Technical Writing

Contexts, Audiences, and Communities

Carolyn R. Boiarsky

Illinois State University

Allyn and Bacon
Boston London Toronto Sydney Tokyo Singapore

Dedication

To my husband Paul, and my son, Michael, who not only supported me, but provided me with *a room of my own.*

And to Bill Powers, whose unfaltering instinct for bringing the right people together gave me the opportunity to write this text.

Copyright ©1993 by Allyn & Bacon
A Division of Simon & Schuster, Inc.
160 Gould Street
Needham Heights, MA 02194

Editor- in-Chief, Humanities	Joseph Opiela
Developmental Editor	Allen Workman
Series Editorial Assistant	Brenda Conaway
Production Administrator	Ann Greenberger
Editorial-Production Service	DMC & Company
Text Designer	Donna Merrell Chernin
Cover Administrator	Linda K. Dickinson
Composition Buyer	Linda Cox
Manufacturing Buyer	Megan Cochran

Library of Congress Cataloging-in-Publication Data
Boiarsky, Carolyn R.
  Technical writing : contexts, audiences, and communities / Carolyn R. Boiarsky.
    p. cm.
  Includes bibliographical references and index.
  ISBN 0-205-12932-9
    1. English language—Rhetoric. 2. English language—Technical
English. 3. Technical writing.   I. Title.
PE1475.B56 1992
808'.0666—dc20   92-35635
                                                            CIP

Printed in the United States of America
10 9 8 7 6 5 4 3 2 1   96 95 94 93 92

CONTENTS

Chapter 3 Readers' Processes and Behaviors   48

Chapter 4 Readers' Characteristics   73

Chapter 5 Purpose and Situation   107

PART II
Processes and Techniques

Chapter 8 Persuading   209

Chapter 9 Rhetorical Techniques   235

Chapter 12 Graphics   328

Chapter 13 Document Design   368

PART III
The Documents

Chapter 14 Correspondence   406

Chapter 15 Correspondence for Employment 432

Chapter 18 Interim Reports   511

Chapter 19 Final Reports   532

Chapter 20 Document Locators and Supplements   572

Chapter 21 Oral Presentations   594

PREFACE

Over the years, I have come to recognize how difficult it is for my students to perceive the complex situations in which they will write in the workplace. Yet, as a teacher I have felt a responsibility to prepare them for the reality of a world that is increasingly multidimensional and multicultural. This text is the result of my efforts to introduce students to the contexts in which they will write as I help them acquire strategies for writing effective documents. Through scenarios and examples based on actual situations, this text shows students the social/political context in which manuals are prepared. It demonstrates how audiences affect the process of writing reports, how organizational and professional communities influence the language and formats of their documents, and how persuasion becomes an integral aspect of technical writing. It also illustrates the ethical and legal responsibilities of technical communicators.

As the text leads students through the corridors of today's workplace, it will help them see how writers use problem-solving skills and strategies to determine appropriate levels of language, tone, visual text, and graphics to reach their audience. Students will also see how writers engage successfully in the composing processes through inventing, planning, synthesizing, drafting, and revising to create effective texts.

Readers of this book will study real-world models of the various types of documents they will be expected to write. The models include a letter supporting McDonald's decision to reduce solid waste, a letter requesting a reduction of a fine for a nuclear utility, a proposal for funding a study on head injuries, documentation for a software program for architects, and excerpts from a report on the effects of lead poisoning on children in cities.

Organization

TECHNICAL WRITING is a fully integrated text. Students are able to perceive how the concepts and processes in Part I and the strategies in Part II are applied to the specific types of documents presented in Part III.

Part I, THE CONTEXT, establishes a framework for writing based on the process approach. The chapters help students recognize how their audience, purpose, situation, and professional and organizational communities affect their decisions concerning content, organizational pattern and sequence, point of view and focus, style, graphics, visual text, and layout.

Part II, PROCESSES AND TECHNIQUES, is concerned with the specific processes, strategies, and skills writers need to master to write effective texts. The chapters provide guidance in gathering and organizing information, using persuasive arguments, writing descriptions and definitions, revising for coherence and readability, integrating graphics with verbal text, and designing documents. All of these discussions are presented in terms of the writer's audience, purpose, and situation.

Part III, THE DOCUMENTS, presents the conventions associated with the major kinds of documents students will be expected to write in the workplace. Examples of actual letters, memos, electronic correspondence, employment correspondence, instructions, documentation, research and marketing proposals, and reports, including progress, trip, feasibility, evaluation, and research and development reports, are discussed. Students have an opportunity to see how each of these must be adapted to the needs of their audience, purpose, and situation.

Finally, the GUIDEBOOK in the Appendix provides brief discussions on the major aspects of grammar, usage, and punctuation to which students may need to refer as they write. These are discussed in terms of the variations found among different discourse communities.

Features

Students in a wide variety of subject areas enroll in technical writing courses. This textbook has been written to apply to a broad cross-section of students, including those in engineering, technology, biology, physics, computer programming, agriculture, sociology, geography, political science, psychology, and professional or technical writing.

The text offers new features in these areas:

Rhetorical Aspects of Technical Writing

- Focuses on audience, purpose, the political/social situation, and the communities in which technical documents are written. Skills, strategies, and conventions are explained and demonstrated in these terms throughout the text.

- Demonstrates through scenarios how invention and decision-making are involved in creating texts.

- Presents persuasion, argumentation, and other rhetorical elements in key chapters.

- Describes real world technical communities through scenarios and running examples (note the text with colored boxes in several chapters).

- Explores legal and ethical issues in discussions of the situation in which documents are written.

Writing Processes and Real-World Examples

- Presents a synthesizing, problem-solving approach to learning the skills of planning, research, organizing, arguing, documenting, and designing.

- Involves visual as well as verbal aspects in decision-making.

- Presents a full range of formats and types of documents organized into four logical categories—correspondence, instructions, proposals, reports.

- Introduces electronic communication, software documentation, manual preparation.

- Introduces multicultural/multinational communication.
- Provides examples covering a broad range of disciplines, including correspondence discussing plans for solid waste reduction by Arby's, the Kemeny Report on the nuclear accident at Three Mile Island, a proposal for legislation to eliminate lead paint, and the patent application for non-fat ice cream.

Pedgogical Features

- Includes sound pedagogical aids, i.e., strategy checklists, chapter summaries, diagrams of models, and visual mnemonic devices.
- Offers a full spectrum of realistic projects, ranging from collaborative work to individual exercises; from long term projects occurring over a period of weeks to those conducive to a single in-class period, from those related to the campus to those involving the community. Some of these projects have been provided by technical writing instructors and content-area faculty involved in writing-across-the-curriculum programs around the country.
- Includes a brief Guidebook of common errors and documentation conventions, discussed in terms of the conventions of various discourse communities.
- Features a fully-loaded Instructor's Manual modeled after those used by industrial trainers. The Manual includes alternative syllabi and weekly lesson plans with accompanying objectives, suggestions for lectures, instructions for conducting additional classroom activities, masters for transparencies and hand-outs, quizzes and accompanying answer sheets, and masters of documents with accompanying explanations. Additional discussions on cognitive processes, multicultural communication, ethics in communication, legal issues, and text grammars are included.

The Beginning of a Dialogue

A textbook is successful if it meets the needs of those who teach. Your responses to TECHNICAL WRITING and those of your students are important to us. We would also like to know of any exercises or projects that you have found valuable in your teaching and would like to share with others in the profession in future editions.

We invite you to enter into a dialogue with us, so that we can provide the kinds of information, activities, and assistance you need in order to offer students effective instruction in technical writing.

Carolyn R. Boiarsky

ACKNOWLEDGMENTS

I would like to take this opportunity to thank those people who have worked with me and shared their knowledge of industry, writing, and teaching, a knowledge which permeates this book as I try to bring the real world to students about to enter the workplace. I would especially like to thank my lifelong friend, Juliet Goodfriend Zimmerman, the President of Strategic Marketing Corp., and my former compatriot, Leah McNeill, Manager of Internal Communications, at Westinghouse Savannah River Plant.

I would also like to express my gratitude to the many people who took the time from their own work to read this manuscript and to offer their expertise to ensure that this text contains information that is accurate and concepts that reflect the most recent thinking in the discipline. To Beth Neman, Wilmington College; Irene Brosnahan, Illinois State University; Donna Goehner, Western Illinois University, and her two assistants, Kathy Dahl and Kate Joswick; Margot Soven, LaSalle University; Emily Thrush, Memphis States University; Barbara Olds, Colorado State University; Mark Moreau, LaSalle University; Lynn Beene, University of New Mexico, Albuquerque; and Leslie Moe-Kaiser, the Asian Group. In addition I would like to thank Deborah Andrews, University of Delaware; Nancy Blyler, Iowa State University; Gregory Columb, University of Illinois—Urbana; Joe Comprone, Michigan Technological University; Dixie Elise Hickman, University of South Mississippi; Debra Journet, University of Kentucky; Carolyn Miller, North Carolina State University; James Porter, Purdue University; Carolyn Plumb, University of Washington; and Diana Reep, The University of Akron.

The support I received from my acquisitions editor, Joe Opiela, and the assistance I received from my developmental editor, Allen Workman, were inestimable in helping me create both a readable and usable text.

Perhaps my greatest debt is to my students. Those in corporations, struggling in complex situations to communicate their plans, programs, and needs have provided me with a wealth of information to pass on to those students getting ready to enter the workplace. And those in undergraduate classes helped me test my models and hypotheses as well as create workable exercises.

Finally, I must thank all of those people who not only permitted me to reprint their documents, but who spent time discussing their organization's document process as well as their own writing processes so that I could provide readers of this text with a broader picture of the communities in which they will be writing. My appreciation goes to Art Curtin, Coca Cola, Port Orange, FL; Tim Curtin, Normal, IL; Alan B. Greenstein, Hingham, MA; Marlise Streitmatter, Elmwood, IL; Meleah Melton, Normal, IL; Rick Schwarzentraub, East Peoria, IL; Leslie Penles and Barbara Pecak, Abbott Laboratories, Abbott Park, IL; Marcia Patterson, ADM, Decatur, IL; Jody Howard, Advanced Technology Services, East Peoria, IL; Stephen Hiles, American Electric Power Corp., Columbus, OH;

Jack Shaver, Appalachian Power Company, Charleston, WV; Handy Truitt, Architech, Chillicothe, IL; Dave Beauarac, Dave Jacque, Joseph Harmon, Karen Haugen, Mary Warren, Argonne National Laboratory, Argonne, IL; Kerry Leifeld, Byerly Aviation, Peoria, IL; Greg Tarvin, David Glore, Merv Rennich, Harry Plate, Mark Sutherland, Caterpillar Inc., Peoria, IL; Russ Watson, Cessna Aircraft Company, Wichita, KS; Janet Attanucci, Chi-Chi's, Louisville, KY; Pamela Parker, Chicago Zoological Society, Brookfield, IL; Greg Heaton, Crawford, Murphy & Tilly, Inc., Springfield, IL; Harry Litchfield, Ed Wahlstrand, Dean Easterlund, Deere & Company, Moline, IL; Albert Manville, II, Chris Sarri, Defenders of Wildlife, Washington, DC; Sach Takayasu, Diamond Star Motors, Bloomington, IL; Michael H. Simpson, Peter Engels, E & A Consultants, Stoughton, MA; Howard C. Kreps, Richard Denison, Environmental Defense Fund, Washington, DC; Michael J. Hoffman, Environmental Science and Engineering, Peoria, IL; Marge Franklin, Franklin Associates, Ltd., Prairie Village, KS; Ron Burling, Greater Peoria Regional Airport, Peoria, IL; Bonnie Noble, Heartland Water Resources Council, Peoria, IL; John Morgan, IBM, Atlanta, GA; Tim Hungate, Illinois Central College, East Peoria, IL; Ralph Anderson, Illinois Department of Transportation, Springfield, IL; Hank Campbell, Steve Eddington, Doug Hesse, Kathleen McKinney, Patrician Monoson, Alan Monroe, Charles Pendleton, Jacqueline Scholl, Mike Sondgeroth, Jill Thomas, Ed Wells, Mike Welsh, Illinois State University, Normal, IL; Chandru J. Shahani, Library of Congress, Washington, DC; Roger Monroe, Methodist Medical Center, Peoria, IL; Robert D. Port, Nalco Chemical Company, Naperville, IL; Michael A. Meador, NASA—Lewis Research Center, Cleveland, OH; Corinne Caroll, NSF Network Service Center, Cambridge, MA; Kevin McCollister, Pennsylvania Resources Council, Inc., Media, PA; John Vigilante, Plessey Semiconductors, Dedham, MA; Robert Moffat, Stanford University, Stanford, CA; Cheryl Asa, St. Louis Zoo, St. Louis, MO; Linda Cooke, USDA, Northern Regional Research Center, Peoria, IL; Wiley Scott, USDA, Soil Conservation Service, Champaign, IL; Phillip A. Garon, U.S. Department of Energy, Washington, DC; Andy Schneider, U.S. Department of Labor, Peoria, IL; Frank Giunta, University of California, Berkeley, CA; Tracy LaQuay, University of Texas at Austin, Austin, TX; Marilyn Hailperin, West Jersey Health Systems, Gibbsboro, NJ; Fukiko Minami, Nara, Japan.

The Context

WHETHER YOU ARE WRITING A PROPOSAL requesting funding for a project, a report evaluating a program for your supervisor, or a manual of instructions on repairing a vehicle, you need to consider the context in which you are writing. Part I concerns the context in which writing takes place in your technical community. It involves your readers, your purposes, and the situation in which you write. In these first five chapters you'll learn how writing occurs in technical communities; how to make textual decisions about the content, organization, and point of view; and how to choose the style, graphics, visual text, and format of a document. You'll learn how to base these decisions on your readers' organizational and professional communities as well as their personal characteristics. You'll also learn how the readers' purposes and your purposes affect textual decisions. In addition, you'll discover how the social and political situation surrounding your work affects your text.

REGARDLESS OF YOUR TOPIC OR THE TYPE of document, you need to consider the context in which you write. If readers aren't knowledgeable in the subject area, you don't want to use terms unfamiliar to them. For example, if your purpose is to help readers use a software program, focus on the aspects of the program that they need to know, not on the inner workings of the program. If your readers may react negatively to your proposal to institute a recycling effort in your organization, build your case so they perceive the need for the effort and are persuaded to accept your proposal.

PART I WILL HELP YOU UNDERSTAND YOUR readers' needs so that, as you begin writing, your decisions concerning the content, point of view, organization, style, graphics, and design are appropriate and will help readers understand and accept your message. ☐

CHAPTER 1

Writing in the Communities

WHETHER YOU'RE CONDUCTING RESEARCH on preserving the wetlands, working with a robotic-powered assembly line, providing rehabilitation for people within the criminal justice system, or marketing a new product, writing will be an integral part of your work. You'll prepare memos, letters, and progress reports, and you'll collaborate with others to draft proposals for new products. Decisions will be made based on reports you will write and present. Whether you are an hourly worker, salaried employee, manager, consultant, engineer, technician, psychologist, marketing analyst, or public information specialist, you will need to engage in a variety of problem-solving skills to plan, write, edit, and revise technical documents.

The ability to communicate your ideas in writing has become an important skill in technical communities. Listen to what several people in the workplace say about the need to write effectively:

> John Nespo, an engineer with Goodyear, comments, "Today's engineers must not only be able to develop an idea, they must be able to promote it...The biggest thing you do is sell [communicate]. They [management] don't just accept an idea; you need to convince them."

> Jody Howard, whose firm, Advanced Technology Services, develops software programs for manufacturers, explains, "Because we're small, we do all the writing. We write the proposals to sell the customer on our services, we write the specs for the systems based on interviews with customers, and we write the documentation to be used by the customer."

> Bill Massih, a manager with Dupont, emphasizes that, "One of your [company's] major problems is communications, especially when you're highly regulated."

Because industry recognizes the importance of written and oral communication, such skills are a priority item in job interviews. One of the first questions a recruiter at a college job fair asks is, "What communication skills do you have?"

Job descriptions often include communication as one of the responsibilities of an employee. Jody Howard notes that this requirement may be even more important to a small firm like ATS, which doesn't have the editorial or technical publications staff a large corporation has. "We're planning our [career] ladders," she explains of the new and rapidly growing firm for which she works. "One of the criteria for moving up into management is good written communication skills."

Included in a job description for a quality control engineer are the following responsibilities requiring skills in written communication:

- Submission of a written monthly report regarding progress on assigned projects and tasks.
- Correspondence with all customers relating to quality matters.
- Preparation of forms and procedures for implementing quality control requirements.
- Preparation of formats for quality audits.

Notice that, in addition to writing reports and correspondence, the engineer is expected to write procedures and to design the formats for various forms and audits. Skills for communicating effectively range from gathering and organizing information to designing layouts and creating tables and charts.

Physicist John Soven emphasizes the need to be able to write about numbers, using correct nomenclature and symbols. Sociologist William Rau comments, "Numbers by themselves are meaningless; you need to be able to interpret them." Karen Haugen, a technical editor with the Argonne National Laboratory, sees skills in desktop publishing as necessary: "Many scientists are now creating their own graphics [using desktop publishing]. They need to learn to integrate visual text [so readers can understand the chart or the table]."

This book will help you acquire the skills that will enable you to communicate effectively in the workplace. In the remainder of this chapter, we'll look more closely at the technical communities in which you'll write. We'll examine your role as a communicator in these communities, and we'll consider the various political and social aspects of the writing situation in these communities.

THE COMMUNITIES

As you begin a career, you enter a technical community defined by the subject field in which you work. Whether you are a chemist for a small consulting firm specializing in landfill reclamation, a marketing analyst for a large international corporation producing plastics, or a technical writer for a state agricultural agency, you become a member of a technical community. Often you are part of several communities. The chemist is a member of a community of chemists as well as a part of a community of landfill reclamation specialists. The marketing analyst is member of a community of marketing analysts as well as a part of a community of plastic specialists. The writer is a member of a community of agricultural specialists as well as of technical writers.

Technical communities are groups of people involved in the technical aspects of a field. Biologists, astronomers, and environmentalists, as well as civil engineers, electrical technicians, and physical therapists, each comprise a distinct technical community. An artist involved in the mixing of paints to create a fresco is as much involved in a technical field as a paint analyst for the automobile industry. Technical communities range from agronomy to zoology; from aviation technology to electrical engineering; from economics to psychology to sociology.

Each of these communities has its own language and conventions. As a member of a community involved in environmental science, you might discuss *polystyrene*, *leachates*, and *recycling*. As a member of a biogenetic community, you might exchange ideas about *recombinant DNA*, *pharmokinetics*, and *placebos*. And as a member of a community involved with earth-moving equipment, you might talk about *fuel efficiency*, *rackshafts*, and *hydrostatic power*. The conventions of one organization's R & D division might permit you to wear jeans and write informal memos, while the conventions of a public utility might require you to wear more formal attire and write more formal correspondence.

To be accepted as an expert in a technical community, you must know more than just the subject matter. You must also know the specialized language and conventions shared by the community. Thus, you are not only a member of a *technical* community, but also member of a *discourse* community.

Consider Jody Howard's situation. Howard, who works at ATS, is an electrical engineer. However, because her work revolves around the design of computer software for manufacturing purposes, she is not only a member of a community of electrical engineers, but also part of a community of computer scientists, and

Figure 1-1
Howard's Discourse
Communities

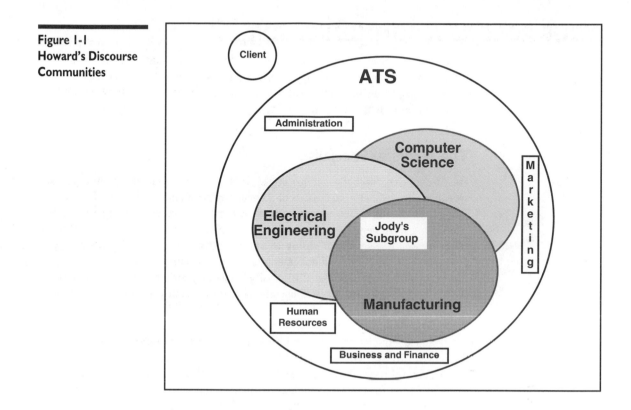

also manufacturing specialists. She is not only familiar with the language and conventions associated with electrical engineering, but also with the language and conventions of those aspects of each of the other discourse communities related to her specialty. You can see in Figure 1-1 how the three communities intersect to form Howard's specialized community.

But you will not only communicate with people in your own discourse community, you will also communicate with people in other communities. For instance, as a civil engineer, maintenance mechanic, or psychologist, you will be writing documents which will be read by marketing managers, clients, and possibly even the general public. You need to remember that these readers are not members of your discourse community and that they may not know the language and conventions of your field. When you write for these readers, you need to "translate" the specialized language used in your community to a language that people in other communities can understand. Instead of using the word "down-sizing," you might use the phrase "reduction of personnel."

As you read this book, you will become more aware of the language and conventions of the discourse communities in which you plan to work. You will learn how to communicate effectively within these communities and with members of other communities, so that, regardless of your readers' communities or the situation in which you write, your readers will understand and accept your message.

YOUR ROLE AS A COMMUNICATOR

You will write for a variety of readers and purposes, usually in one of the following three roles: technical specialist, adjunct specialist, or writer. Let's examine how people in these three roles write, the types of writing they do, and the communities in which they are writing.

The Technical Specialist

As an engineer, biologist, archaeologist, or technician, you may picture yourself working on a project in the office of a small environmental consulting firm, developing a product in a laboratory at a multinational corporation, investigating a field site for a government agency, or operating machinery in a large industry. But you probably don't see yourself sitting at a desk and writing. Yet 25 percent of your time may be spent preparing documents. As you move up the organizational ladder, that percentage may increase substantially. Most of your writing will be concerned with your own field and you will usually write for those in your own community.

Let's take a look at the kind of writing that is done, the communities involved, and the situation in which writing occurs for Patti Haines, a technical specialist.

Haines, a nuclear engineer with Florida Power Corporation, is an Operations Technical Advisor. Her responsibilities include advising operators in the utility's reactor control room during unusual events and accidents, then writing reports on these incidents for both the Nuclear Regulatory Commission (NRC) and her own management. These reports demand a combination of the investigative skills of Jessica Fletcher in *Murder She Wrote* and the political tact of Henry Kissinger.

The reports are required by the NRC which, depending on its reaction to them, may fine the utility or even shut it down, resulting in a loss to the utility of millions of dollars. To avoid this, Haines needs to prove that the company has done everything in its power to correct a problem as quickly as possible. Thus, she needs to learn exactly what has happened and what has been done about it. Further, she is dependent on information given to her by operators who were present at the time the problem occurred. If they don't provide her with all of the data, it is her job to track it down.

Once she has all of the information, she must negotiate between the NRC's purposes and her own company's desire to look good. She must determine the content, tone, and organization of the information in such a way as to persuade the NRC that the company has met its responsibility.

The reports are then read by Haines' supervisor, her supervisor's supervisor, and that supervisor's supervisor, all the way up to the Corporate Executive Officer (CEO) before they are sent out. She is often asked to revise the reports for her supervisors a number of times before they are satisfied with the way in which the information is presented.

If management is dissatisfied with the way in which a problem has been handled in the control room, Haines is also often required to investigate these problems and write up the results of her investigation for management. In writing these reports, she must negotiate between the operators who handle the problem (and with whom she must work) and the management that determines her promotions and salary.

Haines writes as an expert in her field and, because she writes for readers within her discourse community, she uses the specialized language and conventions of her field.

The abstract for a report to the NRC, such as that shown in Figure 1-2, represents the type of writing Haines does. As you read, notice that she uses technical terms because the abstract is written for readers in her discourse community who are familiar with this language. In addition, she follows the conventions of including the power level and the mode of operation, since members of this community not only expect this information to be provided, but also need to know it in order to evaluate the information.

As you can see, writing is an integral part of Haines' job, and, like most technical specialists, she usually writes for those within her own discourse community. While the documents are usually short, two to five pages, she often spends a great deal of time gathering the information for them. Further, because the documents are written within the political context of the organization, Haines is often required to revise them numerous times.

The Adjunct Specialist

You may be a marketing analyst for an organization which produces drugs, a sales representative for a developer of software programs, or an accountant for a

NRC Form 366	U.S. Nuclear Regulatory Commission	APPROVED OMB N. 3150-0104
(6-89) **LICENSEE EVENT REPORT (LER)**		EXPIRES 4:30:92 ESTIMATED BURDEN PER RESPONSE TO COMPLY WITH THIS INFORMATION COLLECTION REQUEST 500 HRS. FORWARD COMMENTS REGARDING BURDEN ESTIMATE TO THE RECORDS AND REPORTS MANAGEMENT BRANCH (P5301) U S NUCLEAR REGULATORY COMMISSION , WASHINGTON , DC 20555, AND TO THE PAPERWORK REDUCTION PROJECT (3150-0104º. OFFICE OF MANAGEMENT AND BUDGET, WASHINGTON, D20503.

ABSTRACT (Limit to 1400 spaces, i.e. approximately fifteen single-spaced typewritten lines) (15)

Unit 3 was operating in MODE 1 (POWER OPERATION) at 98% full power on January 22, 19xx. Reactor Coolant System (RCS) leakage calculations completed earlier that day showed that unidentified leakage was 0.3 gpm. At 1035 Reactor Coolant System leakage calculations indicated that unidentified leakage had increased to 1.4 gpm. This value exceeded Technical Specification limits and a plant shutdown began. The plant entered an Unusual Event in accordance with the Plant Emergency Plan.

Figure 1-2 Abstract for a Licensee Event Report by a Technical

company that generates electric power. Like the technical specialist, you may find as much as 25 percent of your time devoted to writing. Much of your writing will involve both your own community and the technical communities encompassed by the company for which you work.

As an adjunct specialist, you need to be knowledgeable about your own field, and also possess a general knowledge of your firm's field. A purchasing agent for an engine manufacturer is not only expected to know about purchasing, but also about purchasing in relation to the manufacture of engines.

In addition, as an adjunct specialist you need to be "bilingual," capable of using the technical terminology of your own discourse community as well as that of your firm's. For example, the purchasing agent should be able to talk about "on order/on hand inventory" as it relates to "pistons" and "valves."

This bilingual capability is necessary if you are to persuade readers outside your discourse community to accept your message. By using the language and conventions of your readers' field, you indicate that you are not only an expert in your own field, but that you are an expert in your field as it relates to their field.

Let's look at how Leslie Penles, the Associate Manager for Market Research with Abbott Laboratories, communicates within the dual communities of pharmaceuticals and marketing.

Penles' undergraduate degree was in marketing and her master's degree in applied statistics; she had very little scientific or medical background when she took her first job as a marketing analyst with a small marketing research company, Strategic Marketing Corporation.

As a research analyst, she conducted research to determine the marketability of potential new products, and then wrote reports on the results of the studies. These reports were presented to the clients who commissioned the studies.

Because the firm's product is the presentations and reports it produces for its clients, Penles found herself doing a great deal of writing; the presentations often ran to several hundred pages and the reports to 25 or 50 pages. Moreover, these documents concerned a field outside Penles' own discourse community, and were read by people in that community.

"The first thing I had to do was build up a vocabulary of medical terms," she explains. "I found a medical terminology book, a big red book," she remembers, "and studied it." She also spent a lot of time doing background reading. "If I knew we were going to do something on hypertension, I'd pull out everything in the library on hypertension and, with a dictionary at my side, I'd learn about it."

According to Leslie Penles, "We [adjunct specialists] don't know the scientific aspects of the drugs. What we know is the relationship of the drugs to sales, i.e., doctors' preferences for prescribing pills or liquid medication." She also knows the medical language related to the conditions for which the medications are used, e.g., "antihypertension" and "ventricular hypertrophy," as well as the jargon, e.g., "primary care physician" and "monotherapy."

While Penles now uses the vocabulary of the new discourse community in which she works, she continues to use the vocabulary of her own discourse community of marketing analysts, thus assuring her readers that she is an expert in her own field. "I use marketing research language, like 'sample' and 'universe,' in the reports because it lends credibility to them," she explains. "The technical marketing language indicates a level of sophistication with the methodology, and reassures the client that the research has been conducted professionally [so they are persuaded to follow my recommendations]."

By the time Penles moved to Abbott Laboratories, a huge international pharmaceutical corporation, she had become fairly comfortable in the pharmaceutical discourse community. But now she had to adapt to another discourse community, that of the organization. "I had to pick up a lot of organizational vocabulary," she admits. "We use a lot of acronyms here." In addition, she had to learn the vocabulary of the various divisions with which she was working. "Before [at Strategic Marketing] I only had one audience, the product manager, but here I have several," she explains. "If I'm writing to the Vice President of Sales, I need to use sales language, but if I'm writing to the Vice President of Business, I need to use finance language."

In the excerpt shown in Figure 1-3, taken from a proposal Penles submitted to her manager, she requests permission to conduct a pilot study to determine whether or not doctors are interested in a new product for hypertension. As you read, notice that Penles uses the technical language of two discourse communities. Like Haines, she uses terms related to the content area. Terms such as "alpha-beta blockade," "lipid profile," and "titrate," are common terms in a community of pharmacologists, while such terms as "armamentarium," "receptivity," and "exploratory study" represent the language used by a community of marketing specialists. You should also notice that Penles focuses on the product's sales potential to persuade her manager to approve her proposal, since the manager is interested in a product's marketability rather than its scientific aspects.

The Writer

You may be a writer working in the public relations division of a hospital, or in the technical publications division of a utility, or you may serve as an editor in a governmental research facility. All of your writing will be concerned with the communities encompassed by your organization.

Writing in a technical field can be both interesting and challenging. Since you are seldom knowledgeable in that field when you first begin to work, you may feel like a tourist in a foreign country during those first weeks. You don't know the vocabulary or the customs. However, after several years of writing about a specific content area, you acquire much of the knowledge, language, and conventions of that discourse community.

Allan Margolin, who majored in English and who admits avoiding math and science courses, has acquired a knowledge base that includes the economics of lead poisoning in low-income urban housing and the chemistry involved in acid rain. As the Media Director for the Environmental Defense Fund, a private not-for-profit organization, Margolin writes about such scientifically complex subjects as ozone depletion, acid rain, toxic waste, and global warming. His job is to get the message to the general public through the media. This means that the

⊋ **ABBOTT**

DATE: Sept. 19XX

TO: J. S.

FROM: Leslie Penles

RE: Proposal to conduct exploratory research to measure PCPs' reception to a combination product

BACKGROUND

    This proposal outlines an exploratory study to assess primary care physicians' (PCPs) receptivity to a combined product for the treatment of hypertension. Preliminary discussions with physicians suggest that from a scientific standpoint the combination makes sense and could have great appeal. It has inherent features that when combined could possibly produce a synergistic effect in terms of blood pressure, lowering as well as offering some additional benefits over other classes of antihypertensives.

    Despite the intellectual appeal of a combined alpha-beta blockade product, only one similar chemical entity currently exists, and it does not have a strong presence in physicians' armamentarium.

    The proposed product offers a number of advantages, such as a favorable lipid profile and QD dose. In addition, the fixed ratio of alpha and beta blockade makes it impossible to titrate the two activities separately in individual patients. A combined product could offer the flexibility of various formulations with different ratios.

Leslie Penles, Manager, Marketing Research,
Abbott Laboratories

message needs to be presented in a language citizens and legislative leaders can understand. This, in turn, requires providing information in a form that media representatives can use.

    Initial news releases are often written by the scientists themselves and then given to Margolin for editing. Sometimes the releases are eight pages long and full of footnotes. Margolin's first task is to explain to the writer that a release shouldn't be longer than a single page and shouldn't contain any footnotes. His next task is to make certain the text is written so that the ideas will be clear to the general public. He explains, "They (the scientists) come from professions which have their own language. They speak legalese or sciencese or economese. But as an advocate, you need to reach people in a language they understand, not with shoptalk." So Margolin's task is to translate the language of the profession into everyday English.

    Sometimes Margolin drafts a release himself. "I meet with the specialists and we chat about what they want to say or get across. I make them explain their ideas to me. I figure if I can understand it, it will be clear to most folks."

    Margolin then drafts the release and sends it to the specialist to make certain that the facts are correct and that the appropriate scientific terms are used. Sometimes he and a specialist edit a text together.

While Margolin is mainly concerned with news releases, "Op Ed" letters, and "Letters to the Editor," he also provides insights for reports and other documents which may be read by the media. "I try to make sure that there is a clear executive summary that gets across the main idea," he remarks, noting that media reporters prefer to read a summary rather than a full report.

Margolin collaborated with an attorney in writing the Executive Summary for the report on lead poisoning shown in Figure 1-4. The purpose of the report was to persuade congressional representatives to pass legislation to protect children, especially in poor urban areas, from lead poisoning. The report received national media attention, and a number of the recommendations in it were enacted.

Notice that, unlike Haines and Penles, Margolin does not use technical terminology in this report, since many of the readers are not members of a scientific discourse community involved with lead poisoning. However, you can easily recognize that Margolin is knowledgeable about his subject. He provides specific details to support his arguments, specifies the ratio of lead level to dropout rates, and lists the consequences of lead poisoning.

**Figure 1-4
Summary by a
Technical Writer**

Introduction

As a tragic legacy of the decades-long use of leaded products on a vast scale, lead today pervades America's environment. The result is a nation-wide epidemic of low-level lead poisoning, an epidemic that is causing permanent neurologic damage to millions of American children. Recent studies demonstrate that the long-term consequences of this disease are profound: children who had moderately elevated lead levels in early childhood later exhibited seven-fold increases in school dropout rates, six fold increases in reading disabilities, and lower final high school class standing.[1] These effects occurred even though the initial exposures caused no overt symptoms...

The massive amounts of information on lead's toxicity—bolstered by recent findings on low exposure level effects—as well as indications of children's current exposure levels, reveal an urgent need for an aggressive federal program to control America's continuing epidemic of lead poisoning. To be effective, such a program must provide a mechanism not only to **stop adding** lead to children's environment, but also to **remove** it from the areas where they are most heavily exposed: their homes. And, to be politically feasible, it must respond to current budgetary realities the nation now faces.

The Environmental Defense Fund proposes creation of a National Lead Paint Abatement Trust Fund, to be financed by placement of a substantial excise fee on the production and importation of lead. Proceeds from the fund initially would be devoted to the removal of deteriorating lead-based paint from the group of highest-risk homes. In addition, a portion of the monies could be made available for research to develop more lead-removal methods...

EXECUTIVE SUMMARY

The federal government estimates that well over three million pre-school children—more than one in every 6—have dangerously elevated lead levers.

Environmental Defense Fund, *Legacy of Lead: America's Continuing Epidemic of Childhood Lead Poisoning,* (1990).

THE SITUATION

Reader-Based Purposes

When you write in the workplace, you have a reader who actually uses your information. Your readers have a purpose for reading your message: they need to make a decision or do something. In fact, often, you are writing in response to a request by your reader. Thus, your decisions relating to the content, style, and organization of your text rely heavily on your readers' needs.

Haines' managers use the information on control room problems to determine changes for improving the plant's operation, while the NRC uses the information in the reports to determine whether or not to fine the utility.

Leslie Penles writes differently as a manager at Abbott Laboratories than she did as an analyst at Strategic Marketing; most of her writing now is for readers who are also employed at Abbott. "When I wrote for Strategic Marketing, the bulk of the report was the analysis [for clients]. But here [at Abbott], the emphasis is on the recommendations and actions to be taken," she explains. "You have to go directly to the bottom line. When you're writing at the VP level, things need to be kept to a single page because nobody wants to read more."

Instead of writing what you want to write, you write what your readers want to read, and you provide the information they need to know. Your writing is *reader-based*.

Social and Political Contexts

Writing is often social; it involves many other people. It is also political.

Patti Haines is aware that a political situation exists both between the regulatory agency and her utility, and between managers and employees. In order to avoid conflicts, she often meets with managers to make certain her approach is in concert with theirs .

Leslie Penles comments, "Tone is important. You've got to consider the audience. You can't just come out and say, 'Listen, dummy, this is the way to do it.'"

Persuasion

Because social and political factions often exist both within and between communities, you will find you are constantly trying to negotiate between various factions and communities to persuade readers to accept your message. In a letter of application, you want to persuade the reader to hire you for a job; in a manual, you want to persuade the reader to follow procedures; in a set of directions, you want to persuade the reader to repair a mechanism properly; and, in a bread-and-butter note, you want to persuade the reader to provide information again if you ever need it. Even in a research report, such as an Environmental Impact Statement (EIS), which reports on investigations into how proposed projects such as the superconducting supercollider may impact the environment in which they're going to be located, you want to persuade readers to accept your findings.

Bill Vinikour is an environmental scientist with the Environmental Sciences and Information Systems division at the Argonne National Laboratory. Everything Vinikour writes in an EIS is reviewed by a wide variety of readers, ranging from managers of the organization proposing the project and the accompanying environmental research, to local citizens who may be affected by the project. An adverse finding could cost a sponsor tens of thousands of dollars or possibly even force a company to abandon a project altogether. On the other hand, it may save citizens a great deal of tax money in addition to protecting them from a health problem.

The law requires that, before a proposed project is approved, a draft of an EIS must be presented to citizens affected by the project for their response. These citizens belong to a broad spectrum of discourse communities, ranging from the general public to members of environmental groups, such as the Sierra Club, to businessmen with vested interests in the proposed project. Bill Vinikour is responsible not only for writing those sections of the EIS related to his specialty, but also responsible for responding to people and organizations who question or protest his findings. The public's comments and Vinikour's responses are printed in the appendix of the EIS. In these responses, Vinikour must negotiate between the public and the organization which has sponsored the proposed project, and hired his firm to write the EIS. If Vinikour's findings favor the public, he affects the proposed project negatively, and the sponsoring organization may not hire his firm again. However, if his findings favor the sponsoring organization, the citizens may feel he is not telling the truth, and he may lose their trust. He needs to persuade both sides that his findings and recommendations are valid without alienating either.

Vinikour's response to several comments on biotic resources follows. Read the summary in the first paragraph of the public's questions. Then, as you read Vinikour's response, consider how he persuades his readers in both communities that his findings are valid.

CATEGORY 8: BIOTIC RESOURCES

Summary of public's comments.

Vinikour's response counters opponents' argument. Recognizes citizens' concerns. Refers to authorities.

One commentor was concerned that the SSC project would be detrimental to ecological resources in the area while another was concerned about impacts that could occur to aquatic resources if spoils were not adequately contained. Twelve of the 18 comments expressed minor concerns about the status or nomenclature for several of the species listed in Vol. 1, Appendix B.

Category 8, Submission 126, Comment 3 (Detrimental Effects on Ecological Resources) In contrast to the stated concerns, both the U.S. Fish and Wildlife Service and the Texas Parks and Wildlife Department view the SSC as a project that could significantly benefit ecological resources in an area that has been diversely affected by agricultural activities and urban encroachment.*

Presents facts related to policy to support his argument.

The SSCL intends to work with these agencies to establish natural areas (e.g. blackland prairies and wetlands) and preserve high quality areas (e.g. riparian habitats). As stated in Vol. I, Section 4.3.2, about 8,500 acres of mostly agricultural land can be enhanced to create habitats conducive to use by a diverse array of wildlife species.

* Short, R. W., 1990. U.S. Fish and Wildlife Service, letters to T.A. Baillieul, U.S. Department of Energy, Chicago Operation Office, Argonne, Ill., April 3 and May 22.

U.S. Department of Energy, 1990

Collaboration

Many of your documents will be written in collaboration with others. You may collaborate either with those within your discourse community or outside it.

One of the newest forms of collaboration, called concurrent engineering (CE), is being initiated by corporations to stimulate new ideas as well as make the production process better. CE is designed to speed up the manufacturing of new products that have traditionally been a serial or linear process. That is, a product is initiated in the design division, then sent to testing, then manufacturing, quality assurance, and, finally, to marketing, sales, and service. However, by using concurrent engineering, all of the various divisions—engineering, manufacturing, marketing, and services—are involved from the beginning and work along parallel paths. In concurrent engineering the design phase is no longer limited to design engineers. Instead, representatives of testing, manufacturing, quality assurance, and marketing are involved. Even suppliers and customers may be included at various points throughout the process. You can see the difference between the two methods in Figure 1-5.

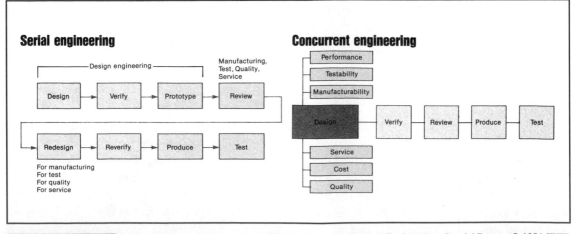

IEEE Spectrum, "Concurrent Engineering, Special Report. © 1991 IEEE

Figure 1-5 Serial vs. Concurrent Engineering

Concurrent engineering significantly affects the communication processes within a company. Instead of communicating only with people within their own discourse community, engineers and technicians must now communicate with people in a variety of other fields. The vertical row in Figure 1-5 indicates the various communities which must communicate with each other.

John Koegan, a software engineer at Digital Equipment Corporation, believes more communication occurs between divisions at DEC since the initiation of concurrent engineering. As changes are made, deadlines met or passed, budgets maintained or missed, division representatives need to be notified so they can change their plans to correspond with the new situation. The constant daily communiqués assume the forms of memos, letters, and progress reports, many of which are simply transmitted by electronic mail.

Concurrent engineering is not the only collaborative effort occurring in industry which affects written communication in a company. At the Argonne National Laboratory a team approach is used to enhance the written products of its researchers. Scientists from many disciplines work in teams, with each researcher assigned a technical editor to clean up written copy for a report.

While Bill Vinikour was working on the Environmental Impact Statement for the supercollider, he was a member of a team that included specialists in geology, hydrology, acoustics, social psychology, urban planning, economics, and paleontology. These researchers each wrote up their own section, which was then put together in a final report by the team leader.

Before Vinikour submits the final draft of a report, he works with a technical editor. Bryan Schmidt, a recent graduate in technical communications, reviews Vinikour's work for grammar, style, format, and consistency with the other sections. Possessing a broad scientific background as well as specific knowledge in linguistics and rhetoric, Schmidt is able to edit Vinikour's copy with a fair degree of content knowledge, spotting inconsistencies in logic and problems in tables or charts. Figure 1-6 is an excerpt from Vinikour's section of an environmental impact statement related to the joint military/civilian use of Scott Air Force Base. You can see where Schmidt makes a last-minute addition after he has checked the map to which Vinikour's text refers. Notice that Schmidt's addition makes the text more accurate and the map to which it refers easier for readers to read .

Day-to-Day Communication

While you may occasionally write a major report or proposal, much of the writing you do as a technical or adjunct specialist in a technical community involves short reports, like the one Haines wrote, and a great deal of day-to-day correspondence. A recent survey of aerospace engineers indicates that they write more memos and letters than any other form of technical document. The engineers estimated they devoted 25 percent of their time to communicating with others.

The memo in Figure 1-7, written by Ron Rothstein, a software engineer with ATS, Inc., is indicative of the kind of correspondence you may write daily.

2.1 SETTING FOR THE PROPOSED ACTION AND ALTERNATIVES

Figure 2.1 shows Scott AFB as it currently exists, and Figure 2.2 identifies many of the base features discussed in this EIS. The base has one runway 7,061 ft long and oriented in a southeast-northwest direction. A system of nine taxiways gives aircraft access to the parking spaces and various facilities. There are 48 parking aprons for fixed-wing aircraft, 13 parking spaces for helicopters, and 4 parking spaces for emergency hazardous cargo. On-base housing for enlisted personnel is provided in CCV, which is centered along the northern edge of the base, and in other housing areas (Figure 2.2). Current plans call for Scott AFB to acquire the Idaho-shaped area between the western edge of the base and Illinois 158 ~~(Figure 2.2)~~ for dormitories and base support facilities, whether or not joint use is approved.

and the rectangular area south of the Mascoutah Gate and west of the Mascoutah Gate Road (Figure 2.2),

---

Figure 1-6 A Corrected Manuscript Resulting from Collaborative Editing

Rothstein writes everything from meeting cancellations to FYI (For Your Information) memos. This memo is simply a notice that a meeting has been changed. Notice that the memo is brief but friendly, and that it follows the conventions of a standard memo format.

Mark Sutherland, a noise analyst with Caterpillar Inc., needs to maintain constant contact with personnel working around the globe for his firm. Problems relating to his specialty occur daily, and he is expected to respond promptly. For this reason, he uses electronic mail, which involves writing a message on a personal computer (PC) and then sending it via computer network to recipients who can call it up on their own monitor from as far away as Brazil and Japan.

Loss of Ownership

In the workplace, a written document is not yours alone; it belongs to the company or to a group. And you are no longer the sole author; many other people have provided input. You must forgo "pride of authorship" and be open to making changes indicated by other people's knowledge and experience.

Most of the documents discussed in this chapter were reviewed and revised by the writers' supervisors and editors. These writers seldom receive a by-line for their text; their documents are usually written under the auspices of the organization for which they work.

Leslie Penles admits to being bothered at first because her name did not appear on the reports on which she worked. "I wanted recognition by the client for what I had done," she explains. Eventually, she realized that clients were aware of the way in which technical documents were put together, and knew she had been involved with the final report.

**Figure 1-7
A Typical Memo-
randum**

ATS   **ADVANCED TECHNOLOGY SERVICES, INC.**

"Setting New Standards In Service Excellence"

```
DATE:   October 31, 19XX

TO:     William Bakker  Ext. 15     James Larson   Ext. 787
        Jim Goble       Ext. 730    Avi Narula     Ext. 511
        Jody Howard     Ext. 43     Jack Rainey    Ext. 746
        Bob Keime       Ext. 37

cc:     Jim Brooks      Ext. 18     Bob Everts     Ext. 21

FROM:   Ron Rothstein   Ext. 33

SUBJ:   **************MEETING CHANGE*********************
```

Our meeting on Thursday, November 2, with Tony Mayo, National Account Manager, Robbins-Gioia, Inc. has been changed
to Tuesday, November 14.

Mayo will be coming to ATS to tell us about Robbins-Gioia's
Job Costing software, Project Management software, and consulting services.

You are invited to attend the formal presentation which will
now be held at 9:30 -11:30 a.m. in the Upper Employee Conference Room at the ATS Main Building.

In addition to the formal meeting, Mayo will be here in the
after on Monday, November 13, to meet with us and learn more
about our requirements. I will contact you individually to
schedule meetings on this Monday.

See you there.

                                        Advanced Technology Services, Inc.

Firm Deadlines, High Expectations, Real Consequences

Documents must be completed on time and written perfectly. Haines has 30 days to complete a report for the NRC. Often she has less than a week to complete a report for her own management.

In addition, you may be expected to revise a document until it is acceptable. Schmidt notes, "At school there's a cut-off point at the end of the term. But here, projects never seem to leave."

One consultant tells of being called into a company to help improve the efficiency of employees' writing. The consultant was shown a document that was marked "REV 104"… it had been revised 104 times before being considered satisfactory for distribution. By acquiring effective strategies suggested in this text, you can avoid that kind of nightmare.

Technical Documents

You will be writing types of documents that you have never before written or even read, such as memos, feasibility reports, and proposals. You need to learn the various styles, formats, and conventions for these documents.

Patti Haines uses guidelines from the NRC to determine the format for her reports to that commission, but she uses a generic format for the evaluation reports she submits to management. Bill Vinikour uses an evaluation report format modified for Environmental Impact Statements.

USING THE TEXT

This text will help you acquire strategies for writing in technical communities by helping you understand the communities, your readers, and their purposes. Once you understand how readers read a text, what their needs are, and the social and political situation in which you are writing, you can begin to develop strategies for writing your text.

Part I helps you analyze your readers, purpose, and situation, and then provides you with strategies for planning your document. In Part II you will learn the processes and techniques for creating effective documents within technical communities. And in Part III you will become familiar with the genres of technical discourse.

CHAPTER SUMMARY

The ability to communicate is a necessary skill for any job. In your job you will belong to a technical community. By acquiring its language and conventions, you will become a member of your technical discourse community.

You will communicate within your organization in one of three roles: technical specialist, adjunct specialist, or writer. In your role you will write to people both within and outside of your discourse community.

Most of the day-to-day communications you write will be in the form of memos and letters. Occasionally you will also write long proposals and reports. Because your writing will have a social and political context, your documents will often involve persuasive elements. You may write as part of a team of content specialists, as in concurrent engineering, or in collaboration with graphic artists and other writers. Your documents will be reviewed and revised by your peers, editors, and supervisors.

Writing in technical communities involves the following aspects:

- Social and political contexts.
- Loss of sole authorship and ownership of texts.
- Firm deadlines; high expectations for quality copy; and real consequences, ranging from rewriting a text to being fired for drafting ineffective documents.
- Documents which are reader-based and used by readers to make decisions or do something.
- Different technical genres and subgenres, including proposals, reports, memoranda, letters, etc.

PROJECTS

Collaborative Projects

1. **Writing in Discourse Communities**

   Whether or not you realize you are doing it, you adapt your language to your audience's discourse community. You use the terms "serve" in a community of waiters and waitresses, the term "love" in a community of sociologists, and the term "volley" in a community of armed services personnel very differently from the way you use those same terms in a community of tennis buffs.

   a. Work in groups of five. Make a list of the different discourse communities with whom you presently talk.

   b. Discuss the differences in your language as you move from one of these communities to another.

   c. Try to think of a time when you used the wrong language in a discourse community. Discuss the incident and the results.

2. **Adapting Language to a Discourse Community**

   Discourse communities often use the same words but define them differently. In a community of computer programmers, the meanings of the terms "Apple," "byte," and "menu" differ from the meanings of these words in a community of restaurateurs. You need to be aware of these differences when you write for a particular community.

   a. Work in groups of five.

   b. Look at the list of discourse communities listed in the far left column of Figure 1-8. Think of terms used in these communities that have different meanings in other communities. For example, "apple" is a personal computer brand name in a community of computer analysts, but it is a fruit in a community of agricultural workers.

   c. In column 2, list each term you think of for a community. For example, for the first community listed in column 1 (computers), you could list the terms "apple" and "bit." Indicate the community's definition of each term in column 3. You would define "apple" as a computer manufacturer and "bit" as the smallest unit of computer memory. Then list the other community in column 4. You could list agriculture or farming for "apple" and manufacturing or even horse breeding for "bit." Finally, in column 5 define the term(s) according to the new communities. "Apple" would be defined as a fruit in an agricultural community; "bit" would be defined as a drill in a manufacturing community and as a the mouthpiece of a bridle in a horse breeding community. The first row in Figure 1-8 is filled in for you.

	Community	Term	Definition	Other Community	Definition
Computers	Apple	PC	Manufacturer	Agriculture	Fruit
Physics					
Biology					
Chemistry					
Genetics					
Marine sciences					
Automotive technology					
Medicine					
Robotics					
Electrical engineering					
Political Science					
Geography					
Agriculture					
Food production					
Transportation					
Others					

Figure 1-8 Conflicting Definitions in Discourse Communities

3. Writing in Your Field

If you plan to enter a content area, follow the instructions under "a. Technical/Adjunct Specialists." If you plan to be a writer, follow the instructions under "b. Writers."

a. Technical/Adjunct Specialists

- Work in groups of five.
- Each person should interview someone in the field in which they expect to work upon graduation. Find out the types and amount of writing being done, and what you should do to prepare yourself to write in that field.
- Meet in your groups and share your findings. Determine similarities and dissimilarities.

b. Writers

- Work in groups of three.

- Each person should interview someone who works as a writer in a technical community. Interview people in different organizations. Find out the types of writing being done, the skills required, and their recommendations for you to prepare yourself for a career in technical communications.

- Meet in your groups and share your findings. Make a list of the skills you will need to succeed in a career in technical communications and rank them.

Individual Projects

1. When you write in a discourse community, you need to use the language and conventions of that community. This is not a new concept. You have been adapting your language to your audience for many years. Figure 1-9 lists four types of audiences. The first row is filled in to demonstrate the differences in the way in which you may greet each type. Fill in the remainder of the chart. Then look at the differences between the various audiences.

	Peer	Parent	Teacher	Unknown older person
Verbal greeting	Hey	Hi	Hello	Nice to meet you/How do you do
Greeting for a letter				
Closing for a letter				
Written request for something important				
Term for policeperson				
Term for female peer				
Term for male peer				
Term for a shirt				
Term for running shoes				

Figure 1-9 Adapting Language to Discourse Communities

2. You have been writing in nonacademic communities for a long time. When you were very young, you wrote invitations to birthday parties to your peers and thank-you letters to relatives who sent you presents. Make a list of the nonacademic writing you have done in the past six months. You have probably written more than you think.

    Place a "T" next to those documents that are technical, a "B" next to those that are business-related, and an "S" next to those that are social. What differences do you notice in the way in which you use language in these different types of documents? Consider why those differences exist.

3. Your technical writing instructor wants to make certain the course meets your needs. Review the case studies in this chapter. Make a list of the skills which Patti Haines, Leslie Penles, and Allan Margolin needed to be effective communicators. Cross out those skills which you already have. Rank the remaining skills according to those you believe to be the most important for you to learn. Then write a brief memorandum to inform your instructor of your needs.

4. Write a letter to someone or interview someone in the field in which you expect to work upon graduation. Learn what types and the amount of writing being done, and the skills you need to prepare yourself to write in that field.

    Based on the response you receive to your letter or during your interview, make a list of the skills you need. Check off those skills you have already mastered. Place an asterisk next to those you plan to learn in this course.

5. Your technical writing instructor would like to find out if you are acquiring the communication skills necessary to enter your field. Either write or meet with a personnel officer at your school, at a nearby employment office, or at an industry in which you'd like to work. Find out the requirements for written and oral communication for an entry-level position in your field. Write a brief memorandum informing your instructor of your findings.

CHAPTER 2

Writers' Processes and Decisions

WRITING IS A COMPLEX TASK. You not only have to keep in mind your ideas, but you also have to think about how your readers will react to them. Furthermore, you need to consider the communities and context in which you write as you make decisions about the content, organizational pattern and focus, point of view, and style of your text. You make these decisions as you engage in the various processes of planning, drafting and revising.

Because writing is so complex, it is often a messy business. Figure 2-1 shows a page from a draft of a technical document written by a professional writer with over 20 years' experience. After drafting a section of the document on the computer, the writer printed out a copy. As she read it, she realized she needed to revise the text so that the reader could more easily follow the discussion. Figure 2-1 shows the results of the revision. Notice how the writer has added and deleted information, moved sentences around, and corrected typos.

a look at the problem. Prior to determining methods for retaining dropouts,

studies ~~will need to~~ *should* be undertaken to determine (1) who the drop-outs are and,

(2) why they have dropped out. Such assessment studies will examine the economic

and solid status and role of minorities in the region to identify aspects

specifically related to minority problems. *Again, the Center for Needs Assessment is uniquely qualified for this study.*

To provide assistance for the *is* wide variety of needs and for the varying

priorities within these concerns among the states, ~~identified during the needs~~

~~sensing activities,~~ (the Laboratory will provide dissemination and technical

assistance strategies involving (a) the development of data bases, (b) analyses

of existing knowledge, programs and methods relating to ~~the~~ various areas, (c)

training programs for educational leaders, and (d) service as brokers between

regional school systems and agencies and other regional agencies, R& D centers

and national agencies and R & D centers which have the expertise, knowledge and

programs available to provide the needed assistance.

In addition, shareholders indicated during a needs sensing meeting a large number of problems.

insert p. 2

Figure 2-1 A "Messy" Revision

Effective technical writers use three writing processes: planning, drafting, and revising. We'll consider these three processes in this chapter. Although we will cover the processes one at a time, good writers do not go through them as three separate steps. Good writers move back and forth among the three—planning a bit, revising their plans, drafting, revising their draft, redrafting, planning further, etc. After looking at each process separately, we'll examine how the three processes work together and why they make writing a messy business.

We'll also take a look at how writers create a text, and we'll see what strategies they use to help readers both understand their documents and accept their message. Finally, we'll consider a scenario that shows how these processes interact as writers make various decisions in their efforts to compose effective technical documents.

PROCESSES OF COMPOSING

This chapter presents an overview that will provide you with a picture of how a document is composed from start to finish. We'll begin by examining how writers engage in the three processes. In the following chapters we'll look more closely at the strategies and techniques involved in each.

All writers, regardless of whether they're professional technical writers, engineers, or college students, and regardless of whether they're writing technical reports, business letters, poems, or short stories, use similar composing processes. While all writers plan, draft, and revise, the amount of time they spend on each process and the ways in which they approach them differ, depending on the contexts in which they are writing, the topic under discussion, the type of document being drafted, and the writer's own style. Some writers spend a great deal of time planning and little time revising. Others spend almost no time planning; instead, they draft everything that comes to mind and then spend time replanning and revising. Technical writers may spend less time on certain aspects of planning and revising than writers composing essays, because technical writers may have to meet deadlines, and therefore have less time to spend in planning and revising.

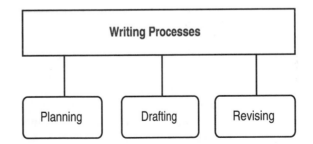

Technical writing is usually done within the context of a social setting that may involve several people whose discourse communities differ from that of the writer. An engineer may hold a conference with a plant operator before writing a report for a vice president whose background is finance. A marketing analyst may work with a dietitian in drafting a proposal to develop a new soup for the company's product line.

Technical writing also occurs within a political context in which writers and readers may have different goals and biases. As a result of a plant evaluation report to the Nuclear Regulatory Commission (NRC), an operator can be fired or a plant fined. While the operator may want to blame the problem on the NRC's procedures, the NRC expects the plant to assume responsibility for its own problems. The writer of the report must therefore negotiate between these two points of view and compose a document persuading the NRC not to issue a fine.

As writers work through the composing processes, they continually consider the social and political context in which a document is being written. Let's look at these processes.

THE PLANNING PROCESS

During this process you engage in numerous activities and use a variety of strategies to make appropriate decisions for composing a meaningful message that can be effectively communicated to a reader. Some planning activities occur only in your head; others involve note-taking, outlining, or interviewing.

The major activities involved in the planning process include determining the context for your document and making verbal and visual decisions about your text. Each of these activities requires you to make a variety of decisions that will eventually determine whether or not your reader understands and accepts your message.

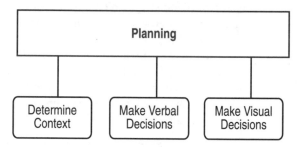

Determining the Context

Technical writing is always done within a context, never in a vacuum. For your writing to be effective, you must constantly keep your readers, your purpose, and your situation in mind.

During the planning process, you need to analyze readers' knowledge of the subject and their attitude toward the information to be presented. You want to consider their organizational and professional communities, and the circumstances under which they read the documents. In addition, you need to determine whether the purpose is to instruct and/or inform readers. Further, because most writing involves some aspect of persuasion, you need to determine how you want readers to perceive your message, as well as the factual evidence and emotional appeals that convince readers to accept your message.

Finally, you need to consider the social and political situation in which the document is being written. You need input from your supervisor and possibly the supervisor's supervisor. You may need to obtain information from your peers or your subordinates. In addition, political aspects may be involved; the document may offend someone, or it may further your career, or it may improve the environment. A deadline may exist or a competitor may need to be outbid.

■ Let's examine how a chemist with the Library of Congress engages in the writing processes. The chemist is drafting a proposal intended to persuade the Stauffer Chemical Company to support a research project to conserve old manuscripts in which the pages are disintegrating.[1] During a telephone call, the product supervisor at Stauffer informs the Library chemist that Stauffer can support the project by supplying some of the chemicals and by training the Library's chemists to work with chemical alkyls. The supervisor suggests that the chemist draft a brief proposal, which will be presented to the supervisor's director for approval of the project.

The chemist begins planning the proposal by trying to get a sense of his readers—the supervisor and the supervisor's director. He recognizes that, while the supervisor at the chemical company may be knowledgeable about the chemical aspects, he has little knowledge of the problem of manuscript deterioration. This is probably also true of his supervisor. In addition, the readers are members of a discourse community of product developers, not researchers.

Next the chemist considers his purpose, which is to persuade the readers to support the project.

Finally, the chemist considers the situation in which he is writing the proposal. Chemical companies often want to support good public relations projects that may also provide them with knowledge for improving their products. By basing his argument on the claim that the research on manuscript deterioration may also improve the company's product lines, the chemist should be able to develop an effective argument. ■

Making Verbal Decisions

These decisions relate to the information you include, the way in which you organize and sequence your information, the point of view you take in relation to your material, the content on which you focus, and the style in which you write.

Let's look at how you make these various decisions in relation to the context of your document.

Gathering Information   In writing term papers, you probably gathered much of your information from written sources such as general news magazines like *Time*, specialized popular magazines like *Home Office Computing*, and academic research journals. In writing a technical document, written sources are usually only one of several forms of information; field sources, such as observations, surveys, interviews, and experiments, often provide information for technical documents.

---

[1]Information is drawn from letters and memos from the Library of Congress, Stauffer Chemical Company, "Preservation of Libraries and Archives," *American Scientist*, 1987, by Chandru J. Shahani and WIlliam K. Wilson, and "The Self-Destructing Book," *Encyclopedia Britannica*; *Science and the Future Yearbook*, by John Dean.

■ The chemist at the Library of Congress has already gathered much of the information for his proposal by reading books on the topic, articles from professional journals and bulletins, and papers presented at conferences. In addition, he has read patents for several related procedures, and lab notes for related experiments conducted at the Library.

The chemist's field research includes listening to a consultant who is visiting the Library of Congress, interviewing another consultant over the telephone, and observing work at a laboratory already engaged in paper deacidification. In addition, he recalls what he knows about this topic from previous projects, or from research in which he's been involved. He also thinks of his own personal experiences related to the topic, including memories of crumbling pages of leather-bound volumes in his grandmother's library. ■

Selecting Content   The information included in a document depends on readers' needs. A writer's task is to determine what those needs are. Administrators may need to know that problems exist in their plants, but they don't need to know how to repair the heating system; the maintenance crew needs that information.

A scientist may need to know what specific types of chemicals in a food may be useful in lowering blood cholesterol; consumers probably do not.

Reader	Text
Scientist (researcher)	It is advantageous to use thermostable a-amylases referred to as 1,4-a-D-glucan glucanhydrolases and having the essential enzymatic characteristics of those produced by the Bacillus stearothermophilus strain ATCC Nos. 31,185; 31,196; 31, 197.
Consumer	Oatrim contains beta-glucan, one of the active components in lowering blood cholesterol.

■ While the chemist at the Library of Congress has gathered a great deal of information, he can only use some of it in his proposal. As a writer, the chemist is aware that if Stauffer can use the results of his proposed research to improve its products, those improvements will more than make up for the amount of money the company donates to the Library. Therefore, he needs to include information that supports his claim that the results of the proposed research can be used by the company. He also recognizes the need to provide background information on manuscript deterioration and on the chemical treatments for preserving paper. His reader is not familiar with this information, but needs to know it to understand and accept the proposal. On the other hand, the chemist doesn't need to include long explanations for the chemical treatments because his reader, a chemist, will understand. ■

Organizing and Sequencing Information Once you gather your information, you need to organize it so readers can easily understand and accept your message. You need to decide whether to put the most important information at the beginning or to lead up to it; how the various pieces of information are related; and how to sequence various chunks of related information.

In planning the organization and sequence of a document, you need to consider readers' reading behaviors as well as attitude and knowledge. If readers have requested certain information, they'll want to see it at the beginning of a text. However, if you're trying to persuade them to do something they may not want to do, they may be turned off if you start out with your request. You may need to lead up to your point instead.

■ The chemist at the Library of Congress synthesizes information about pH-values with the effect of acid on certain elements. He adds this data to the other information he has been collecting. As all of this data comes together, he begins to formulate ideas for categorizing his information. Eventually, he's ready to communicate his message—the need to find a successful and efficient method for preventing manuscripts from being destroyed by acidification. Because his reader knows little about the problem of decaying manuscripts, he decides to begin with an explanation of the problem so the reader can understand the need for the research. Once the manuscript deterioration problem is described, he can discuss the problems involved in finding a satisfactory chemical solution.

The Stauffer reader will expect the Library chemist to follow the conventions of a proposal, so the chemist knows he must organize the information in a problem/solution pattern. He plans to provide an historical overview of the problem and will therefore sequence the information chronologically. He then plans to follow the conventions for the solution section by presenting first an overview of the solution, and then subsections on the procedures, equipment and materials, and cost. ■

Determining Point of View and Focus The point of view of a text is twofold:

- *The way readers view what is being described or discussed.* Readers' points of view are based on their organizational roles as users, decision makers, or builders, as well as on their experience and knowledge.
- *The way writers views a topic.* Writers' points of view may range from unbiased informers, to instructors, to biased persuaders.

The view of a vice president of a company differs from that of a regulatory agent overseeing a company, just as a manager's view differs from that of an employee's. Writers need to determine how readers will use the information.

■ The chemist at the Library of Congress recognizes that the reader's point of view is that of a manager and decision maker whose main purpose in reading the document is to determine whether or not to support the proposal. ■

The focus, the main idea around which the information in a text is organized, should be related to the reader's point of view. If a reader is a manager, the focus should be on decisions a manager needs to make; if a reader is a consumer, the focus should be on how a consumer uses a device.

For a newly-developed pesticide, a document for scientists who may use the information in their own experiments might focus on the chemical composition of the product, while a document for farmers who may use the pesticide on their crops might focus on the results of the product.

Reader's Pt. of View	Text	Focus
Scientist (researcher)	Biological control agents such as pathogenic bacteria and viruses have been encapsulated in a protective starch matrix without the use of clerical cross linking agents.	Results of experiment
Farmer (User)	Insect pests of crops can be killed with a deadly "dessert" made of cornstarch and laced with microbes.	Solution to problem

■ The proposal from the Library of Congress needs to focus on the aspects that the director at Stauffer needs to know to agree to fund the project. Thus, the chemist plans to focus on how the research can lead to new scientific information, since this aspect is directly related to the interests of the chemical company. ■

Selecting Style   Style relates to how a text "sounds." It involves language, sentence structure, and tone.

Language.   Language relates to the words you use and the way in which you use them. Of course you may use technical terms when you know your readers are familiar with them. However, if you use words readers do not know, your language is considered jargon. You need to avoid this kind of language.

When you're writing for scientists who not only know what "dietary fiber compositions" are, but also expect you to use appropriate technical terms, the following sentence is acceptable:

Scientists   The dietary fiber compositions are colorless and devoid of inherent undesirable flavors and are, therefore, uniquely suitable for use in a variety of foods.

However, consumers probably don't know that a food product "devoid of inherent undesirable flavors" is tasteless. They may consider the sentence above to be pretentious and filled with jargon. If you're writing for consumers, the following sentence would be more appropriate:

Consumers	The product can be used to prepare foods without affecting their taste, texture or appearance.

■ The Library of Congress chemist is aware that the supervisor with whom he has been talking is a member of the general discourse community of chemists, and he may therefore use general scientific terminology. However, because the reader is *not* a member of the more specialized area of manuscript deterioration, the chemist should avoid technical terms related to that specific topic. ■

Sentence Structure   Sentence structure is related to the audience's professional and educational background as well as to their specialization. The less familiar readers are with a topic, the more they tend to become confused if complicated sentence structures are used. The simpler and more straightforward the sentences, the easier it is for readers to understand a text.

Syntax	Text
Complicated	I have, through a series of 12 tests ranging from the effect of process pII to low temperature treatment to sensory evaluation, discovered that water soluble dietary fibers, considered to be among the components of food not digestible by enzymes in the human gastrointestinal tract, can be recovered from milled products of oat such as oat bran and oat flour, which are useful in lowering cholesterol as observed in experiments with rats, after enzymatic hydrolysis of these substrates with a-amylase.
Straightforward	A series of 12 tests were conducted on water soluble dietary fibers, including milled products such as oat bran and oat flour which are useful in lowering cholesterol as observed in experiments with rats. These fibers are not considered to be digestible by enzymes in the gastrointestinal tract. The tests ranged from assessing the effect of process pH to low temperature treatment to sensory evaluation.
	I discovered that water-soluble dietary fiber compositions can be recovered from these milled oat products. The recovery occurs after hydrolysis of these substrates with a-amylases.

■ Realizing that his field is a complex one, even for another expert, the chemist uses a clear, fluent style so that the text can be read easily. ■

Tone   Tone relates to the writer's attitude toward a subject. A tone can be ironic, defensive, or gracious; formal or informal.

Tone	Text
Formal	We would like to express our appreciation for your assistance.
Informal	Thanks for your assistance.

■ Since the chemist at the Library of Congress is writing a proposal to decision makers whom he does not know, he knows he should use a formal tone. ■

Making Visual Decisions

In addition to decisions about verbal text, you need to make decisions about visual text. You may not have recognized that readers "look" at a document as well as read it. In fact, the visual aspects of a document, such as charts and illustrations, headings and subheadings, and the way in which a text is printed on a page can help readers locate information more easily and read text more fluently. Let's look at the visual decisions related to graphics, visual text, and format you need to make.

Selecting Graphics Graphics involve tables, charts, illustrations, diagrams, and other visuals. These are used to help readers visualize the concepts being discussed in the text. The use of graphics depends on the topic and the readers' familiarity with it. Readers unfamiliar with the process of generating electricity for commercial use can easily see how electricity is transmitted to homes and industries by looking at the diagram shown in Figure 2-2.

Figure 2-2
Visual Presentation
of a Complex Process

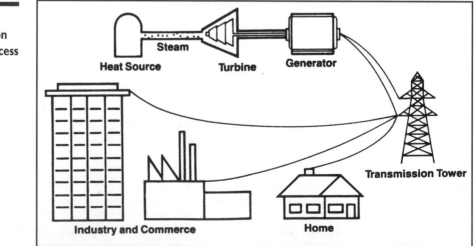

U. S. Department of Energy

■ The chemist at the Library of Congress does not believe that his topic can be graphically depicted in a way that would help the reader, so graphics are not included. ■

Determining Visual Text Visual text refers to such aspects as headings, lists, and special type like boldface and italic. Visual text is used to catch readers' attention, to help readers locate information, and to make text legible.

■ Because the proposal will be several pages long, the Library chemist decides to use bullets (such as that used to highlight the lists on pages 34-35) to list information, and special headings to introduce major sections. ■

Designing Format The format for a document determines the way text looks on a page. Formatting involves the placement of graphics and number of columns. The format for a letter involves placing a date and inside address at the upper left of the paper while the format of a report involves placing a title at the center top of the paper.

■ The Library chemist combines a letter and proposal format. The document begins like a letter, with the date, inside address, and salutation, but is then divided into sections with headings like a proposal (See Chapter 14). ■

THE DRAFTING PROCESS

Once you have made your verbal and visual decisions, you are ready to draft. However, as you draft, you may find that some of your decisions don't work and you need to replan. Don't be discouraged. Remember, writing is a complex task. During this process you actually put pen to paper, fingers to typewriter or computer keyboard, and write. If you have planned properly, drafting is much easier and goes much faster.

■ Let's consider how the chemist at the Library of Congress might begin to draft his proposal. Based on his sense of his reader and purpose, he concludes that his reader needs to understand the extent of the problem of manuscript deterioration and what it means to society, so he decides to begin the background section with a discussion of the problem of decaying manuscripts. Since his audience's bias is toward scientific knowledge rather than toward the preservation of old manuscripts, he realizes that in order to persuade the agency to fund the research, he must discuss previous attempts to solve the problem. He drafts the document. Consider how he applies his decisions to the draft:

Problem related to
manuscript deteriora-
tion.

Our histories, the visual images of our times, and the sounds of our civilization are all recorded on materials which are organic in nature and therefore inherently impermanent. They are subject to deterioration and decay as they age…

Books throughout the world are decaying at an alarming rate, placing in jeopardy much of the accumulation of human knowledge. Approximately 25% of the collections housed in the Library of Congress are in brittle, possibly unusable, condition and many more are highly acidic. Approximately 77,000 more become brittle every year. Libraries throughout the United States are reporting similar problems. Half of the Western European literature collections of both Yale University and the University of Michigan are also brittle. The problem is not limited to the United States but is being experienced in libraries throughout the world…

Problem of finding
chemical solution.

The problem of deteriorating paper is complex, massive and growing. While several methods for preserving manuscripts exist today, none can be accomplished quickly, easily, and inexpensively. At present we do not know of a practical, commercially feasible vapor phase deacidification process in operation or existence. ■

THE REVISING PROCESS

This process involves rereading and reviewing your text and probably changing it. During this process you assume an additional role: You become a *reader*, putting yourself in your reader's shoes. You read the document to determine whether you are saying exactly what you want to say, and to see if you have said it so readers can understand and accept it. If not, you must revise until you're sure you have communicated the message to the intended readers.

■ After completing the first draft of his proposal, the chemist at the Library of Congress decides the introduction does not focus sufficiently on the chemical problems involved with deacidification; his reader will be more interested in the chemical aspects than the manuscript problem. He revises his draft accordingly. As you read, try to decide if the revision is more effective than the first draft.

Problem related to
manuscripts.
Problem related to
chemical solution.

The problem of deteriorating paper is complex, massive, and growing. While several methods for preserving manuscripts exist today, vapor phase deacidification (VPD) is the most favored one.

VPD is an ideal method because:

- The whole book may be treated.

- Many books can be treated at once.

- Autoclaves for ethylene oxide fumigation of books are available.

- VPD is not likely to make an ink run, though it may change colors.

However, several aspects of present methods are questionable.

- The book is exposed to ammonia and then to formaldehyde. A great deal of formaldehyde polymer is formed in the books and this process is questionable.

- The use of an epoxidamine system provides some protection but the paper is severely yellowed.

At present we do not know of a practical, commercially-feasible process in operation or existence. ■

We'll discuss these composing processes in more detail in the following chapters.

A COMPLEX PROCESS

This step-by-step discussion of the three processes may make you think that composing is linear, that the three processes occur in sequence: first you plan, then draft, and then finally revise your writing. In reality, you move between the three processes any number of times and in different patterns. Figure 2-3 depicts this complex process.

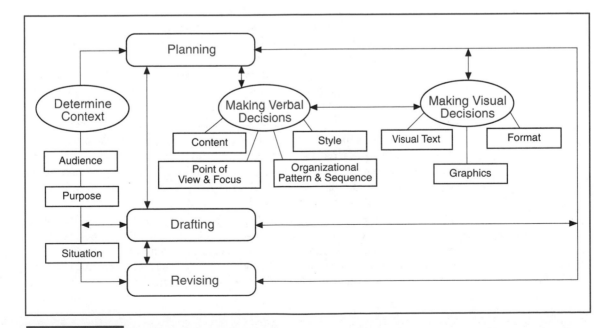

Figure 2-3 The Complex Writing Process

You may revise before you ever draft anything. You may think of one way to present your information, but before you draft, you may revise your plans and decide to present your ideas in a different way. As you revise after drafting a document, you may replan an entire section because you don't think readers will understand your discussion unless you begin the section with some background. You may even be drafting paragraph five when you think of a word you couldn't remember in paragraph two, so you go back to paragraph two to revise your language, then return to paragraph five to continue to draft.

A THINKING PROCESS

The writing processes are complex because *writing involves thinking*. Writing is like putting together a 2,000 piece jigsaw puzzle without ever seeing the picture on the front of the box. It's a process in which pieces of information, like the pieces of a puzzle, are put together to create a new meaning or picture greater than the sum of all the parts.

When you begin to work on a puzzle, one of the simplest ways to organize the pieces is by color, putting the reds, the browns, and the blues together. However, during this sorting process you begin to discern other criteria for determining relationships among the pieces. You find that some of the red pieces are not related. Instead, you notice that there are light reds and dark reds, light browns and dark browns. Some of the pieces have straight lines painted on them as if they were the sides of a barn, and others have curving lines like the petals and stalks of rose bushes. You not only realize that some of the dark red pieces are related to some of the light red pieces, but you also recognize that some of the red pieces are related to the brown pieces. You define these new relationships by line shapes rather than by color.

Just as you may fail to see the more complex categories when you first begin to organize the pieces of a puzzle, you may not perceive the intricate and subtle relationships between pieces of information when you begin to plan a draft of a technical report. Only after you begin drafting do you perceive these relationships and realize you need to revise. Thus, writing is always messy.

You can feel assured that, messy as your draft may appear, it is similar to that of other writers, including professional writers. You can also take some comfort from the fact that professional writers often need to make just as many revisions as you do.

Each time you write you are dealing with an entirely new set of information in which the pieces relate to each other in a unique way, so no one set of rules can accommodate the complex thinking process that comprises writing. However, you can learn some strategies to make the process easier, and you can develop criteria for determining whether you're on the right track.

STRATEGY CHECKLIST

1. During planning, think about your topic, audience, purpose, and the situation in which you're writing.
2. Determine the context for your document by responding to your readers' needs.
3. Adapt your point of view to that of your readers. Relate the focus of your document to the readers' point of view.
4. Organize your text to meet your readers' needs, and sequence the information according to their reading behaviors.
5. Use language which is easily understood by your readers.
6. Don't use complicated sentence structure, but don't write simplistically.
7. Match your tone with your readers' attitude.
8. Draft your text, based on your plans.
9. Revise your focus, content, etc., as you discover better ways to present your message so readers understand and accept your message.

INVENTING THE TEXT

When you compose a technical document, certain decisions are usually made for you. You don't need to decide on a topic; it's determined by your work. You write about topics in which you are or have been involved, or in which you plan to become involved. Your audience is also usually predetermined. You write for someone who has requested the information or who can provide you with needed resources. And, finally, your purpose is determined. You write to provide people with information or to persuade them to think in a certain way. But there are still many decisions which need to be made. As you make these decisions, you invent or create your text. You must decide on the information to be included and excluded, emphasized or subordinated. You must determine a point of view—that of an instructor or of a narrator. You must decide how to organize the information—in a narrative or in a problem/solution or cause/effect pattern, and how to sequence the information—chronologically, sequentially, or by importance. And you must decide on the style—formal or informal, technical or non-technical.

There are no rules to help you make these decisions, but there are three key questions that you can ask yourself to help you make the appropriate choice. These questions are based on the context (audience, purpose and situation) in which you write:

- How will readers react to this?
- Will this help achieve the readers' and my purposes?
- How will this affect the situation in which this document will be read?

By asking specific questions related to these three general ones, i.e., "Will readers understand technical terminology?", "Will readers have time to study the entire document?", you can begin to determine the content, shape, and tone of your text. The answer to the question "What does the audience know?" can help you eliminate unnecessary information because you don't need to include information the audience already knows. The answer to the question "What are the audience's biases?" can help you determine the sequence of information for a request. If the audience is biased against your topic, you may want to lead up to your request rather than present it at the very beginning of your document. The answer to the question "Will using a courteous tone achieve my purpose?" can help you determine the tone, and the question "Will my supervisor want this information included?" can help you determine the content of a document.

Let's look at a situation that occurred when a magazine printed a recipe containing an error. The recipe called for wintergreen oil, which may cause a bad reaction in some people if they ingest it. By the time the problem was discovered, the magazine had already been sent out. Subscribers needed to know about the error as quickly as possible. Management decided to send a letter informing subscribers of the error. With the letter they enclosed a sticker containing the revised recipe, which subscribers could paste over the incorrect one. One of the magazine's writers was assigned the task of drafting the letter to be signed by the editor.

Let's examine the decisions the writer may have made about the text. While the primary purpose of the letter is to prevent subscribers from using the ingredient, there are several secondary purposes. Because this country has become a litigious society, the magazine needs to preempt a possible suit over the ill-effects of eating cookies containing the oil. In addition, the editor needs to uphold her job and reputation since she is ultimately responsible for the contents of the magazine. The writer must negotiate among these multiple purposes to persuade subscribers to revise the recipe, to continue subscribing to the magazine, and to refrain from suing the magazine if the cookies were made and eaten.

Before the letter can be drafted, the writer needs to determine strategies that will persuade readers to accept the message—we made a mistake but don't be angry with us; and carry out the instructions—use the sticker and revise the recipe. These strategies are determined collaboratively during a series of meetings between the writer and the editor, the editor's supervisor, the firm's legal staff, and several peers.

Let's examine the kind of debate that may have occurred as the magazine staff attempted to decide on the claims they should make to persuade readers to do as they wished:

Strategies for Evidence/Appeals	**Readers' Projected Responses**
Do readers need supporting evidence to accept the claim that the oil can cause a problem?	**Issue.** Should the information concerning the health effects be included or eliminated as supporting evidence to persuade readers to revise the recipe?

Strategies for Evidence/Appeals	Readers' Projected Responses
	Discussion. Including the health information may scare readers. This could result in (1) cancellation of subscriptions by readers who stop trusting the magazine, knowing it may include harmful ingredients; and (2) suits by people who may have made the cookies before they received the letter. If the information is omitted, however, people may not pay sufficient attention to changing the recipe and may use the ingredient at a later time.
	Resolution. Include the information.
What is the readers' emotional response to the health problem? To the correction of the error?	**Issue**. Should the focus be on the recipe correction or on the health problem?
	Discussion. If the focus is on the recipe, readers may perceive the magazine as trying to correct a situation and therefore view it positively. If the letter focuses on the health issue, readers may perceive only the negative aspects of the problem.
	Resolution. Focus on the correction.
What is the readers' emotional response to old-fashioned things? To explanations of errors?	**Issue**. Should the information that the wintergreen oil was actually used by the grandmother of the person who submitted the recipe be used as supporting evidence to appeal to readers' values of family? Would such an effort reduce readers' anger with the magazine?
	Discussion. By explaining that the person who submitted the recipe actually used the ingredient in its natural form many years ago, before it was synthetically produced, may exonerate the magazine. On the other hand, readers may think the magazine is being defensive and form a negative reaction to the letter. In addition, readers will probably not be persuaded to change the recipe by that information. Furthermore, the explanation, which would have to include information concerning the difference between the natural oil and the more toxic synthetic oil now on the market, may increase the length of the letter, and many readers may not read a long message.
	Resolution. Don't include an explanation.
Will readers accept the magazine's policy as a valid claim?	**Issue**. Should the magazine claim that its policy not to test recipes is one reason the error occurred?
	Discussion. Readers will probably perceive this policy negatively. Most readers believe all recipes printed in a magazine are workable and safe, and that it is the magazine editor's responsibility to make certain that this is so. They may believe the magazine is not fulfilling its responsibility if they learn of this policy, and may very well cancel their subscriptions, feeling they can't trust the recipes which are printed there.

Strategies for Evidence/Appeals	Readers' Projected Responses
	Resolution. Don't include information about the policy.
What is the readers' emotional response to letters containing instructions? Will readers follow the instructions and the line of reasoning without supporting evidence explaining the reason?	**Issue**. Should the letter begin with the problem—the incorrect recipe—or with the solution—paste on the corrected version?
	Discussion. Readers will want to know why the magazine is sending them a letter, so they should be told to change the recipe immediately. However, readers may need background information to understand why they need to make the change. If they don't understand the reason, they may not think they need to make the change, because they don't plan to use the recipe. However, if they decide to use it at a later time, they won't remember the letter and the change won't be there. In addition, if readers are not told of the problem in the beginning, they may stop reading before they get to the explanation. Without knowing the reason for the change, they may not make it.
	Resolution. Begin with the problem and lead up to the solution.
What is the readers' emotional response to attention-getting devices? To form letters?	**Issue**. Should the letter call attention to the problem by using attention-catching phrases with large lettering on the envelope or at the top of the letter?
	Discussion. While large lettering probably catches a reader's eye, readers may not pay attention to phrases, such as "ATTENTION" and "IMPORTANT INFORMATION INSIDE," because they are often used by advertisers to introduce sales pitches rather than truly important information. On the other hand, phrases such as "WARNING" or "ERROR" may scare them. However, if the lettering and phrasing is not made to appear important, readers may toss the letter away without ever reading it, imagining it's a form letter for early renewal.
	Resolution. Use something to catch readers' attention.

The resolutions by the editorial staff are based on their judgment of the context in which the document is being written. For some decisions, such as selecting a word to catch the reader's attention, there does not appear to be a "good" resolution, so the word which appears to be the least negative and the most eye-catching is selected. Using these decisions as guidelines, the writer drafts the final letter. As you read it, consider how the writer applies the decisions to the actual text:

Attention-catching head.	ATTENTION SUBSCRIBERS
Focus on recipe.	We have been advised that some persons may unknowingly be allergic to an ingredient used in the recipe for cookies on page 90 of the September issue of our magazine. The recipe includes among several alternatives 1/4 teaspoon of wintergreen oil which contains an ingredient that may cause some people to have an allergic reaction. Furthermore, wintergreen oil is generally not intended for internal consumption. In light of this, we strongly urge you not to use wintergreen oil in preparing the cookie recipe and not to eat cookies that have been made with wintergreen oil. The other ingredients in this recipe, including the alternatives to wintergreen oil, are acceptable.
Begins with problem.	
Information on health effects.	
No explanation for cause of problem.	We urge you to mark out the recipe immediately so that neither you nor anyone else inadvertently uses this ingredient in preparing the recipe. We are also enclosing a sticker containing a revised recipe which we suggest you place over the original recipe.

Had the writer arrived at some of the alternative solutions to her decisions, the letter's message may have been entirely different. Consider the differences between the following letter and the previous one.

No attention-getting head.	Dear Subscriber:
Focus on recipe.	There has been an error related to one of the ingredients included in the cookie recipe on page 90 of the September issue of our magazine. The recipe includes 1/4 cup of wintergreen oil which should not have been listed among the alternative ingredients. Please cross it out. We are enclosing a sticker containing a revised recipe which we suggest you place over the original one.
Begins with problem.	
Omits information on health effects.	

Different decisions create different texts, as the two texts above indicate. In the remainder of this chapter we will look at the various decisions you make in writing technical documents, and examine some of the problem-solving strategies writers use to make effective decisions.

STRATEGY CHECKLIST

1. Determine readers' reaction to a decision.
2. Determine if your decision will help you achieve the readers' and your own purposes.
3. Determine the effect of a decision on the situation in which a text will be read.
4. Relate decisions to specific aspects of the context in which you write a document.

Scenario

This scenario describes the context in which a technical writer at an agricultural research laboratory writes, the processes in which she engages, and the decisions she makes as she drafts a document. You can follow her through the complex processes of composing to create a technical document. As you do, think about the various decisions the writer makes and the number of times she plans and revises her decisions.

An entomologist with the U. S. Department of Agriculture Southwestern Regional Laboratory has developed a biological engineering technique to eliminate the screw worm, a parasite that kills livestock. Linda Conley, a technical writer with the Laboratory, has been asked by her supervisor to write an article on the technique for an agricultural magazine read by the region's farmers. The purpose of the article is to convince farmers that the Laboratory is helping them, so that they in turn will persuade their Congressional representatives to continue funding the agency.

Conley's supervisor gives her an abstract and a report by the entomologist about the project. However, Conley has difficulty understanding the report, which was written for the scientist's peers, and she telephones the entomologist for further explanation. Later in the day, she picks up several copies of the magazine for which she is writing and scans the articles in it, noting the style, point of view, and types of visuals used. She also considers the information presented and from it she develops an idea of the kinds of knowledge her readers have and the topics in which they are interested. The articles serve as models for her own article.

The next morning she begins to *plan* the organization, point of view, and focus of the article. At first she plans to use the same organizational structure and focus as the entomologist has used in his report. She will start with a background description of the problem, describe earlier attempts to eliminate the parasite, and the biogenetic methods used in developing the present technique. She plans to conclude with the results of the new method. But as she thinks about this approach, she realizes it doesn't match the point of view of the various articles she read in the magazine. The local farmers probably aren't interested in the details of the discovery of the technique. What they want to know is how to get rid of this parasite. So she *revises* her plans. This time she *plans* to start with the results.

When she arrives at work, she turns on her computer, pulls up a word processing program, and *drafts* the following two paragraphs:

The government has initiated a program to sterilize about 500 million blow flies every week with Cesium 137. They are dropped over ranches where livestock have been infected with screw worms, the larva of the blow fly. The female is inseminated but fertilization does not result and, since the female can only mate once in her lifetime, there are no further screw worms to infect the cattle.

The drop is part of a program of biological control, sponsored by the U.S. Department of Agriculture's Animal and Plant Health Inspection Service.

By the time she begins the third paragraph, she has a sinking feeling that the readers will find the article dull. The articles she scanned in the magazine were written in a light, entertaining style rather than in a straight news format, indicating that the readers like to be entertained as well as informed. After *reviewing* her draft, she confirms her feeling and *plans* an entirely new beginning, which she *drafts* as follows:

Every morning, when the visibility is good, a squadron of 16 Cessna 206s take off with a cargo of sterile male blow flies to be dropped over ranches where a bevy of unsuspecting females wait below.

Happy with this introduction, she *revises* the next two paragraphs so the tone and style match that of the new paragraph. Here is her revision:

The government has initiated a program to zap about 500 million blow flies every week with Cesium 137 to sterilize them. They're dropped over ranches where livestock have been infected with screw worms, the larva of the blow fly. Since the female can only mate once in her lifetime, she mates and dies, leaving no larva. And, therefore, no screw worms.

The drop is part of a program of biological control, sponsored by the U.S. Department of Agriculture's Animal and Plant Health Inspection Service. The government has discovered a way to use the sexual habits of blow flies to eradicate the screw worm.

After *rereading* the new draft, Conley realizes that if she wants the reader to know what the article is about from the beginning, then the information in the last sentence of the third paragraph needs to precede the information in the second paragraph. She moves the sentence and makes several additional *revisions* to reflect the new organization. The resulting draft is as follows:

Every morning, when the visibility is good, a squadron of 16 Cessna 206s take off with a cargo of sterile male blow flies to be dropped over ranches where a bevy of unsuspecting females wait below.

Science has discovered a way to use the sexual habits of blow flies to eradicate the screw worm. The government has initiated a program to zap about 500 million blow flies every week with Cesium 137 to sterilize them.

The sterile blow flies are dropped over ranches where livestock have been infected with screw worms, the larva of the blow fly. Since the female can only mate once in her lifetime, she gets her once-in-a-lifetime thrill, but no larva. The result—the screw worm is being virtually exterminated in the Southern area of the United States.

The drop is part of a program of biological control, sponsored by the U.S. Department of Agriculture's Animal and Plant Health Inspection Service.

Conley returns to her initial *plan*—to list the areas involved in the program next—and *drafts* this fourth paragraph:

This program was begun 15 years ago in Flor., Texas and Mexico. Last Sept. air drops began in Guatamala and they are now being conducted in Belise. It is estimated that the program has saved the U.S. livestack industry nearly ten million dollars.

She then *reviews* and *proofreads* what she has written, *revising* the final paragraph for spelling and typographical errors, and grammatical and mechanical problems:

This program was begun 15 years ago in *Florida*, Texas and Mexico. Last *September* air drops began in *Guatemala* and they are now being conducted in *Belize*. It is estimated that the program has saved the U.S. *livestock* industry nearly *$10 million*.

The flow chart in Figure 2-4 depicts Conley's composing processes as she writes the article. As you follow Conley's decisions, you can see the complex nature of the process.

Conley plans and then revises her plans a number of times as she considers her readers' needs and her own purposes. Sometimes she plans and revises without ever drafting; other times she revises and returns to drafting without replanning. By making some major revisions in terms of point of view and focus in the beginning, before she drafts, she avoids revising an entire document.

The better you define the context in which you write, and the more carefully you make your initial decisions, the less revision you may need to do after you begin to draft. As you draft you may find that your plans need to be revised, but your revisions may not be as massive as they would have been if you had not planned. Planning can be especially important if you have a tight deadline.

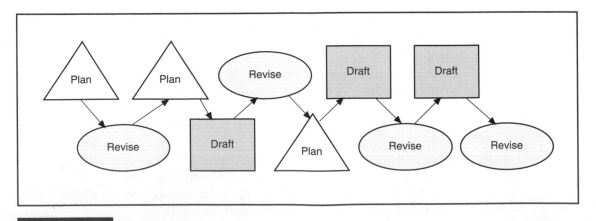

Figure 2-4 Conley's Writing Process

**CHAPTER
SUMMARY**

Writing consists of three processes: planning, drafting, and revising. These processes are complex; the writer moves back and forth between them.

Writing always occurs within the context of a reader and a purpose and in a social and political context.

Writing is a complicated process in which you invent a text. You make decisions based on your audience, purpose, and situation. These decisions involve content, point of view and focus, organizational pattern and sequence, and style, including language, sentence structure, and tone.

PROJECTS

Collaborative Projects

1. **Determining how you engage in the writing processes**

   a. Work in groups of three to five people. Take turns talking about the following aspects of your writing processes:

   - The planning process—Do you consider your audience and purpose? How do you usually gather your information? How do you determine your organizational pattern and style?

   - The drafting process—Do you sometimes begin to draft, then tear it up and start over? How many drafts do you usually write? Do you write very fluently on some topics and hesitantly on others? What do you think is the cause of the differences?

   - The revising process—Do you revise your drafts? When you revise, do you add or delete or reorganize information? Do you proofread your final drafts?

   b. As the others in your group discuss their processes, listen for new strategies that you can use to improve your own writing. Make a list of these strategies on the last page of your notebook. Try one of these new strategies each time you have a writing assignment. At the end of the assignment, return to the list and put an asterisk next to the strategy if it was helpful, cross it out if it was not.

2. **Making decisions**

   It usually takes time for freshmen to discover places to eat off campus. Your class has decided to help by publishing a booklet recommending places.

   Work in groups of three. Your group should select a place you would recommend. Write a paragraph explaining why you believe it's a good place for freshman to eat. Before you draft, make the following decisions:

   - Content
   - Organizational pattern and sequence

- Point of view and focus
- Style: language, sentence structure, tone

3. **Determining how writers write in technical communities**

   a. Work in trios. Each of you should interview a professional technical writer, or someone in your respective fields, to determine how the person writes. Ask the interviewees about their processes for writing and the various decisions they make.

   b. Compare the answers each of you receive from your respective interviewees. What similarities do you find? What differences?

   c. Make a list of the similarities and differences on a transparency. Report your trio's findings to the class, using the transparency.

Individual Projects

1. Think about the last time you wrote a paper for a class. What were your writing processes? As you think about how you write, consider the various activities and strategies discussed in this chapter.

   a. List the activities and strategies in which you engage.

   b. Compare your responses to the strategies in which Conley engaged in the scenario in this chapter. What differences do you find? Why do you think these differences occur?

2. Think about the last time you wrote a letter to a friend or relative. What was your process? As you think about how you write consider the various activities and strategies discussed in this chapter.

   a. List the activities and strategies in which you engage.

   b. Compare your responses with your responses to question 1. What differences do you find? Why do you think these differences occur?

3. Write a memo of three to five paragraphs explaining your writing processes to your instructor. Your instructor will use the information to decide the strategies you still need to learn to become an effective writer. The instructor will then develop a syllabus based on this information.

4. This chapter discusses different ways writers write. Think about the way you write. Do you spend a lot of time planning and revise very little, or do you dump everything on the page and then revise a lot? Do you do some planning and some revising, or do you skip the revising or planning processes altogether?

   Do you think that if you changed some of the activities and strategies you use you could write better? What changes would you make? Write a list of resolutions for improving your writing processes.

5. Study the letter in Figure 2-5.
   a. Consider the decisions the writer makes in terms of content, organizational pattern and sequence, point of view, and style. Why do you think these decisions were made?
   b. The writer also makes decisions about the format. He uses a block style (everything is flush left; see Chapter 13) and lists the points to be emphasized. Why do you think he makes these decisions?
   c. The writer has also made revisions. Study each one and see if you can determine why he made them.

Germantown Hills Fire Department

R. R. #2, Metamora, Il 61548
Chief Wayne Cox

RR #8 Ten Mile Creek Rd.
East Peoria, IL 61611
August 26, 1991

Emergency 116 Ambulance Service
Metamora, IL 61548

Dear EMTs:

I want to address a serious safety situation that has appeared in the auto accidents we respond on together. We have had people from your department as well as mine who get into these wrecked automobiles without any protective gear on. From this date on I am going to ~~reinstate~~ enforce the basic safety rules we in the fire and rescue service have had set down for us for many years.

1. Everyone who approaches or enters an automobile that has been in an accident must wear protective clothing.

2. ~~Upon the arrival of my department~~ anyone who is in or near the automobile without protective clothing will be asked to step back or leave the car.

3. ~~After~~ Once the victim or victims have been extricated, they will be moved to the ambulance gurney and turned over to the other EMTs for further evaluation and transport.

4. At all times we will provide protection with a charged fire line an any stabilization necessary ~~to protect all rescuers.~~

5. The fire department will try to set up a safety zone around the car and maintain it to keep spectators out of the work area.

I am not initiating these changes for ~~any reason other th~~an the personal safety of all the responding emergency persons. We all work together for the good of our community and I ~~feel a friendship and mutual respect for all our EMTs and firemen, and I~~ will not let any of them be injured or killed doing some unsafe act while I am chief of the fire department.

If you have any questions ~~on this subject~~ feel free to call me and I will attend your meeting to address these concerns. ~~As I stated earlier my people will be ordered to back away from the scene also if hey do not have the proper protective gear on.~~ Thank you.

Wayne Cox, Chief

Germantown Fire Department

Figure 2-5 A revised letter

Readers' Processes and Behaviors

YOU WRITE TECHNICAL DOCUMENTS for specific readers who need to understand your message to make decisions, use or repair a product, or follow procedures. In addition, your own job may depend on your ability to persuade readers to give you a raise, fund your project, or take your advice. The remaining chapters in Part I will help you learn strategies for writing so that your readers will understand and accept your message. In this and the following two chapters you'll learn how to get a sense of audience so that you can make textual decisions based on your readers. This chapter will help you understand your readers' reading and thinking processes and behaviors and will suggest strategies for facilitating them.

Let's examine your own processes by looking at the following article in *Tracks*, a quarterly magazine circulated to all employees of the John Deere Dubuque/Davenport Works. The purpose of the magazine is to maintain good morale by informing employees of events and people associated with Deere.

Here is the first line of the article's headline:

AN OUNCE OF PREVENTION...

You probably recognize these words as the first part of the Ben Franklin proverb: "An ounce of prevention is worth a pound of cure." Knowing that this headline appears in a magazine related to John Deere, you may predict that the article will be about something Deere does to prevent a problem.

Here is the entire headline:

AN OUNCE OF PREVENTION...
Preventing Oil Spills and Devising Effective Emergency Countermeasures

Our prediction after reading only the first line of the headline turns out to be correct. The full headline indicates that the article will indeed be concerned with John Deere's ability to prevent oil spills, and with countermeasures that can be taken if a spill occurs.

Read the first sentence of the article to see if it discusses oil spills as we are predicting:

What if the John Deere Dubuque Works experiences an accidental oil spill?

It does, and we can predict that the next sentence will begin to answer the question posed in this sentence. Read the following paragraph to discover if, as we have predicted, the writer answers the question posed in the sentence above.

Because of the large volume of petroleum-based products we use in the manufacturing process, a plan for dealing with accidental spills or leaks of storage facilities is needed to protect our land and waterways in the vicinity of such storage areas. Personnel need to be trained to deal with such emergencies, and a considerable amount of funds need to be invested.

The writer does answer the question posed in the first sentence. Apparently, Deere has a good chance of an oil spill occurring, and the company needs to plan for such an emergency. This is good "proactive" planning.

What do you predict the next paragraph will discuss? Read on to see if your prediction is correct:

Currently, if a spill should result in the north two thirds of our plant property, oil can be contained in skimmer ponds where it can be effectively cleaned up. The southern one third of the factory, however, could be a problem. Storm drains exist that discharge directly into the Mississippi River. Any chemical or oil spill is serious business; however, an oil spill in an area south of W-3 Building is extra critical.

And one such incident did occur this past winter with the accidental spill of some "tramp oil" from a malfunctioning containment tank. The overflow of an estimated 20 gallons escaped onto the ground, and a portion of that oil found its way into a storm drain.

But plant security personnel, who have been trained to deal with such incidents, were on the scene immediately and the discharge point was quickly secured by oil booms, absorbent devices that contain the floating oil. The amount of oil that escaped to the discharge pipe was actually quite small thanks to our emergency preparedness...

Prevention is our first line of defense and our employees are our best and most valuable resource in this war on pollution.

If you predicted you would read about Deere's emergency plans, then you guessed correctly.

Even though you don't work for John Deere and may not be knowledgeable about oil spills, you have been able to predict what you will read. That's because the information, organizational pattern and sequence, and style of the article reflect the way in which you read.

To write effectively for readers, you need to understand how readers read and interpret a text. Now that you've had some insight into the way you read, you should be ready to examine the reading process in general. After a discussion of the process, we will examine the thinking processes used by readers to understand a technical document, readers' reading behaviors in the workplace, and barriers that prevent readers from understanding a text. In each section you'll learn strategies for writing to meet your readers' needs.

THE READING PROCESS

Reading, like writing involves three processes: predicting, reading, and aligning.

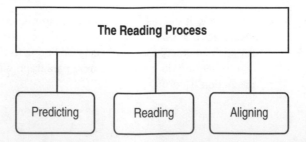

Predicting

As a reader you predict what you're going to read. Usually you make these predictions subconsciously, and you base them on clues, such as a document's appearance, titles, headings, and lists, and also on your previous knowledge and experience.

For example, when the employees at John Deere picked up the magazine with the article on oil spills, they predicted they would read about events occurring at Deere because all the previous issues dealt with topics related to Deere. Based on the headline, they predicted, just as you did, that they would read about a problem.

The predicting process in reading is similar to the planning process in writing. Readers "plan" what they'll read just as writers plan what they'll write.

Reading

Once readers predict what they'll read, they begin to read. This is the actual act of looking at the letters on a page and creating meaning from them. Again, this process of reading is parallel to the drafting process of writing in which a writer puts pencil to paper and creates meaning by translating thoughts into words.

Aligning

After reading, readers align their predictions with their actual reading experience. If they read the information they predicted they'd read, they continue reading and make their next prediction. But if the information they read isn't aligned with their prediction, then their reading is interrupted. They may reread the previous paragraph to make a different prediction, or they may reread the same paragraph to try to interpret the information differently. If they still can't align their prediction with their understanding of the information, they may stop reading.

If John Deere employees had begun to read about a child's school problems instead of a problem related to Deere, they would probably have stopped reading the article after the first paragraph. They might then have turned to the cover of the magazine to make certain that they had the right magazine, or they might have decided to continue reading to see if the writer eventually indicates a relationship between the child's problem and John Deere. Regardless of which method they use, their reading will have been interrupted. For the text to make sense, readers must be able to align their predictions with the information they actually read.

Fluency

Readers of technical documents need to read fluently (without stopping). They should be able to read through a text without going back and forth, engaging in what amounts to a revision process in reading. *You, as a writer, need to engage in revision to keep the reader from revising.*

You can help readers engage successfully in the reading process so that they read fluently by helping them make accurate predictions, and by ensuring that the text they read is aligned with their predictions. You can help readers make accurate predictions by providing cues to the information they will read, and aligning the text with the cues.

Providing Cues The following devices provide cues that readers can use to predict the information they will read:

- **Titles**. A title, whether of a whole document or a chapter, cues readers to the main topic of a text. In letters and memos, a subject heading indicates the topic (see Chapter 14). In the Deere article, the full headline provides a general statement of the topic to be discussed.

- **Tables of contents**. A table of contents should be included in a long report to indicate the various topics covered and the sequence in which each is discussed.

- **Headings and subheadings**. Headings and subheadings, like titles, give readers cues to the information they will read by indicating the main idea of a chapter section and its subsections.

- **Forecasts**. A forecast is comprised of one or two sentences in which the topic and subtopics of a document, section, or chapter are stated and the sequence in which the subtopics will be discussed is listed. A forecast should be placed at the beginning of a text, usually either at the beginning or end of an introductory section, chapter, or paragraph.

  By providing a forecast of the information to follow, writers help readers accurately predict not only what they will read, but also the order in which the various subtopics will appear.

  The first paragraph of the Deere article establishes the main topic. If, at the end of the first paragraph, the writer had written, "Let's look at Deere's plans to handle spills at both the north and south ends of the plant," the reader could have accurately predicted that the article would go on to describe, first, Deere's plans for the north end of the plant and, second, Deere's plans for the south end.

Aligning Text You need to be certain that your text reflects the information contained in your titles, tables of contents, headings and subheadings, and forecasts, because readers use these devices to predict what they will read. If the information following such a device is not related to it, the text won't align with readers' predictions, and readers' fluency will be interrupted.

In addition, the information following a forecast should be sequenced in the same order as it is presented in the forecast. This aligns the text with the readers' predictions.

All of the information in the Deere article relates to the main topic—preventing and cleaning up oil spills. If the forecast had included a sequence of the information, the writer would have had to list the north end of the plant first and the south end last to reflect the sequence in the text.

STRATEGY CHECKLIST

1. Help readers predict what they will read by using titles, tables of contents, headings and subheadings, and forecasts.

2. Align the information following titles, tables of contents, headings and subheadings, and forecasts with these cues. Make certain information is sequenced in the same order it is presented in the forecast.

A THINKING PROCESS

Like writing, reading is a thinking process. Whether you're trying to communicate an idea in writing, or trying to understand an idea you're reading, you're involved with a process that has to do with meaning. Just as you *create meaning* through your writing, you *create meaning* from your reading. Readers attempt to understand a writer's message by relating the information in a text to previous knowledge; categorizing related pieces of information (chunking); sequencing information in logical order; and integrating visual memory with verbal memory.

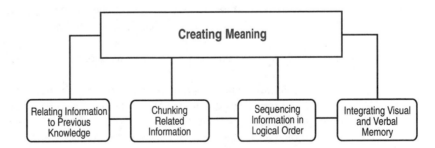

Previous Knowledge

Readers get meaning from a document by relating the information in it to their previous knowledge. Read the section below, which is excerpted from an annual report of Genentech, Inc. As you read, consider the information you must know to understand this excerpt.

Readers must know about	Text
Caesarian sections.	According to the National Center for Health Statistics, nearly three out of every 100 women giving birth in the United States suffer complications which may lead to delivery by costly and sometimes risky caesarean section. The overall number of caesarean sections may be on the rise, considering that more women are having children now.
Pregnancy complications: toxemia, diabetes.	In February, 1990, Genentech initiated Phase I human clinical trials on a recombinant form of the human hormone, relaxin. It will be evaluated for its usefulness in facilitating safe and natural childbirth in women who near the end of pregnancy suffer complications of toxemia or diabetes, or who are more than three weeks past due. These initial trials involve non-pregnant women volunteers and are being conducted at three centers: The University of North Carolina at Chapel Hill, University of California, San Francisco, and the University of Utah, Salt Lake City.
The birth canal and the cervix. Definition of "efficacious," "non-invasive".	In humans, relaxin appears naturally and is elevated throughout pregnancy to aid in reshaping the birth canal and relaxing, or softening, the cervix in preparation for delivery. If proven to be safe and efficacious in child-bearing women, relaxin may provide a non-invasive and cost-efficient alternative to caesarean section delivery.

Genentech, Inc.

Readers refer to four aspects of their memory to comprehend a document: facts, storyline/format, sentence structure, and previously-read text.

Facts To understand a message, readers need to relate it to the facts already in their memory.

Reading and *understanding* can be two different experiences. Reading is the act of translating the alphabetic code to create words. Understanding is the act of getting meaning from those words. To understand a sentence, you not only need to read each word in it, but you need to put the words together to make sense.

Read the following sentence:

Bill sat on the couch and leaned his head on the *antimacassar*.

You may be able to *read* the sentence above, but unless you are familiar with the word "antimacassar," you don't completely *understand* the sentence. You've probably concluded that a man is sitting on a couch, leaning in some direction against something. However, you don't know what his head is leaning against because you don't know what an antimacassar is.

You can use several techniques to determine the meaning of the word without looking it up in a dictionary.

- **Analyze it**. You can break it down into its root, prefixes, and suffixes which, in this case, are "anti" and "macassar." While you probably know that "anti' means "against," it's doubtful that you know what "macassar" means. Even if you were to learn that "macassar" is a kind of oil, named for a city in the Celebes Islands of Indonesia from which it came, you probably still won't arrive at the conclusion that an antimacassar is a piece of material placed on the arms and back of upholstered chairs and sofas to prevent oil from a person's head or arms from dirtying the upholstery.

- **Relate it to its context**. Based on the sentence, you can deduce that whatever an antimacassar is, it can be found on a couch. However, that clue can refer to a pillow, a child's toy, a small blanket, or even a painting above the couch. Furthermore, unless you've seen couches with extra pieces of upholstery draped across the back, you won't think of these pieces of cloth when you read the sentence.

As you've probably noticed, neither of these methods is satisfactory, because each requires that you have some previous experience or knowledge of an antimacassar. To understand a word, readers need to relate it to information already in their memory. If they can't, they don't understand the message. Readers who don't know that toxemia in a pregnant woman is life-threatening and often causes doctors to take a child by caesarean section won't understand why the drug in Genentech's annual report is being evaluated for women with this problem.

Readers' previous knowledge not only includes facts, but also *attitudes* toward facts, which are based on their prior knowledge of the topic. If they have had previous negative experiences with your topic, they will approach your text with a negative attitude. Readers' previous knowledge of the dangers of caesarean section births may cause them to perceive this section in the annual report positively.

By referring to readers' prior knowledge and attitudes, you can help them understand the new knowledge you are presenting. This is called the "given-new contract" between the writer and the reader. Under this contract, the writer agrees to relate "new" information to previously-learned (given) information. In the third paragraph of Genentech's annual report, the author relates the effects of the new drug to knowledge readers already have about childbirth.

By adhering to the given-new contract, not only in terms of the information you present, but also in terms of the language you use, you help readers understand your message. By defining words your readers may not know in terms familiar to them, and by using analogies to compare new ideas with objects or concepts your readers already know, you help your readers understand your message. You wouldn't define an antimacassar as "A doily placed on the back of a sofa to prevent the Vitalis from men's hair from soiling the upholstery," because your readers may not know what a "doily" or "Vitalis" is.

If your readers do not have knowledge about a topic or do not know a term, then provide background information, explanations, descriptions, or definitions so they'll understand your discussion. Genentech's discussion of the new drug includes background information that explains how the hormone being researched occurs naturally in humans and increases during a woman's pregnancy.

Because readers' prior knowledge influences their attitudes toward a topic, avoid references to topics your readers might perceive negatively. If you cannot avoid the reference, approach it positively. Readers may have negative attitudes toward experiments using humans, especially if a fetus is involved. The writer of the annual report assures readers that the first trials do not involve pregnant women, thus preempting negative responses.

Story Lines/Formats In trying to understand a message, readers not only refer to facts they've previously acquired, but they also refer to previous story lines and formats. If you know you're going to read a story, you expect it to have a beginning, a middle, and an end, because that's the story line you're accustomed to reading. If you're reading a letter, you know you can discover who wrote it by turning to the end of it, because you're accustomed to that format.

Technical documents need to follow appropriate formats so that readers find the information they expect, and find it where they expect it to be. By using standard formats with which your readers are familiar, you facilitate your readers' reading processes. (See Part III for standard formats for various documents.)

Sentence Structure Your knowledge of sentence structures also influences your understanding of a text. One of the reasons it is so difficult to read Shakespeare is that much of his syntax differs from today's sentence structure.

Readers expect sentences to follow the sequence of subject, predicate, object. Most people are used to reading instructions in this straightforward style. When these syntactic elements are inverted or interrupted, readers have difficulty understanding the meaning of a sentence.

Complex	The board, which has the two holes, one at the top and one at the bottom on the left side as you look at it, should be placed on its side.
Simple	The board has two holes in it. As you look at it, one hole is at the top and one is at the bottom on the left side. The board should be placed on its side.

Notice that three sentences were required to present clearly the information contained in the single sentence in the first example. The first example is difficult to follow because the subject of the independent clause (the board) is separated from its simple predicate (should be placed on its side) by the dependent clause (which has the two holes...). By using the straightforward sentence structures with which your readers are familiar, you can help them follow your discussion more easily. (Chapter 11 discusses more fully writing clear sentences.)

Previously-Read Text Readers use their short-term memory to understand text they are reading by relating it to text they have read recently. In reading the third paragraph of Genentech's annual report, readers can relate the effect of the drug (preparing the cervix for delivery) to the information in paragraph two (that the drug will be used to facilitate safe childbirth) and conclude that the drug will help a pregnant woman by softening her cervix so that she doesn't need a caesarean section.

By providing readers with signals, such as headings or subheadings, you can indicate to readers that you are about to discuss a topic noted in the forecast. Readers can then easily align what they read with the predictions they made when they read the forecast. Subheads and transitions between paragraphs also help readers align their predictions with the text by indicating the relationship between the new information and the old.

Chunking

Readers not only comprehend by relating information they're reading to previous knowledge, but also by relating various pieces of information to each other. Readers are able to follow a writer's discussion more easily when related ideas are grouped together in "chunks." They also remember the main gist of the ideas better. After reading a chapter in a book, you probably remember the main concepts of the chapter, but not all of the details.

Think back to the excerpt from Genentech's annual report. You probably remember that it concerns a new drug to reduce the necessity for caesarean section deliveries. You remember this because you have related all of the information into a single chunk—drug for caesarean deliveries. Now think about the testing of the drug. You probably remember something about it because there was an entire chunk of information about it.

By chunking related information together in your documents, you help your readers understand information and remember it. In the excerpt from the annual report, the information related to the testing of the drug is chunked in paragraph two, while the background information is chunked in paragraph three.

Sequence

The order in which information is presented also helps readers construct meaning. Readers tend to remember information presented first or last in a book, chapter, section, or paragraph.

Thinking back to the excerpt in the annual report, you probably remember that there are relatively large numbers of risky births, even though you may not remember the exact number. That information is presented in the very beginning of the report. You may also remember that this drug could provide an alternative to caesarean section deliveries, a statement made at the conclusion of the text. However, you may not remember that this new drug is a hormone, information provided in the middle of the text.

To help readers recognize and remember your main points, the most important information should be placed in the beginning or end of a book, section, chapter, or paragraph. In addition, the main topic or point to be emphasized should be placed in an independent clause and, if possible, should serve as the subject of a sentence.

Visual Memory

Readers attempt to create meaning from visual as well as verbal text. Approximately one-third of the brain is devoted to visual memory. Readers tend to remember more information when they combine their visual and verbal memories than when they use only one or the other.

When people read a text, they usually create visual images of the text in their minds. Some people actually have photographic memories; they remember the information on a page because they can visualize it. In addition to the visual image of the actual text, readers often create visual images of the content. For example, when you read a novel, you imagine how the main characters look and act. Even when they're reading a technical document, readers often imagine a person involved with the topic. While you read the instructions for setting a digital clock, you may picture yourself working on the clock, or you may picture the clock dial itself, before you actually follow the directions.

By providing your readers with pictures, graphs, etc., you help them visualize the content of your text. In addition, by providing a well-designed page, you help readers visualize the actual text.

STRATEGY CHECKLIST

1. Use terminology, sentence structures, and document formats with which readers are familiar.

2. Adhere to the given-new contract in terms of information, sentence structure, language, and document formats.

3. Provide readers with background, explanations, descriptions, and definitions if they do not have prior knowledge about a topic, or are not familiar with a term.

4. Avoid topics for which readers may have previously developed negative attitudes. If you can't avoid the topic, approach it positively.

5. Signal the relationship between your text and the forecast, between chapters, sections, paragraphs, and sentences by using titles, headings, subheadings, or transitions.

6. Keep related information together.

7. Place your main ideas in titles, headings, subheadings, and the first and last sentences of a text.

8. Place your main idea in the independent clause of a complex sentence. (See the Guidebook for a definition and examples.)

9. Use visuals, i.e., graphs, photographs, or line drawings, to help readers visually comprehend your message.

10. Present the text attractively so readers can visually perceive the page as a chunk of information.

READING BEHAVIORS IN THE WORKPLACE

You not only need to consider the reading and thinking processes of your audience, but you also need to consider the reading styles and patterns. These are often influenced by the audience's purpose for reading a document, and the conditions under which it is read.

Styles

Readers use a variety of reading styles, depending on their needs and situation. Huckin (1983) identifies the following styles:

- **Skimming**—reading quickly to get a general idea of the content. Administrators often skim material since they are busy and need only a broad overview of a topic.
- **Scanning**—reading quickly to find specific information. Supervisors or persons who have initiated a document often scan a document to locate specific information.
- **Searching**—reading quickly to find specific information which is then read more thoroughly. Readers who will use specific information within a document often use this style. This method basically combines skimming or scanning with one of the following two styles.
- **Understanding**—reading for thorough comprehension. Readers use this style when they want to know specific information and when they have sufficient time to study a document.
- **Evaluating**—reading to evaluate the information. Readers use this style when they will use the information to make a decision and when they have sufficient time to study a document.

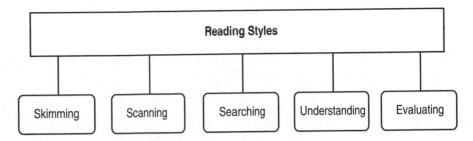

Patterns

Readers not only have different styles, but they also follow different patterns in reading a document. These patterns depend on readers' positions and roles in an organizational community, their purpose, and their situation. A vice president (VP) may only read the cover letter of a report before passing it to a division manager, who will study it. Or a VP may read the cover letter and then the executive summary to discover the major recommendations of a report before passing it on to the division manager who will be expected to carry out the recommendations. Division managers may read just those parts of a report which are of specific interest to them, or they may read a report in its entirety. In some cases readers, such as news reporters, only need to read the introduction and conclusion and can skip everything between.

Readers may read a document hurriedly, along with a number of others, if their purpose is simply to obtain some information, or they may set aside time to study a document if they need to make a difficult decision based on the information in it. If multiple readers receive a document, one of them may skim the text while another may read it thoroughly. Readers may or may not expect your document. If they do expect it, they may be anticipating it with anxiety or with pleasure.

Regardless of their behaviors, information should be organized, sequenced, and formatted so readers can understand and accept your message.

In writing an article for elementary school principals, an editor helped a writer perceive her readers' reading behaviors by suggesting the following:

Think of a person who has been at school since 7 a.m. It is now 3:15 p.m. and she has just finished seeing the last school bus leave. She has a staff meeting in 15 minutes. The magazine with your article is lying on the top of a pile of things to do. She picks it up.

Obviously, the writer needs to catch her reader's attention first and then get to the point quickly. She can't make the article too long: this reader has only a few minutes. If the writer uses subheads, the reader can easily locate those sections that are of special interest. The sentence structure needs to flow easily: this reader won't have the time or patience to wade through difficult prose. And unfamiliar ideas or terms should be defined: the reader isn't going to take the time to look anything up in a dictionary.

A person's reading behavior helps determine the length of a document, the parts of a report, the content, the sequencing of information, and the opening of a text.

STRATEGY CHECKLIST

1. Keep documents short and to the point for busy readers.

2. Provide administrators with cover letters that summarize documents, and/or with executive summaries.

3. Begin with background information or explanations to allay anxieties or defuse hostility.

4. Begin with background, explanations, or descriptions to familiarize readers with a topic if they aren't knowledgeable in the subject, or if they're not expecting a document, or if they are expecting a different response from the one you are making.

5. Use a table of contents, and headings and subheadings so readers can easily locate information. Separate different chapters and sections with extra spacing.

BARRIERS TO READER COMPREHENSION

No matter how hard writers work to facilitate their readers' processes and behaviors, readers may still misunderstand a text. The causes for this failure include the lack of two-way communication, differing frames of reference (points of view) and biases, and the knowledge effect.

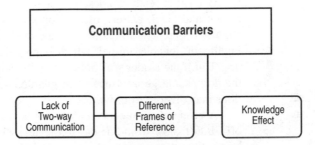

Lack of Two-Way Communication

Conversations involve two-way communication. You can convey a message more accurately in two-way communication than you can in one-way because you and the listener can see and hear each other. Your voice conveys as much about your message as your words. Listeners can tell by your intonations, pauses, and stress patterns whether you're angry, sad, or happy. In addition, by watching your face, body, and hand motions, listeners get clues about your message. If your listeners are not sure they understand what you're saying, they can question you and receive immediate clarification. By listening to your listeners respond to you, you can determine whether or not they have understood your message, and you can correct any misinterpretations on the spot.

However, when you write, you don't have these opportunities. You can't hear your readers' questions or their responses. Nor can your readers get clues about your message from your gesticulations, body movements, and facial expressions, or by hearing your intonations, pauses, and stress patterns.

When readers study written documents, they don't have an opportunity to talk to the writer. They can only reread the text; they can't get additional clarification. For this reason writers need to think of the questions their readers may ask and then provide answers. Writers need to *predict* how their readers will interpret their text and then solve any problems they think might arise.

Differing Frames of Reference

Readers approach a document with different frames of reference and biases based on their background and experiences. These frames of reference and biases influence their interpretation of a text, sometimes causing them to misinterpret a writer's message.

If a writer writes, "A dog can prolong the life of a person," a reader's frame of reference may be in terms of Lassie rather than the writer's dog, a mutt who is a mixture of a Labrador and a German Shepherd. If the writer describes his dog by explaining its mixed breeding, readers may be closer to visualizing the dog, but their perception is still not the same. They may picture a dog with the brown coloring of a German Shepherd and the floppy ears, blunt nose, and sweet personality of a Labrador. But the writer's dog may actually be predominantly black like a Lab, and have the pointy ears and sharp nose of a German Shepherd.

Even after reading a description, readers may not be able to picture the exact brown and black pattern of the dog's hair because everyone perceives things slightly differently. In addition, if the reader is a dog lover, he may perceive an intelligent, lovable animal, which is probably the way the writer perceives the dog. But if the reader was bitten by a dog when he was a child, he may perceive the dog as a large nuisance to be guarded against. Once again, readers' perceptions may diverge from writers'.

A group of engineers employed by an organization undergoing massive layoffs didn't know from one day to the next whether they would continue working. Every time they went to their mailboxes they expected to see pink slips telling them they were being let go. In a workshop on writing, these engineers were asked to revise the following statement:

In those establishments where suspension of labor is possible, direct those parties of management to the termination of the illumination.

Excerpted from Murray, Donald. 1968. *A Writer Teaches Writing*. Boston: Houghton Mifflin.

The statement was actually written during World War II by an official who was trying to tell people in government offices to "turn off the lights when you're finished working." But the engineers in the workshop didn't perceive the message this way. Instead, several of them thought the writer was saying that, wherever it is possible to stop work (suspension), management should lay off (terminate) the workers (labor). This exercise had been used in numerous workshops and no one had ever interpreted the message in this way before. The engineers' interpretation was influenced by their precarious situation in the organization. Had the message been written more clearly, the employees' interpretation may have been closer to the writer's meaning.

When you write, you need to remember that your readers' frames of reference are different from yours, and that their biases may cause them to misinterpret your message. You need to determine their frames of reference and biases and then consider these as you determine the information to include in your document. If the writer of the dog description had known the reader was afraid of dogs, he could have included a description of the dog's gentleness, or, as an example of the animal's good nature, described an incident in which the dog took care of someone.

You should also consider readers' frames of reference in adhering to the given-new contract discussed earlier. New ideas should be related to something within readers' frames of reference and should not conflict. If the writer of the memo about the lights had recognized the readers' mindset, he might have known the term "labor" could be confusing.

Because it is often difficult to know readers' biases, you should use neutral terms to avoid offending anyone.

Knowledge Effect[1]

Writers are saddled with the "knowledge effect;" they have knowledge their readers don't have. Because writers already know details and background, they have difficulty realizing that readers may not have this information, and that without it they may not understand the information.

The knowledge effect is one of the major causes of poor computer documentation. Documentation writers, who are often programmers, have difficulty understanding that anyone may not know such a mundane fact as how to turn on a computer. Yet, many new users do not.

You need to be careful that you don't assume that your readers have the same knowledge as you do. If at all possible, someone else should review your document before it goes to your readers. Because peer readers don't have your knowledge about a document, they can often perceive problems the actual readers may have in understanding your text. Peer readers are not limited by the knowledge effect.

STRATEGY CHECKLIST

1. Predict your readers' interpretation of your text, and then solve problems readers may have *before* they read the document.

2. Don't assume your readers have the same frames of reference, biases, or knowledge that you do.

3. Consider readers' frames of reference in determining the details, explanations, and descriptions to include in a text.

4. Consider readers' frames of reference in adhering to the given-new contract.

5. Use neutral terms so you don't offend your readers.

6. Have a peer review your document for problems readers may have, and then make the necessary revisions before submitting the document to the actual readers.

Scenario

Let's study a person actually reading a document to see how a reader responds to a text.

Janet W. wrote the letter in Figure 3-1 to a mail order flower nursery after discovering that some bulbs she ordered from the firm were not good. As you read the letter, think about the kind of response Mrs. W. wants from the nursery.

---

[1]Schriver, Karen. Adapted from a presentation at the Conference on College Composition and Communication, 1988.

**Figure 3-1
Janet W's Letter
of Request**

Dear Sir:

I ordered ten exotic Iris from you last fall. They cost $32.44 and I sent you a check in this amount on October, 3, 19xx. I planted them on the north side of my trailer on high land and covered them with straw, as you advised, in late December after the ground was frozen.

I uncovered the straw yesterday to check on them. They are all soft and moldy. We have had a very mild January. Very little snow and in the twenty's at night and 40 degrees to 60 degrees during the day. It has rained most of the month. Should I have left the straw off of the plants? Is it possible to get replacements this late?.

I'm no stranger to this type of Iris. I ordered some from Jackson and Perkins in 19xx in Central Illinois. I had wonderful luck with them.

I lived in a Senior Housing Apartment after my husband died and I was excited about being able to have my favorite flowers again.

I moved to Ohio last summer and usually their weather is the same as it is here.

I'm very upset about this and wonder if you could help me. If I did something wrong, please let me know. I'm enclosing the names of the plants I ordered. I am hoping to hear from you soon.

Sincerely,

Janet W.

Fredericktown, OH 43019

In response, Mrs. W. received the letter in Figure 3-2. As you read it, think about what your reaction to it would be if you were Mrs. W.

**Figure 3-2
Response to W's
Letter**

Dear Ms. W.:

Thank you for advising us of the difficulty you have experienced with your order #8207210531300.

We have issued credit to your account in the amount of $32.44 and applied it to your order #82210071274800.

This credit has brought your account to a zero balance and resulted in a refund check in the amount of $7.30. You may wish to use this check to reorder next season.

If we can be of further assistance, please let us know.

Sincerely,

Mary Doe

Customer Service Representative

Unfortunately, the replacement merchandise you requested is not available at this time.

Your care of the iris appears to have been correct. Perhaps it is possible that the soil in your iris bed is not well-drained. This would result in rotting iris rhizomes. The straw, if placed around the plants, should not be harmful. You may also want to check the area for insects as they also may be a cause of disease.

The above mentioned refund was sent prior to this date. Please advise if you have not received it.

Let's consider how the reader may perceive this letter. Compare your response with hers in Figure 3-3. The actual letter is in the lefthand column; the reader's thoughts as she reads the letter are recorded in the middle column; and an analysis of the reader's responses to the text appears in the righthand column. The reader's responses were obtained by having her record her thoughts into a tape recorder as she read the letter. Notice that the reader is interpreting, not editing, the text.

As you can see, the reader is not satisfied with the response and, in fact, doesn't understand some of it.

Using the strategies discussed in this chapter, the document is revised to solve the problems indicated in the reader's response, and get the reader to accept the writer's discussion and continue to order from the firm. Let's examine the revision in Figure 3-4.

The writer uses the reader's response to determine strategies for revising. Using readers' responses to documents is almost like having two-way communication. It permits readers to respond to the writer, and allows the writer to discover the readers' interpretation of a text. (We will talk more about reader response in Chapter 10.)

As a writer you need to learn to perceive your text in the same way another person reads it.

CHAPTER SUMMARY

Readers engage in a reading process similar to the writing process in which writers engage. The reading process is comprised of three phases: predicting, reading, and aligning. Writers can help readers read fluently by helping them accurately predict what they will read, by providing cues to what they will read, and by aligning their text with the cues.

Reading, like writing, is a thinking process in which readers create meaning from a text. Readers try to understand a text by relating their previous knowledge and attitudes to the information in the text, chunking related pieces of information in a text together, considering the sequence in which information is presented in a text, and combining visual and verbal memories of a topic.

You can help readers interpret their message correctly by avoiding information and language with which readers are unfamiliar; by providing details, explanations, and background information for readers who are not knowledgeable about their topic; and by using familiar document formats.

Readers' organizational and professional communities influence how they read and interpret a text. You need to consider the readers' reading styles and patterns in making textual decisions.

Readers misinterpret texts because of the lack of two-way communication, differing frames of reference, and the knowledge effect.

Text	Reader's Response	Analysis
Dear Ms. W.: Thank you for advising us of the difficulty you have experienced with your order #820721053 1300.	I don't want them to thank me, I want them to apologize.	The reader becomes agitated because the writer fails to respond with the appropriate etiquette the reader expects, based on her previous experience.
We have issued credit to your account in the amount of $32.44 and applied it to your order #8221007 1274800.	This number looks different from the one in the previous paragraph. Let me check...It is. I'm confused. Why is it different?	The number in sentence #2 differs from the number in sentence #1. It's probably a typo.
This credit has brought your account to a zero balance and resulted in a refund check in the amount of $7.30. You may wish to use this check to reorder next season.	Wait a minute. I don't understand. How can I have a credit if I have a zero balance? Let me read this again. I still don't understand.	The refund information is confusing because, based on previous experience, the reader does not expect a credit with a zero balance.
If we can be of further assistance, please let us know. Sincerely, Mary Doe Customer Service Representative	But what about the bulbs I have in the ground now? They didn't answer my questions.	The letter closes before the reader's questions have been answered, further upsetting the reader.
Unfortunately, the replacement merchandise you requested is not available at this time.	Oh, here it is. Why did they put it down here? I was afraid of that.	The response is provided as a postscript because the writer uses a "boilerplate," a generic letter with spaces to fill in names, addresses, account numbers, and order. Instead of integrating the response with the letter, the writer adds it in a postscript. But, based on her previous knowledge of letter formats, the reader expects the response in the body of the letter.
Your care of the iris appears to have been correct. Perhaps it is possible that the soil in your iris bed is not well-drained. This would result in rotting iris rhizomes. The straw, if placed around the plants, should not be harmful. You may also want to check the area for insects as they also may be a cause of disease.	Oh, good. I didn't think of that. I'll have to look the next time it rains. Rhizome? Do they mean bulb? I didn't think it would be. That's ridiculous. Not in winter.	The reader doesn't know the technical term for bulb.
The above mentioned refund was sent prior to this date. Please advise if you have not received it.	Oh, yes, I did get it. So that's why I only received $7.30. I put it with my bills so I wouldn't forget it because I wanted to ask for the rest of my money. I couldn't figure out why they had only sent me $7.30. I don't think I'll order from them next season. I wonder if I can find my address for the other place I used to get my bulbs.	The reader's agitation apparently began before she received this letter—when she received the check for $7.30 rather that $32.44 and no explanation was enclosed.

Figure 3-3 Response and Analysis of W's Response

Begins courteously.
Refers to flower, not
numbers. Opens with
apology. Uses "bulb,"
not "rhizome." Clarifies
credit problem.

Concludes courteously
with offer of further
assistance.

Dear Ms. W.:

We have received your letter notifying us of the problem you have had with your order for exotic irises and apologize for the difficulty.

Unfortunately, we cannot replace the bulbs at this time. We are, therefore, issuing you a credit of $32.44, the cost of the plants. We have applied the $32.44 to your account in which you owed a balance of $25.14. This leaves you with a credit of $7.30 which we are sending to you. We hope you will use this to reorder next season. Please let us know if you do not receive our check.

Your care of the iris appears to have been correct. The straw around the plants should not have been harmful. However, you might want to check that the soil in your iris bed is well-drained. A bed with too much moisture could cause your plants to rot. You may also want to check the area for insects which might be a cause of disease in your plants.

We appreciate your ordering your plants from us and hope you will continue to do so. If we can be of further assistance, please let us know.

Sincerely,

Mary Doe

Customer Service Representative

Figure 3-4 Revised Response to W's Letter of Request

PROJECTS

Collaborative Projects

1. Learning how people read your documents

a. This exercise requires that you work with a partner. You each need to select a short paper written recently for a class. Redo the paper in the following format:

- Leave a margin of about 4 inches on the right side of the page.
- Put each sentence on a separate line.
- Skip an extra line between paragraphs.
- Number the sentences consecutively.

b. Trade papers with your partner and respond to your partner's paper by writing down what you are thinking as you read. Record your thoughts in the right margin next to the sentence you are reading. Remember, you are interpreting, not editing, your partner's paper.

c. Exchange your papers when you are finished. Look over your reader's comments. What problems did your reader have in understanding your paper? What strategies can you use to solve these problems?

2. **Learning how you read documents**

    a. Work in groups of five. Assign one person to each of the following tasks:

- Read the first 25 pages of a novel.
- Read three newspaper articles.
- Read three letters.
- Read three articles in a magazine.
- Read a chapter one of your other textbooks.

    b. As you read, determine your reading style and process. Do you scan, skim, search, understand, evaluate? Why do you think you use this style? Do you predict what you will read? What kinds of predictions do you make? Did you read fluently? If not, why do you think you were unable to read fluently? Did you reread sections? Did you stop to look up information? On the chart in Figure 3-5, place an X in the appropriate boxes to indicate your reading style and processes for your reading assignment.

    c. Meet with the others in your group and discuss your reading styles and processes for your respective assignments. Fill out the remaining boxes on the chart, based on each person's style.

    d. Compare your chart with the charts of other groups in your class. Can you make some generalizations about reading styles for different types of documents? If so, what are these generalizations?

3. **Learning how people in technical communities read**

    This assignment is comprised of two parts:

    a. Interview someone in the workplace—a supervisor, technician, nurse, etc. Learn about the kinds of documents this person reads. Consider the following questions in determining the styles and patterns this person uses in reading the various documents.

- Does the reader read a lot of documents or only a few? Why does the reader read these documents? For what kinds of information is the reader searching ?

	Scan	Skim	Search	Understand	Evaluate	Predict	Read Fluently
Novel							
Newspaper							
Letter							
Magazine							
Textbook							

Figure 3-5 Reading Styles and Processes

- Does the reader have different styles and patterns for different documents?

- Does the reader set time aside to read, or does the reader just fit the reading in?

- Does the reader read a document in its entirety or only parts of it? What parts?

- Does the reader like to have an abstract or executive summary if it's a long document?

- Does the reader read appendices?

b. After you have completed your interview, compare your notes with other students' notes. Work in groups of five.

- What similarities and differences do you find in the reading styles and patterns among the people interviewed in your group? What do you think are the reasons for these similarities and differences?

- Do readers on the same organizational levels (administrators, technicians, professionals) have the same reading styles and patterns?

- Does their role (decision makers, users, producers) in the organization affect their reading behaviors? If so, how?

c. Your group should write a one-paragraph summary of its findings. Make and distribute enough copies for each member of your class.

d. Compare your group's findings with those of the other groups.

Individual Projects

1. The entire article for the oil spill was not printed, as the ellipsis (three dots) at the end of paragraph 3 on page 50 indicate. Here are the two paragraphs following paragraph 3 and preceding the final paragraph. Read each sentence and see if you can predict what the next sentence will be.

Paragraph 4
    Oil and petroleum products are but one among many chemical hazards which may affect emergency response.

    While petroleum or oil floats are fairly easy to round up and capture, some liquids which are highly flammable pose special safety hazards.

Paragraph 5
    The Dubuque Works has embarked on an ambitious training program to deal with all accidental spills to prevent environmental impact.

    In fact, Dubuque Works' training and preparedness is held in such high regard by area government officials, that recently, Dubuque County emergency personnel were trained right along with factory employees at Dubuque Works.

2. Examine one of your textbooks or a set of instructions that came with something you bought recently. Determine how the writer helps you predict accurately what you are going to read. Identify which of the strategies listed on page 53 is used by the author of the text you are examining. Note the strategy in the margin.

3. Examine a section of a chapter in one of your textbooks or a set of instructions you may have from something you bought recently. Determine how the author uses prior knowledge, chunking, sequencing, and visuals to facilitate readers' thinking processes. Mark the strategy in the margin.

4. Examine a paper you have recently written for a class to determine what strategies, if any, you used to help your instructor make accurate predictions about your text. What additional strategies could you have used to help your reader read your paper fluently? Make a list of these additional strategies. Before you write your next paper, look at the list. Try to use at least one of these strategies as you write your paper.

5. Mrs. S. from California placed an order with the same mail order firm as that discussed on pages 63 and ff. However, she only received part of her order and the rose she received as a bonus arrived in poor condition. She wrote to the firm, inquiring about the remainder of her shipment. She received the reply shown in Figure 3-6.

Figure 3-6
Response Letter

1/14/19xx
Mrs. Susan S.
3250 Sherman Drive
Los Angeles, CA

Dear Mrs. S.:

Thank you for your recent inquiry regarding order no. xxxxxxxxxxx.

The only part of your fall order that could be shipped is the following:

15 Ranunculas	$4.99
10 Freesia	4.99
+ Postage	4.55
	$14.53

This also included your bonus. The remaining items were not available. The balance was first applied to your bankcard. But for some reason, the charges were rejected. The charges were then applied to Visa no. xxxxxxxxxxx. You now have a zero balance.

We have replaced your bonus for shipment in the spring. The merchandise will be guaranteed through next fall. We hope this meets with your approval.

We sincerely appreciate your interest and look forward to being of service again. Should you have any further questions regarding these matters, please let us know. We are more than happy to be of assistance.

Sincerely,

C.D.
Customer Service Representative

Mrs. S' response to the letter appears in Figure 3-7. What problems does it show she has with the letter from the mail order firm?

What changes should the writer make in the letter to solve the reader's problems?

Text	Reader's Response
1/14/19xx Mrs. Susan S. 3205 Sherman Drive Los Angeles, CA 95608	
Dear Mrs. S.:	
Thank you for your recent inquiry regarding order no. xxxxxxxxxxxx.	Which order was that? I guess that's my original order.
The only part of your fall order that could be shipped is the following:	
15 Ranunculas $4.99	
10 Freesia 4.99	
+ Postage <u>4.55</u>	
$14.53	Yes.
This also included your bonus.	Yes, I got the bonus.
The remaining items were not available.	So that's why I didn't get them.
The balance was first applied to your bankcard. But for some reason, the charges were rejected.	That's strange. I don't understand. Which card did they charge originally?
The charges were then applied to Visa no. xxx-xxxx-xxxxx.	A zero balance where? How could I have a zero balance if it's been charged? Do they mean with them? But I don't have a charge with them. That's why they're using my Visa card.
You now have a zero balance.	
We have replaced your bonus for shipment in the spring.	Good.
The merchandise will be guaranteed through next fall.	What merchandise? Do they mean just the bonus? I would have thought all of it was. What about the stuff I haven't gotten? Are they going to ship me that in the spring? If not, I want the credit. I'm not going to pay for something I haven't gotten.
We hope this meets with your approval.	No. I either want my flowers or credit for them. I guess I'm going to have to write to them again.
We sincerely appreciate your interest and look forward to being of service again.	My interest! I didn't know why they hadn't sent me all of my order. They haven't been of much service.
Should you have any further questions regarding these matters, please let us know. We are more than happy to be of assistance.	Not very much.
Sincerely, C.D. Customer Service Representative	

Figure 3-7 Reader's Response to the Response Letter

Readers' Characteristics

YOU WILL WRITE FOR PEOPLE in diverse communities. As members of organizational communities, they will hold a variety of positions, sometimes several at the same time. They may be hourly workers, managers, consumers, clients, or citizens. They may also assume various roles in these communities, such as decision makers, regulators, and performers. You may try to persuade the marketing director of your organization to use a certain angle in marketing your product, or perhaps you will inform your supervisor of your work on a project during the past month, or you may instruct members of a maintenance crew to perform repairs on a snow blower manufactured by your company.

Readers will also be members of various professional communities, which may be the same as yours—or completely different. They will be specialists in fields ranging from oceanography to civil engineering, from economics to psychology.

You will also write for readers with a variety of backgrounds, who may or may not be familiar with your topic. The more you keep your readers in mind, the better your textual decisions will be, and the closer you'll come to achieving your purpose.

This chapter will help you learn strategies and skills for making textual decisions based on characteristics of readers in their various communities.

CHARACTERISTICS OF ORGANIZATIONAL COMMUNITIES

One of the communities to which readers belong is that of the organization for which they work. Each organizational community has a structure in which readers assume specific roles, such as hourly employees, professionals, and managers. Each community also has its own characteristics, standards, and conventions, which readers often adopt. Some have a dress code, others do not. Some are proactive, others are reactive. Some are conscious of environmental issues, others are not.

Your readers' organizational characteristics often determine their knowledge and attitudes toward your topic, as well as their use for your information. In making textual decisions, you need to consider the following organizational characteristics of your readers: relationship to your organization, position in the organizational hierarchy, and role in relation to your topic.

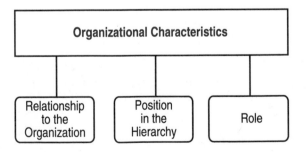

Relationship to the Organization

Some of your readers will be *internal*, working for the same firm as you; others will be *external*, working outside your organization. External readers are not limited to persons outside your firm. They can also be readers who work for the same company as you, but who work at plants or divisions different from yours. If your firm is multinational, you may need to consider readers located in another country. In a national firm, you may consider as external those readers who work in a plant in Dubuque, Iowa, because they have little in common with your plant in San Diego, California. You may even consider as external a reader in the Department of Information Services at your company if you work in a different division.

Because they are not part of your organization, external readers may not know as much about your topic as internal readers. When you write for someone who is external to your division or organization, you often need to provide more details, concrete examples, and background information than when your audience is internal to your organization.

On the other hand, don't take for granted that internal readers have the same knowledge or point of view as you do, especially if they are in different positions or departments. Project managers often perceive projects differently from those who are actually working on the project, and fiscal managers often perceive projects differently from either the project manager or the workers.

The two excerpts in Figure 4-1 describe a software program, "History of Charges," for accounting departments of small utilities. The first description is included in a sales brochure for external readers, managers of utility accounting departments. The second description is included in an orientation booklet for internal readers, the firm's programmers. As you read, compare the decisions made for each of the descriptions.

Both descriptions include information that readers need to know, and exclude information they already have or won't use. In the paragraph from the orientation booklet, the writer simply provides a general statement of the program's purpose, and includes the fact that this information is required by the federal government, information external readers would know but internal readers would not. Internal readers need to have this information because they will be expected to include all data related to governmental requirements when they write programs.

In the sales brochure, the writer gives a detailed description of the data that the user views on the monitor. The writer also provides the specific aspects of the program, since external readers will need the information to decide whether or not to purchase the program for the employees in their departments. Internal readers do not need to know this much detail, since they will not use the program.

Position in the Hierarchy

You will write for people in positions above, equal to, or below you in the organizational hierarchy. Those readers who are your supervisors usually have a

**Figure 4-1
Writing for Internal
and External Readers**

Sales brochure
(external readers).
Detailed description.

> **DIGITAL SYSTEMS**
>
> **History of Charges**
> The detailed on-line billing history file holds all the pertinent data about an account for each billing period and includes time-stamped payments and adjustments. Users can view four consecutive periods of billing data at a time, and they can scroll through the entire billing history maintained by your system so customer inquiries about previous billings can be answered quickly and accurately.

Orientation booklet.
Overview.

> **DIGITAL SYSTEMS**
>
> **History of Charges**
> REACT's History of Charges program provides an accurate response to customers' questions by maintaining details of the past 15 billing periods as required by the federal government. The program allows users to respond quickly and accurately to customer inquiries.

Digital Systems Inc.

broader view of a topic than you do. They perceive your project as only one aspect of a marketing or production plan. However, they may not be as knowledgeable about a particular topic as you are, since they are involved with it in a more general way.

Your supervisors are also often more knowledgeable about a political situation in an organization than you, and therefore may not share your point of view. Having evaluated a section of a plant, you may feel you want to write an evaluation strongly urging the foreman of that division to correct the problems you found. However, your supervisor may have evaluation results for over 20 areas of the plant, and your findings, in comparison to others, may be relatively minor. Your supervisor may suggest that you word your report less strongly, so that findings for those problems which are worse can be worded more strongly.

Administrators often have little time to read. They want information summarized, and they want to be able to find information in which they're interested quickly and easily. Therefore, your information needs to be organized efficiently for readers above you in the organizational hierarchy. You should present the most important information first. Long documents for administrators should include a cover letter and/or an executive summary to provide a brief overview of the information. A table of contents and headings and subheads should be used so that readers who are scanning a document are provided with cues for locating information in which they're interested. Documents should be kept as short as possible.

While readers above you may want only summaries, your subordinates may need detailed information to carry out specific tasks. A manager whose job is to oversee several projects may only need to know that an improper weld exists and what major actions are required to correct the problem, but operations personnel need to know exactly where the weld is located and what needs to be done to fix it.

Let's consider the documents in Figure 4-2, which include a letter to managers of maintenance facilities and an attachment to the letter for maintenance crews at these facilities. The letter from Cessna headquarters informs readers of an electrical problem found in one of the manufacturer's models. As you read the two documents, consider the different decisions the writer makes to compensate for the two different hierarchical levels.

Figure 4-2
Writing to Readers in Different Positions

To managers.

Purpose: To ensure that certain mechanical parts are correctly installed.

When maintenance should be done.

Requirements met.

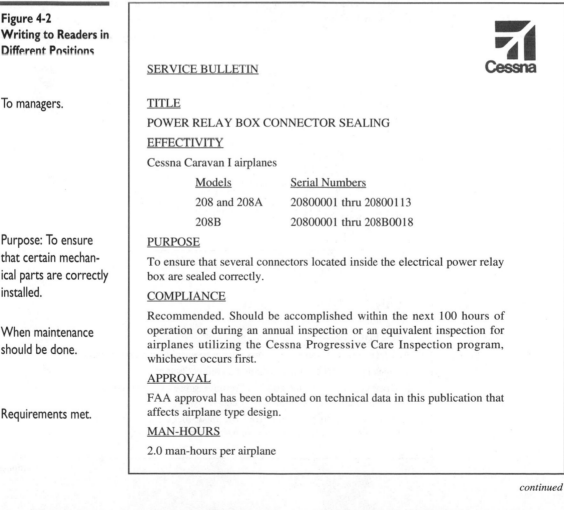

SERVICE BULLETIN

Cessna

TITLE

POWER RELAY BOX CONNECTOR SEALING

EFFECTIVITY

Cessna Caravan I airplanes

Models	Serial Numbers
208 and 208A	20800001 thru 20800113
208B	20800001 thru 208B0018

PURPOSE

To ensure that several connectors located inside the electrical power relay box are sealed correctly.

COMPLIANCE

Recommended. Should be accomplished within the next 100 hours of operation or during an annual inspection or an equivalent inspection for airplanes utilizing the Cessna Progressive Care Inspection program, whichever occurs first.

APPROVAL

FAA approval has been obtained on technical data in this publication that affects airplane type design.

MAN-HOURS

2.0 man-hours per airplane

continued

Figure 4-2
Writing to Readers in
Different Positions
continued

Labor required.
How to get materials.

Instructions attached.

Cost.

Cessna

MATERIAL

Required parts are available from the Cessna Supply Division through an appropriate Cessna Service Station for the suggested list price shown.

Part Number	Description	Qty/Airplane	Price
Q3-6077KT	Sealant Kit	1	$22.00 (PS) ea

ACCOMPLISHMENT INSTRUCTIONS

Power Relay Box Connector Sealing instructions are attached.

CREDIT

Cost Parts credit and a labor allowance credit of 2.0 man-hours per airplane will be provided to accomplish this service bulletin if the work is completed and a quick claim submitted by an appropriate Cessna Service Station by February 6, 19xx.

To maintenance
crews.
Purpose: To perform a
task.

List of materials to be
used.

Detailed instructions
for carrying out pro-
cedures.

ATTACHMENT TO SERVICE BULLETIN

Cessna

TITLE   POWER RELAY BOX CONNECTOR SEALING

The following procedures provide instructions to clean and seal the power relay box connectors which may not have been properly sealed.

EFFECTIVITY

Cessna Caravan I airplanes

Models	Unit Numbers
208	20800001 thru 20800113
208A	20800001 thru 20800113
208B	20800001 thru 208B0018

MATERIAL REQUIRED

Quantity	Part Number	Nomenclature
1	Q3-6077 KT Ablative RTV, Dow Corning	Sealant

CHANGE IN WEIGHT AND BALANCE

Weight Increase   Negligible

PROCEDURES

A. Clean and seal power relay box connectors as follows; refer to figure 1.

1. Open upper RH engine cowl and disconnect battery

2. Open upper LH engine cowl and remove screws (3) and cover (2) from power relay box (1); retain cover and screws.

3. Remove existing sealant from the base of connectors (4 & 5) and around wires which feed into connectors; refer to Detail A.

continued

Figure 4-2
Writing to Readers in
Different Positions
continued

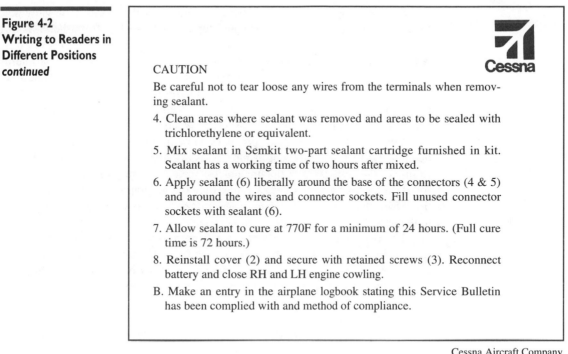

CAUTION

Be careful not to tear loose any wires from the terminals when removing sealant.

4. Clean areas where sealant was removed and areas to be sealed with trichlorethylene or equivalent.

5. Mix sealant in Semkit two-part sealant cartridge furnished in kit. Sealant has a working time of two hours after mixed.

6. Apply sealant (6) liberally around the base of the connectors (4 & 5) and around the wires and connector sockets. Fill unused connector sockets with sealant (6).

7. Allow sealant to cure at 770F for a minimum of 24 hours. (Full cure time is 72 hours.)

8. Reinstall cover (2) and secure with retained screws (3). Reconnect battery and close RH and LH engine cowling.

B. Make an entry in the airplane logbook stating this Service Bulletin has been complied with and method of compliance.

Cessna Aircraft Company

The managers of maintenance facilities and their maintenance crews are internal readers; all are employed by Cessna. The managers need to know the general problem and the overall cost for correcting it in terms of time, labor, and parts so that they can schedule sufficient time for the repair, assign the correct number of crew members, and order parts if necessary. The maintenance crew does not need this kind of general information. They need specific, detailed instructions for repairing the problem. They can learn about the general problem by scanning the original document, and then use the attachment for making the actual repairs.

Role

Readers can assume a variety of roles relating to the organization, the writer, and the topic. If readers are within the writer's organization, they may assume the role of subordinate, peer, or supervisor to the writer. If they are external to the organization, they may assume the role of client, subcontractor, regulator, or attorney.

They also assume roles in relation to the topic. They may be decision makers, producers, users, consumers, or regulators. Some of these roles are related to readers' jobs, others to their avocations, and still others to their personal lives as homemakers, spouses, parents, travelers, citizens, and consumers.

Readers' points of view are often determined by their roles. The view held by a decision maker toward a particular topic usually differs from the view of a laborer toward the same topic, and the view of a designer probably differs from both the consumer's and the salesperson's. While a product designer is mainly interested in how a product looks, a consumer wants to know how to use it, and a service representative wants to know how the product is better and differs from the competitor's. In addition, the views of members of legal and accounting staffs often differ from employees of other divisions.

You need to keep readers' points of view in mind as you write, relating a document's focus and content as closely as possible to your readers' roles. For example, if you're writing documentation for a software program, you need to focus on the user's point of view rather than the programmer's. The documentation writer for a word processing program needs to provide instructions related to the user's needs rather than to the computer's capabilities.

The two excerpts in Figure 4-3 are from a tutorial and a reference manual for *Codebuster*, a software program for architects. The tutorial is read by readers in the role of learners who want to discover what the program does. The reference manual is read by readers in the role of users who want to know what to do to perform a task. As you read the excerpts, consider the differences in decisions made by the writer.

Because readers of the two documents assume two different roles, they perceive the documents from two different points of view, and therefore the writer focuses on each text differently. The tutorial is focused on the *reasons* for marking or not marking an item, whereas the reference manual focuses on *how* to mark an item.

The content, too, is different. The information in the tutorial is more general and explanatory because the reader wants to understand how the program works. The information allows the reader the flexibility to try out different commands. However, the information in the reference manual is more detailed, and provides specific functions in which the user must engage to accomplish a task.

The style of the two texts also differs. The tutorial follows the conventions of expository prose, while the reference manual follows the conventions of instructional prose. Almost all of the sentences in the manual tell the reader to do something and use the second person, imperative mood, i.e., strike the <enter> key. In contrast, only the sentences in the last paragraph in the tutorial use this style.

CHARACTERISTICS OF PROFESSIONAL COMMUNITIES

Your readers are not only members of an organizational community, they are also members of a professional community comprised of people within a field of study. These communities may be as specialized as robotics engineering or as broad as biology. Some of your readers will be in your professional community, others will be in related communities, while still others will be in entirely different communities. You may write to an administrator of your company who spe-

**Figure 4-3
Writing for Readers
in Different Roles**

Tutorial (for learners).
Simulated project
(building a restaurant).

▲
Architech

QUESTION MODE

Codebuster starts out assuming that the entire [building] code applies [to your project to build a restaurant]. Thereafter, it relies on you to indicate what isn't involved. This lets you trim the code database down to more accurately reflect the scope of your restaurant. The more you know about your project, the greater your ability to tailor the code database.

> Example: If revolving doors are not involved in your project, but ramps are, you would strike the <enter> key when you came to the item for revolving doors and would pass over the item for ramps. Later when you print out code requirements which are applicable to your project, your list will include requirements for ramps but not for revolving doors.

If you look down the list on the pull down menu, you will see the item for revolving doors. Strike the <enter> key and watch the computer place a check next to the item. Select three or four other items and strike the <enter> key. Then go back to one of the items you marked and strike the <enter> key again to erase your mark. Go through the list. Mark off about ten items.

Reference manual
(for users).

How to use the
screen.

QUESTION MODE

▲
Architech

A pull down menu appears on the screen. At the top of the menu you're asked "Which of the aspects below are NOT involved in your project?"

Determine whether or not the first item is involved in your project. If it is not, strike the <enter> key. A check mark will appear next to it. If the item is involved in your project, don't do anything.

Move on to the next item by using the arrow key to move down the menu.

Strike the <enter> key whenever you come to an item that is not involved in your project. If you change your mind, strike the <enter> key again. The tag will disappear.

Architech, Inc., Chillicothe, IL

cializes in emergency preparedness, a field very different from yours as a chemist. Or you may write to a chemist at a university in another country who is engaged in a project similar to your own, and who is, therefore, a member of your professional community even though you are separated by thousands of miles.

Each of these professional communities has its own knowledge base, terminology, and conventions. Readers who are in the same community as you can discuss similar topics using the same language. Moreover, they often judge your knowledge and authority in a field by your ability to use the appropriate technical terminology. When you are writing for people in your own field, you should use terms, organizational patterns and sequences, conventions, and formats that are recognized by your professional discourse community. However, when readers are not members of your professional community, you need to avoid using terms with which they may be unfamiliar or which are defined differently by their field. When engineers use the term "noise," they refer to "sounds which do not exhibit clearly defined frequency components, but which comprise a frequency spectrum of energy." However, when communication specialists use the term "noise," they refer to "an electric disturbance that interferes with or prevents reception of a signal or of information." And when elementary school teachers use the term, they simply mean a hubbub.

Discourse communities also have their own conventions. Industrial technologists write in the passive voice, sociologists write in active voice.

Based on their knowledge of your topic and their proximity to your professional discourse community, readers can be classified into three categories: expert, generalist, and novice.

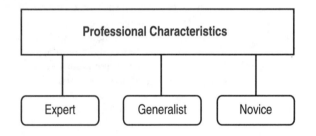

Expert

Readers who are members of your professional community are experts. They possess much of the same background information, use the same terminology, and have much of the same knowledge as you do. A nuclear engineer in one BWR (Boiling Water Reactor) utility can write to a nuclear engineer in a different BWR plant and consider the reader an expert. In addition, a nuclear engineer in a BWR plant can write to a nuclear engineer in a PWR (Pressurized Water Reactor) plant and consider this reader an expert, since the same basic concepts of nuclear power govern both types of plants.

When you write for experts, you should include technical details and use proper technical terminology. You can omit common background information and knowledge.

The following document abstract (Figure 4-4) for a scientific paper is written by a biologist who is the prime investigator of a research project with the U.S. Department of Agriculture. The abstract is read by other biologists who are experts in the same field as the writer. As you read, notice the terminology and focus of the paper.

Members of scientific communities expect information that discusses the results of experiments to include the categories and follow the organizational pattern of a classical report: hypothesis, procedures, results, and consequences. The abstract does so.

Figure 4-4
Abstract for
Biologists

Technical terms.
Begins with objectives.

Detailed description
of methods.

Results in terms of
increased intake.

 Agricultural
Research
Service

Cholecystokinin Octapeptide (CCK-8) Immunization: Stimulation of Feed Intake and Growth of Swine.

The objectives were: 1) to determine if immunization against cholecystokinin (CCK) would elicit production of CCK antibodies and 2) to investigate the hypothesis that CCK antibodies would enhance voluntary nutrient intake, growth and lean accretion.

Twenty-four castrated weanling pigs were randomly assigned to two groups (12 pigs/groups; 2 pigs/pen). The control group of pigs was immunized by intra-dermal-subcutaneous injections of human serum globulin (hSG); the CCK-8-immunized group by injections of the C-terminal octapeptide of CCK-8 conjugated to hSG. The same feed was offered to all pens ad libitum. Daily feed intake and weekly gain were measured. The animals were slaughtered for carcass dissection after 82 days. Serum antibody titers were confirmed in all CCK-8 pigs; mean binding at 1:200 was 37.9% + 1.19 SE. The group means, kg. hSG vs CCk-8 (probabilities) were as follows...

In conclusion, active immunization resulted in antibodies to CCK-8 and increased intake 5.4%, growth 8.2%, and lean accretion 7.2%. The growth and carcass responses to increased voluntary intake in this study are consistent with the greater responses obtained in a previous study where intake was increased 20% by experimental superalimentation (Pekas, Growth 49:19-27, 1985).

Since most experts know about CCK, the abstract does not define CCK, nor does it explain CCK's effect on pigs, or describe the composition of CCK. In addition, because experts are interested in details, the paper provides a precise description of the administration of the injections. It also presents the results quantitatively and in detail, as experts in the field expect it to do.

The language of the text is technical and contains long words without roots, e.g., superalimentation, ad libitum. Because experts are familiar with the subject matter, they have no difficulty understanding the long sentences (two of the sentences exceed 30 words). The document has a serious and formal tone; if they are to accept his conclusions, the author needs to persuade readers that this is a serious research study.

Generalist

Readers who are in a related field, but only partial members of your professional community, are generalists. They may know some, but not all, of the principles, terms, and formulas in your specialty, and they have no in-depth understanding of them. Often these generalists hold administrative positions in service divisions. The manager of the evaluation division of a software firm may have a degree in education rather than in computer science. The business managers of many firms have MBAs rather than bachelor of science degrees in chemical or industrial engineering. Donald Fites, the CEO of Caterpillar Inc., has a background in marketing, not engineering. If engineers want approval to work on a new design, they need to explain their ideas so that Fites, a generalist, understands.

Even in technical divisions, administrators often manage employees who are in fields other than their own. The manager of a research and development division in a papermaking firm may have been an expert in one aspect of chemistry, and then moved up to an administrative position in which she manages people in a variety of chemical areas. An employee who writes to this manager would need to be careful not to include technical terms related to a particular specialty. However, he would need to include information necessary for understanding a concept that is not common knowledge to those outside his area of expertise.

When writing for generalists, err on the side of clarity. You're better off giving too much information than failing to include enough. Define terms that are specific to particular fields, provide needed background information, and avoid including detail that is irrelevant to the reader's point of view.

The article shown in Figure 4-5 is based on the report of the CCK experiment described in the previous example (page 83). It was written by a technical writer as a brief article for a magazine for farmers (generalists). As you read, compare the technical terminology of this article with the previous one on CCK.

Figure 4-5
Article Written for Farmers

Begins with results. Discusses results in terms of pigs' intake, weight.

Background. Description of compound. Provides definition of technical terms. Future expectations.

Heavier Pigs with Less Fat Now Possible

The injection of a special compound into pigs appears to produce leaner pork. By injecting pigs with cholecystokinin (CCK) antibodies, scientists have been able to increase the amount of food ingested, thereby increasing overall weight but decreasing the amount of body fat. Over a period of 82 days, the injected pigs ate 5.4% more feed and grew 8.2% more than those who were not injected. Carcasses of the injected pigs contained 6 pounds more lean meat than the control pigs. They ate an average of 22.5 pounds more feed.

The new compound immunizes the animals against a natural hormone in their bodies, cholecystokinin (CCK), which inhibits pigs from eating once they're full. The compound consists of a fragment of CCK chemically bonded to a harmless foreign protein, serum globulin, found in human blood. The pigs were injected at 11 weeks of age and received three booster shots over an 82-day period. Scientists, however, believe the pigs probably need only one booster.

CCK-8 immunization of pigs requires federal approval. USDA scientists are pursuing a patent on using it to control appetite in animals.

Unlike the abstract for the expert, this document includes background information on various technical aspects, since readers need to have this information to understand the importance of the discovery to farming. The text includes a definition of CCK and a description of its composition, as well as an explanation of its effect on pigs.

While the background information has been added, many of the scientific details from the experts' document have been summarized or eliminated. These readers won't understand the statistical methodology, nor do they need it to understand the message in terms of their purposes. They simply need to know the statistical results.

Because the article is not written for a scientific community, the organizational pattern differs from that of the abstract. The pattern for this article is the one used by most magazines and newspapers, the inverted pyramid. The information is sequenced according to importance, with the most important information presented first.

The language is less technical and contains fewer Latin roots than the abstract for the expert. Those technical terms that are included are usually defined, e.g., the writer explains that CCK is a natural hormone, and that serum globulin is found in human blood. Sentences are also kept short, since readers are not familiar with the topic. No sentence is over 25 words. The tone of the article is serious. The author, whose purpose is to persuade farmers to consider using CCK if the USDA approves it, wants farmers to accept the conclusions of the research.

Novice

Readers who know very little, if anything, about a subject, and who are outside a professional community, are novices. A novice is not familiar with the technical terms of a particular field; does not have much, if any, of the basic knowledge or background information common to that field; and is unaware of recent findings in that field. A member of the chemistry department at a nationally-recognized drug company, who has a Ph.D. in biogenetics, would still be considered a novice as a reader of a report on automobile safety.

When Jeannette Goodfriend, the owner of a successful marketing firm, was looking over a twin-engine airplane that she was considering purchasing for her company, the salesperson commented that the plane was ready for a "hot section." Seeing the buyer's puzzled expression, a mechanic standing nearby explained, "At certain intervals, an area of an engine known as the 'hot section' needs to be checked for cracks or other problems occurring from abnormal wear. It's a routine check."

Though Jeannette is a college graduate and the owner of a successful business, her area of expertise is marketing, not avionics. While the mechanic's explanation is a simple one, the sentence structure and language are sophisticated. He doesn't "talk down" to the customer who is, after all, an educated person. The mechanic also realizes that the buyer needs to be reassured that this is a routine check, not an unexpected problem.

When writing for a novice, use as few technical terms as possible. When you do use technical terminology, define the terms or use an analogy that relates the technical, unfamiliar item to the reader's world. In addition, include whatever background information and details are necessary for the novice reader to understand your point or argument.

The article shown in Figure 4-6 is also based on the CCK report (p. 83). This article is a news release for the general public, which may range from experts to novices. The technical writer who drafted the article wrote it for a novice. Again, as you read, compare the terminology and focus of this article with the two previous ones.

As in the article for the generalist, this article contains more background information and fewer technical details than the abstract for the expert readers.

The general public expects information about experiments to be presented in the form of a news story, with the results up front, followed by a discussion of what has occurred. Therefore, this article, like the one for the generalist, uses an inverted pyramid as its organizational pattern.

The vocabulary in the news release is fairly simple and almost devoid of technical terms. When the writer cannot avoid using a technical term, such as CCK, she relates it to the reader's experience, i.e., CCK "tells them [pigs] when to lay off the feed." The sentences are also kept short. Unlike the previous two documents, the tone for this article is fairly light. The author is more interested in informing readers than in persuading them to do anything. She uses a light tone to catch the readers' attention.

**Figure 4-6
News Release for
General Public**

Begins with results in terms of readers' health.
Background.

Relates to common knowledge.
Description of process.

Results in terms of pigs' intake and weight.

RESEARCH NEWS

United States Department of Agriculture

Information Staff
1815 N. University St.
Peoria, IL 61604

Compound Makes Pork Less Fatty

The latest health trends have turned consumers away from fatty foods, reducing the demand for pork. Less fatty pigs could once again put pork back on the table.

Getting more lean pork to the marketplace faster might mean getting pigs to "pig out" more at the feed trough. Ordinarily pigs know when they're full and don't really "pig out." However, when injected with a special compound, pigs keep eating though they actually produce less fatty tissue. That means Miss Piggy can eat to her heart's delight and stay as slim as Vanna White.

Pigs produce a natural hormone, CCK, that tells them when to lay off the feed. But when injected with a compound that immunizes them against this hormone, the pigs eat more and put on more lean tissue. The pigs "think" they're hungry, so they eat more.

On the average, carcasses of the pigs injected with the new compound contained six pounds more lean meat than pigs who had not received the injection. The leaner pigs consumed an average of 22.5 pounds more feed and gained 11 pounds more weight than the control pigs over the 82-day treatment period.

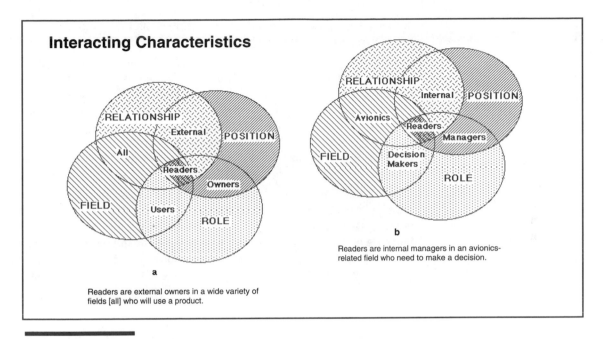

Interacting Characteristics

Readers are external owners in a wide variety of fields [all] who will use a product.

a

Readers are internal managers in an avionics-related field who need to make a decision.

b

Figure 4-7 Interaction of Readers' Characteristics

CHARACTERISTICS OF INTERACTING COMMUNITIES

The categories of readers' organizational characteristics —relationship, position, and role—interact with readers' professional characteristics. Readers possess the characteristics from each of the intersecting categories shown in Figure 4-7.

You need to be careful in making assumptions about your readers. Their needs will differ as different aspects of the various categories interact.

- DON'T ASSUME that an expert in your field, but external to your plant, knows all of the technical terminology you do. Some of your terms may be plant-specific, related to your organizational rather than professional community. An architectural engineer in one area of the country may use the term "vomitory" to refer to an exit area, while an engineer with expertise in the same field, but located in another region, may never have heard the term.

- DON'T ASSUME that an expert in your general field, but a specialist in an area different from yours, is also an expert in your special field. A mycologist who is an expert in one type of fungi, such as mushrooms, will not necessarily be an expert in other types of fungi, such as yeasts. Because both mycologists belong to an extended discourse community involved with mycota (fungi), they have some technical terms and conventions in common. However, as specialists in two different areas, they are members of two different subcommunities, and may differ in their use of various terms and conventions specific to their subcommunity.

- DON'T ASSUME that an expert in a field other than yours is also an expert in your field. A doctor learning to fly may be an expert in cardiopulmonary disease, but a novice in avionics: he is not an expert in your professional community and is not familiar with the technical terminology of its discourse. If you are writing an owner's manual for a small, single-engine airplane, you need to remember that the reader may not be an expert any more than the driver of a car is an expert in auto mechanics.

- DON'T ASSUME that readers above you in the organizational hierarchy are experts in your professional community, and that those below you are novices. Persons above you in the hierarchy may come from different branches of your specialty, or they may not even be in your field. Therefore, they may not be familiar with the technical terminology of its discourse. The Director of Research at the Library of Congress is a librarian, not a chemist, though chemists are employed in that division.

- DON'T ASSUME that persons associated with your field, but in different positions in the organization, are experts in your field. They may be members of your organizational discourse community, but not your professional discourse community. The director of training at a chemical company may be a chemist, or a former high school chemistry teacher or a specialist in personnel development.

STRATEGY CHECKLIST

A. Content

1. Provide more details, concrete examples, and background information for your external readers, as well as for internal readers outside your division, internal readers outside your field, and internal readers in hierarchical positions different from yours.

2. Omit or summarize information you're sure your reader knows; omit common background when writing for someone in your discourse community.

3. Include technical details when writing for an expert.

4. Provide a more general, overall description for those higher in the organizational hierarchy than you. Provide more detailed descriptions to those on lower levels.

B. Point of View and Focus

1. Relate a document's focus to readers' points of view.

C. Organizational Pattern and Sequence

1. Organize your information efficiently for busy readers. Present important information first. Provide a cover letter and/or executive summary, as well as a table of contents with long reports. Include

headings and subheadings. Place unnecessary information in an appendix.

D. Style

1. Use proper technical terminology when writing for experts.

2. Use as few technical terms as possible and define those you do use when writing for a novice.

3. Err on the side of clarity. Include details and define terms if you're unsure of readers' knowledge.

4. Adopt a formal, mannerly tone for readers above you in the organizational hierarchy, and also for external readers.

E. Assumptions

1. Be careful to consider the following:

   • Internal readers do NOT necessarily have the same knowledge as you do.

   • An expert in your field, but in a different organization, plant, or geographic location may NOT know all of your technical terms.

   • An expert in your general field, but in a different specialty, is NOT necessarily an expert in your specialty.

   • An expert in a field other than yours is NOT necessarily also an expert in your field.

   • Readers above you in the organizational hierarchy are NOT necessarily experts, and those below you are NOT necessarily novices in your area.

   • Readers associated with your organization, but in different positions, are NOT necessarily experts in your field.

PERSONAL/SOCIAL CHARACTERISTICS

Regardless of the type of work you do, you'll find yourself writing for readers of differing educational backgrounds and cultures. While these aspects may not be as important as readers' organizational and professional characteristics, they can be factors in making textual decisions. In addition, you will, more and more often, be writing for readers of different cultures. You need to be aware of the linguistic rules and conventions of these various cultures to make appropriate decisions.

In this section we will examine situations in which the factors of education are significant. We will also consider some linguistic rules and conventions of different cultures.

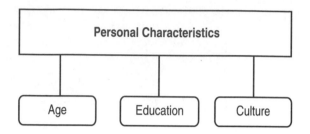

Education

Your reader's educational background can provide some guidelines for determining a document's style. You should gauge your syntax and language to readers' levels if they're not familiar with a topic. Write straightforward sentences. However, do not write simplistically.

Culture

We have entered an international economy. Many firms have clients throughout the world. Furthermore, because many firms have plants located around the globe, employees in a single company may come from a large number of nations. Caterpillar Inc. has plants in Scotland, Australia, India, Brazil, and Japan. Digital Equipment Corporation has clients in 82 countries from A(ustralia) to Z(ealand). In addition, an increasing number of firms are forming partnerships with firms from other countries. In the Diamond Star plant in Bloomington, Illinois, citizens of this country work side-by-side with Japanese workers in positions ranging from technicians to top level administrators. Mark Southerland, a noise analyst with Caterpillar Inc., finds himself sending several electronic-mail messages daily to readers at Caterpillar's various plants in Japan, Brazil, Germany, and Belgium.

The linguistic rules and conventions of these countries often differ from ours. If you are to communicate successfully with people of other cultures, you need to be aware of differences in their semantics (which relates to the meaning we give to words), sentence structure, and conventions.

Semantics The way in which we perceive objects and ideas often depends on our culture. We perceive dogs as pets and cows as food, while some Asian cultures perceive dogs as food, and the Indian culture perceives cows as sacred. When we read the word "cow," we visualize a bovine creature which is good for milk and steaks; however, an Indian visualizes a creature with far more stature. When we write, we need to be certain we're using words that enable the reader to perceive our ideas the same as we do.

Vocabulary differences abound. We have all heard stories about tourists in a foreign country who are misunderstood because a tourist's meaning of a word differs from that of a native speaker's. These differences exist even between England and America. A tourist tells the story of ordering a screwdriver[1] in a

London pub, and being told by the waitress that tool boxes were kept in the basement. When you are writing to readers of other cultures, you want to be extremely careful that they don't misunderstand you.

To avoid such misunderstandings, use technical terms whenever possible. Unlike most words, technical terms do not change from language to language. Regardless of your readers' culture, their definition of a technical term will not differ from yours. A "byte" is equal to eight bits anywhere in the world, and FORTRAN refers to the same computer language in Japan as in America.

Furthermore, you should avoid using idioms.[2] Think of the difficulty you had remembering idioms when you studied a foreign language. Readers from other cultures will not be familiar with many of your idioms and may even misinterpret them, translating them at face value.

Finally, don't include jokes that revolve around the connotations of words in your correspondence. Sometimes it's difficult to "get" a joke even in your own language. It is much more difficult in a foreign one. In addition, topics that one culture finds funny may be taken very seriously in another, and you may inadvertently insult your reader.

Sentence Structure If you have studied a foreign language, you know that linguistic aspects, such as sentence structure, differ from country to country. Differences in sentence structure can cause readers of other cultures to have difficulty reading a text we consider to be simple.

Neither the Chinese nor the Japanese language have sentences or paragraphs as we know them. Thus, the way we use paragraphs may actually create difficulty for Asian readers. Here is how an American letter was written by a Japanese translator:

The season for cherry blossoms is here with us and
everybody is beginning to feel refreshed. We
sincerely congratulate you on becoming more
prosperous in your business.[3]

The German language, which usually places the simple predicate at the end of a sentence, can be dense with modifying phrases and appositives. Readability formulas may not be appropriate for German readers who are used to reading dense sentences.

Conventions Just as the vocabulary and sentence structure of countries differ, so do conventions. To communicate effectively with people of other cultures, you need to know the conventions they follow in communicating within

[1]A drink composed of orange juice and liquor.
[2]An idiom is an expression peculiar to a language that may not make sense literally. From *The Glossary of Literature and Composition* by Arnold Lazarus and H. Wendell Smith, NCTE, 1983.
[3]Hamedo and Shima. *Journal of Business Communication.* 1982. p. 21.

their own culture. Asian convention demands an indirect approach in business correspondence; Latin American conventions require an extremely mannerly and formal approach; and German conventions require an historic approach.

The indirect approach of the Asian and Arab cultures is related to the concept of "saving face." Both Arabs and Asians go to great lengths to avoid embarrassing another person. For example, when a well-to-do Arab wishes to help out a relative with financial problems, he may say, "I have some money that I don't have time to put in the bank. Could you keep it for me, say for two months?"[4] The Japanese do not have a word for "no." If they do not agree with you, they simply don't say anything.

If you are writing to someone in an Asian or Arab country, you need to keep this face-saving concept in mind. In responding to a request for a replacement of a product that didn't work because the user hadn't followed the instructions properly, you might say, "We are sending you a new product and would like to suggest that if you will turn off the lock located at the lower side of the handle, you should have no difficulty using it." You have acknowledged the problem, and the reader, by not being blamed for failing to follow the directions correctly, has saved face.

To avoid embarrassing or irritating your readers, and to be certain that your readers understand and accept your message, you need to learn the various conventions they follow. You can do this by contacting the American embassy in that country or that country's embassy here in America. You can also contact a university faculty member who specializes in that culture, or you may interview someone from that country who lives locally. And, of course, you can always look up recent information about the culture in the library. Once you learn the conventions, you need to adapt them to your text.

STRATEGY CHECKLIST

A. Content

1. Include detailed information and explain all technical terms when writing on a topic with which a person may not be familiar.

B. Style

1. Gauge your syntax and language to your readers' level. Use straightforward sentences and few polysyllabic and Latinate words.

2. Learn the conventions of your readers' culture and adapt them to your text.

3. Use technical terms whenever possible when writing to readers of other cultures.

4. Avoid idioms and do not use jokes when writing to readers of other cultures.

[4]From an interview with M. Cherif Bassiouini, MidAmerica Arab Chamber of Commerce, by Charles Leroux, Chicago Tribune, 1989.

CHARACTERIZING UNKNOWN READERS

If you're writing to your own supervisor or a client with whom you deal regularly, then you probably have a sense of your audience. However, if you're writing to someone with whom you have little contact, you may have more difficulty analyzing your readers.

How can you go about getting a sense of your readers if you haven't even met them, let alone gotten to know them? You can refer to previous knowledge and experience with similar persons, and you can also obtain information about readers from secondary sources. It may also be possible to contact them prior to writing your document. In addition, try to visualize them.

Secondary Sources

To get a general sense of your readers, you may be able to get some specific help from your supervisor or a mentor. Before beginning to write, ask about the reader's knowledge, attitude, biases, etc.

Visualization

To get a sense of your readers, try to picture them, regardless of whether they're known or unknown. Write out a brief description of the primary reader, draw a picture or locate a photograph of someone similar, or fill out an Audience Analysis Chart (page 97). Keep it near you as you write.

Personal Contact

If it's at all possible, talk to your readers prior to drafting your document to get some idea of their attitude toward your topic, what they know, and what they need or want to know. These discussions can be very helpful if you're writing for readers whom you don't know, or with whom you are only slightly familiar, especially if they're in divisions other than your own or external to your organization. If you can't arrange a face-to-face meeting, then use a telephone. You may need your supervisor to arrange the meeting if the person is a client, subcontractor, or someone higher up or in another division of your own organization.

CHARACTERIZING A GENERAL AUDIENCE

As a content specialist, much of your writing will be for specific readers whose work is related to yours. However, you will occasionally need to write for the general public. For instance, you may be asked to prepare a news release about a project on which you're working, or to speak to the local Kiwanis club about some aspect of your field. On the other hand, if you are a writer for a technical organization, most of your writing will probably be for the general public. Often your task will be to "translate" a document written for experts into one the general public can understand.

The general public is comprised of readers who range from experts in your field to novices; from people sympathetic to your topic to people who are hostile toward it; from people who read your entire piece to understand it to people who simply scan it or read only the opening to determine whether or not they're interested.

How do you make textual decisions when readers have such a broad set of behaviors and characteristics? How do you determine what they know and don't know, or what their biases are?

One way to characterize the general public is to apply the characteristics of the majority of readers to the entire audience. However, in doing so, you may lose readers who do not fit the norm. Therefore, as you make textual decisions, you may also want to consider factors that may prevent individual readers from understanding and accepting your message.

Writing for the Majority

Many readers comprising the general public are probably novices in your field. Therefore, your textual decisions relating to content and style should be the same as those you would make in writing for a novice. You should provide background necessary for understanding a topic, and avoid technical language. Newspapers often aim at a ninth or tenth grade level in determining vocabulary and sentence structure level, while some magazines, the *Atlantic Monthly*, for example, aims at a twelfth grade level.

General readers often scan an introduction to decide whether or not they're sufficiently interested in a topic to read the remainder of a document. Therefore, introductions not only need to present the main idea but also catch readers' attention. Readers may also skim a document, searching for a specific topic or subtopic in which they're interested. Visual text, such as tables of contents, headings and subheadings, and boldface type, help these readers easily locate the information they want.

Considering the Individual

In writing for the majority, you may exclude or antagonize some readers who do not possess the characteristics you have assigned to them. If you want to write so that all readers can understand and accept your message, you need to consider the least knowledgeable reader as well as the nonsympathetic reader. This is known as writing for the "lowest common denominator."

In writing instructions for a computer program available to the general public, you may be aware that the majority of users know how to prepare a backup disk, but some may not. To be certain that all users use the program properly, you will want to include information on preparing a backup. Those readers who are knowledgeable can skip this section.

Some readers may even be hostile to your topic. If you want to persuade them to consider and perhaps even accept your view, you will want to maintain an objective and positive tone. For instance, if you are trying to persuade readers of the financial benefits of investigating alternative power sources, you might want to begin the document with a discussion of the low cost of such energy once it is developed, rather than with a statement about the amount of time and money required for the research. Such a preliminary discussion will reinforce those who are already sympathetic to the topic, and may provide the data necessary to persuade those who are not to at least read further and possibly accept your claim.

STRATEGY CHECKLIST

1. Write for the majority of readers when writing for the general public.
2. Write for the reader with the least knowledge of the subject, if you need to be certain *all* readers understand your message.
3. Write for the hostile reader, if you need to persuade *all* readers to do a specific activity or think in a certain way.

Scenario

You need to know as much as possible about your readers so that you can make informed textual decisions involving the content, organizational pattern and sequence, point of view and focus, and style of a document. Let's observe someone making these decisions.

To obtain information on traffic patterns for controlling traffic, a city is investigating the feasibility of hooking up with a supercomputer located at a nearby university. Dr. Susan Thomas, the Manager of Information Services for the city, has written the letter in Figure 4-8 to Dr. Carol Rose, Director of Computer Services at Midwestern University. Dr. Thomas has sent copies of her letter to two other city employees, Gordon Jones, the Director of Traffic Control, and Melissa Morgan, the City Treasurer, since both of them will be involved in making the final decision. In addition, if the project goes ahead, Mr. Jones will be involved in the use of the supercomputer and Ms. Morgan will be responsible for the hook-up. As you read, consider how Dr. Rose might respond to the request.

Before answering the request, Dr. Rose needs to get a sense of her audience to determine the best response. She recognizes Dr. Thomas as the primary reader, and Mr. Jones and Ms. Morgan as secondary readers. In addition, Dr. Rose has an internal reader, her boss, Dr. Joseph O'Reilly, the Vice President for Business and Finance. While he doesn't review her letters before she sends them out, she does send him a copy of everything she writes to keep him apprised of possible additions to the university's computer services.

Figure 4-8
Letter of Request to
Dr. Rose

City of XX
XX, KY
October 2, 1990

Dr. Carol Rose
Director, Computer Services
Midwestern University
XXXXX, IL

Dear Dr. Rose:

I would like some information concerning a hook-up with your Cray. Our city is considering the possibility of using the supercomputer to provide information on traffic patterns in an effort to alleviate some of our problems. We have a ZZ mainframe here and so could serve as a node if that is required.

I'd be interested in whatever information you can provide concerning how we would make the hook-up, what kinds of equipment we would need, what our responsibilities would be and how it would be operated.

I appreciate your assistance in this matter.

Yours truly,

Susan Thomas

Susan Thomas, Ph.D.

Manager, Information Services

Because Dr. Rose has never met any of these readers, she uses her intuition and previous experience in dealing with people in similar administrative and decision-making positions to determine the characteristics of these particular readers.

Dr. Rose arrives at the following conclusions:

- Dr. Thomas' title indicates she will be in charge of the node. The titles for the secondary readers indicate that Mr. Jones is responsible for the traffic flow and will use the supercomputer to control traffic, and that Ms. Morgan will be responsible for financing the hook-up, but will not be involved with the computer. Based on her previous experience with persons in accounting, Dr. Rose assumes that Ms. Morgan may be concerned with the cost differences between a mainframe and a supercomputer.

- Because Dr. Thomas is in the computer field, she can be assumed to be an expert concerning micro-, mini- and mainframe computers. However, she is probably closer to a generalist when it comes to supercomputers, since she admits in her letter that she does not know the technological details of supercomputers. On the other hand, both Ms. Morgan and Mr. Jones (though he has probably talked with his peers and read some information about the use of supercomputers in controlling traffic flow), can be considered novices since their positions are not directly involved with computers and neither probably knows much about them.

Audience Analysis Chart

While most people analyze their readers in their heads, you may find that using an audience analysis chart helps you get a better idea of your readers so you can make appropriate textual decisions. Some of the information may not be relevant for making decisions for this particular document but, if you get in the habit of thinking about all of these factors every time you write, then when one of them is relevant, you won't overlook it. It's like the checklist a pilot uses before taking off. If Dr. Rose were to fill out a chart indicating her sense of her audience, it would look like the chart in Figure 4-9. Look at this chart to see how it helps you get a sense of your readers.

	S. Thomas, Mngr. Information Services	G. Jones, Dir. Traffic Control	M. Morgan City Treasurer	J. O'Reilly V.P. Business
I. Reading Behavior				
Amount of time to read?	Sufficient	Little	Little	Little
Scan? Skim? Search? Study? Evaluate?	Evaluate	Evaluate	Search	Scan
Expect/Not expect?	Expect	Expect	Expect	Expect
Expect with anxiety? Pleasure? Skepticism? Neutrality?	Neutral	Neutral	Neutral	Neutral
II. Position in Organizational Community				
Internal? External?	External	External	External	Internal
Decision Maker? Operator? Technician? Service Person? Professional? User? Producer?	Decision Maker, Operator	Decision Maker, User	Decision Maker, Financial Services	Decision Maker
Primary? Secondary? Intermediary? Peripheral?	Primary	Secondary	Secondary	Intermediary

continued

Figure 4-9 Audience Analysis Chart

	S. Thomas, Mngr. Information Services	G. Jones, Dir. Traffic Control	M. Morgan City Treasurer	J. O'Reilly V.P. Business
III. Knowledge of topic				
Expert? Generalist? Novice?	Expert	Generalist	Novice	Novice
Knows what about topic?	Technical knowledge of computers, general knowledge of super-computers.	General knowledge. Knows what it can do in general terms but not how it works. A user's knowledge.	Very little. May have heard of them but doesn't know anyone who uses one and has never used one herself.	Very little. Only aspects related to cost.
Needs to know what about topic?	What she will need to do to hook up and what she will have to do to keep it going.	Whether or not it can really help improve traffic control.	How much it will cost and is it cost effective.	Cost.
Use of information?	To make decisions based on what she will need to do to operate it.	To make decisions based on how he will use it.	To make decisions based on how she can fund it.	To make long range decisions on computer usage.
Biases?	Don't know	A supercomputer can solve all his department's problems.	Don't know.	Don't know.
IV. Personal Characteristics				
Culture?	American	American	American	American

Figure 4-9 Audience Analysis Chart

Based on her sense of her audience, Dr. Rose uses the various strategies discussed in this chapter to make the following decisions:

Content

- The letter should be directed to Dr. Thomas as the primary reader.

- Dr. Thomas' position indicates she knows a lot about computers and therefore doesn't need an explanation of what they are.

- Since Dr. Thomas has never been involved in a hook-up, but will be responsible for the technical aspects involved in running the node to the supercomputer, she needs detailed information.

- Because Dr. Thomas is probably extremely busy, the letter should be kept fairly short, and a list of general responsibilities should be appended rather than included so that she can refer to it when she has time.

- The letter should provide a general discussion of the ways in which the Cray can help Mr. Jones, and also include general budget figures for Ms. Morgan.

  Mr. Jones needs specific information in terms of procedures for scheduling the supercomputer and the amount of delay to expect.

  Ms. Morgan needs information on the cost of the hook-up as well as the cost in relation to alternative services, such as a mainframe. The letter should provide a general estimated total budget figure with a breakdown of the major costs appended; a detailed budget isn't necessary at this stage of the project. Ms. Morgan may also need an explanation of the difference between a supercomputer and a mainframe.

- The specific information Mr. Jones and Ms. Morgan need should be appended to the letter, since Dr. Thomas is not interested in it.

Point of View

- Since Dr. Thomas will use the information to make a decision based on her operation of the supercomputer, the letter should reflect the operator's point of view and emphasize the hook-up itself.

- Since Mr. Jones will use the information to make a decision based on his use of the supercomputer, and Ms. Morgan will use the information to determine whether the supercomputer is cost effective, the letter should also recognize the user's and accountant's points of view respectively, and should include information concerning the use of the supercomputer to control traffic as well as a general estimate of the cost of the project.

Organizational Pattern

- Because Dr. Thomas is a busy administrator, the information should be presented as efficiently as possible. Dr. Rose will use the inverted pyramid pattern, which presents the most important information first and the remaining information in descending order.

Style

- Because of readers' high educational level, the language and syntax can be sophisticated.
- Since Dr. Thomas is the primary reader and is a member of Dr. Rose's extended discourse community, Dr. Rose can use technical terms.
- Since Dr. Thomas is not knowledgeable about supercomputers, but will eventually become involved with them, Dr. Rose needs to define technical terms specifically related to supercomputers.
- Because Dr. Rose doesn't know Dr. Thomas, she plans to use a formal, objective, and mannerly tone.
- Since Dr. O'Reilly does not appear to have any needs or knowledge different from those of the primary and secondary readers, the decisions Dr. Rose makes based on her analysis of Dr. Thomas, Mr. Jones, and Ms. Morgan includes Dr. O'Reilly as well.

As with an audience analysis chart, most people make decisions in their heads. However, you may find that a decision chart forces you to think about your reader's needs. If you fill out a chart, you will probably be better prepared to draft your text. If Dr. Rose were to fill out a chart to indicate her decisions, based on an analysis of her audience, it would look like the chart in Figure 4-10. As you look at the chart, consider how Dr. Rose can use it to draft her letter.

Based on these decisions, Dr. Rose writes the letter in Figure 4-11. As you read, relate the letter to the responses in the audience analysis and decision charts.

The letter reflects the decisions Dr. Rose made based on her sense of her readers. The letter is short, but it provides all the necessary information. The language and syntax are mature. Such technical terms as "Cray," "vector processor," and "node" are common to the general discourse community to which both Dr. Thomas and Dr. Rose belong. The tone is mannerly and formal. In addition, Dr. Thomas is addressed by her title and last name in the salutation. A block format and the formal closing "Yours truly," is used.

CHAPTER SUMMARY

Your textual decisions should be based on your readers' organizational characteristics. You need to consider your readers' relationship to your organization; position in the organizational hierarchy; and roles as decision makers.

Your decisions are also affected by your readers' professional characteristics. Your readers will be either experts, generalists, or novices in relation to the topic of your document.

The organizational and professional characteristics of readers intersect.

In addition to readers' organizational and professional characteristics, you should get a sense of readers' personal and social characteristics, including cultural background, when making textual decisions.

As the economy becomes increasingly international in scope, you'll be writing for readers from other countries. Be aware of the conventions these cultures follow and use them in your documents. Also recognize the linguistic differences between your language and your readers'.

	Manager	Engineer	City Treasurer
Information	Hook-up, responsibilities for running it.	How it helps in traffic control.	Cost and comparison with main frame.
	Fairly detailed.	Some detail on the computer in terms of traffic control	Very little detail on the computer, somewhat detailed on cost.
	Be specific.	General level of information.	General level of information.
Pt. of View	Decision maker/ Operator	Decision maker/ User	Decision maker/ Accountant
Org. Pattern	Inverted pyramid	Inverted pyramid	Inverted pyramid
Style Tone	Objective Factual Impersonal Neutral	Objective Factual Impersonal Neutral	Objective Factual Impersonal Neutral
Language	Technical College level	Somewhat technical College level	Non-technical College level
Syntax	College level	College level	College level
Format Length	Short	Short	Short
Opening	To the point	To the point	To the point

Figure 4-10 Decisions Chart

You will be writing for readers you have never met. You will need to use your prior knowledge and experience and communicate with your supervisor to gauge the characteristics of these readers. You will also be writing for the general public, which encompasses a wide range of characteristics.

Figure 4-11
Rose's Letter of
Response

Date

Dr. Susan Thomas, Director of Information Services
City of X
Address
City, State Zip

Assumes the persona
of the institution.
Uses first person plur-
al. Responds to Dr.
Thomas' questions.
Relates to treasurer's
concerns.
Provides details as
addenda for studying
later. Indicates infor-
mation for treasurer,
traffic manager in sep-
arate documents.
Closes on a note of
courtesy.

Dear Dr. Thomas:

We are pleased to hear of your interest in becoming a node for our supercomputer. It would appear that the kinds of problems you are attempting
to solve can be expedited by hooking up with our Cray.

Because the supercomputer is able to process information in far less time
than any main frame presently on the market, your hook-up with us will
not only allow your planners to get their results more quickly but in the
long run will save you money.

We are enclosing information on the hook-up as well as on the training
sessions you will need to attend to acquire the skills we require of anyone
hooking up with the X-MP vector processor. A general breakdown of
costs is included. We are also enclosing an article concerning a hook-up
by another city to facilitate traffic flow. If we can answer any further
questions, please feel free to call on us.

Yours truly,

Carol Rose

Carol Rose, Ph.D.

Sends copy to her
supervisor.

cc: J. O'Reilly

PROJECTS

As you work on the following projects, you may want to use the Audience
Analysis Chart (used in Figure 4-9, page 97) and the Decisions Chart (used in
Figure 4-10, page 101).

Collaborative Projects: Short Term

1. Analyzing readers of a college recruiting brochure

Work in trios. You have been asked to serve on a college committee to
recruit high school students in your home town to enroll in your major field.
The committee has decided to develop a brochure to be distributed to three
different audiences: (1) students, (2) their parents, and (3) their teachers.

Work together to fill out an audience analysis and decisions chart for these three types of readers. After you have filled out the charts, compare your decisions for the three readers. Discuss the similarities and differences you find. Why do you think these exist?

2. **Determining conventions of different cultures**

   a. Write to an embassy of a foreign, non-European country and request a copy of a business letter that will enable you to learn that country's conventions. When you receive the letter, study it, and make a list of the similarities and differences between that letter and our business letters.

   b. Meet in groups of five. Take turns explaining the differences you found between the conventions of the other country and those in the United States.

   c. What similarities and differences did you find among the various countries? Make a chart indicating the conventions for the different countries.

3. **Determining readers of organizational documents**

   a. Work in pairs. One person should interview a manager and the other person an employee of the same division of a firm. Find out the types of documents each person writes and who their readers are. Determine if readers are internal or external to the organization, higher or lower in the organizational hierarchy, and the role their readers assume, i.e., decision makers, users, clients, etc.

   b. Compare the responses. Can you make any generalizations from their responses? If so, what are they?

Collaborative Projects: Long Term

The following projects can be continued through Chapter 13. As you complete each chapter, you will find activities for these projects that are directly related to the chapter you are reading. In this chapter, you'll engage in activities related to analyzing your readers.

1. **Communicating about a campus/community problem**

   Work in groups of five. Consider a campus or community problem that you want to alleviate, i.e., lack of local entertainment facilities, inadequate security, need for student evaluations of faculty, missing issues and pages in magazines and journals in the library, inadequate library hours, failure to provide a site to dump household toxic wastes, need for recycling trash collection, inadequate landfill for solid waste, water/air pollution, etc.

   a. Determine to whom you should write to get the problem solved. You will probably have several readers. For example, if you would like something done about a problem in the library, you might write to the Reference Director, the Director of the Library who is the Reference Director's supervisor, and the Provost who oversees the Library as part of the academic program on campus. If you are concerned with the disposal of toxic waste, you might contact the state environmental protection agency.

b. Determine how you would communicate to your readers. Would you write a single letter for all the primary readers, a separate letter to each of the readers, or would you combine several readers in a single letter and write separate letters to other readers?

c. Relate your decisions to strategies included in this chapter.

d. Fill out an audience analysis chart for your readers. Then fill out step 1 on the worksheet in Appendix B.

Exercise adapted from *Introduction to Technical Writing*, Catherine Peaden, Department of English, Purdue University, 1989.

To be continued, Chapter 5, Purpose and Situation, Collaborative Projects: Long Term.

2. Proposing microcomputer laboratories

Three alternative projects are listed below. Select the most relevant project for your campus. Once you've determined which project to work on, read the instructions that follow the list of projects.

Project 1: *Proposal for a Microcomputer Laboratory.* A group of students, including yourself, would like the university to establish a microcomputer laboratory for student use on the campus. A committee has been organized to write the proposal for the laboratory. You are a member of the committee.

Project 2: *Proposal for a Desktop Publishing Laboratory.* A group of students, including yourself, would like the university to establish a desktop publishing laboratory for student use on the campus. A committee has been organized to write the proposal for the laboratory. You are a member of the committee.

Project 3: *Proposal for Microcomputers for a Desktop Publishing Laboratory.* The Student Council recently requested that the university's Vice President for Academic Computing establish a microcomputer lab for desktop publishing. Yesterday, the Provost met with the Director of Computer Services, the Director of the Library, and the Vice President for Business Affairs and approved the project. The lab is to be set up in the Library's basement, which will be renovated to accommodate 100 microcomputers. The vice president has requested the Student Council to recommend the brand or brands of microcomputer to be purchased for the lab. This morning the president of the Student Council appointed a subcommittee to determine which microcomputer brand or brands should be purchased. You are a member of that subcommittee.

Instructions for Engaging in Projects 1, 2, and 3

   a. Call a meeting of the subcommittee.

   b. List all the people who should be involved in the decision, and determine which person would be most likely to make the final recommendation. This person should be your primary reader.

   You may want to include some or all of the following administrators who are on your campus in your list:

   • The President of the University

   • The Vice President for Business

   • The Provost

   • The Director of Computer Services

   • The Dean of Undergraduate Instruction

Once you decide on the location of the laboratory, you may also want to include the administrator in charge of that building. You should consider the president of the Student Council as one of your readers. You may also think of other readers who should be included, such as the Director of the Physical Plant, who will be in charge of renovating the room where the lab will be located, and possibly the editor of the student newspaper, the *Daily Skimmer*.

   c. You need to get a sense of the audience that will receive your memo. Fill out step 1 of the worksheet in Appendix B. Also fill out an Audience Analysis Chart, like the one on page 97. Then make decisions for writing your proposal based on your sense of your audience. Fill out a Decisions Chart, like the one on page 101.

   To be continued, Chapter 5, Purpose and Situation, Collaborative Projects: Long Term.

Individual Projects

1. Revise the following sentences so they are more appropriate for the intended reader.

   a. Reader: Your parents

   In essaying to augment my socialization in this institution of higher learning, I did not sufficiently consecrate my cognitive energy toward cogitating the effects of Hamlet's dilemma and thereby failed to elucidate his Gordian Knot in a recent interrogation.

   b. Reader: One of your professors

   The topic was boring so I never finished getting the information. That's why I didn't hand in the paper that was due today.

   c. Reader: Your local congressional representative

   I want to buy a new car so don't raise tuition any more.

    d. Reader: Director of Public Relations:

      The car radio your company makes broke after I used it twice. I want another one.

    e. Reader: Your little brother or sister

      The gentleman who dons the crimson raiment and has the expansive calcimine whiskers is a chimerical figment of your right brain.

2. Write a short letter or memo to one of the following readers:

    a. Your technical writing instructor. Explain why you can't hand in your paper this coming period (you thought the assignment was due the following week) and request an extra day.

    b. The head of student services. Request that the cafeteria in the dorms include a greater variety of ethnic foods.

3. People are often called upon to talk to community groups about work in which they're engaged, and companies encourage their employees to make these presentations because it's good public relations. Select one of the groups below to present a talk about a project on which you're working either in or out of school.

    a. The local Kiwanis Club

    b. The Association of University Deans of Undergraduate Instruction in Colleges of Engineering and Technology

    c. A professional organization related to your project

    d. A sorority or fraternity

To prepare for your speech, write a description of your project. Before writing the description, fill out an audience analysis and decision chart.

4. The most recent issues of the weekly *Official Gazette of the United States Patent and Trademark Office* can be found in the index section of your library. You'll find abstracts of patents issued each week in the *Gazette*. Choose any **chemical** patent and order a copy of it from the library. The circulation desk can assist you with the patent request. The patent can be ordered from the closest patent depository and is free. Since it can take as long as two weeks to receive the patent, you should make the request as soon as possible.

The abstracts in the *Gazette* are not necessarily written for a chemistry audience. However, you have some colleagues who you think would be interested in this patent. Revise the *Gazette* abstract so that it is more appropriate for other chemists with your expertise.

Exercise developed for Chemistry 380, Chemical Literature
by Mike Welsh, Ph.D, Illinois State University, Normal, IL

CHAPTER 5

Purpose and Situation

YOU MAY HAVE MANY PURPOSES for writing a document. Some of them are related to your reader, others to yourself, and still others to the situation in which you're involved. You may write to inform colleagues of the results of an experiment in which you're engaged, or to persuade your supervisor to allow you to work on a new project, or to provide consumers with instructions for using a product your firm produces.

Your readers also have purposes. Your colleagues may read your research report to decide whether they can integrate your results into their own work; your supervisor may read your memo to decide whether to budget additional monies for your project; and consumers may read your instructions to learn to hook up a CD player.

Your purpose is further affected by the situation in which you write your document. If you write to persuade your supervisor to increase your budget on a project at a time when funds are limited, you may also need to persuade the supervisor to reduce funds on another project, but not a project on which you're working.

When the Nuclear Regulatory Commission (NRC) issues a written *Notice of Violation* (NOV) to a utility violating a regulation, the utility must file a written report, explaining why the infraction occurred and what has been done to correct it. The purposes for writing these reports were cited by a group of engineers:

- keep the utility's managers out of jail
- keep the utility from being fined
- keep the NRC busy
- make the utility's CEO happy
- because the NRC said to
- let the NRC know we corrected the problem
- keep the NRC happy
- because my manager told me to

These responses are probably all correct. However, only one response relates directly to the *reader's primary purpose*—to tell the NRC that the utility has corrected the problem. The NRC's purpose in requiring the response is to determine whether or not the problem has been corrected, so that the Commission knows whether to shut down the plant, fine it, or allow it to continue operation. All the other reasons are related to the *writer's purpose* and the *situation* that involves readers whose organizational relationships, positions, and roles differ from that of the writer. The manager of the division in which the violation occurred wants to stay out of jail, and the utility's CEO wants the utility to avoid a fine. The vice president expects the responsible division to respond, while the engineer, who was told by his division manager to write the response, wants to keep his job.

Of course, the reader also has additional purposes, including:

- determine the violation
- determine if requirements have been met
- determine whether or not to impose a fine
- keep the manager happy
- keep the Congress happy
- because it's part of the job

Notice that some of these are directly related to the writer's purpose and situation. The writer is drafting the report to keep the utility from being fined, while the reader is reading the report to determine whether or not to impose a fine. In addition, the situation for both the reader and writer requires that the report be written and reviewed; it is mandated by federal law.

You need to keep all of your purposes in mind as you write your document. However, you need to concentrate on the readers' purpose, just as you need to concentrate on your primary audience while simultaneously making provisions to fulfill your secondary purposes.

By identifying your readers' purpose, you add one more facet to your sense of audience. Not only do you need a sense of what readers already know, but you need a sense of what they *need* to know to achieve your purpose and theirs.

In the remainder of this chapter you will look more closely at purpose to see how it affects textual decisions, and you'll learn strategies for achieving readers', as well as writers', purpose.

TEXTUAL DECISIONS

Just as your readers' processes, behaviors, and characteristics directly influence your decisions concerning content, point of view and focus, organizational pattern and sequence, and style, so do the purposes of a document and the situation in which you write.

Let's look at an actual situation to see how purpose and situation affect textual decisions.

An engineer is assigned by his supervisor to write a letter persuading the Nuclear Regulatory Commission (NRC) to reduce a fine levied against the utility for failing to correct a problem promptly. The reason for the delay was the utility's inability to find the cause of the problem. While plant engineers suggested the cause was related to a calculation error, management disagreed. A number of weeks and several additional problems later, the engineers' diagnosis proved correct, and the utility finally fixed the problem.

The NRC's guidelines indicate that penalties can be reduced if a problem is "identified and corrected in a timely fashion." Therefore, the primary purpose of the engineer who has been saddled with the chore of writing the letter is to persuade the NRC that the utility moved as promptly as it could and that the fine should be reduced. In addition, the utility's managers do not want the NRC to think that the delay was caused by their failure to listen to the engineers. So the writer's second purpose is to keep management from being blamed. A committee of NRC administrators, familiar with the utility and with nuclear problems in general, but not necessarily with the particular problem under discussion, will review the letter to determine whether or not to reduce the fine.

The excerpt in Figure 5-1 is from a section of the letter relating to mitigating factors, such as prompt identification and reporting of the cause of the problem. Assume you are a member of the NRC board. As you read this section of the engineer's letter, consider whether you would lower the fine imposed on this utility.

If you were a member of the NRC Board, you probably wouldn't reduce the utility's fine after reading this section. In fact, you might even consider raising it. In the very first sentence you discover that five trips associated with the problem occurred before the problem was corrected. There is no mention of prompt identification of the problem, the NRC's criterion for determining whether or not to reduce the fine.

Figure 5-1
Letter of Request for
a Fine Reduction

RESPONSE TO PROPOSED IMPOSITION OF CIVIL PENALTY

NOS. 650-327/88-35, 650-328/88-35

650-327/88-55, 650-328/88-55

FEBRUARY, 19XX

Mitigation Factors (Section V.B)

A. Identification and Reporting

Shortly following restart of unit 2 from the extended shutdown, five reactor trips[*] occurred—May 19, 19XX: May 23, 19XX; June 6, 19XX; June 8, 19XX; and June 9, 19XX. Immediately after the first trip, personnel involved in performance of the posttrip shutdown margin (SDM) calculations noticed an unexpected reduction in excess SDM, and discussions commenced both internally and with the Nuclear Steam Supply System (NSSS) vendor regarding the cause and possible resolution. It was initially thought that the observed reduction was solely the result of additional conservatism recently added to the xenon tables. Although calculated SDM was well above the technical specification (TS) requirements, a questioning and safety-conscious attitude prompted additional review. However, because adequate margin existed to the TS limits, this was not an issue that should have prevented restart of the unit.

Following the first reactor trip, management was also questioning the cause of observed reactor coolant system (RCS) overcooling from an operational standpoint. Discussions were held among management, the site reactor engineers, and personnel from the corporate fuels organization relative to possible effects of low decay heat. Telephone inquiries were made to other plants to determine if RCS overcooling existed or had existed and what actions may have been taken to resolve the problem. Early management concern was evidenced by an operator aid issued on May 24, 19XX, regarding isolation of main feed pump governor valves that had been found to be leaking. The operator aid directed isolation of high- and low-pressure steam to both main feed pumps on a unit trip to prevent excessive cooldown of the RCS. At this point, the RCS overcooling was viewed as an operational concern; the tie had not yet been made to the SDM issue and the cooldown "event" itself was considered bounded by the main steam line break accident analysis.

[* A trip halts a nuclear reaction and basically shuts down a utility's operation.]

This letter reads like a mystery novel, as the reader follows the utility's efforts to discover the culprit behind the trips. But the purpose of the document is not to entertain readers with a detective's deductive reasoning or, in this case, the utility engineers. Rather it is to persuade the NRC to reduce the fine. To achieve this, readers need to know that the utility identified the cause as promptly as possible. However, the writer doesn't focus the information around the identification of the problem, and in failing to do so he fails to achieve his purpose.

To be effective, the "Identification and Reporting" section of the letter needs to be revised. This revision appears below in Figure 5-2. As you read it, consider how the writer's purpose affects the textual decisions.

Let's examine the writer's decisions in light of his purpose.

**Figure 5-2
Revised Letter of
Request for a Fine
Reduction**

Relates to readers' criteria for determining a fine reduction.
Provides evidence to support claim that utility took aggressive action.
Provides evidence to support management's actions.
Provides details so readers understand evidence.

Mitigation Factors (Section V.B)

A. Identification and Reporting

The shutdown margin (SDM) issue was identified as a result of aggressively pursuing the root cause of an unexpected condition. Management addressed the reactor coolant system (RCS) overcooling as an operation concern while reactor engineers simultaneously investigated the possibility of a correlation between the cooldown and the SDM safety issue. Once this correlation was established, the utility promptly documented the issues using its approved corrective action program and reported the issues to NRC in accordance with 10 CFR 50.73.

The utility initiated a two-pronged investigation: one into the RCS overcooling and one into the SDM. The RCS overcooling was viewed by management as an operation concern. In investigating the cause of observed RCS overcooling, management held discussions with site reactor engineers and personnel from the corporate fuels organization relative to possible effects of low decay heat. Telephone inquiries were made to other plants to determine if RCS overcooling existed or had existed and what actions may have been taken to resolve the problem.

Simultaneously, personnel, who had noticed an unexpected reduction in excess SDM during posttrip SDM calculations, commenced discussions both internally and with the Nuclear Steam Supply System (NSSS) vendor regarding the cause and possible resolution. Although it was initially thought that the observed reduction was solely the result of additional conservatism recently added to the xenon tables, an additional review was undertaken, despite calculated SDM being well above the technical specification (TS) requirement.

Content. The information you include in a document depends on what readers need to know to achieve their purpose, and what you need to achieve yours. If readers need to decide whether or not to continue a project, they need to know the particular aspects of the project that either recommend it for continuation or tag it for dumping. If the purpose of the NRC Board is to determine whether or not to reduce a utility's fine, then the Board needs to know whether or not the utility has met the criteria for a fine reduction, i.e., prompt identification and correction of a problem. To persuade the utility to lower the fine, the writer must be able to claim that the utility identified the problem as quickly as possible, and then support the claim with evidence demonstrating the utility's efforts to correct the problem.

Once you've determined what your readers need to know, you can eliminate irrelevant information and information which doesn't advance your cause. The engineer at the utility eliminates the information related to the five trips, not only because it doesn't advance the utility's cause but because it is unnecessary. The material has been included in previous reports, and the Board is aware of it.

In addition to including information readers need to know to achieve their purpose, you want to include information that fulfills your purposes. The writer of the request for a fine reduction includes information on management's attempts to determine the cause of the problem. The writer thereby demonstrates that management had been involved and had reasonable grounds for following its path, even though it was the wrong one. Thus, the writer fulfills the secondary purpose.

Point of view and focus. The point of view and focus of a document is directly related to the purpose. If a reader's purpose in reading a document is to assemble a tractor, then the reader's point of view is that of a user, and the focus should be how something is done. If the reader's purpose is to determine what can and can't be done in a situation, then the reader's point of view is that of a follower, and the focus should be on rules and regulations. If a reader's purpose is to decide whether or not to purchase a product, then the reader's point of view is that of a decision maker, and the focus should be on the general aspects of the product. However, if a reader's purpose is to decide on a particular brand to purchase rather than whether or not to purchase a product, the focus changes and should be on the specific aspects of each of the different brands.

Since the NRC's point of view is that of a decision maker who must decide whether or not to reduce the fine based on the timely correction of the problem, the second letter focuses on the utility's identification of the cause, rather than on the search.

Your focus will also depend on whether or not you have a vested interest in readers' decisions. If you are simply providing readers with information and you don't care how they use that information, then your focus may encompass all facets of a topic. However, if you want to persuade readers to use the information in a specific way, then your information should focus your view as closely as does the second letter to the NRC.

Organizational pattern and sequence. As the two drafts to the NRC show, the different ways in which information can be organized and sequenced depend on the purpose of a document. The writer of the first draft uses a *chronological* pattern, focusing the reader's attention on the history of the incident rather than on the aspects of the incident related to the criteria for reducing fines. The revised letter uses an *analytical* pattern, listing the activities related to the criteria. By presenting the solution before discussing the problem, the writer responds to the reader's main concern—prompt solution of the problem.

Sometimes you can achieve your purpose better if you use an indirect approach in which you begin with background to a problem and lead up to your main point or solution. If you're writing a letter to persuade your parents to buy you a computer, you may want to start your letter by building a case rather than just coming out and asking for a PC. You might begin with a hard luck story about how you have to stay up late writing papers for class.

Style. Readers often infer the personality of a writer from the tone of a document. To achieve your purpose, you want to be certain that the tone of your document presents you in the best possible light. The tone of the revised letter to the NRC is more formal and factual than the original. By deleting such adjectives as "questioning" and "safety-conscious," the writer avoids the cavalier attitude of the first letter.

READERS' PURPOSES

Your purpose for writing is inextricably related to your readers' purposes. Readers read a documents for the following purposes:

- **To acquire information for future use**. Readers may read about personal computers to keep current on what is available so that they can eventually make an informed decision on when and what to buy. Or they may read catalogs, ads, or patent announcements to learn what their competition is doing.
- **To make a decision**. Readers may need to decide whether or not to purchase a cellular phone, to send employees to a conference, or to order more parts for a machine.
- **To do something**. Readers may need to repair a diesel engine, put together a barbecue, replicate an experiment, or repair a light socket.
- **To follow regulations**. Readers may need to know whether or not state legislation requires the use of plastic bags for dumping grass clippings after mowing a lawn, whether plant procedures require that a supervisor be informed prior to repairing a turbine, or at what level of radiation federal regulations require employees to undergo a health check.

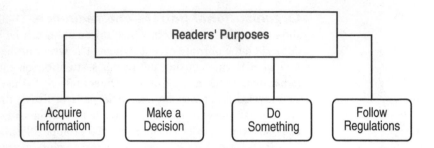

In industry much of the writing is reader-initiated: the same person who reads a document has assigned it and has specified the purpose. Your supervisor may assign you to write a periodic report to keep her up to date on your projects so she can make certain you're doing your job satisfactorily. A client may ask you to provide information about your product so that he can decide whether or not to purchase it. Or a governmental agency like the Office of Safety and Health Administration (OSHA) may require you to write a report documenting a violation so that they can determine whether or not to fine you.

WRITERS' PURPOSES

Initiating the Document

Your purposes can relate to either reader-initiated or writer-initiated documents.

Reader-Initiated Documents Your primary purpose in writing a reader-initiated document is to fulfill your reader's purpose. If your supervisor expects you to write a periodic report, your purpose is to inform her of the work you've been doing during a particular period of time. If clients request information about your product, your primary purpose is to provide the information, though your secondary purpose is to persuade them to purchase your product. If a state government asks you to determine if archaeological artifacts exist on a potential construction site, your purpose is to inform the reader of the results of your study.

The following paragraph from a memo is a response to a supervisor who has requested a recommendation for a laptop computer. Notice that the writer not only provides information that meets the reader's purpose, but also uses the information as supporting evidence to persuade the reader to accept the recommendation.

Lists the best features of the recommended model.
Because the reader is biased toward low cost, price is the first item discussed.

The Toshiba XXX offers one of the best combinations of price, weight, and battery life of laptop computers on the market. The 6.4 lb. laptop, which costs $999, has a battery life of approximately 3-1/2 hours. In addition, its resume feature saves your place when it's necessary to shut down the system for a battery change. The laptop has 512K RAM and a reflective supertwist LCD screen.

Writer-Initiated Documents When documents are writer- rather than reader-initiated, the primary purpose is related to writers' needs rather than readers'. The purpose of most writer-initiated documents is usually to persuade the reader. You may write a letter to persuade someone to hire you, to persuade your supervisor to allow you to work on a new project, or to persuade a governmental agency to fund a program with which you're involved.

Even in these cases, your purposes are tied closely to your readers'. Because your readers' purpose is to make a decision concerning your topic, you need to persuade them that your needs are related to theirs. If you're applying for a job, you need to persuade the personnel officer that your qualifications fit the company's requirements. If you're requesting permission to work on a new project, you need to convince your supervisor that the project enhances the company's aims. If you're seeking funds, you need to persuade the agency that your project meets their funding criteria and deserves to be funded.

The following paragraph is part of a writer-initiated memo drafted by an employee to persuade the supervisor to purchase a laptop computer. As you read, consider how the writer indicates that she and the reader share the same needs and purposes.

Lists the best features. Explicitly states organization's needs and relates information (color) to them.

An important feature is the screen's readability. The liquid crystal displays (LCD) of the new laptops are now comparable to cathode ray tubes, although none provide color. However, color is irrelevant to our purposes.

The Persuasive Element

As you have probably noticed, almost all of the documents we've discussed in this chapter have had a persuasive purpose. Most documents are written to persuade someone to do something or to think in a certain way, though they may also be written to provide information, to offer instructions, or to regulate actions.

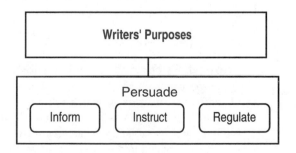

Sometimes the persuasive purpose is explicit; the purpose is openly acknowledged in the document. But at other times the persuasive purpose is implicit; it is a hidden agenda that is never discussed openly.

Explicit Purposes In documents in which the purpose is explicit, readers are aware that you are trying to persuade them to do something or think in a certain way. Readers automatically assume that, in such documents as proposals and letters of application, you are being persuasive. Reader-initiated documents may also be persuasive. Readers may ask you to draw conclusions or make recommendations in various reports. Documents in which you explicitly try to persuade readers may involve justifications, evaluations, and requests. These documents usually assume the form of feasibility and evaluation reports, and complaint letters and letters of request.

You may write a memo to persuade your supervisor to promote you, or a letter to persuade a client to purchase an upgrade of a computer program you've developed.

The following excerpt is included in a proposal to the federal government for funds to establish a project to retrain electricians. To persuade the reviewers to approve funding for the proposed project, the writers make three claims. As you read, consider how the writers' claims and the information they use to support their claims helps them achieve their purpose.

Relates claim to individuals affected. Relates claim to needs of organization submitting proposal. Relates claim to needs of the country. Uses statistical facts from written sources respected by the readers to support claims.

First, large numbers of individuals need to be retrained because their skills are eight to ten years behind what they should be. Second, the Company's manufacturing technology is becoming more and more sophisticated, requiring more highly trained individuals. Currently, Caterpillar has almost 1,000 industrial electricians and/or electronic technicians providing support for electronic equipment throughout the U.S. Third, the industry faces an aging workforce...According to the 19xx edition of *Statistical Abstract of the U.S. Department of Commerce*, the need for electronics technicians will grow from 37 to 51 percent between 1986 and 2000. The XXX Edition of the *Occupational Outlook Handbook* says, "Employment of commercial and industrial equipment repairers is expected to increase faster than the average for all occupations through the year 2000 as the volume of electronic equipment to be installed and maintained grows (p. 295).

Electronics Technology Training for Central Illinois

Implicit Purposes Often implicit purposes are related to hidden agendas, i.e., purposes which are not openly acknowledged. Implicit purposes are usually related to *the situation* in which you write a document, and for which a reader reads it. The engineer who requested a fine reduction from the NRC had an implicit purpose—to keep management from being blamed for the delay—which was directly related to the situation in which he worked.

The following paragraph from a reader-initiated progress report on a construction project was written by a manager to a supervisor who requires monthly reports to determine whether a project is on schedule. The manager who oversees the subcontracting work by Progressive Factors, Inc., is coming up for a job review that may result in a salary raise or even a promotion. As you read, consider how the writer achieves the explicit purpose of informing the supervisor of the project's progress, and the implicit purpose of persuading the supervisor that he is doing a good job and should be rewarded accordingly.

Places blame for delay on another office. Takes credit for getting the project going.

Progressive Factors is not satisfied with the water flow test obtained in October. The Chicago office gave approval to proceed but the Atlanta office refuses to approve the drawings until a second test is obtained. I've forced the issue with the Atlanta office and they will run a test by 11/22 with the sprinkler contractor so that both will have the same information. I've impressed on Progressive Factors that we cannot stand a delay and that the work must be done as soon as possible.

By assuming the responsibility for Progressive Factors' scheduling the test, and by indicating that everything possible has been done to see that the project is on schedule, the writer implicitly attempts to persuade the reader that he is doing a good job.

Other Purposes

In addition to persuasion, writers write documents for three other purposes: to inform, to instruct, and to regulate.

To Inform You write a variety of documents to provide your readers with information for making decisions. These documents, which involve explanations and descriptions, include such forms as responses to requests, lab journals, research reports, and minutes to meetings.

A supervisor may require that you submit a report explaining a delay in completing a project so that whatever is necessary can be done to prevent similar problems from occurring in the future. A client may request information about an aspect of a product she's purchased so she can repair it. You may keep a notebook of your work on a design so that you can use the information later in applying for a patent.

The following excerpt from a brochure informs local residents about the reason for regulations against burning leaves. Notice that the information not only informs residents why leaf burning is banned, but also tries to persuade them that the policy is a good one.

Relates to readers' physical safety and attitudes toward dirt.

When leaves are burned, smoke and soot are released into the air. Smoke and soot are particulate matter which become suspended in the air. When airborne particles are inhaled, they may irritate the respiratory system or damage the clearance mechanism of the lungs, thereby contributing to acute respiratory illnesses. Particulate matter can also be a nuisance when it settles on cars, homes, patios, etc., creating a film of soot and causing the surface to appear soiled.

Illinois Environmental Protection Agency

To Instruct Depending on the type of work in which you're involved, you may find yourself writing instructions to help readers put together, use, or repair a product.

If you're involved in developing computer programs, you need to provide readers with documentation for using a program. Professional technical writers are often requested by their firms to write instructions that help users put together and use the firm's products, while engineers and mechanics involved in designing products may need to write instructions to help users repair products.

The following instructions help consumers use automatic coffee makers. As you read, notice how the writer tries to persuade readers not to do something that could ruin the mechanism.

Uses typography (all capitals) to warn people.

Fill carafe with cold water and pour into reservoir. Markings on water level indicator show the amount of water needed to make desired number of cups. (Note: You'll get fewer mugs than cups since mugs hold more coffee than cups.) DO NOT EXCEED 10 CUP LEVEL. Place a coffee filter into the filter cone. Use a fresh coffee filter for each carafe of coffee...

To Regulate You may find you need to write regulations so that readers know what to do under certain circumstances. If you work for a regulatory agency, such as the NRC or the Federal Aviation Administration, or if you work in a service department, such as information or computer services, you may find you need to draft rules and procedures for readers to follow.

The State of Illinois issued the environmental regulations reprinted below to halt the use of plastic bags for collecting lawn cuttings. Notice that the regulation is implicitly persuasive.

Implies that if regulation is not followed, persons will be punished.

Beginning July 1, 19xx, no person may knowingly put landscape waste into a container intended for collection or disposal at a landfill unless such container is biodegradable.

The Illinois Environmental Protection Act
Section 22.19b

STRATEGY CHECKLIST

1. Determine necessary information according to readers' purposes for reading a document and your purpose for writing it.
2. Include all information readers need to know, including information related to secondary purposes.
3. Eliminate information which isn't relevant, which may not advance your cause, or which may create a negative attitude on the part of the readers.
4. Focus on information related to the readers' purposes.
5. Focus on information representing your point of view if you have a vested interest.

6. Organize information to meet both your readers' and your own purposes.

7. Build your case before making your recommendation if you're trying to persuade your readers, who may be unsympathetic to your topic, to make a specific decision.

8. Consider your readers' purposes when you make decisions about style.

THE SITUATION

You don't write in a vacuum. Rather, you write within the social and political environments of an organization. Your documents involve a company that competes against other companies for business, and which contends with outside forces, such as regulators and interveners, that attempt to influence the organization's operations.

Your documents are also written within the broader context of society. You may want to improve the environment for wildlife or the quality of life for the physically handicapped. However, you may be trying to preserve land that other organizations may want for development of a mall. Or your own organization may not have the funds to implement your architectural plans for clients in wheelchairs. As a writer, you need to negotiate between these conflicting forces.

The following paragraph, from a report by the Illinois State Water Plan Task Force, is written for government representatives and interested citizens. Charged with proposing an action plan to improve the state's waterways, the task force has to negotiate between citizens who will not approve further tax hikes, and environmentalists who want nothing less than complete rehabilitation of the area's rivers. As you read, consider how the writer negotiates between the conflicting needs of these readers to persuade both sides to approve the proposed plan.

Relates to values of environmentalists.
Relates to attitudes of taxpayers.
Relates to shared values of citizens and environmentalists.

Before a course of action is recommended, it must be pointed out that the Illinois river *cannot be returned to its original pristine condition.* Moreover, some areas of the bottomlands and backwater lakes have undergone almost irreversible changes and can not be altered or revitalized without significant cost and effort. It is essential that a thorough evaluation of the backwater and bottomland lakes be made to determine the area or areas of these resources that are of significant value to Illinois citizens. Once this determination has been made, efforts should be concentrated to revitalize only these high-value areas.

Illinois River Action Plan:
Special Report No. 11

In this section we'll examine the dilemma writers face when they write for readers who have different and even conflicting purposes. We'll also look at the review process for company documents, and we'll learn strategies for writing company, rather than personal, documents. Finally, we'll consider some of the legal and ethical factors associated with writing in technical communities.

Multiple Readers

You'll constantly engage in one-to-one correspondence with your supervisors and peers. However, you'll often find yourself writing for more than a single reader. Even if you're simply keeping a laboratory or engineering journal, you may have readers other than yourself. While engineers may keep journals for their own records, managers may insist on reading those journals to check on what employees are doing or to keep up on their progress. Such notes may also be read in the future as historical documents, by people who want to know something about what has been done, or by lawyers who want to use the information in a trial.

Each time a new product is developed at John Deere, an operator's manual must be written to provide information on operating the new piece of equipment. The manual is written by a technical writer in the technical publications division. Before the document is sent out, the manual must be reviewed by at least four persons: a representative of the safety committee, the engineering team that worked on the product, a field service representative, and a senior editor. The manual is written for the operator of the equipment. However, because the manual contains information for maintaining and servicing the equipment, the manual may also be used by technicians who work for the operator. In addition, if a problem with the equipment results in a suit by the operator, attorneys for both sides will read the manual. Figure 5-3 depicts all the readers whom the writer must consider.

In the remainder of this section, you'll examine the types of readers for whom you'll write, and then you'll learn strategies for meeting the needs of multiple readers as well as for meeting the needs of readers with conflicting purposes.

Types of Readers A single document may be read by four types of readers: primary, secondary, intermediary, and peripheral.

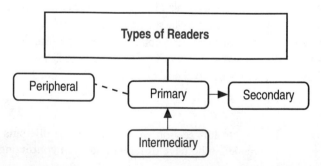

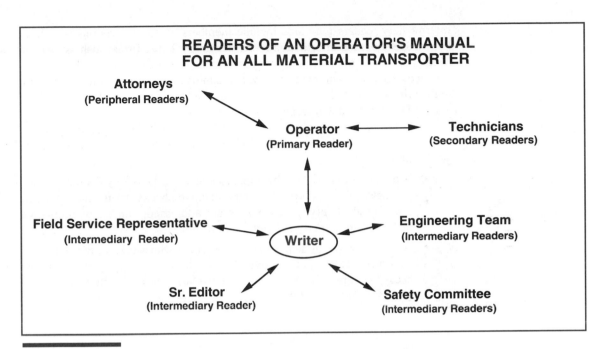

Figure 5-3 Multiple Readers of an Operator's Manual

Primary readers. The primary reader is the person or persons for whom the document is being written. You need to keep the primary reader in mind at all times, regardless of how many other types of readers read your document. Your decisions should be based on your primary reader's needs.

Secondary readers. Secondary readers are people affected by the document, like marketing managers and comptrollers, who need to develop a marketing plan or a budget based on the information in a document.

Secondary readers need to be considered, since they are often the ones to use the information to engage in a project or to repair a problem. Often, secondary readers need more information than the primary reader. By placing in an appendix information that is irrelevant or unnecessary to the primary reader, but important to a secondary reader, you can meet the secondary readers' needs.

You can also provide for secondary readers' needs by using a table of contents, headings, subheadings, and forecasts so they can quickly scan for information they specifically need.

Technicians are the secondary readers for the Operator's Manual. The information they need is included in the main section of the manual, and listed in the table of contents where they can find it easily and the primary reader can skip it.

Intermediary readers. Intermediary readers are usually supervisors who are responsible for approving a document before it is sent to the primary reader.

Intermediary readers may also include members of other divisions who have knowledge about the information in the document. Large firms, such as General Motors or Apple, often require many intermediary readers, such as the supervisors and executives who read, review, and approve a document before sending it to the primary reader.

At Deere, the manual receives four reviews.

- A technical review by the engineering team that developed the product. The team checks the manual for accuracy and adequacy in presenting the information.

- A product safety review by a representative of the Safety Committee and attorneys (if necessary). The representative checks the manual to ascertain that all safety-related issues are addressed, and all warnings are included and properly worded.

- A reader review by a service representative who works in the field with customers. This representative checks that all information the customer needs to operate the product is included, and that it is written so the customer can understand it.

- An editorial review by a senior writer. The writer checks that the manual conforms to the firm's standards, and that it is grammatically and stylistically correct.

Peripheral readers. Peripheral readers, like lawyers or news reporters who are not directly affected by a message, may read a document after the primary reader receives it. Peripheral readers are the readers on the sidelines. There is always a possibility that they will read a document. You always need to remember that the peripheral reader may be looking over your shoulder. If there is a problem with the product that results in a suit or a personal injury, lawyers and even reporters may read the manual.

If you're not careful to consider peripheral readers as a potential audience for everything you write, you can wind up in an embarrassing situation, as one nuclear utility did: A poll commissioned by the utility indicated that the less people knew about nuclear energy, the more negative they felt about it. As a result of the poll, one of the VPs circulated a memo stating, "We must make these people understand nuclear energy." Somehow national columnist Jack Anderson saw the memo and wrote a scathing editorial against the company's "fascist" tactics, based on his interpretation of the word "must." But the VP hadn't meant that the company should force the public to accept its point of view, only that if it was ever going to get the public behind it, it needed to find a way to educate the public about nuclear energy. If only the VP had written, "We *need* to help people understand nuclear energy," instead of "We *must* make people understand," Anderson might not have felt the memo had fascist overtones.

Multiple Readers' Needs

If you're writing for a single primary reader, it's relatively easy to make decisions. But when you're writing for several readers in different organizational relationships, roles, and positions, the decisions are more difficult, because their needs often differ and even conflict. You must negotiate between them to achieve your purposes.

The Director of the Heartland Water Resources Council needed to develop a brochure to persuade citizens in a midwestern county to approve a bill in an upcoming referendum establishing a river district. Potential readers included citizens who did not want to pay the additional tax the bill would require; members of the Sierra Club who felt the bill did not provide enough help; and farmers who didn't want to be assessed to help pay for the district. While many citizens had little knowledge of the environmental issues involved, members of the Sierra Club were well versed in the topic.

The problem: How to draft a single brochure which would meet the needs of all of these constituencies.

The solution: Use a question-and-answer format.

As you read the questions and answers in Figure 5-4, consider how the writer relates to each of the various types of readers and their respective needs to persuade each group to approve the bill.

**Figure 5-4
Q & A Format to Satisfy Multiple Readers**

Relates to values of citizens, environmentalists, farmers.

Relates to environmentalists' criteria for effective conservation.

Relates to citizens' attitudes.

HEARTLAND
WATER RESOURCES
COUNCIL OF
CENTRAL ILLINOIS

RIVER CONSERVANCY DISTRICT

Q. Why do we need a Heartland River Conservancy District?

A. Over the past 90 years, a major disaster has befallen the Illinois River and Peoria Lake. Over 800,000 tons of soil enters the River each year from Marshall, Peoria, Tazewell and Woodford counties. In 10 to 15 years a mud flat with a transportation channel will be all that is left of the Illinois River. A River Conservancy District will provide effective, unified management of the River and lake system.

Q. What are the purposes of a River Conservancy District?

A. A River Conservancy District can effect soil conservation by augmenting erosion control and instituting streambank and bluff stabilization. The statute allows control of lakes, river and stream systems, prevention of stream pollution, protection of water supply, preservation of water levels, prevention of floods, reclamation of wetlands, and construction of bridges, roadways and streets.

Q. What is the tax rate of a Conservancy District?

A. The formation of a Conservancy District authorizes a maximum tax rate of $.083 cents per $100 assessed valuation.

1. This rate amounts to $16.60 per year for the owner of a $60,000 home.

2. The average farmland may be assessed at $.10 to $.15 per acre, depending upon productivity and assessed valuation.

continued

Figure 5-4
**Q & A Format to Sat-
isfy Multiple Readers**
continued

Relates to citizens'
values.

Relates to values of all
readers.

HEARTLAND
WATER RESOURCES
COUNCIL OF
CENTRAL ILLINOIS

Q. What can I expect for my tax dollar?

A. Your dollar will provide the local share necessary to secure state and federal support for a comprehensive, balanced plan of prevention and cure. A unified local strategy will focus attention on the needs of the four counties. Every local dollar is expected to bring in $7 to $40 to our Central Illinois area for erosion control, streambank and bluff stabilization, and management of sediment already in the river.

Q. What will happen if the referendum doesn't pass?

A. Unfortunately, much of the River from Chillicothe to Pekin will become a mud flat. Flooding will be severe. With the further destabilization of streambanks and bluffs, bridges and roads will fall. Prime agriculture land will be lost with the resulting loss in revenues. Fishing and hunting revenues will dwindle with the choking of aquatic habitat. Both the economic and ecological well-being of our Central Illinois communities will severely decline.

Heartland Water Resources Council of Central Illinois

The brochure contains information that each constituency needs to know to determine whether or not to vote for the bill. The questions serve as subheads, allowing readers to scan the questions and then read only those answers which are relevant to their needs. The question and answer format allows the writer to focus on several issues, based on the readers' varying points of view. The writer follows a problem/solution format, providing sufficient information on the problem for those readers who are unaware of it to understand it. Those who are aware of the problem can simply skip the section. The writer also uses comparison/contrast, comparing the beneficial results of the district with the deleterious effects if the referendum isn't passed. The style is aimed at the novice who is not a member of an environmental discourse community. While some of the language is technical, the terms are familiar to the general populace in this age of ecological awareness. Sentences are simple and easy to follow. The tone is factual.

Let's consider the textual decisions that are made in writing documents, and examine some additional strategies that can meet the needs of multiple readers.

Content. When multiple readers are involved, conflicts may arise concerning the information that should be included in a document.

A primary reader may not need the information secondary readers need. Either write an executive summary so the primary reader doesn't have to read all the details in the report, or place these details in an appendix. Perhaps there is information a primary reader doesn't need, but the writer wants an intermediary reader to have it or feels it should be kept for the future. Such information should be placed in a separate document to be sent to the intermediary reader, or it should be filed in the writer's records. It should not be included in the document to the primary reader.

Sometimes a primary reader needs to have information that the writer doesn't want peripheral readers to see. Information like this should be eliminated from the document, and either communicated orally or placed in a separate, confidential document.

The writer may also have information that a primary reader probably should have, but an intermediary reader wants withheld. The writer can eliminate the information entirely or send it in a separate document. However, doing so means going over a supervisor's head and, if the intermediary learns of such actions, the writer may be demoted, fired, or denied a raise or promotion.

Point of view. Multiple readers usually require multiple points of view. Your primary reader may be the manager of a division who reads your document from an administrator's point of view, while your secondary readers may be technicians within a division who read your document from an operator's point of view. Multiple readers become even more problematic when their points of view conflict.

If your primary reader is a government regulatory agency, then your document will be read from an external evaluator's viewpoint, but, since your intermediary readers are your supervisors, your document will also be read from the point of view of an internal administrator. If you are writing to the Office of Safety and Health Administration (OSHA) to request an extension on a deadline for correcting a problem, your manager is not going to want you to state that you are requesting the extension because nothing has been done.

You need to negotiate between your readers to find a compromise solution that relates to both their needs.

Organizational pattern and sequence. Multiple readers may need information presented in different patterns and sequences. A primary reader who is an administrator may only want an overview of a mechanical problem at a plant, while a secondary reader, who is an engineer, may want sufficient information to locate and repair the problem. You can provide for each of these readers' needs by including a cover letter or executive summary for the administrator, or by placing the detailed information in an appendix for the secondary reader.

If you have several primary readers whose positions, roles, or fields differ, and whose needs are not parallel, you should write a separate document for each reader.

Style. Multiple readers may require different styles. For example, your primary reader may be a subcontractor who specializes in one aspect of your field and is, therefore, familiar with technical terms used in that field. Your intermediary reader may be a supervisor who is not familiar with your subcontractor's specific field, and who is unfamiliar with many of those terms. How do you communicate effectively with both?

- Whenever possible, use language that all of your readers understand.
- Though familiar to the primary reader, define any terms you use which may be unfamiliar to secondary readers.

- Place definitions in parentheses in the main text or include them in a glossary. The primary reader can then skip over them, but the secondary reader can refer to them.

- Provide a glossary for intermediary readers who are unfamiliar with technical terms well-known to your primary or secondary readers.

Your sentence structure should be easy to read regardless of who your reader is.

You may find that different readers require different tones. Your supervisor may suggest you use an assertive tone in your correspondence to a governmental regulatory agency; but the agency, your primary reader, might interpret that tone as antagonistic. You need to find a compromise whereby you can be straightforward without being offensive.

For example, a chemist, in drafting an evaluation report, wants to begin by stating, "X division has failed to correct the problems cited in the last report." However, the chemist's supervisor suggests a more tactful approach, e.g., "There continues to be opportunity for improving the division." The chemist sees this phraseology as soft-pedaling, but the supervisor believes the original wording will aggravate, rather than motivate, the division's manager. Because the supervisor's argument is politically sound, and because the supervisor has the power to recommend raises and promotions, the chemist will revise the sentence.

You should try to use neutral tones and language to avoid non-sympathetic responses, especially from peripheral readers. For example, if you're writing a report on a chemical spill, you'll want to phrase your report very carefully, since a reporter may quote the report in a newspaper article, or a lawyer may use the report in a suit if someone has been hurt at your company.

Readers with Conflicting Needs A major problem occurs when the purposes of your primary and intermediary readers conflict.

The Nuclear Regulatory Commission's purpose in requiring that a utility file a written report in response to a Notice of Violation is to determine if the problem has been resolved. However, engineers responsible for writing these reports may be faced with incorporating the hidden agendas of those above them in the organizational hierarchy. The engineer's supervisor may want to rationalize the cause of the problem, so that his capability as a supervisor is not called into question. Furthermore, the CEO may want to minimize the significance of the problem so that the utility is not perceived as unsafe. The CEO may also want to avoid making expensive or time-consuming corrections.

According to procedures of the NRC, all supporting evidence for test results are to be filed with the results of a test. Realizing that the supporting evidence for an instrumentation test had been lost, a supervisor assigned an engineer to write a report identifying the violation. The supervisor recognized that the proper way to correct the problem was to rerun the test, but this would result in the loss of about a day's work. Therefore, the supervisor told the engineer to write the report in such a way as to preclude the agency's requiring a new test.

Conflict:
- The engineer knew the problem should be corrected.
- The NRC (primary reader) wanted the problem corrected.
- The supervisor (intermediary reader) didn't want to correct the problem.

Result:
- The engineer, recognizing that the supervisor was the one who determined his salary, did as the supervisor requested.
- The NRC recognized the smoke screen for what it was and required the test.

The results of minimizing the significance of such a problem can be much worse. Consider such classic cases as the Three Mile Island (TMI) and Chernobyl nuclear accidents, when the public completely lost trust. In the long run, telling the truth is beneficial to both the company and the public. Fudging or throwing up a smoke screen can eventually cost a company far more than telling the truth from the beginning.

The Review Process

You will almost always have at least one intermediary reader. Your supervisor will want to review all major documents before you send them to the primary reader, and your supervisor's supervisor, or representatives of other departments involved with the information, may want to review your documents from time to time.

Often, because a written report must go through so many people in a review process, writers forget who the primary reader is, and instead, write for the intermediary reader who reviews the report. This is most likely to happen when intermediary readers require a large number of revisions. While these revisions may appear arbitrary, they are usually based on sound reasoning. Sometimes revisions are suggested because a text is unclear to the reader/reviewer. At other times revisions must be made for political reasons, as in the following example, excerpted from a report explaining the reason for a problem to an external reader.

Original   The plant never suspected that the vendor who made the modification to the mechanism would deviate from the original indexing letters.

The reviewer felt the reader of the original text would believe that the plant was at fault. He revised the sentence so that the reader would accurately perceive that it was the vendor who caused the problem.

Revision   There was no intent for the modification to change the indexing.

To make appropriate revisions and avoid making mistakes in the future, you need to understand why a reviewer has made certain changes. If you don't understand, you should ask. Don't make the mistake the engineers at the Institute for Nuclear Power Operations (INPO) made:

INPO sends teams of engineers to various nuclear plants to evaluate them. Following their observations, team members collaborate to write an evaluation report, which includes the results of the investigation as well as recommendations. After the engineers draft the report, it is submitted to at least five reviewers, including the immediate supervisor, the supervisor's manager, the vice president of the plant, and the chief executive officer (CEO). The document is not sent to the plant until all the reviewers approve it. Figure 5-5 depicts this cumbersome review process.

**Figure 5-5
A Nuclear Utility's
Review Process of a
Report to the NRC**

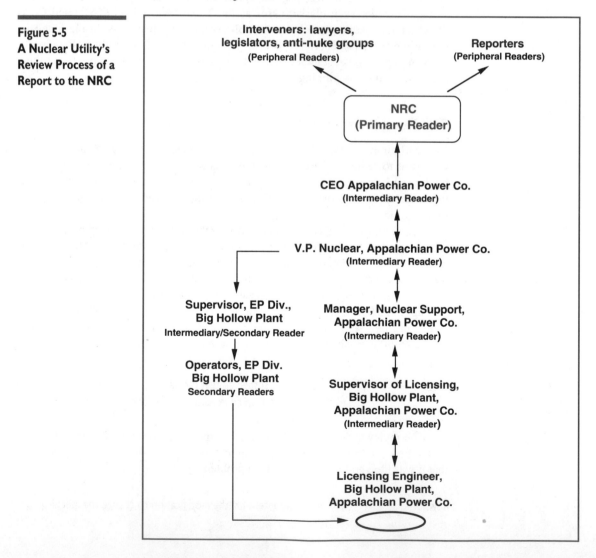

Often, the reviewers make so many changes that the engineers who write the reports have difficulty recognizing their own documents. Noticing that the last sentence in the first paragraph of the recommendation section is often changed by a reviewer to read "X needs improvement," the engineers began using the phrase in their documents. Sometime later, a consulting firm, hired to improve INPO's writing, suggested replacing this too-general phrase with a more precise one so readers would know what needed to improve. Instead of the sentence "Contamination control practices need improvement," the firm recommended using "New procedures need to be written for contamination control."

When the engineers explained why they'd been using the more general phrase, the consultants questioned the intermediary reader. Admitting to making the changes, the reviewer explained that the engineers often included several different aspects under the same problem and, therefore, the only phrase that seemed to cover all their points was "X needs improvement."

In this case, the engineers had written for their intermediary reader, and had been mistaken in their perceptions of what the reader wanted. The intermediary reader wanted the reports to be more specific, as the consultants had suggested, so that the primary reader would understand the requirements for improving conditions at the plant. The consultants provided the engineers with a new format which separated the various aspects and thus permitted the reports to be written more specifically.

Writers often become frustrated by intermediary readers, perceiving changes as personal affronts to their writing capabilities. As a result, friction often exists between writers and intermediary readers. Such attitudes are counterproductive, since intermediary readers can provide useful suggestions for improving a document.

Antagonism between yourself and your intermediary readers can be reduced by collaborating with your intermediary readers, keeping lines of communication open, and using objective criteria for evaluating documents.

Collaborating with intermediary readers.
While you may perceive the intermediary reader's revisions as a reflection on your ability to write, this is usually not true. These changes are simply part of the collaborative process discussed in Chapter 1.

Authorship is collective. Each person involved with a document may have knowledge the writer doesn't have, but which needs to be considered. The higher up in an organization people are, the more they are aware of the context in which documents are read and, therefore, more able to perceive readers' needs and biases.

Keeping lines of communication open.
To reduce the frustration that is often caused by intermediary readers, you should discuss your document with them as you engage in the various composing processes. The more you discuss your plans with your intermediary readers, the more you'll be able to satisfy their needs by integrating their perceptions with yours, or by persuading them

that your decisions are correct. By discussing your plans before you begin drafting, you may avoid selecting an inappropriate point of view or focus, or using an inappropriate tone.

You should also discuss intermediary readers' editing changes with them so you can understand and learn from these changes. In addition, you can clarify the changes so you don't misunderstand the reviewer's suggestions as the engineers at INPO did.

Furthermore, you should use intermediary readers as editors. The higher up in the hierarchy they are, the more they know about the primary reader and, therefore, the better they can gauge readers' reactions to your drafts. The intermediary and the primary reader should both read a report or memo in similar fashion. Realistically speaking, however, this isn't always the case. A supervisor may have some pet phrases or some peeves about certain words or style. You need to consider these as well as the needs of the primary reader. After all, your supervisor is responsible for determining whether or not you receive a raise or a promotion.

Using objective criteria.   One way to reduce frustrations caused by intermediary readers' making changes related to their personal preferences is to use generic checklists. By emphasizing the aspects of a document which interfere with communication, the checklist can eliminate, to a large degree, personal stylistic preferences. The sheet can also provide both you and your intermediary readers with a basic vocabulary for discussing writing.

Writing in the Company's Name

As you have seen, documents can be either writer- or reader-initiated. Some documents relate to you personally, such as requests for vacation leave; others relate to the company in general, such as annual reports to clients.

While you may assume your own persona (identity) as an employee of a company, you may be asked to assume the persona of your supervisor or even of the company itself. Employees may be asked to write letters that their supervisors sign. Many reports go unsigned, indicating that they are company documents.

At times a company itself is responsible for submitting a report or responding to a request. A company may also want to propose or request something. These reports, responses, and proposals are drafted by writers who speak for the company. They assume not only the company's persona, but also its purpose. The request to the NRC for a penalty reduction is written by an engineer, but signed by the CEO. The engineer who wrote the request assumes the persona of the utility. He uses only third person, never first person, and he never identifies himself as either an engineer or a manager.

The following paragraph is excerpted from a proposal written by a member of a college faculty to persuade readers to support a project. Notice how the writer assumes the persona of the two agencies sponsoring the proposal—the college and Caterpillar Inc.

Use of first person plural (we) to indicate that the writer is assuming the persona of the two sponsoring agencies.	This proposal seeks to retrain a group of 90 workers who have the potential for holding electronics positions at Caterpillar Inc. ... We estimate that at least ten percent of the students will be members of racial ethnic minority groups, women, handicapped persons, or the elderly. Although some of the instruction for the project will be provided by full-time college teachers, much will be provided by retired Caterpillar electronics personnel who will receive in-service training to become teachers.

Electronics Technology Training for Central Illinois

Writing Legally and Ethically

Regardless of the situation in which you work, you are expected to maintain legal and ethical standards in your documents. You are expected to conform to all laws related to your profession and organization, reporting facts honestly and completely. Concern for the consumer and citizen should remain foremost in relation to everything you write. In this section we will study three aspects of legal and ethical standards related to written documents: product liability cases, the Code of Federal Regulations, and professional ethics.

Product Liability Cases

Product liability refers to civil cases involving injury to consumers from defective or dangerous products.

Over the past decade manufacturers have been held liable for the products they produce. The documents accompanying these products, including warnings and instructions, are considered an integral part of a product. Even if a product meets all production and design requirements, it may be judged defective if warnings and/or instructions are considered inadequate.

More and more often, the courts have ruled in favor of consumers in product liability cases. Therefore, you need to be especially careful in writing instructions and other information concerning products. Most organizations have specifications for writing instructions and warnings to ensure their own, as well as the user's, protection.

In addition, any documents concerned with a product may be introduced in a suit as evidence that indicates the manufacturer's attempt or failure to design a "reasonably" safe product. All memos, minutes to meetings, letters, etc., should be meticulously maintained. The Consumer Product Safety Act (CPSA) requires manufacturers to keep records, make reports, and submit information related to product-related injuries. Furthermore, the Consumer Product Safety Commission has the right to inspect records related to the manufacture of consumer products. You should remember that whatever you write may be used in a legal situation. Keep a copy of everything you write for possible future use.

Code of Federal Regulations

All industries that work for the federal government, as well as industries regulated by the federal government, are subject to the Code of Federal Regulations (CFR), which requires that not only must all documents contain accurate information, but they must also contain "complete" information. Writers who fail to include all information relevant to a situation are guilty of making a "material false statement" and are in violation of the code. Depending on the severity of the situation, the firm and/or individual can be fined and/or prosecuted.

For example, if, in responding to a violation by the Environmental Protection Agency, a firm admits dumping paint thinner down a drain on the day the agency conducted the inspection, but fails to admit dumping it daily for the past two years, the firm has made a material false statement.

Professional Ethics   As a writer, you are not only responsible for adhering to the letter of the law, but also for considering the ethics of a situation. Schimberg[1] suggests three aspects of communication that may create unethical texts: imprecision and ambiguity; understating the negative; and semanticism.

Imprecision and ambiguity.   If a problem occurs at a plant, administrators may prefer not to admit it. They may "throw up a smoke screen" or place the blame elsewhere. In the following excerpt, a utility responds to a violation cited by the NRC. Notice how the utility rationalizes the cause of the problem:

Failure to notify and obtain permission from Shift Operations personnel prior to troubleshooting the recirculation pump control circuit...deviated from normal practices of Instrumentation and Controls (I&C) personnel. On this occasion the technicians were working under the direction of the system engineer to clarify a problem with the Reactor Recirculation Flow indication. Although their activities resulted in the recirculation pump transient, the technicians had no reason to believe that installing the recorder to test jacks would have that effect.

The utility never admits the technicians should have obtained permission. Instead, it simply explains the technicians never expected the problem to occur, and therefore didn't see any need to inform shift operations personnel about their work. This explanation resembles that of a child who has just dropped and broken the cookie jar, and explains he didn't see any need to tell his mother he was going to take a cookie because he didn't expect to break the jar.

Understating the negative.   By downplaying a problem, you may inadvertently mislead consumers or employees about the consequences. Following the gas leaks at the Union Carbide plant in Bhopal, India, a company statement was made that "The probability of the kind of accident that happened in India happening here [in the United States] is just not the same." In fact, within a year a Union Carbide plant in West Virginia experienced a gas leak.

[1]Schimberg, H. Lee. 1989. "Ethics and Rhetoric in Technical Writing" in *Technical Communication and Ethics*. Ed. R. John Brockman and Fern Rook. Washington, D.C.: Society for Technical Communication. 59-62.

Semanticism. Semantics refers to the meanings words have according to legal definitions and in readers' minds. In relation to ethics, semanticism refers to the art of using words to avoid assuming responsibility for a product or situation.

In the following sentence, excerpted from a response by a utility to an NRC violation notice, the writer uses the term "reorder," which connotes a change, instead of the term "correct," which implies an error. As you read the two sentences below, consider the impressions a reader gets from the two different words.

The utility has revised the procedure for transfer rod groups to *correct* the sequence of steps.

The utility has revised the procedure for transfer rod groups to *reorder* the sequence of steps.

Many professional societies, such as the Society for Technical Communication, have developed codes for their members. You should become familiar with your professional society's code.

STRATEGY CHECKLIST

1. Keep your primary reader and purpose in mind at all times.

2. Base your textual decisions on your primary reader.

3. Write a separate letter for each primary reader if you have more than one and they are in different relationships to your organization or in different organizational or professional communities.

4. Make provisions for the secondary reader by placing information irrelevant to the primary reader, but relevant to the secondary reader, in an appendix. Place definitions for terms the secondary reader may not know in parentheses or in a glossary.

5. Place information which you want to record, but which is irrelevant to the primary reader, in a separate document, and file it or send it to your supervisor.

6. Use cover letters, executive summaries, and tables of contents so primary readers get a general overview of the subject of a document. Use headings, subheadings, and forecasts so secondary readers can locate information in which they're interested.

7. Keep lines of communication open with your intermediary readers. Discuss editing changes with them. Learn from the editing changes.

8. Use your intermediary readers as editors.

9. Use a generic checklist when discussing document changes with your intermediary readers.

10. Use neutral tones and language if peripheral readers may read your document. Phrase your report carefully.

continued

11. Adapt your style and tone to your persona.

12. Use the editorial "we" to include the reader as part of your organization or ideas, or to persuade the reader to agree with you.

13. Use first person or second person to speak directly to the reader to create a tone of intimacy and immediacy.

14. Maintain legal and ethical standards.

Scenario

Let's observe how purpose and audience may affect a writer's decisions:

Arby's Franchise Division received a telephone call from Tom Giles, the owner of an Arby's franchise in Eau Claire, Wisconsin. Giles was disturbed because Arby's had "missed the boat" by not being at the forefront of a move to switch from plastic foam products to paper packaging. The person who spoke with Giles promised to convey his concern to the packaging division, and to have them get back to him. He subsequently sent a memo to Debra Irwin, the manager of new products at Arby's, requesting that she respond to Giles.

Irwin knows that Arby's has been aware of the environmental problems associated with polystyrene (styrofoam) for some time, and that the company has been investigating alternatives, including changing its packaging from polystyrene to paper. However, based on its study of the problem, the company has decided that the best way to handle plastic waste is to reduce and recycle it, rather than replace it with paper.

Irwin needs to respond to Giles in such a way that she *persuades* him that paper is not the answer, and that recycling is.

Before starting to draft her response, Irwin needs to get a sense of her reader, and to determine her purpose. She has never met Giles, and so she has to make some guesses about him. She goes about getting a sense of her reader by considering the following:

- Since Giles is sufficiently concerned with the issue to make a long-distance telephone call, he probably plans to read her letter thoroughly. Therefore, the letter can be as long as necessary so all of the relevant details can be presented.

- Because Giles' call was related to using paper packaging, he will probably be disturbed that Arby's is not going to follow this route. He needs to be persuaded that Arby's decision is valid. The letter will need to relate Arby's decision to Giles' values, and support the claims with factual information.

- While Giles appears to be knowledgeable about plastic waste, he is not an expert and may not even be a generalist. He probably acquired his information from newspaper and magazine articles. The letter should not be technical.

Responding to letters such as this one is part of Irwin's responsibility. She has kept in touch with her suppliers and a number of organizations, such as the Council on Plastics and Packaging in the Environment (COPPE), and has acquired quite a bit of knowledge about polystyrene and plastic waste.

She knows that her manager, her internal reader, will read through the entire letter to determine whether she has represented Arby's in the best possible light.

Irwin considers all of the aspects discussed in Chapters 3, 4 and 5. If she were to fill out an audience analysis chart, it would look like that in Figure 5-6.

Figure 5-6
Audience Analysis
Chart

	Tom Giles, Owner	Manager, Arby's
I. Reading Behavior		
Amt. of time to read?	Sufficient	Little
Scan? Skim? Search? Study? Evaluate?	Evaluate	Scan
Expect/Not expect?	Expecting	Not expecting
Expect with anxiety? Pleasure? Skepticism? Neutrality?	Skepticism	Neutral
II. Position in Organizational Community		
Internal/External?	External	Internal
Decision Maker? Operator? Technician? Service Person? Professional? User? Producer?	Decision Maker, Proprietor	Decision Maker, Administrator
III. Knowledge of Topic		
Expert? Generalist? Novice?	Novice	Novice
Knows what about topic?	Polystyrene hurts the environment; it contains CFCs. Nothing about Arby's efforts.	The general results of Arby's testing and some economics and politics behind the decision.
Needs to know what about topic?	How Arby's will take care of the problem.	How Arby's is positively taking care of the problem.

continued

Figure 5-6
Audience Analysis
Chart *continued*

	Tom Giles, Owner	Manager, Arby's
III. Knowledge of Topic (cont.)		
Use of information?	To feel better about continued use of of Arby's products. Possibly to decide to participate in pilot recycling project.	Make a decision about writer's performance.
Biases?	Against plastic packaging. For paper.	For plastic packaging.
IV. Personal Characteristics		
Culture?	American	American

Irwin next considers her purpose. Her main purpose is to *persuade* her reader to change his attitude toward the use of paper and plastic for packaging. If she were to fill out a purpose chart, it would look like the one shown in Figure 5-7.

By keeping her sense of audience and her purpose in mind, Irwin can make the following textual decisions:

Content
- Giles needs background information to persuade him that Arby's is aware of the problem and isn't ignoring it.
- Giles needs concrete evidence that supports Arby's solution. The evidence should be factual, and it should persuade the reader that the writer is knowledgeable and the argument is valid.

**Figure 5-7
Purpose Chart**

Reader's purpose: To be informed about Arby's position on the use of plastic rather than paper packaging.

Writer's purpose:
(x) persuade   ( ) inform   ( ) instruct   ( ) regulate
To persuade the reader to approve of plastic recycling as a viable alternative to paper.

Initiated by whom? Reader

Situation: Part of writer's job. Supervisor reviews letters.

- Since the reader is a novice in his knowledge of polystyrene, he probably has ideas about styrofoam and plastic waste that need to be demythologized.
- Giles is expecting the letter and probably will read it thoroughly. Therefore, it can be long enough to include all the points and details needed to support Arby's decision.

Point of View and Focus

- As the owner of an Arby's franchise, Giles provides users with the packaged food. Therefore, his point of view is that of a provider. To persuade him to change his mind and accept plastic recycling as an alternative method of eliminating plastic waste, the focus needs to be on the good aspects of recycling plastic.

Organizational Pattern and Sequence

- Since the reader is interested in a solution to the problem of solid waste, the problem/solution pattern appears obvious. However, since Giles is already aware of the problem, the information should begin with the solution.
- Since the reader is skeptical and possibly even hostile toward the solution, the letter should open by conceding that plastic waste is a problem as the reader has indicated.
- The reader should be left with a positive point about recycling.

Style

- To alleviate any hostility on the part of the reader, the tone should be friendly, using first person plural (we) to include the reader as part of the company.
- The tone needs to be positive in proclaiming Arby's solution.
- Since the reader has addressed a company issue, the writer should assume the persona of the company.
- Since the reader is a novice in the field of plastics, technical terms should be used only when necessary, and then they should be explained, though the language and syntax should not be simplistic.
- As a novice, the reader won't want a lot of scientific detail. If he's interested in further information, a suggestion for finding additional sources could be appended as a P.S.
- The information should not be too abstract; it should be specific enough for the reader to understand that Arby's solutions are good ones.

**Figure 5-8
Decisions Chart**

Information	Background	EPA's view as related to Arby's
	Arby's solution	Complexity of problem
	Problem with myths	
Point of View	Arby's solution is right	
Organizational Pattern	Solution/problem	
	Begin with concession	
Style		
Persona	The company	
Tone	Positive, conciliatory, friendly, factual	
Language	Non-technical	
Syntax	Clear, straight forward	
Amount of detail	General	
Level of abstraction	Specific	
Conventions	Friendly address and salutation	
	Block format	
Format		
Length	Short	
Opening	Concession, catch attention, to the point	
Report parts	NA	

If Irwin were to fill out a decision chart, it would look like Figure 5-8.

Based on these textual decisions, Irwin drafts the letter in Figure 5-9. As you read, consider how the text is based on Irwin's decisions.

The letter reflects the textual decisions Irwin makes based on her sense of her reader and her purpose. Rather than stating openly that Giles' suggestion is not valid, she tactfully implies that he should reconsider his decision in light of her argument.

Was Irwin successful in achieving her purpose? Yes. Her supervisor noted on a copy of the letter he returned to her, "Excellent follow-up," and Giles was convinced, commenting that, after reading the facts, recycling seemed to be the proper decision. He requested a pilot recycling project at his franchise.

CHAPTER SUMMARY

To communicate your message successfully, you need to think carefully about your purpose for writing a document. You need to keep the reader's primary purpose in mind at all times.

You may have multiple purposes for writing a document. These purposes may relate to both the reader and the writer. Your own purpose may be closely related to that of your reader.

**Figure 5-9
Arby's Letter of
Response**

Begins with problem. Indicates shared values. Opens on positive note. Assumes company's persona. Moves to solution section. Provides background.

Relates evidence to authority accepted by reader. Begins by discussing most important solution. Provides factual evidence.

Relates evidence to reader's attitude toward recycling.
Refutes counter-argument for paper.

Refutes other counter-arguments. Tone is friendly, conversational.

Relates to reader's needs as a restaurant owner. Uses quantitative evidence to support claims. Firm tone. Recognizes reader's biases as a small business owner toward the free market. Ends on a positive note. Short and punchy conclusion. Easy to remember.

Provides reader with further sources of information rather than write a longer letter.

May 10, 19xx
Mr. Thomas Giles
1412A South Hastings Way
Eau Claire, WI 54701

Dear Mr. Giles:

Arby's appreciates your interest in packaging and the environment. We, too, here at Arby's are very concerned with the environment and what the real solutions entail.

We began solving the problem in 1988 when the use of CFCs in the manufacture of polystyrene foam was phased out. Today none of our containers include CFCs.

After several years of studying the problem of plastic waste, we believe the best solution lies in a combination of source reduction and recycling, two of the solutions approved by the Environmental Protection Agency. As a step toward that solution Arby's began a Pilot Recycling program in June, 1990 in Connecticut and the Portland, Oregon, area.

Two polystyrene plastic recycling plants are in operation nearby: Plastics Again in Leominster, Massachusetts. and Denton Plastics in Portland, Oregon. The National Polystyrene Recycling Council has named Los Angeles, San Francisco, Philadelphia, and Chicago as locations for four other recycling plants.

Recycled plastics can be made into a number of different products, such as construction material (boards/insulation), office supplies, flower pots, video tapes, plastic trays and holders. However, once a paper container is contaminated with food, it is unrecyclable. Switching from a polystyrene food container to a paperboard container merely replaces one material with another.

Let me briefly set you straight about some common misinformation concerning foam and plastic packaging. One common misconception is that packaging in general, and plastic food packaging in particular, comprises an inordinate amount in the waste stream. This myth suggests that if we ban plastic food packaging, we will create additional capacity in our landfills and other waste disposal systems.

This view ignores the fact that takeout containers are a small part of the total trash deposit in solid waste landfills. Excavations of landfills have shown that ALL "restaurant" containers comprise less than 1/3 of 1 percent of total landfill contents. Obviously, banning plastic food containers will have very little effect on the total material in the solid waste stream.

Banning plastic food containers ensures that the growing ability to recover and recycle plastics will never reach your area. Product bans deny sellers, buyers and customers the convenience of free market choices and, worst of all, they do not solve the problem. Product substitution is not the answer. Product reduction and recycling is.

Sincerely,

Debra A. Irwin
Manager, New Products
cc: Will Fisher [Manager]

P.S. You can call this toll free number for additional information—Mobil Chemical Company 1-800-333-0124.

Purpose affects your textual decisions involving information, organization, focus, and style.

Documents can be reader-initiated or writer-initiated. Most writer-initiated documents are persuasive.

Readers read documents to acquire information, make a decision, learn to do something, or determine what can and can't be done.

Writers draft documents to inform, persuade, instruct, and regulate.

The situation in which you write affects your purpose. Sometimes you may be faced with conflicting purposes. You need to use your judgment to navigate them. Some documents relate to you personally, others are company related. You may be expected to assume your own persona, that of your supervisor, or even that of the company itself.

You may also have a hidden agenda when you write. The focus of your information should relate to your hidden agenda.

PROJECTS

As you work on the following projects, you may want to use the Audience Analysis Chart (used in Figure 4-9, page 97), the Decisions Chart (used in Figure 4-10, page 101), and the Purpose Chart (used in Figure 5-7, page 136).

Collaborative Projects: Short Term

1. **Writing in a complex situation: paper or plastic**

    You are on the product packaging staff of a firm that makes toys. Your company has just developed a new toy, and you have been debating whether to use paperboard or plastic to package it. Yesterday, management made the decision to use an inexpensive plastic. However, because of the public's interest in degradable materials, management is worried that some retailers may not carry the product and some customers may not buy it. You have been assigned to a team to determine the best approach for the sales force to take in relation to the packaging.

    In your discussion, you may want to refer to the notes (shown below) you took during the packaging debate, and, of course, you will want to consider any legal or ethical consequences.

    NOTES

    - There is a difference between biodegradability and photodegradability. Many products are photodegradable, not biodegradable.
    - Photodegradability refers to the ability of a product to deteriorate in sunlight. In photodegradable products, ultraviolet rays break down the molecules.
    - Biodegradability refers to the ability of a product to degrade in the ground. In biodegradable products, microorganisms break down the molecules.
    - Paper is biodegradable. Plastic is photodegradable.

- Photodegradable products can only degrade in sunlight. Litter left along the country's highways degrades fairly quickly.

- Biodegradable products can often take many years to degrade. Research in landfills has unearthed whole newspapers which never degraded. A copy of the *Cincinnati Enquirer*, unearthed in a landfill ten years after being dumped, had degraded so little that the headlines could still be read. *

- Photodegradability can be enhanced by adding impurities.

- Increasing photodegradability increases cost by 5-10%.

- Photodegradable products only break down if exposed to direct sunlight, not buried in landfills.

- Increasing degradability reduces the ability to recycle materials.

- Biodegradation can cause air and water pollution. The liquid filtering through biodegrading waste, called leachate, can contaminate groundwater. The gaseous by-product is methane which is unhealthy to breathe. Landfills that have been properly constructed in the last few years eliminate these problems.

- Both paper and plastic degrade slowly in landfills.

- Plastic can be safely incinerated.

* Data obtained from paper presented by Dr. Riley Kinman, Degradable Plastics and the Environment Symposium transcript, 1988, Solid Waste Management Solutions, New York.

   a. Work in groups of five. Determine an approach. You may want to consider one or more of the following suggestions, or you may have a different idea.

   - Don't say anything about the packaging and hope the retailers don't care.

   - Write a letter to the retailers that describes all of the good aspects of plastic and all the bad aspects of paper.

   - Write a letter to the retailers telling them that the product is degradable, but not specifying that it is photodegradable rather than biodegradable.

   - Write a letter to the retailers telling them that the packaging is photodegradable, and why this is beneficial to the environment.

   - Print information on the wrapper on how to dispose of the product in the most environmentally-beneficial way.

   - Print the following information on the wrapper—degrades in sunlight, landfill safe, no groundwater contamination, nontoxic when incinerated, recyclable, meets FDA requirements.

   - Establish an 800 number to provide consumers with information about plastic packaging. Print on the wrapper, "For information about the degradability of this plastic wrapper, please call 800..."

b. Each group should present its approach to the class.

c. The class should vote on the approach the company should adopt.

Option: Depending on your decision, write the letter and/or the text to be printed on the wrapper.

2. **Writing to persuade**

Work in trios. You are part of a project dedicated to improving the environment. Your team has targeted an industry. Select the industry and a company within that industry that you would like to see improve its impact on the environment. Write a letter to the company, suggesting the actions the company should take to improve the environment. Before drafting your letter, fill out audience analysis, purpose, and decision charts.

Your team may decide to gather information and write the letter together, or you may assign each member one of the following tasks:

a. Gather information about the company's efforts, or lack of efforts, to solve a specific environmental problem.

b. Gather information about strategies for improving the specific environmental issue you are addressing.

c. Use the information you have gathered to write the letter.

3. **Writing to inform**

You have just recently been hired by the National Geographic Cartographics Laboratory. You will be part of a team to design a new atlas for the local elementary school system in which your college is located. The atlas will be titled, "Sample Atlas of Map Types Used in Cartography."

You need to take into consideration your audience, how the atlas can be used in classroom projects, teacher usage and effects, the types of maps represented, and the design of the atlas.

You and your team are ready to begin planning the atlas. You need to engage in the first two steps of the planning phase—getting a sense of your audience and determining your purpose. Working with your team members, fill out the following charts: (1) sense of audience, (2) purpose, and (3) decisions.

Exercise adapted from Geography 300, Introduction to Cartography
Jill Thomas, Illinois State University, Normal, IL

Collaborative Projects: Long Term

The following projects can be continued through Chapter 13. As you complete each chapter, you will find activities for these projects which are directly related to the chapter you are reading. In this chapter, you'll engage in activities related to analyzing your purpose and situation.

- **Communicating about a campus/community problem** continued from Chapter 4, Collaborative Projects: Long Term Exercise 1.
- **Proposing microcomputer laboratories** continued from Chapter 4, Collaborative Projects: Long Term Exercise 2, Projects 1, 2, 3.

Work with your group or subcommittee to fill out the Purpose Chart (used in Figure 5-7, page 136). Then fill out Step 1, "Determine Purpose," on the Planning Worksheet in Appendix B.

To be continued, Chapter 6, Gathering Information, Collaborative Projects: Long Term.

Individual Projects

1. The letter in Figure 5-10 was written to Arby's. You have become Irwin's assistant and she has asked you to respond. Based on what you know from

Figure 5-10
Letter of Request to Arby's

Dear Sir:

Allow me to introduce myself. I am a supervisor at store #5338 in Lincoln, Nebraska.

I am a junior attending a small liberal arts college here in Lincoln. I am majoring in biology, and I am working toward an endorsement to teach secondary science. In my biology classes this semester we are stressing environmental care, e.g., recycling, pollution control, conservation, etc.

This brings me to the main point of this letter. As my "activity project" for this class I am attempting to learn about Arby's attitude about such things as conservation, pollution, etc.

An area of debate here on campus was a proposal by the student senate to ban all styrofoam. This issue caused me to notice all of the styrofoam we use at Arby's. Much has been said lately about how styrofoam is linked to the current problem of ozone depletion over Antarctica...

Basically, I would like to ask if this is more or less the practice in all Arby's stores or is localized to the stores here in Lincoln. I would like to ask though if Arby's as a whole is planning a switch away from styrofoam and other non-biodegradable products. Are alternate sandwich box materials being researched?...Any information you can give to me about product research and development and cost analysis would be a great help and I would appreciate it very much. I thank you for your time and effort, and I anxiously await your reply.

Sincerely,

Marc Kroger

her letter on page 139, write a response. Before you do, you should fill out an audience analysis chart (page 97), a purpose chart (page 136), and a decision chart (page 101). If you have a computer, make a template for the charts.

2. You have read Irwin's letter on page 139. If you disagree with her findings, based on your knowledge of environmental issues, write a letter to convince Arby's to use a different solution. Before doing so, fill out audience analysis, purpose, and decision charts. If you have a computer, make a template for the charts.

3. You probably have some classes besides this one in which you are required to write papers. You may also have some classes requiring another use for a computer. Although you have access to your school's computers to do this work, the access is limited and you realize it would be a lot easier if you had your own computer.

    The following assignment has three parts:

    a. Write a letter to your parents requesting either a laptop computer or a PC.

    b. Now that your parents have agreed to buy you a computer, they want to know which one you want. You may already know the model and brand. If you don't, you will need to do some investigating to make a decision. You can get information from computer magazines or computer stores, and your school probably has a computer expert who can help you. Once you decide, write a letter to your parents telling them the computer you want.

    c. Compare the two letters. What are the differences in the following:

    • the information included

    • organizational pattern

    • point of view and focus

    • style

    How has your purpose influenced these differences?

4. Read the memo in Figure 5-11 written to a supervisor by an employee. The memo is writer-initiated; the supervisor did not request it.

    What do you think is the writer's purpose? Write a sentence, stating the purpose of this memo, to insert as the first sentence. By stating your purpose up front, you help readers understand from the very beginning why they are receiving a document.

    Does the writer of this memo have a hidden agenda? If so, what is it? Put brackets around words or phrases that you believe indicate a hidden agenda.

Figure 5-11
A Persuasive Memo

INTER-OFFICE MEMO

TO: H. B. Supervisor

FROM: R. W.   RW

SUBJECT: Energy Conservation by Automatic Light Switching

DATE: February 3, 19xx

The trial installation of an automatic light control in the cafeteria has gone well. There have been no problems with the equipment in the nearly two months it has been in service and employees have had no complaints.

My calculations indicate that this unit, which is saving about $150.00 per year in electricity costs, will yield a simple pay-back in about 1.57 years.

Although this type of savings is low in terms of total dollars, it is worth following up on.

To this purpose I had the entire facility surveyed to identify areas which would yield reasonable savings. As you can see in the resulting report from the vendor (attached), a "complete" installation would cost $3,032 for materials and another $850 would be required for installation. Depending upon the savings percentage used, the payback would be 1.5 to 2.0 years if all areas are done.

However, I would recommend installation only in those areas with a very good potential either due to the nature of the area or due to the amount of lights involved.

Area	Annual Energy Use $	Savings Factor	Savings $
'B' Lab	$288	50%	$144
'C' Lab	$384	50%	$192
'D' Lab	$288	50%	$144
Elec Shop	$288	50%	$144
Cons. Lab	$382	40%	$153
Lab Stor. Room	$192	80%	$153
Gross Savings			$930
Material Required=(6) x $186 = $1,116			
Installation= (6) x $ 50 = $300			
Total Cost $1,416			

The simple pay-back for these selected areas is 1.5 years, which, I feel is realistic. I would recommend a limited installation as outlined.

5. You've been working in the local frozen yogurt store, "Frozen Rainbows." Yesterday, a man stopped in for a cone. He's an engineer from China, and he is visiting the United States to attend a class at your college on innovations in electrical engineering for telecommunications. He'll be in your city for a month while he attends the class. He's staying at the big dorm on the college campus. He told you he comes from a city similar to the one in which your college is located. While he's in the United States, he wants to learn about the kinds of small businesses that flourish in these cities,

because he's hoping to set up his son, who graduates from college at the end of this year, in business in China.

You've spoken with your boss. She thinks the company would be interested in having a franchise in China. The firm, which has 30 franchises, located mainly in the northeastern United States, has several in England and Germany. The franchise would cost approximately $100,000, and the son would have to attend a six-month training program at the firm's headquarters in Maine.

a. Write a letter to this father suggesting he contact the Director of Franchises, Mr. John Dirkson, at the firm's national headquarters in Waterloo, Maine, to discuss purchasing a franchise for Frozen Rainbows.

b. Since you first talked about this with your manager, she has discussed the Chinese man's interest with a representative from the company's national office. The representative has agreed to give you a point (1% of the franchise fee) as a bonus if the man buys the franchise. Your manager gives you this information just in time; you have not yet mailed your original letter. Write a new letter to this father, persuading him to purchase a franchise for Frozen Rainbows.

c. When you have finished your new letter, compare it with your original one. What differences do you notice in terms of the information you've included, your organizational pattern, focus, and style? If you find differences between the documents, how do you think your purpose affected them?

Processes and Techniques

IN PART I YOU LEARNED STRATEGIES TO help you make decisions based on the context in which you write. That context includes your readers, your purpose, and your situation.

IN PART II YOU'LL LEARN STRATEGIES FOR presenting information once you have analyzed your context and made your decisions. These strategies help you communicate your message to your audience to achieve your purpose and meet your readers' needs. You'll also learn techniques for designing documents that facilitate readers' reading processes and behaviors.

AS YOU ENGAGE IN THE PLANNING PROCESS, you need to gather and organize information in relation to your audience, purpose, and situation. Because all writing involves some form of persuasion, you need to acquire strategies for persuading readers to accept your message.

You also need to learn various techniques, such as defining, describing, analyzing, and summarizing, to achieve your purpose. Once your document is drafted, you need to evaluate it to determine if you are communicating your message to your readers. If not, you need to use effective revision techniques.

THROUGHOUT THESE PROCESSES YOU NEED to consider graphics that will help your reader comprehend your discussion. You also need to help readers quickly and easily locate information. You should provide them with a "road map," using effective visual text and a format that will help readers perceive where they are going and the route they will take to get there.

IN PART II YOU WILL ACQUIRE STRATEGIES for engaging in composing. These chapters will also provide you with strategies for effectively carrying out your decisions. □

Gathering Information

THROUGHOUT YOUR ACADEMIC CAREER, you have gathered most of the information for a report from a library. In many cases you know very little about a subject until you look it up in a reference book, magazine, or journal. However, the situation in industry is different. You'll often know a lot about a topic, and your information will be derived from a variety of sources, not just the library.

Some documents, like progress reports, require very little, if any, research because the information is all in your head. However, other documents, such as feasibility studies, proposals, and research reports, may require gathering information from a number of sources. These sources might include personal experience, journals, industrial reports, interviews, observations, and other forms of field research.

This chapter begins by looking at some of the methods used for planning to gather information, including listing, clustering, questioning, visualizing, and collaborative debating. The chapter also considers the major sources from which information is obtained, including personal, written, and field sources. Finally, we'll observe a writer using the various planning strategies and sources to gather information for a proposal.

PLANNING TO GATHER INFORMATION

As you have seen in previous chapters, the information you include in a document depends on your sense of audience, purpose, and situation. Once you decide on the content of a document, you need to determine information you already have, information you need to get, and how and where to get information you don't have. You can determine a plan for gathering the information you need by using one or all of the following methods: listing, clustering, questioning, visualizing, and collaborative debating.

The method you select depends on the type of document you're writing, the topic, and your own thinking processes. Some people like to use listing, while others prefer visualizing. You may want to try each of the methods to see which is most effective for you. Often a combination of methods can be helpful.

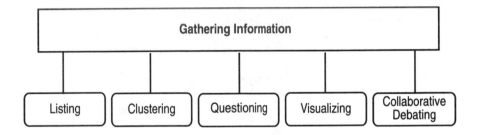

Listing

This method is simple. Just follow these steps:

- List information your readers want to know based on their organizational, professional, and personal characteristics, as well as their purposes, and situation.
- List information your readers need to know to achieve their purposes.
- List information you want your readers to have to achieve your own purposes.
- List information you need to support your claims.
- Go back and look at your list. Put a check next to the information you already have.
- Next to each of the unchecked items on your list suggest where or how to get the information.

■ Let's look at how Jody Howard, a design engineer for a software firm, may have used this method in preparing to write a letter to a client closing out a completed project. Her purpose is twofold: to notify the client that the project is completed so the client can begin using the system; and to prevent additional requests for services so the client can be billed. The client will read the letter in the role of a decision maker (determining whether the program has been completed satisfactorily), and also as a future user. Since the client has been involved in the project, background information and details are not necessary. Notice how Howard considers information the reader needs to achieve his purposes (what reader wants to know), as well as the information she needs to achieve her purposes (I need to tell reader). ■

Reader wants to know

Needed for legal identification.

a. Name of project, project no. (Check contract)

Needed for legal identification.

b. Date when project completed. (Check calendar)

Permits reader to compare with contract.

c. Summary of project. (See proposal)

Needed for legal identification.

d. Person present when project completed. (x)

Needed by user.

e. Warranty info. (Use boilerplate[1])

Needed by user.

f. Referrals for questions. (Use QA people, check with supervisor who will be responsible)

I need to tell reader

Needed for legal purposes.

a. Statement of completion. (See boilerplate[1])

Appeal to reader's self esteem.

b. Thank you

Clustering

This method, like listing, involves jotting down information. However, rather than writing out the entire list and then going back to look at each item, work with each item as you think of it. As you list each item, write down all of the

---

[1]A generic document or document excerpt which can be used for a variety of purposes or sent to a variety of persons. Specific information, such as readers' names, addresses, and other relevant information, can be easily inserted to individualize the message.

information you know about that item in a cluster around it. If you don't know anything, list what you want to know and indicate where and how you can get the information.

■ If Howard had used this method, she would probably have thought first of the statement of completion. In thinking of the statement, she would have recognized that readers should have a record of the name of the person present at the time of completion, as well as the date. Because she doesn't remember the exact date, she makes a note to check her calendar. As she thinks about things to check, she remembers readers need information about the warranty, along with the name of the company's contact for the project in case of problems with the software program. Again, she notes that she needs to find out who the contact will be. At this point, she remembers she needs to include a "thank you" in the letter. She then goes back to the completion statement and decides the reader may need a brief summary statement of the project for the record as well as a list of the services the company has provided.

Howard's planning sheet looks like Figure 6-1. Notice the similarities between the results of this method and listing. ■

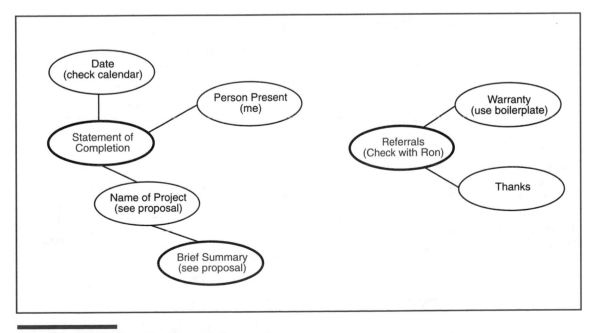

Figure 6-1 Clustering to Plan for Gathering

Questioning

This method only concerns information you *don't* have and is often good to use in combination with one of the others.

List all the questions your reader may have about your topic. Next, list questions you have about your topic. Go back and, based on your prior knowledge and experience, try to answer the questions. Next to the questions you can't answer, or for which you don't have complete answers, indicate where and how you can get the answers.

■ If Howard had used this method, she might have come up with the following three questions. Notice how they relate to the list of items or the items in the clusters:

Client's questions

1. Has everything been done for which we contracted? (check original proposal and contract)

2. What about a warranty? (use boilerplate)

3. Who should be contacted when questions arise? (check with supervisor)

Engineer's questions

None ■

Visualizing

It's easier for some people to make a drawing, sketch, or flow chart (a sequence of steps for solving a problem) than a list.

■ If Howard had used this method, it might have looked like the chart in Figure 6-2. Again, notice the similarity between this chart and the previous results. ■

**Figure 6-2
Charting Your
Information-
Gathering
Activities**

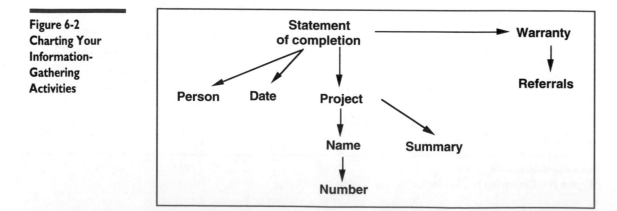

Collaborative Debating

This method is a good one to use when a document involves a number of people.

1. Hold a meeting with all persons involved. Arrive at a consensus on the purpose for the document and a sense of audience.

2. Draw up a list of the information needed in the document.

3. Indicate whether or not the information is already known, and if so, by which person at the meeting, or if not, how and where the information can be obtained and who will be responsible for gathering it.

The College of Engineering at a university needed to draft letters to U.S. airlines requesting funds for a research project involving wind shear. To plan the list of information required for the proposal, the Director of Research spoke with the head researcher and several members of the project to determine the information that should be included in the request. Since the readers were familiar with the problem, background information was not needed.

The following list resulted from that discussion. Notice that some of the information will come from other reports written on the topic; some from the researcher, George, or his assistant; and some from a boilerplate statement:

1. Name of research (see recent report, George's assistant will send it).

2. Goals and objectives of research (see recent report).

3. Specific aspects to be investigated (George will list them in a memo to me).

4. Researcher's credentials and track record (bio at end of report).

5. Summary of work completed to date (enclose recent report).

6. Amount of funding needed and length of time needed (see memo George sent last month); grant stipulations (boilerplate).

7. Present support (see recent report).

8. Referrals (to either researcher or director of research, George agreed at meeting).

STRATEGY
CHECKLIST

Use one of the following methods for planning the information you need to gather and how you will gather it.

- Listing
- Clustering
- Questioning
- Visualizing
- Collaborative Debating

USING SOURCES OF INFORMATION

Once you know what information you need, you can begin to gather it. Information can be derived from three major sources: personal knowledge and experience, written sources, and field sources.

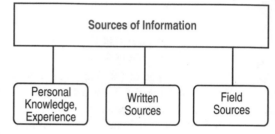

Personal Sources

On the job, much of the information you include in a document comes directly from your own knowledge and experience. To gather information for a progress report on a project on which you're working, you can simply review what you've done over the past week or month; the information is already in your head. Knowing that you will need to write a report, you may also have kept field notes to which you can refer.

Informal lab reports are based on the research you've been doing, just as minutes to meetings are based on decisions made at meetings you've attended. You may refer to the notes you made while working on a project or attending a meeting, or you may refer to your lab journal to jog your memory, but you don't need to do library research to obtain your information.

■ Howard knows most of the information for the close-out letter because she worked on the project. However, she needs to verify certain information. She checks her calendar and the contract to be certain she has the correct date for the closing, and the correct project name and number. The abstract for the project proposal includes a good summary statement that she can copy directly from her word processing file. In addition, the company has boilerplate for warranties and referrals that she can copy into the letter. Also, her supervisor recently sent her a memo listing the Quality Assurance (QA) people responsible for responding to questions on this particular project.

Let's take a look at the letter in Figure 6-3 that Howard drafts, once she has gathered her data. As you read, consider her sources. ■

STRATEGY CHECKLIST

1. Keep good, detailed notes of your activities so you have all of the information you may need. Make sure you specify names, dates, times, model and part numbers, etc.

2. File information so you can find it easily.

3. Maintain a list of boilerplates for easy reference.

**Figure 6-3
Howard's Letter**

ADVANCED TECHNOLOGY SERVICES, INC.

—ATS

January 11, 19xx
Rod Simpson
Bldg. A2 Systems
Caterpillar Inc.
East Peoria, IL 61611

Dear Mr. Simpson:

Project name, number from files. Acknowledges cooperation.

   Thank you for choosing Advanced Technology Services, Inc. (ATS) as your systems integrator for the host communications link to your Vehicle Sound Intensity Testing System (P.O. 11175) and for allowing us to participate in this project (Job 500725). I would like to extend my appreciation to Caterpillar Inc. and especially to you for your cooperation and assistance during this project.

Date of closing, person present, from experience. Summary of work completed from contract. Statement of completion from boilerplate. Warranty from boilerplate.

   On January 3, 19xx, I was present when the updated Sound Intensity Testing System program, updated Sound Intensity System main menu program, and the Procomm Plus script file developed by Advanced Technology Services successfully met the acceptance criteria defined in the Project Proposal. Advanced Technology Services considers this job to be complete and acceptable to Caterpillar Inc.

   Warranty will be in effect for one year from January 3, 19xx for Advanced Technology Services supplied materials and workmanship as provided in the Project Proposal. Work requested beyond the warranty is subject to ATS normal service charges.

References from supervisor. Statement of thanks.

   If there are any questions in reference to the above, please contact Mark Rottman or Ron Isaia (Quality Assurance).

   Thank you for helping make this project a success and please extend my thanks to those at Caterpillar Inc. who were involved in this effort.

Sincerely,

Jody L. Howard
Design Engineer

Written Sources

In the industrial world most documents are written to suggest *new* ideas. Most of the information you gather from written sources provides background data, but it comprises only a small part of your document. While the amount of space in a document devoted to background information may be relatively short, the data may provide essential evidence to persuade readers of the need for your proposal, or to prepare them for your solutions or recommendations.

   For example, if a feasibility study concerning the conversion of a utility from nuclear to coal power is to be presented to the utility's officials and stockholders, you need to include a background section related to the problems presently being experienced by the nuclear industry, as well as information about other plants that have already made the conversion. The audience needs this information to persuade them to accept the claim that the conversion will be cost-efficient in the long run.

Your information needs to be the most recent information available on a topic. You will find much of the information in journals, on computer databases, and in reports from other utilities that have investigated similar projects. Though numerous books may be concerned with problems of nuclear energy, books on coal conversion may not yet be published because the topic is too new.

In the following example, you will read a one-page introductory section from a manual on leaf and yard waste composting. The manual is sent to municipalities interested in changing methods of landscape waste disposal. Notice the writer cites the sources in the text.

WHY COMPOST?

From a solid waste management perspective, leaf and yard waste composting can reduce the amount of solid waste which has to be sent to a landfill or incinerator. Less waste going to these traditional disposal options results in a decrease in a municipality's cost for transport and disposal.

Source 1

Burying leaves in the ground at landfills usually occurs over a relatively short time period during specific times of the year. Although leaves and yard waste may make up only 17 percent (by weight) of a municipality's annual waste stream (USEPA, 1988), during the fall season this waste may reach as much as 35 to 45 percent of a community's waste stream. A second peak for the material is in the spring and early summer. As a result such material will end up concentrated in one area of a landfill...

Source 2

Composting has become more prominent as people recognize its role in solid waste management. In addition, composting is a method for renewing a dwindling resource: soil. Where the solid waste disposal crisis has been considered by many as the crisis of today, soil loss in this country has been termed the "quiet crisis." In the United States 1.7 billion tons per year of soil is lost to erosion; this loss has direct impact on food production and the economy (Brown, 1984). Composting is one of the few methods for quickly creating a soil-like material which can help mitigate this loss...

Produced by E & A Environmental Consultants, 1990 for
National Corn Growers Association, Copyright © 1990.

Notice that the writer refers to two written sources—USEPA and Brown—in the background discussion (the use of yard waste composting to solve the soil erosion problem). Because composting on a large municipal scale is a relatively new topic, the sources are limited to a paper presented at a conference and an update from the U.S. Environmental Protection Agency. The sources provide specific quantifiable data concerning the percentage of yard waste in the waste stream, and the amount of soil lost to erosion. In addition, the two sources were written within the past decade, and the consultants and NCGA are considered authorities in the field by their professional community. Thus, the data provides evidence readers will accept as support for the writer's claims.

With the introduction establishing composting as a viable form of waste disposal, the remainder of the manual is concerned with procedures for selecting an appropriate form of composting.

If you use written sources, you need to acquire reference skills. These include locating relevant and current information, using computer databases, and taking relevant notes.

Locating Relevant and Current Information Because as long as three years can elapse between the writing and publishing of a book, much of the information contained in books may be dated. In addition, if you are researching a very recent topic, a book may not yet have been published on the subject.

The most recent in-depth technical information may be found in organization reports, such as those submitted to governmental agencies, stockholders, etc.; papers presented at various meetings; and even newsletters distributed to interested parties, including employees. These sources can be found in a variety of locations.

On microfiche.[2] This procedure allows the relatively rapid reproduction of reports and presentations which, prior to this technology, could not be made available quickly on a mass scale.

In journal, magazine, and newspaper articles. These may be available at a local university library. However, many publications are so specialized that libraries do not order them. You may need to request a library to obtain a copy of an article for you. In addition, as much as a year may elapse between the writing and the publication of a journal or magazine article. While current information can be found in newspaper articles, news items seldom cover technical topics in depth.

At private libraries. Your own and other companies often have their own libraries for storing information that they have generated, as well as information specifically related to their industry. In addition, many professional groups and public agencies maintain their own libraries. These organizations often keep in-house documents, such as brochures, newsletters, and conference proceedings, which are not catalogued in the large outside information databases. However, these documents may contain information you need. You can often write to these organizations directly, or you can request that your interlibrary loan department obtain them for you.

In sources cited in texts and bibliographical reference lists in documents you have read. These sources can often provide you with an historical perspective of the topic. You can use the references noted in these sources to read even earlier sources.

---

[2]A microfiche is an index-card-size sheet of film on which are reproduced very reduced photographs of pages of text. These must be viewed on a special machine available at most libraries. An entire book can be reprinted on several pages of microfiche. Government documents are often printed on microfiche and then indexed in special indices. Librarians can help you locate the reports.

Using Computer Databases   A variety of documents are catalogued on the library's computers, either in online databases or CD ROMs (compact disc read-only memory). This computerization allows the computer, rather than your fingers, to "do the walking." You simply ask the computer to locate the information you need. A list of sources, including books, journal and newspaper articles, and reports and presentations related to your topic appears on the monitor.

You can request information about a topic as broad as nuclear energy, or as narrow as the conversion of utilities from nuclear to coal power. The computer limits the search for you once you specify your topic.

While some libraries are more computerized than others, and while they use a variety of systems, most provide the services described here. Many organizations also subscribe to systems that provide databases specifically related to their respective industries. They may or may not elect to provide such on-screen services as abstract and full-article availability, since these can be expensive.

In those libraries where these services are available, you can call up either the abstract of an article in which you are interested, or the full article. By scanning the abstract or article online, you can determine whether or not it includes relevant information. If it does, you can print it out and make notes directly on the hard copy.

Some of the major database suppliers include DIALOG, BRS, and WILSEARCH, which provide indices in the social sciences, humanities, and general sciences. More specialized databases include BPO (*Business Periodicals Ondisc*), which provides indices for business periodicals; and PAIS (*Public Affairs Information Service*), which includes reports, newspaper articles, etc., from federal, regional, and local governmental agencies. NTIS (*National Technical Information Service*) and *Engineering Information* provide specific databases in technological subjects. STN (*Scientific Technical Network*) offers databases ranging from *Bioquip*, the biotechnology equipment suppliers' databank, to *COMPENDEX PLUS*, containing citations from the world's engineering and technology literature, to *Energy*, containing abstracts referencing energy research and technology, to *Ulidat*, a German language file containing abstracts on environmental subjects.

Depending on the subject matter, some compact discs are updated monthly, others quarterly or annually. Some databases, like those for medical and legal information, are updated daily. Those for newspapers, such as the *Wall Street Journal* and the *New York Times*, are updated hourly. You can actually read an article in the *Times* on the computer before the edition carrying it comes off the presses.

Computers not only search the sources for you, but, via a modem[3], bring the information into your office, thereby saving you a trip to the library. Using services such as DIALOG, you can search for information from your own personal computer (PC). In addition, your office or university can subscribe to such services as WILSEARCH or relevant databases from Infotrac, which provide updated CDs on a variety of topics.

---

[3]A device that transmits a computer's digital signals over telephone lines.

Taking Relevant Notes   You need to take notes for your list of references as well as for the content of your document.

Reference notes.   Take reference notes on index cards, legal pads, or printouts of database reference lists. Not only can you use these notes for locating books, journals, etc., but you can also use them when you are ready to cite your references in your documents. You should be sure to include:

- the name of the book, magazine, journal, newsletter, conference report, etc.
- the name of the relevant article or chapter
- the name(s) of the author(s), editor(s), or organization responsible
- the name of the editor(s) if the information is contained in a chapter of an edited book
- date of copyright
- place of publication
- name of publisher
- pages on which the article or chapter are located.

Content notes.   As you read, make notes in the margins. Indicate relationships between information you are presently reading and information you have read previously, i.e., similarities, contrasts or contradictions, explanations or support for information you've previously read, or cause/effect or problem/solution relationships. Indicate your personal response to ideas, whether you agree, disagree, or question them. Put asterisks next to ideas you especially want to remember and include in your document because they support your discussion. These notes make it easier for you to find and discuss specific information once you begin to draft your document.

Write a brief summary of each article or chapter on index cards as soon as you have finished reading. The summary helps jog your memory when you are considering which information to include in your document.

STRATEGY CHECKLIST

1. Use relevant, current sources.
2. Don't limit your sources to books.
3. Refer to reports, presentations, conference proceedings, newsletters, journals, magazines, and newspapers.
4. Use online databases to locate and reproduce current texts.
5. Use sources cited in texts and bibliographies.
6. Search for information in private, company, organizational, and governmental libraries.
7. Take reference notes on index cards, legal pads, or printouts of databases, articles, etc.
8. Make notes in the margins of texts you have read as sources of information.
9. Summarize briefly each source.

Field Sources

In many cases, written sources will not provide sufficient information to meet your readers' needs. To gather necessary information, you'll need to spend time doing field research—*interviewing* people, *observing* something being done, *surveying* groups, *conducting* focus groups, and *researching, developing, and testing* your own products or projects.

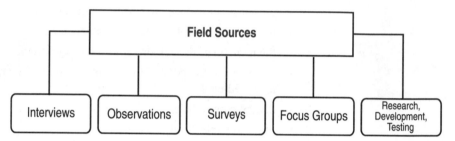

To acquire information necessary to the technical manuals for John Deere farm equipment, writers literally tear down the machines and then put them back together.

An engineer wrote the following evaluation report of a plant based on his observations during a visit:

The plant's informal labeling program is inadequate.

- Formal label tags are often missing or not used.
- Many pieces of equipment are not labeled.
- Some equipment has incorrect labeling, causing operators to disagree on equipment designations.
- Some labels in the turbine building have missing digits.
- Chain-operated valve labels are inaccessible.

The engineer concluded in the opening sentence that, based on his observations, equipment was either mislabeled or not labeled at all.

Let's look at four types of field research in which you may find yourself involved: interviewing, observing, surveying, and conducting focus groups. We will not discuss researching, developing, and testing, since these aspects are related to your specific field of study, and you will learn the necessary techniques in your specific content courses.

Interviews   Sometimes people working on a project or activity have information that hasn't been reported in a journal or presented in a conference paper, but which is relevant to your document. Sometimes, too, people involved in incidents can provide the descriptions that you need.

Interviewing is a form of oral communication that involves one person who asks a preconceived set of questions, and another person who responds to the questions. Interviews can be formal or informal; they can be conducted in person or over a telephone; and they can be structured or unstructured.

Structured interviews are often used when more than one person is to be interviewed about the same topic. All questions are prepared ahead of time, and they are the same for each person. Often, even the introduction and conclusion of the interview are written out and presented to each interviewee.

Unstructured interviews are more informal. Although you should usually begin with a set of prepared questions, you can ask questions you did not originally consider but you now believe to be relevant, or which follow a line of reasoning introduced by the interviewee.

Phrase your questions so the interviewee provides a full discussion or description, not just a yes-or-no answer. You may want to use such words as "why," "what," "how," "explain," or "describe." Start out with open questions, such as those below. These are usually broad in nature and simply specify the topic:

How do you think the plant's efficiency could be improved?

Tell me how work assignments are made.

Open questions allow the interviewee to determine the amount and kind of information to give, and are not threatening, since there are no right or wrong answers. The problem with open questions is that they can be time-consuming; the interviewee may not give you the information you need; and they are difficult to record.

To make certain you receive the specific information you want, you may need to follow up open questions with closed ones such as the following:

When was the last time you did preventive maintenance on the car?

To whom do you report on the back shift?

While closed questions take less time and are easier to record, interviewees may respond only in monosyllables. They may fail to give you much information, or they may leave out vital data, as in the following:

Q: Did you look at the manual before you did the repair?

A: Yes. (Unspoken, "...but I didn't follow it.")

Use both primary and secondary questions. Primary questions are used to introduce a topic. Secondary questions are used to probe or follow up responses to primary questions which seem incomplete, superficial, vague, irrelevant, or inaccurate.

What do you mean by....

Could you give me a specific example.

There are three phases to an interview: the preparation, the interview, and the follow-up.

Preparation.   Do your homework well in advance. Read whatever you can about the person you are interviewing and the topic you will be discussing.

Plan and write out your questions ahead of time. Notify interviewees of the topic you want to discuss so they can prepare for it. They may want to jot down some notes, or bring some documents or other relevant materials.

Select a neutral place and time to hold the interview. Find a place where you will not be interrupted and a time that is convenient to you and the interviewee. Give yourself enough time. If you're not sure how long the interview may take, err on the side of more time; that keeps you both from feeling rushed.

Ask for permission to record the interview if you plan to use a video or tape recorder. It is illegal to record a telephone conversation without informing the party on the other end Check the audio or video tape recorder if you plan to record the interview. Make certain your batteries are good, that you have enough tape, and that it is working.

The interview.   Try to make the interviewee feel comfortable. Spend the first few minutes simply discussing neutral topics.

About 70% of the interview should be spent in listening. Use good listening skills. Ask one question at a time, and wait for the answer. Interviewees may not respond immediately. Sometimes they need time to consider a response or to recall an answer. While you wait, you may feel awkward and want to break the silence with a comment. Be patient; try counting to thirty before you jump in.

Maintain eye contact with the interviewee, and avoid visual distractions. If you are taking notes, look up periodically. Let the interviewee do the talking; you should stick to the questions. Don't overtalk or interrupt.

Listen to the entire answer; don't jump to conclusions. Be open to new or divergent ideas. Don't allow your biases to color your perceptions. Try to get all of the information by following up on incomplete answers or answers that appear to purposely avoid certain topics. Use such probes as "Anything else?", "Any other examples?", or "Why?"

Evaluate each response for logic, clarity, and accuracy.

Remain in control of the interview. Don't let interviewees get off track. If they do, bring them back tactfully, but firmly. To get back on track, you can refer to your list of questions and comment:

"If we could return to…"

"Could you tell me about…"

"Since we have a lot to cover, let's turn to the next area…"

Try to keep the interviewee's questions to a minimum. If you find you're not getting the answers you need, use closed questions.

Take notes even if you are using a tape recorder. You never know just how clear the tape will be. Don't just note "bad" things. You may want to discuss good aspects in your report. In addition, you may inhibit the interviewee, who can easily see that you write notes only if you are recording problem areas.

Follow-up.   Write down everything you can remember immediately after the interview. In addition, jot down impressions, ideas, and follow-up questions immediately after the interview, so you don't forget the details and what you've learned.

STRATEGY

CHECKLIST

1. Do your homework about the topic and person you are interviewing well in advance.
2. Plan and write out your questions ahead of time.
3. Phrase your questions so the interviewee provides a full discussion or description.
4. Use both primary and secondary questions and probes.
5. Select a neutral place and time.
6. Inform the interviewee of the topic ahead of time.
7. Request permission to record the interview.
8. Check the audio or video recorder works.
9. Make the interviewee feel comfortable.
10. Use good listening skills.
11. Remain in control of the interview.
12. Take notes.
13. Note good as well as bad observations.
14. Jot down everything you can remember following the interview.

Observations   To support an idea, you may need to describe a situation you've observed or a mechanism you've seen.

Observations can be conducted in several ways. You can observe something once or several times. A machine probably performs the same way no matter how many times you watch it, so once should be sufficient. However, people perform differently, depending on their moods, the situation, etc. You may want to observe a project or a person several times. You may decide to conduct a single, lengthy observation, or you may decide several short observations would be better.

To observe effectively, you need to be thorough, analytical, attentive, selective, receptive, and creative.

*   **Thorough**. Give thorough consideration to the important aspects of the activity, person, or object you are observing. Don't be superficial. You not only need to observe what *is* said or done, but also what *is not* said or done.
*   **Analytical**. Look below the surface. If everything isn't going as it

should, you need to determine why it isn't. Your conclusions should accurately reflect what you have seen.

- **Attentive**. Give your undivided attention to whatever you are observing.

- **Selective**. Select areas that are relevant to your purpose and which will provide the information you need. Focus on those aspects.

- **Receptive**. Be open to understanding exactly what is happening. If you perceive from a biased perspective, you'll only see those things that reinforce your opinion.

- **Creative**. Consider alternatives. As you observe, think of better, safer, more efficient or effective ways the activity can be done.

After observing a method of leaf collection for yard waste, two environmental consultants wrote the following description. The consultants' purpose for the observation was to learn how leaves are collected. Notice the details the writers include.

The vacuum truck crew consisted of three workers, two rakers, and a driver. The driver advanced slowly while the two rakers pushed leaves toward the vacuum hose. When the leaf box was full, the vacuum apparatus was detached and the truck driven to the compost site to deposit leaves. Collection resumed when the truck returned and the vacuum was reattached.

By Lori Segall and Ron Smith
Biocycle, 419 State Ave., 2nd floor, Emmaus, PA 18049 (215) 967-4135

The consultants would have made completely different observations and written a different description if their purpose had been to learn how the truck worked. Rather than describing the crew and the comings and goings of the truck, they would have described the leaf box, the way in which the vacuum hose was attached, and the mechanism used to detach and re-attach the hose.

Like interviewing, there are three phases to conducting an observation: preparation, the observation, and the follow-up.

Preparation. Do your homework. Read whatever documents are available about the person, project, or object so you have some basic knowledge of what you are seeing. Note anything about which you specifically want information so you'll be sure to watch for it.

Request permission to make an observation from the supervisor in charge.

The observation. Put everyone at ease so they relax and operate as they normally do. People tend to be anxious when they know someone is watching them, and their anxiety level rises as they perceive the observer taking notes. You can alleviate some of their anxiety by making certain everyone understands why you're there, and by notifying them ahead of time that you will be taking notes to remind you of what you've seen.

Ask everyone to follow their normal patterns and procedures, performing as if you didn't exist. This can be difficult; your presence invariably influences people's performance.

Try to get as close to the activity as possible without interfering. Focus on three aspects:

- The individual or object involved

- The situation or surroundings in which the observation is occurring

- The overall process

To avoid influencing the observation, be careful when you ask questions. Don't imply a response or offer a prompt. The old question, "When did you stop beating your wife?" implies that the respondent has been beating his wife. The question "When will you do the post-maintenance test?" implies that such a test should be performed. Without the question, the test might have been skipped.

Don't ask leading questions that could cause a defensive reaction, such as "Isn't it part of an operator's job to check bearing temperatures?" Obviously, you perceive that it is part of the operator's job and, since the operator didn't do it, the operator is in the wrong. A better way to phrase the question might be "Do you check the bearings?"

Note questions and comments you may have about an activity. When you are not certain you heard or saw something, put a question mark next to the item. When you're on site, keep your questions to a minimum; only ask for information when you will not interfere with the activity.

Take notes. The following details should be included in your notes as references to your observation:

- Date, time, and place of observation

- Persons' names and/or positions involved

- Equipment components, rooms, etc.

Include good as well as problem aspects in your notes.

Follow-up. Write up your observations immediately. Even with notes it is difficult to recollect accurately what you have seen. You should try to get out a rough draft of your report before you go to bed.

STRATEGY CHECKLIST

1. Do your homework about the person, object, or event you are observing.
2. Note specific aspects you want to observe.
3. Request permission to make the observation.
4. Put all those involved at ease.
5. Request those involved to follow normal patterns and procedures.
6. Get as close to the activity as possible.
7. Keep questions to a minimum.

continued

8. Take notes, including date, location, persons and/or equipment involved, questions, and comments.

9. Focus on individual, object, or event; situation or surroundings; overall process.

10. Avoid influencing the observation by implying a response, offering a prompt, or asking leading questions.

11. Write up your observation immediately.

Surveys You may want to conduct a survey to obtain information that will support your ideas and help readers make a decision.

Surveys can be done over the telephone, in person, or by mail. To conduct a survey you need to take the following steps:

- Determine the goals.
- Identify the population to be surveyed.
- Write the questionnaire.
- Conduct the survey.
- Tabulate the data collected from the survey.
- Analyze the data.

A number of software programs are available for facilitating the latter two steps.

Surveys can be either formal and scientific, in which a statistically-derived, random sample of people is surveyed, or they can be informal collections of opinions. Formal, scientific surveys require special methods related to the selection of a sample. We will not get into sampling techniques in this book. As with research, development, and testing, you will learn applicable techniques in your content areas.

We are concerned here with developing a good questionnaire for either a formal or an informal survey. Good questionnaires are brief and attention catching as well as unbiased.

Brief. Questions need to be kept short and simple to prevent the respondent from becoming confused and from responding inappropriately. Only one subject should be dealt with in a question.

Confusing   Would you be willing to purchase a special container for a $20 deposit and a $5 monthly fee from the local trash collector which you could use for dumping your landscape waste?

Brief & simple   Would you be willing to use a special container from the local trash collector for dumping your landscape waste?

                 Would you be willing to pay a $20 deposit and a $5 monthly fee for the container?

Attention-catching. When you ask people to fill out a mail survey questionnaire, you are asking them to take time away from their own work. If there is no incentive for them to do so, they may very well toss your questionnaire in the waste basket.

Not motivating   We would appreciate your taking the time to fill out the following survey...

Motivating   As the warm weather approaches and your grass begins to grow again, you need to think of how you can dispose of your grass clippings in the most efficient and environmentally-safe way.

Unbiased. Your questions need to be worded neutrally so participants respond truthfully to the questions rather than as they think you want them to respond.

Biased   Do you find it bothersome and expensive to place your lawn clippings in a special container furnished by the local trash collector for curb pickup?

Unbiased   Do you prefer to place your lawn clippings in a special container furnished by the local trash collector or in special bags purchased at the local grocery stores?

In addition to the data about the topic being surveyed, you need to collect two types of data. You need to have a name or code to identify individual respondents in case you need to contact them to clarify a response, to provide further information, or to prove that they actually do exist and did respond to your survey. You also need to collect demographics (gender, education, socioeconomic status, etc.) about respondents. Sometimes people's age, education, or current product usage affects their responses to your questions. You need this information to interpret participants' responses, and to compare responses by different types of participants.

Questions assume two forms: closed and open.

Closed questions. Closed questions require respondents to select from a predetermined list of answers.

What do you do with your grass clippings?
  a.   Place them in plastic bags for curbside pickup   ( )
  b.   Leave them on the lawn   ( )
  c.   Gather them and use them as mulch   ( )
  d.   Burn them   ( )
  e.   None of the above   ( )

Closed questions should be used to obtain respondents' demographics. Closed questions are also easier to tabulate, and assure that you will get answers you want.

Open questions. Open questions allow respondents to write in whatever response they wish. For example, the following question is not followed by a list from which to select a response.

What do you do with your grass clippings?

When constructing open-ended questions, consider how much information you want from the respondents. If you are concerned with only two or three issues, limit the discussion to these. If there are many possibilities, ask the respondents to discuss the three major items so that you are not overwhelmed and can arrive at some consensus.

Survey questions need to be constructed carefully so participants' responses are valid. Use familiar, commonplace words whenever possible. In addition, avoid using words with several meanings or with highly-emotional connotations.

Invalid   Are you willing to place your landscape waste in a special container furnished by the local waste management firm?

Valid   Are you willing to place your grass clippings in a special container furnished by the local trash collector?

Avoid leading questions which imply a correct response.

Invalid   Are you willing to help maintain an environmentally-sound community by separating your trash into recyclable and non-recyclable bins?

Valid   Are you willing to separate your trash into recyclable and non-recyclable bins?

In multiple-choice items, develop realistic alternatives and allow for all possible responses. Use specific, quantitative criteria, rather than qualitative criteria, when possible, i.e., use "daily, twice a day, three times a day," rather than "seldom, sometimes, often."

Invalid   Should recyclable items be picked up
a. often   ( )
b. frequently   ( )
c. seldom   ( )

Valid   Should recyclable items be picked up
a. as often as regular garbage pick up   ( )
b. more frequently   ( )
c. less frequently   ( )

Sequence questions in correct psychological order. General questions should be asked before specific questions. Questions concerning respondents' general impressions of their working conditions should precede questions concerning specific aspects of those conditions, such as hours, lighting, etc. Objective questions should precede subjective questions. Questions concerning specific data should precede questions related to opinions of the data.

Study the wording of other surveys in your field and borrow those phrases that have already been tested for validity. Figure 6-4 shows a page from a survey conducted by the U.S. Department of Labor, Occupational Safety and Health Administration (OSHA). The survey includes both open and closed questions.

WORKER ERGONOMICS QUESTIONNAIRE

OSHA would like your help in its attempt to determine the extent of work-induced physical discomfort or pain so that we may begin to plan to reduce the problem. Please fill out the following questionnaire. Use the back of the questionnaire or a blank sheet for extended answers. Information in the box is confidential.

1. Name_____ Date _____
2. Home Address _____
               (Street)

          (City)   (State)   (Zip Code)

3. Telephone Number (Home) _____
4. Sex: Female _____ Male _____
5. Date of Birth   _____ / _____ / _____ 6. Weight _____ 7. Height _____
            (Month)   (Day)   (Year)
8. Job Title_____
9. Department _____ 10. Line _____
11. How long have you held your present job? _____/ _____
                              (Years)   (Months)
12. How long have you worked for this company? _____/ _____
                               (Years)   (Months)
13. Previous job titles at this company? _____
14. Does your job require repetitive arm, hand, shoulder, or back motions?
    Yes _____ No _____
15. Describe your job _____
    _____
    _____
    _____

16. Are you: Right handed? _____
           Left handed?_____
17. Which hand do you use most at work? Right? _____
                          Left? _____
                          Both? _____
18. How many hours do you work per week on the average?
Less than 10 _____ 11-20 _____ 21-30 _____ 31-40 _____ More than 40 _____

Figure 6-4 Questionnaire for OSHA Survey

continued

Figure 6-4 Questionnaire for OSHA Survey *continued*

19. In the last three (3) years, have you had pain or discomfort in your:
    Neck? _____ Shoulder(s)? _____ Arm(s)?_____ Elbow(s)? _____
    Wrist(s)?_____ Hand(s)?_____ Back?_____*
* If you did not check any item in question 19, please go to question 34.
20. For the body part(s) that gave you pain or discomfort, please mark an "X" on the pain degree scale
    below. 1 = no discomfort, 10 = severe pain.

Neck   No discomfort _____ Severe pain
                           0 1   2   3   4   5   6   7   8   9   10

Shoulder(s) No discomfort _____ Severe pain
                           0 1   2   3   4   5   6   7   8   9   10

Arm(s)   No discomfort _____ Severe pain
                           0 1   2   3   4   5   6   7   8   9   10

Elbow(s)   No discomfort _____ Severe pain
                           0 1   2   3   4   5   6   7   8   9   10

Wrist(s)   No discomfort _____ Severe pain
                           0 1   2   3   4   5   6   7   8   9   10

Hand(s)   No discomfort _____ Severe pain
                           0 1   2   3   4   5   6   7   8   9   10

Back   No discomfort _____ Severe pain
                           0 1   2   3   4   5   6   7   8   9   10

STRATEGY CHECKLIST

1. Keep questionnaires brief, attention-getting, unbiased, and in-depth.
2. Use familiar, unequivocal, and neutral language.
3. Use short, simple sentences.
4. Avoid leading questions, questions implying a response.
5. Provide for all possible responses.
6. Use quantitative responses.
7. Sequence questions in psychological order.

Focus Groups A focus group is a semi-structured group interview in which a small group of people are brought together during a two- to three-hour session to respond to a series of open-ended questions.

The object of a focus group is to obtain the opinions of participants on a topic. Focus groups are especially effective for obtaining information concerning "why" people believe as they do. They are often used to develop marketing strategies. You may want to convene a focus group to obtain information on how people feel and what they believe about a specific topic. For example, a pharmaceutical company may want to determine whether consumers prefer pills or a skin patch for asthma or allergy medication. To obtain this information, a marketing researcher for the firm may set up several focus groups, each representing a different population, i.e., doctors, nurses, medical technicians, patients under the age of 16, between 16 and 65, and over the age of 65.

A focus group is comprised of seven to ten individuals who represent a homogeneous population. The group is led by a facilitator who asks the questions and keeps the discussion on the task and the topic.

Conducting a focus group involves:

- Developing a plan.
- Developing the questions.
- Identifying and recruiting the participants.
- Pilot--testing the questions.
- Conducting the session.
- Analyzing the tape recordings or transcripts of the sessions.

This section is concerned with developing the questions which fall into two categories: key questions and probes. A session involves approximately five or six key questions that are designed to stimulate ideas about a subject. Each key question is follwed by a number of "probes," that are developed to "probe" specific aspects of each general question.

Key   • What kinds of experiences have you had taking pills?

Probes   • What difficulties do you have swallowing pills?

           • How do you remember to take them?

           • What problems do you have if you need to take one while dining out or at a business meeting?

The key question above is open-ended, allowing respondents freedom to explore a topic. Probes, such as those above, dig a little deeper. Questions toward the end of a session, allowing the facilitator to close in on a certain aspect, e.g., "How would you feel about using a skin patch instead of taking a pill?"

Because discussions in focus groups often elicit unexpected but important information, questions need to be phrased to encourage discussion. Avoid dichotomous questions that can be answered with a simple "yes" or "no" answer; e.g., "Do you think allergy sufferers will use the product?" Such questions effectively cut off discussion.

Use an indirect approach rather than a direct "why" question to flush out responses. "Why" questions often put respondents on the defensive as they attempt to determine a rational reason for something they may have done on the spur of the moment, or because of someone's influence. Instead, specify the aspects of the "why."

Why   Why do you prefer pills?

Alternative   What three things do you like/dislike about taking pills?

             What features do you like about the pills?

             How do you feel about swallowing pills?

Begin a session by explaining the general purpose, e.g., "We are interested in your opinion on the various types of drugs on the market." Place the discussion in a context, e.g., "As we age, we find ourselves taking drugs in the form of pills more often."

To warm up participants before beginning the questions, you may ask them to think about a situation in which they have been involved with the product; e.g., "Think about a time when you were taking pills and you forgot to take one on time."

Sequence the questions in an order that will appear logical to the participants. Often questions are sequenced from general to specific.

Following are key questions for a focus group conducted by a firm interested in developing a portable 72-inch screen that could be rented along with a video cassette and hooked up to a TV for viewing a program.

- How do you feel about going to the movies?
- Why do you go to a theater instead of watching a movie on a VCR?
- What would make you watch a program on a VCR rather than going to a theater?
- What kinds of video cassettes do you rent?
- How do you feel about the cost of cassette rental?
- If you could rent a six-foot screen along with a cassette, what would you be willing to pay?

STRATEGY CHECKLIST

1. Develop five or six main questions and probes for each question.
2. Use open-ended questions.
3. Sequence questions from general to specific.
4. Avoid using dichotomous questions.
5. Use an indirect approach.

RESEARCHING A DOCUMENT

Let's observe how a writer uses various planning strategies and information sources to gather information for a document:

To be in compliance with a recently-enacted state law to reduce yard waste, the assistant manager of a municipality needs to determine a new plan for the city to dispose of landscape waste, such as grass and leaves. The assistant manager is expected to present his plan to the city council in the form of a written proposal. Using the questioning method to determine the information he needs, he arrives at the following list:

- What is the law—local, state, federal? (research this)
- What are we already doing? (I know this)
- What are other cities doing, esp. similar cities in relation to size, socioeconomic status, etc.? (research)
- What alternatives exist? (research)
  Problems?   Positive aspects?
  Cost?
  Logistics
- Which alternative is best? for us? (I decide)
- What would we have to do to institute the best method? (focus group)

Because the writer has little background in the area on which he is writing, he needs to use a variety of sources to gather the necessary information for his document.

Personal knowledge and experience. The assistant manager studied waste disposal in at least one college course, but never studied yard waste specifically. He has done some recent reading about the problem and is aware that even though leaves and grass are biodegradable when bagged in plastic, they do not deteriorate. His main experience with yard waste is bagging the grass clippings he mows around his own home. He is aware that if clippings can't be bagged, he either has to cut grass more often or modify the mower into a mulcher. Neither pleases him.

Written sources. To become as knowledgeable as possible about the problem, the assistant manager goes to the local university library and runs a search for books, articles, etc., containing information on landscape waste.

Computer search. He begins by searching the library's online local network. The search indicates there are two relevant sources, one on residential recycling and one on permit requirements for setting up a composting operation.

He prints out the bibliographic information, figuring he may need the second one later if he decides to propose a composting operation. The first one appears relevant and he pulls up the summary on it.

It reads as follows:

A planning guide for residential recycling program in Illinois: drop-off, curbside, yard waste/ by Gary Mielke and David Walters. [Springfield, Ill.] : Office of Solid Waste and Renewable Resources, Illinois Office of Energy and Natural Resources, [1988] vii, 36 p; ocm17-601279.

He decides this source appears promising, and when he gets back in his office, writes for it.

Scientific sources. Next he uses WILSEARCH, which includes databases for the *Applied Science and Technology Index*, the *Biological and Agricultural Index*, and the *General Science Index*. The results indicate three articles from the *Applied Science and Technology Index* may be relevant. The articles include references to a Cornell study on yard waste composting; a paper discussing the Kirkland pilot recycling yard waste program; and a municipal yard waste composting handbook for Wisconsin communities. He prints out the list of these sources and later looks them up in the library. Since the library does not have any of them, he orders them through the interlibrary loan department.

Popular sources. Next he searches through *InfoTrac's* Academic Index, which is an index to popular periodicals, including the *New York Times* and some scholarly journals, and discovers two articles in *Consumer Reports* that may be relevant. He prints them for future reference.

Government sources. He next searches through *GPO on a Silver Platter*, the monthly catalog of United States Government Printing Office publications, which contains information about publications of the federal government. However, this time the only source he finds is related to defense scrap yards.

Academic sources. Finally, he searches through *ERIC,* which contains studies, reports, and journal articles of projects at academic institutions. This search elicits eleven sources. Once again he requests abstracts for the eleven. He quickly scans these and decides that four of the sources are relevant to his needs. He adds these to his list to search for in the library. All in all, the searches took no more than an hour's work.

Note-taking. As he reads the various articles, reports, and so on, the assistant manager makes notes of towns and cities already involved in alleviating the problems of yard waste disposal; of people who are either mentioned in articles, or who have authored articles, with whom he wants to speak further; and of issues to cover if he conducts a survey or focus group of citizens in his municipality.

Field sources. The next phase of his investigation involves field research. He observes leaf collection operations at a number of sites, including Bristol, Connecticut, and Lincoln, Nebraska. While at these sites, he interviews several persons associated with the projects, including the truck drivers and crew members. Finally, he conducts a mail survey in his area to determine citizens' reactions to several alternative plans. In the end, the assistant manager determines on a plan involving the use of special biodegradable bags for yard waste. These can be picked up by the local garbage disposal company in a separate collection each week, and disposed of in special composting facilities.

Figure 6-5 depicts the large number of sources, as well as the various types, used by the assistant manager to obtain the information.

Drafting. With this decision made, the assistant manager writes his proposal. Aware that many of the members of the city council know as little as he did at the beginning of this investigation, he includes a background section in which he refers to several of the articles he read that support his decision. In addition, he describes several of the projects he saw, and quotes several of the people whom he interviewed. Finally, he uses the results of the mail survey to indicate that local citizens approve of his decision. He then outlines his proposed method for yard waste collection and disposal.

The following is an excerpt from the introductory section. Notice the writer uses both written and field sources. As you read, consider whether he provides sufficient background information and evidence to support his claim that composting is a good method for disposing of yard waste.

Field sources. Written sources. Quantitative evidence. Authoritative evidence.	You can collect leaves and yard waste in two ways: bulk and bagged. Our field observations indicate that the most cost-effective method is bagging as long as debagging is not required. Total bag collection costs are 45-65 percent lower than total costs for bulk collection, according to studies in two New England communities (E & A Environmental Consultants, 1989; Division Solid Waste Management, 1988)...

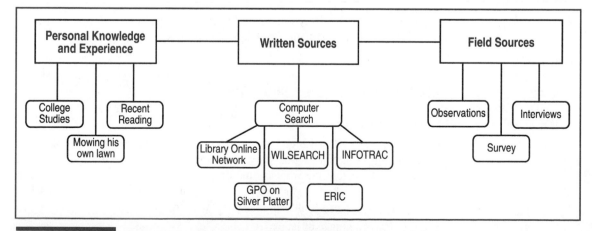

Figure 6-5 Gathering Information for Landscape Waste Disposal Proposal

> The first generation of biodegradable cornstarch bags are currently being field-tested. Preliminary results indicate that residents find the bags perform as well or better than traditional plastic yard waste bags (Hanlan, 1989; Darling, 1989). Laboratory analysis at the University of Missouri indicates that the bags are degrading relatively quickly. In one study, plastic film strength is reduced by an average of 26 percent and as much as 60 percent after 12 to 24 weeks (Iannotti, Tempesta, et al, 1989). In another laboratory analysis a cornstarch pro-oxidant polyethylene film lost 94 percent of its original toughness after six months of soil burial (ADM, 1989).

The writer argues for bag collection, based on the data from his field research. He then uses written sources to support his claims.

CHAPTER SUMMARY

There are several strategies for gathering information. These include listing, clustering, and visualizing. Collaborative debating is a good strategy to use when a group is involved in writing a document.

You get information from three major sources:

1. **Personal sources**—your own knowledge and experience.
2. **Written sources**—books, magazines, journals, conference proceedings, research studies, newsletters, catalogs, and databases.
3. **Field sources**—interviews, observations, surveys, focus groups, and research, development, and testing.

Many of your documents require you to use no more than your own knowledge and experience. Most of the information you obtain from written sources is used as background in documents. Much of the information you need comes from field sources.

PROJECTS

As you work on the following projects, you may want to use the Audience Analysis Chart (page 97), the Decisions Chart (page 101), and/or the Purpose Chart (page 136).

Collaborative Projects: Short Term

1. **Surveying campus trends**

    Your technical writing class has been asked to put together a booklet for incoming freshmen so that they are aware of some of the "in" things on campus. To determine what is "in" on your campus, you need to conduct a survey of the habits of students attending your university. You should work in pairs. Each pair should select one of the items below.

    • Favorite make of jeans
    • Favorite make of sneaker

- Favorite TV program
- Favorite TV personality
- Favorite movie
- Favorite eating place
- Favorite pastime
- Favorite soap opera
- Favorite male movie star
- Favorite female movie star
- Favorite movie
- Favorite sport

Survey 26 students. You and your partner can survey them together or you can each take 13 students. Use the strategies discussed earlier in this chapter to create your questionnaire.

When you have finished the survey, combine your results with your partner's, then tabulate and rank your combined results.

Option A: You and your partner should create a graph to help readers understand the results more easily (See Chapter 12).

Option B: Create a booklet.

2. **Gathering information for a feasibility report: diaper laundry service**

A client is considering opening a diaper laundry service. Because of the problems with plastic waste disposal, the client believes that an increasing number of parents are purchasing cloth diapers. The client wants to know if this belief is true, and if these parents would use a laundry service. Your firm has been asked to conduct the market research, and your supervisor has asked you to be a member of the project team.

a. Work in groups of six. Divide each group into three pairs. One pair should conduct a survey of parents. A second pair should conduct three extensive interviews with parents. A third pair should conduct a focus group for parents.

b. Share your findings with the other team members. Based on your team's findings, your team should determine whether or not to recommend that the client establish a diaper service.

Option A: Your team should write a summary of your team's findings for the supervisor (See Chapter 9).

Option B: Make a presentation to the client of the team's findings. Continue this project in Chapter 7, Organizing Information, or go directly to Chapter 21, Oral Presentations.

3. **Gathering data for an information report: cleaning supplies**

You work for a small company which is very conscious of environmental issues. The contract for the present janitorial service is up for renewal. The owner is interviewing both the present service and others in an effort to find the one which uses the most environmentally-safe supplies. The owner has asked you to work on a committee to gather information on the effect of cleaning supplies on the environment.

a. Work in groups of three. One member of the committee should gather information from magazines and journals, the second member should gather information from catalogs, advertisements, etc., and the third member should gather information from labels on various cleaners.

b. Share your findings with the members of your committee.

Option A: With your committee, create a table illustrating various cleaning supplies and their respective environmental impacts, based on your committee's results (See Chapter 12).

Option B: Collaborate with the members of your committee to write a summary of your committee's findings for the owner (See Chapter 9).

Option C: Collaborate with the members of your committee to make a report for the owner, presenting your committee's findings. Continue this project in Chapter 7, or see Chapter 19.

Collaborative Projects: Long Term

The following projects can be continued through Chapter 13. As you complete each chapter, you will find activities for these projects which are directly related to the chapter you are reading. In this chapter, you'll engage in activities related to gathering information.

- **Communicating about a campus/community problem** continued from Chapter 4, Long Term Exercise 1.

    To obtain information to support your point, your group has decided to conduct a survey among students on the campus to determine their feelings about the problem. You should survey at least 50 students. Use the strategies discussed in this chapter to develop your questionnaire, then conduct the survey. When you have completed your survey, tabulate the results. Fill in step 3, "Gathering and Listing Information," on your worksheet.

    To be continued, Chapter 7, Organizing Information, Collaborative Projects: Long Term.

- **Proposing microcomputer laboratories** continued from Chapter 4, Long Term Exercise 2, Projects 1, 2, or 3.

    You are now ready for step 3 of the worksheet, "Gathering and Listing Information." Using collaborative debating, you and your subcommittee should determine the information you need to include in the proposal. The information should be based on your sense of audience, purpose, and situation. Then you need to indicate whether or not the information is already known, and if so, by which person at the meeting, or if not, how and where the information can be obtained and who will be responsible for gathering it.

Your next task is to gather the information. Once you have gathered it, fill out step 3: "Gathering Information," on your worksheet.

To be continued, Chapter 7, Organizing Information, Collaborative Projects: Long Term.

Individual Projects

1. You believe it is important to recycle plastic waste. You decide to present a proposal at the next meeting of your residence hall, sorority or fraternity, or apartment house to enlist all students in a recycling project. Determine the information you need based on your audience, purpose, and situation and then gather information that you can use to support your oral proposal. Note your information on index cards.

2. You are interested in starting a project to recycle plastic food packaging. Interview someone, either in your school's cafeteria or at a nearby fast food restaurant, to determine whether this could be done within the next semester. Write a memo to inform your instructor of the interviewee's point of view on the topic.

3. Your technical writing class has been asked to write a handbook for incoming freshmen to help them adjust to college life. You are going to write the section on "Meeting People." Based on your sense of your audience, purpose, and situation, determine the information you need to include. Use either listing, clustering, or visualizing.

4. You are concerned with a problem on campus, and you plan to write to the person in charge to request that something be done to alleviate it. If you have a specific problem on your campus, you may want to select it. If you can't think of a problem, you may want to consider one of the following:

    *   Lack of entertainment facilities
    *   Inadequate security
    *   Inadequate course offerings
    *   Missing issues and pages in magazines and journals in the library
    *   Inadequate library hours

    You should determine the information you need to achieve your purpose and then determine how to gather that information. Finally, conduct your own field research to obtain information to support your point of view. Keep notes on your research and your findings in a notebook.

5. Select a topic in your major field on which you would like to obtain more information. Run three different computer searches to determine what sources you might use to obtain this information. Read and summarize on a 3x5 card the information from a source that is either a report or a presentation.

CHAPTER 7

Organizing Information

▶ ORGANIZING FOR READERS' NEEDS
▶ USING PREFORMATTED DOCUMENTS
▶ REFERRING TO DOCUMENT GENRES
▶ SYNTHESIZING

    Planning
    Drafting

▶ ANALYZING
▶ USING COMPUTERS
▶ PROJECTS

    Collaborative Projects: Short Term
    Collaborative Projects: Long Term
    Individual Projects

ONCE YOU'VE GATHERED YOUR INFORMATION, you face the challenge of putting it into an organized, coherent document that will meet your readers' and your own purposes. The more information you have, the more difficult the task may seem.

You may simply start writing. However, if you do, the results could wind up looking like the following paragraph from an evaluation report by a catering operation. As you read, try to determine the specific problems the writer is discussing, and the recommendations the writer makes for solving the problems.

Staff.
Staff knowledge.
Levels of management.
Need for cooperation.
Functions.
Ordering.

Reform needs to begin with the management staff. While staff members are educated and knowledgeable in their respective fields, they are not knowledgeable in our standards of operation. There are three levels of management: the manager of operations, the manager of catering, and several assistant managers. Because of this large number of managers, total cooperation is required. Functions range from catering to small gatherings of 10 persons to luncheons for 1000, requiring a wide variety of supplies which are ordered cumulatively, often resulting in one division running out before another. Each manager should be responsible for ordering supplies for his/her division.

Because the writer has mixed several ideas in the paragraph—need for reform in management, lack of knowledge of company standards, management levels, company functions, and need to decentralize supply orders—you may have some idea of the problems found and the recommendations suggested, but you're probably not sure exactly how the recommendations solve the problems. To avoid this helter-skelter style, the writer needs to organize information so the text is focused and readers can follow the arguments.

The more planning you do, the better chance you have of creating an organized, coherent text your readers can easily follow. This chapter begins by discussing how the organization of information is affected by readers' processes and behaviors, as well as the purpose and situation. The chapter then discusses four strategies for organizing information to meet readers' needs, and to achieve both your readers' and your own purposes. These strategies include using preformatted documents, referring to document genres, synthesizing information, and analyzing information.

ORGANIZING FOR READERS' NEEDS

As you sequence information, you need to keep your readers' reading processes and behaviors in mind. You also want to focus your information on your purpose and situation.

The memo shown in Figure 7-1 is a response to a request for information to determine which brand of computational software should be purchased for a small engineering consulting firm. The writer sequences the information according to his own thinking processes, beginning with a comparison of the major differences between the programs and concluding with his decision.

If you were the reader, would you accept this writer's recommendation?

As you read, you may have had trouble trying to guess which brand the writer would recommend. And, when you finally got to the last paragraph, you may have wondered why the writer arrived at the conclusion he did.

The problem with the memo is that it is writer- rather than reader-based. The writer uses an organizational sequence based on his own thinking process rather than on the reader's reading process. Because of the sequence of information, readers cannot accurately predict what they will read; without knowing the writer's recommendation, they can't understand how the information in the first three paragraphs relates to the writer's decision. In addition, the writer has failed to recognize that the convention for a reader-initiated recommendation report is to discuss the recommendation first and the reasons second, since the reader's primary concern is the recommendation.

Furthermore, because the information focuses on the features of the programs, the text jumps back and forth between the three programs. The reader has difficulty remembering which feature belongs to which program because the writer has failed to recognize that the organizational pattern needs to reflect the purpose—to choose a program rather than features.

Figure 7-1
A Response Memo

MEMO
TO: Richard Mullee
FROM: John Doe
RE: Purchase of an engineering computational software package
DATE: May 24, 19xx

Mathcad for engineering and scientific calculations allows you to do complex equations and formulas as well as exponentials, cubic splines, FFTs, and matrices. It can accommodate Postscript printers and HPGL files. Xmath can also accommodate Postscript and HPGL. In addition it can accommodate EPS. While both Xmath and MATLAB have open architecture, allowing you to include your C or Fortran, there is little reason to expect us to need this advantage. Mathcad has over 120 functions built in and access to additional ones in packs specifically for mechanical and civil engineers. MATLAB has 300 built in functions and Xmath has over 200. While Mathcad, Xmath, and MATLAB all provide 2-D and 3-D graphics, XMath provides WYSIWYG graphics.

Mathcad and MATLAB are available for Macintosh and PC Dos systems. XMath is designed specifically for X-windows workstations. MATLAB can be used with Sun, VAX, and other workstations. In addition, XMath is distributed by Integrated Systems which has a good reputation for its programs. However, MATLAB is developed by The Math Works which is a leader in software for data analysis and mathematics.

All three programs are user friendly. However, because it is compatible with Macintosh PCs, Mathcad would allow our employees to work on problems at home and then transfer their work to their workstations. Xmath cannot be run on a Macintosh.

For these reasons, I recommend Mathcad as the program to buy. If you have any questions, I'll be glad to answer them for you.

If the writer had organized the information according to the reader's needs, the memo would have looked like the one in Figure 7-2.

With the recommendation presented in the first paragraph, the reader understands the details that follow. The entire memo focuses on the recommendation, which is presented in the first sentence of the introduction. By relating all of the following information to the recommendation, the reader can understand the writer's decision. In addition, because information is chunked together according to the program, the reader can easily discern the differences between the three programs.

Consider readers' processes and behaviors, as well as purpose and situation, as you determine the organization of your information so that your message can be understood and accepted.

**Figure 7-2
Revised Response
Memo**

Recommendation.
Mathcad features.

MEMO
TO: Richard Mullee
FROM: John Doe
RE: Engineering computational software purchase
DATE: May 24, 19xx

I would like to recommend that we buy Mathcad rather than XMath or MAT-LAB. Mathcad is compatible with Macintosh PCs. It would therefore allow our engineers to work on a project, using the program, at home and then transfer their work to their workstations with little difficulty. While Mathcad has only 120 built-in functions, the applications packet will allow us to add functions which are specific to our areas of engineering; thereby giving us more than enough. While the other two programs have far more functions built in, many of those would not be applicable to our needs. Thus, Mathcad provides us with a small but applicable set of functions. Mathcad also accommodates Postscript printers and HPGL files which we would need. XMath can also accommodate EPS files but we have no need for this.

Mathcad as well as the other programs provide 2-D and 3-D graphics, and all three have been developed by reputable software firms. All three have also been designed with the user in mind and are relatively user friendly.

Features that are
equal in the other
programs.
Features that are
better in the other
programs.

Xmath and MATLAB both have open architecture, allowing you to use C or Fortran to make changes, but again, I believe the functions which come with the program will be sufficient as they are.

Let's look at several strategies for organizing information.

- Use preformatted documents
- Refer to document genres
- Synthesize information
- Analyze information

All of these strategies are based on recognizing your readers' needs.

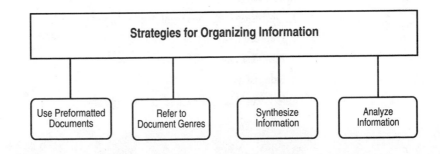

USING PREFORMATTED DOCUMENTS

Certain types of documents are preformatted; the major categories, and often subcategories, into which information should be organized are predetermined, usually by the reader. In documents such as insurance claims, job applications, and car accident forms, the organization for the information is often indicated by a form which the writer simply fills in. For other documents, such as proposals or progress reports, the organization for the information may be outlined in special guidelines issued by the reader.

The following guidelines were issued to writers who were engaged in drafting reports for a state Department of Transportation. Notice that the guidelines dictate much of the content, as well as organization and sequence of information.

I. Report Types
   • Research Reports: An ITS publication documenting the completion of a substantive area of work. A "stand alone" document that defines the research undertaken and presents the results. It can refer to other technical documents for the details of technical data. It should normally have an introduction and conclusion section, in addition to the body of the report...

II. Elements of Reports—Each final report written needs to contain the elements listed below.

A. Cover

B. Technical Report Documentation Page

C. Table of Contents

D. Table of Illustrations and Tables

E. Disclosure Section

F. Disclaimer

G. Foreword, Preface, Acknowledgments

H. Introduction

I. Technical Discussion

J. Conclusions and Recommendations

K. Appendices

    1. Details on Test Equipment, Methods, etc.

    2. Detailed Drawings, Specifications

    3. Raw Test Data

    4. Side Issues

L. References, Literature Cited, or Bibliography...

California Department of Transportation

Even with such detailed guidelines, you might still need to organize information within the various sections. The remainder of this chapter provides strategies for organizing your information within sections of preformatted documents, or in documents for which you do not have guidelines.

REFERRING TO DOCUMENT GENRES

Genres are categories of documents. Technical documents are categorized into the genres of correspondence, instructions, reports, and proposals, with information in each of these presented in a specific organizational pattern. If you are familiar with a particular technical genre, then you know the basic pattern for organizing information in that genre.

A proposal is usually comprised of two major sections. The first concerns background, involving discussion of a problem to be solved or a need to be met. The second revolves around a solution, including the methods or procedures for solving the problem over both the short and long term. A scientific report invariably includes the following four sections: a discussion of the question to be studied; procedures for investigating the question; results of the investigation; and a discussion of the results.

By referring to documents written in the genre in which you plan to write, you can discover the basic organizational pattern for a text.

SYNTHESIZING

The introductory sentence for the revised memo recommending the engineering computational software is the main organizing idea. The text is organized around it; all of the information in the memo is focused on the recommendation.

How does the writer of the software memo develop a main organizing idea, and then chunk related information together so that it all focuses around the main organizing idea? In actuality, he first chunks the information, based on the reader's needs, and then, from those chunks, he develops a main organizing idea.

The writer uses an inductive approach. He synthesizes the pieces of information he has gathered into related chunks, which he then further synthesizes into a single statement, the main organizing idea. In the process, he moves from very specific information to a general statement (the main organizing idea), which encompasses all of the pieces of information. A main organizing idea is constructed from pieces of information just as a house is constructed from bricks and mortar and plumbing and wiring.

Planning

You determine your main organizing idea during the planning process. To develop a main organizing idea you need to engage in five steps. You have already learned three of them:

1. Getting a sense of your audience
2. Determining your readers' and your own purpose and situation
3. Gathering your information

In this chapter you will examine the remaining two:

4. Categorizing your information
5. Developing a main organizing idea

You will also learn two additional steps to take during the planning process that will meet your reader's needs and achieve your own and your reader's purpose. These involve:

6. Determining relevance
7. Sequencing information

Figure 7-3 depicts the planning process as part of a recursive writing process that occurs within the context of audience, purpose, and situation.

**Figure 7-3
The Planning
Process**

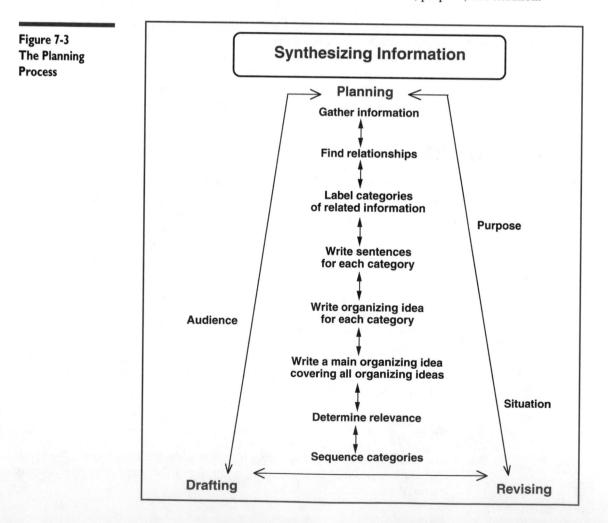

Let's observe a writer using these steps to write a proposal.

An engineer with the Earthquake Engineering Center at the University of California, Berkeley, is responsible for writing the "Statement of Need" section of a proposal to request funds from the National Science Foundation (NSF). The engineer is one of three persons involved with the project. The object of the proposal is to study structural problems found in buildings and bridges after the Loma Prieta earthquake. The study is specifically aimed at examining ways to strengthen structures using supporting columns.

Step 1: Get a sense of your audience.
The writer considers his audience and arrives at the following conclusions.

Audience	Analysis
Primary	Members of the NSF staff who make the final decision on funding.
Intermediary	Reviewers who evaluate the proposal and recommend whether or not to fund it.
Position	Primary readers: Administrators at the NSF.
	Intermediary readers: Professors at universities or research organizations.
Role	Primary readers: Decision makers.
	Intermediary readers: Evaluators.
Field	Primary readers: Scientists (generalists in writer's field at NSF).
	Intermediary readers: Specialists (expert consultants in writer's field).
Readers' knowledge	Primary readers: General knowledge of the topic and of the need for the research.
	Intermediary readers: Expert knowledge of the research and need.
Readers' needs	Both audiences want to know the methods involved in the research, the capability of the investigators, the importance of the results to the scientific community, and the usefulness of the research.

Step 2: Determine your purpose and situation.
After considering his own and his readers' purpose, as well as the situation in which he is writing, the writer makes the following determinations:

Readers' Purpose	To determine whether to fund the project.
Writer's Purpose	To obtain funds for conducting research.
Situation	Studies associated with earthquakes are considered high priority.

Step 3: Gather your information.
Because the three engineers involved in the project are specialists in the field, they already know what information they want to include in the proposal. During a collaborative debating

session between the writer and the other two engineers, a list of information to be included in the "Statement of Need" section is drawn up. As the information is listed, some of the items, such as structural types, examples of structural problems, strengthening efforts, and research needs, fall into clusters. As you read the following list of items, try to find relationships among the data.

- 5 years
- What deformations and loads can columns sustain
- How can we strengthen columns efficiently
- If global stiffening elements are used, how can elements be added to columns to achieve the effect
- Structural frames prior to 1970 are inadequate
  Columns on structures are bad
- Structures
  - Pre-seismic code building with no lateral force resisting system
  - Pre-seismic code building with effective lateral force resisting system
  - Moment frame buildings designed for lateral forces with weak columns
  - Shear wall buildings with inadequate strength
- Loma Prieta earthquake illustrates problem especially in elevated structures
    Examples: Cypress Viaduct, Central Freeway, Southern Freeway
    Hospital in Palo Alto
    Hotel in downtown Santa Cruz
- Column failures due to lateral and vertical load
- Structures typical of this problem prevalent in state
    High occupancy in these structures
- Steel bracing retrofitting systems used before earthquake only partly effective
    Using steel bracing now
- To strengthen existing columns now
    Concrete infills or external buttresses
    Heavy and difficult installation
    Steel bracings
- Need to understand behavior of deficient columns
- Need to understand effectiveness of steel bracing elements for retrofitting existing structures
- Need to understand behavior of retrofit on shaking structures
- Project experimental

- Review literature and experience on similar problems
- Test isolated columns
- Static test of structures
- Dynamic test of structures
- Use researchers from university and professional practice together since research is applied

Step 4: Categorize the information. The writer is now ready to organize the information. Keeping in mind the purposes for the document, he tries to put related information together.

In looking at the list, the writer notices that several items relate to building codes. However, neither his nor his readers' purpose is related to building codes. Then he notices some items related to background. As he thinks about it, he feels the building codes actually provide background information, so he places a *B* next to all background information. Next he notices items related to the problem itself, and places a *P* beside these. Other items seem to relate to the experiment, and he marks these with an *E*. The remaining items all appear to relate to present conditions; he marks these with a *C*.

The writer then rearranges the information so that all related items are chunked together. Next he labels each category of related information. See if these categories are similar to those you considered when you read the list of data:

Problem

P   What deformations and loads can columns sustain

P   How can we strengthen efficiently

P   If global stiffening elements are used, how can elements be added to achieve effect

P   Need to understand behavior of deficient columns

P   Need to understand effectiveness of steel bracing elements for retrofitting existing structures

P   Need to understand behavior of retrofit on shaking structures

P   Column failures due to lateral and vertical load

Background

B   Structural frames prior to 1970 are inadequate

   Columns on structures are bad

B   Four types structures

- Pre-seismic code building with no lateral force resisting system
- Pre-seismic code building with effective lateral force resisting system
- Moment frame buildings designed for lateral forces with weak columns

- Shear wall buildings with inadequate strength

B   Loma Prieta earthquake illustrates problem

Especially in elevated structures

Examples: Cypress Viaduct, Central Freeway, Southern Freeway

Hospital in Palo Alto

Hotel in downtown Santa Cruz

B   Structures prevalent in state

B   High occupancy in structures

B   Steel bracing retrofitting systems used before earthquake only partly effective

Present Conditions

C   Using steel bracing now

C   Strengthen existing columns now

Concrete infills or external buttresses

Heavy and difficult installation

Steel bracing

Experiment

E   Project experimental

E   5 years

E   Review lit and experience

E   Test isolated columns

E   Static tests of structures

E   Dynamic tests of structures

E   Researchers from university and professional practice for applied research

Step 5: Develop the main organizing idea. Up to this point, the relationship between the pieces of information has been somewhat vague. Now the writer needs to become more precise. He looks at the various pieces of information in a category and composes a sentence relating several of them. Developing this sentence forces him to state precisely the relationship between some of his ideas. The writer creates several sentences, relating two or more pieces of information for each category. Consider the pieces of information the writer has combined in each sentence from the category that contained the information.

Problem

- We need to understand how columns behave if we are to determine how to strengthen them.
- We need to study columns under static and dynamic conditions.

- Column deficiencies can lead to lateral as well as vertical load failures.

Background

- The Loma Prieta earthquake illustrates the inadequacy of columns, especially in elevated structures.
- Most of the structural frames built prior to 1970 have high occupancy.
- Four types of structures built prior to 1970 are affected.

Present conditions

- Present methods for strengthening columns are heavy and difficult.
- Two methods are being used now to strengthen existing columns.

Experiment

- The experimental project will take five years.
- The major experiments include (1) testing isolated columns, (2) static testing of structures, (3) dynamic testing of structures.

The writer doesn't write a sentence for every piece of information in a category, just enough to determine if the data really are related. Two to four sentences for each category are usually sufficient.

Once you feel comfortable that all the data in a category is related, you are ready to synthesize the information into a more general statement, called an "organizing idea." An organizing idea is a complete sentence that serves as an umbrella statement for all of the information in a category.

An organizing idea can be derived in three ways:

- Copying an existing sentence
- Revising an existing sentence
- Writing a new organizing idea

The writer of this research proposal is ready to compose an organizing idea for each of the categories in the proposal. After studying the sentences in the "Problem" category, the writer concludes that none of them could serve as an organizing idea, or be revised to serve as one, because none includes all of the information in the category.

- Sentence 1 doesn't encompass the information contained in sentence 2, nor does it include the information about the stiffening elements.
- Sentence 2 is limited to a study of static and dynamic conditions. References to studies related to stiffening elements or retrofits are not included.
- Sentence 3 is limited to deficiencies, and doesn't consider such problems as retrofits or stiffening elements.

Since none of the sentences in the "Problem" category can serve as an organizing idea, the writer develops one:

Problem
We need to understand the factors controlling the behavior of deficient columns.

None of the sentences for the second category can serve as an organizing idea either, so the writer develops another new one:

Background
Of those structures built prior to 1970, those with frames having inadequate columns are the most critical.

In the category for "Present Conditions" the second sentence can serve as an organizing idea:

Present Conditions
Two methods are being used now to strengthen existing columns.

Finally, the writer drafts a new organizing idea for the fourth category:

Solution
The proposed research is of an experimental, but applied, nature.

Once an organizing idea is developed for each category, the writer can further synthesize the information by composing a *main* organizing idea. The main organizing idea serves as an umbrella statement for all of the organizing ideas.

As with an organizing idea, a main organizing idea can be developed in one of three ways:

- Copying an existing organizing idea
- Revising an existing organizing idea
- Writing a new main organizing idea

With some revision, the organizing idea for the "Problem" category can serve as the main organizing idea.

Main Organizing Idea
It is essential to develop an understanding of the factors controlling the behavior of columns with various types of deficiencies so that the efficacy of various retrofitting strategies can be determined.

Step 6: Determine relevance.   Once the writer determines the main organizing idea, he is ready to return to the list of information to determine whether all of the information is relevant to the readers' needs, purpose, and situation. If it isn't, he crosses it out.

The information on the four types of structures does not appear relevant. Readers do not need this information to understand the proposal, and the information is not related to the purpose.

Step 7: Sequence information. Information should be sequenced in logical order, not necessarily the order in which the information happens to be listed. The pattern you use is determined by your readers' needs and reading behaviors as well as by certain genre conventions. Most information fits one of the following organizational patterns: alphabetical, chronological, analytical, sequential, comparison/contrast, problem/solution, cause/effect, and most/least important/effective.

Table 7-1 provides a list of the main organizational patterns used in technical documents. To select an appropriate pattern, consider your purpose and situation. (See Chapter 9 for more information on some of these patterns.)

Pattern	Use	Purpose
Alphabetical	List names, develop glossaries and indices.	Provides readers with easy methods for quickly locating information.
Chronological	Describe events, present a history, narrate a situation.	Provides readers with background to understand events, trends.
Analytical	List the parts of an object, concept, or event.	Provides readers with an understanding of the parts of an object, concept, or event.
Sequential	Instruct or describe something in which one object or activity logically precedes another.	Provides readers with ability to use information, perform a function, or engage in an activity.
Comparison/contrast	Compare two or more objects, concepts, or events. Often appears with one of the others.	Provides readers with alternatives for making decisions.
Problem/solution	Describe a solution to a problem. Often used in proposals, feasibility studies.	Provides readers with information on solving a problem.
Cause/effect	Explain the reason or the results of an occurrence. Often used in research and evaluation reports.	Helps reader understand reason for something.
Most/least important/ effective	Emphasize important information. Often used in correspondence, it is known as the inverted triangle, and is the basic organization of a news story.	Provides readers with important information first.

continued

Table 7-1 Organizational Patterns

Table 7-1 Organizational Patterns *continued*

Pattern	Use	Purpose
Inductive/deductive	Present/delay main organizing idea/purpose of document.	Gives reader information immediately, leads reader to understand background/ causes before presenting concept/request/proposal.

The information in the earthquake research proposal fits a problem/solution pattern as well as a chronological pattern. The purpose of the proposal is to provide a solution to a problem. In addition, since readers are not necessarily familiar with the specific topic, the "Problem" section needs to include information on background and present conditions so readers can understand the proposed solution.

The categories for background and present conditions are included in the "Problem" section. The "Experiment" category, furthering the chronological continuum by discussing future work, comprises the solution section.

The writer sequences the categories as follows:

1. Problem
2. Background
3. Present Conditions
4. Experiment

Once the major categories are sequenced, the writer organizes the information in each category. Because an organizing idea provides an overall statement of the information in a category, an organizing idea serves as the *topic* sentence of a paragraph. The writer begins with the organizing idea, and then determines information within that category that helps readers understand and accept the organizing idea.

The result is a viable outline in which the main organizing idea is Roman numeral I, and the organizing ideas are the capital letters. The Arabic numerals are the sentences, and the lower-case letters are further supporting details.

The sequence of information for the earthquake proposal appears as the following modified outline. As you read, consider how the writer created this outline from the main organizing idea, organizing ideas, and sentences in their respective categories.

Main organizing idea (M.O.I.).

Organizing idea (O.I.).
Sentences.
Supporting details.

I. It is essential to develop an understanding of factors controlling the behavior of columns with various types of deficiencies so that the efficacy of various retrofitting strategies can be determined.
  A. We need to understand the factors controlling the behavior of deficient columns.
    1. We need to study columns under static and dynamic conditions.
      a. We need to determine what deformations and loads columns can sustain.

    b. We need to understand retrofit on shaking structures.
   2. Column deficiencies can lead to lateral as well as vertical load failures.
 B. Of those structures built prior to 1970, those with frames having inadequate columns are the most critical.
 C. Two methods are being used now to strengthen existing columns.
 D. The proposed research is of an experimental but applied nature.

Had the writer tried to outline before synthesizing the information, he may or may not have come up with these categories.

Drafting

Once the main organizing idea is determined, irrelevant information eliminated, and categories sequenced, you are ready to draft.

The Introduction The introduction can be an entire section, or just a paragraph. Because of its importance, the main organizing idea is usually the first or last sentence of the introduction. It serves as a forecast to help readers predict what they will read. Since the introductory paragraph contains the main organizing idea, it is called a forecast. In addition to the main organizing idea, the forecast paragraph contains the purpose of a document if the purpose is not implied in the main organizing idea. If it is appropriate, it also contains a list of the major categories in the sequence in which they will be discussed.

Because a list of categories is not appropriate in the earthquake proposal, it is omitted from the introduction. However, the following introduction to a report on a riverfront plan in a midwest area contains the purpose of the document as well as the sequence of topics to be covered. As you read, consider the sequence of information you expect the writer to follow.

The objective of this project is compare trends over the past twenty years (since the prior Tri-county Riverfront Plan) which reflect both the condition of the river and the way people in the Tri-county area relate to the river. The following four aspects have been selected for discussion in this report: (1) riverfront acquisitions for public use, (2) water quality, (3) sedimentation, and (4) public knowledge and interest in the waterway.

                                     U.S. Army Corps of Engineers

The Body If the seven steps for synthesizing and sequencing your information have been followed, the body of the document should be easy to draft. The first category becomes the first paragraph after the introduction. Because an organizing idea provides an overall statement of the information in a category, the organizing idea for that category serves as the topic sentence of the paragraph. The remainder of the paragraph is comprised of the assertions and supporting detail from that category. Following the sequence established during the planning process, the remaining categories are drafted in a similar fashion.

As a result of this planning process, the information is organized logically, and automatically focuses around the main idea because the main organizing idea has been developed from it. The steps in the process are based on inductive logic, moving from specific pieces of information to general statements.

In drafting, the process is reversed. The writer leads the reader from general statements to the specific details supporting the general assertions. The reader has no difficulty following the information within each category because all of the information is related.

If you begin looking for relationships among your pieces of information while you are gathering information, you may do many of these steps in your head. However, you may want to go through each step in a methodical fashion to make certain you have chunked your information logically. A worksheet, with each step listed, can be found in Appendix B.

The final draft for the "Statement of Need" section of the earthquake proposal appears in Figure 7-4. Notice that organizing idea for each category is placed at the beginning of a paragraph, and that the information in each paragraph is directly related to the organizing idea.

**Figure 7-4
An Organized Section of a Proposal**

Main organizing idea.

Organizing idea.

Organizing idea.

Recent earthquakes have repeatedly demonstrated that column deficiencies can lead to the catastrophic failure of the lateral as well as vertical load carrying systems in frames with inadequate columns. To improve seismic safety, it is essential to develop an understanding of the factors controlling the behavior of columns with various types of deficiencies, the consequences of these deficiencies on the overall integrity of a frame, and the efficacy of various retrofitting strategies.

Of the many types of seismically inadequate existing building structures, the reinforced concrete frame constructed prior to the 1970s stands out for its prevalence, its high occupancy, and its complex retrofitting requirements. Within this broad class of structures, the frame having inadequate columns is the most critical for research at this time. The recent Loma Prieta earthquake, though of modest magnitude, clearly illustrates column deficiencies in several structures. The deficiency was most clear in elevated freeway structures, notably the Cypress Viaduct, the Central Freeway, and the Southern Freeway. Damages were also apparent in several building structures, including a hospital in Palo Alto, a hotel in downtown Santa Cruz, and many buildings south of the Mission in San Francisco. Fortunately, strong ground shaking was localized and of short duration, so that catastrophic collapses of the buildings generally were not observed...

Two retrofitting schemes currently in use for frames with deficient columns include concrete infills or external buttresses and steel bracings. However, concrete is often heavy and presents installation difficulties. Steel bracing retrofitting systems in place before the earthquake were only partly effective and many of those being placed today are likely to perform similarly...

Adapted from a proposal by Jack O. Moehle, Stephen A. Mahin, and A. Astaneh
University of California, Berkeley.

This method can be used to organize information for a variety of documents, ranging from long reports to one-page letters or memos. In long reports, you can determine the major sections into which the report is divided by learning the format for the genre. Then you can synthesize the information in each section, as the writer of the earthquake proposal did for the information in the "Statement of Need" section.

This method is also helpful if you cannot decide on an appropriate format. By synthesizing categories of information into large chunks, you can determine major sections for a document. All of the categories in the earthquake "Statement of Need" section are chunked into a single category related to a need for research.

STRATEGY CHECKLIST

1. Engage in the seven steps for synthesizing information.

2. Place the main organizing idea in the beginning or end of the introduction. The main organizing idea should reflect the purpose. There should be one, *and only one*, main organizing idea.

3. Include a forecast in the introduction. A forecast should include:

   • the main organizing idea

   • the purpose of the document

   • the sequence of categories if appropriate

4. Use the organizing idea of each category as the opening sentence of a paragraph. The organizing idea usually begins a paragraph. There can be one, *and only one*, organizing idea for each category.

5. All related information should be chunked together and should relate to the organizing idea of that paragraph.

6. Sequence categories in a logical order, using one or a combination of the following organizational patterns: alphabetical, chronological, analytical, sequential, comparison/contrast, problem/solution, cause/effect, most/least important/effective.

ANALYZING

If you are aware of the overall format and organizational pattern for the genre to which your document belongs, you may want to use a deductive approach, analyzing your data to determine which items belong in the various categories of a particular organizational pattern. This approach contrasts with the previous inductive approach in which you synthesized your information.

You may use any one of the following analytical methods for organizing your information. However, you still need to follow the first three steps for planning a document: getting a sense of your audience; determining your purpose and situation; and gathering the information.

You also need to engage in the last two steps, determining if any information is irrelevant and sequencing the information to reflect your readers' needs.

The strategies for drafting, after analyzing your information, are the same as those for drafting after synthesizing your information. Begin by drafting an introduction that includes a forecast paragraph. Then draft the body of your document, sequencing your categories in a logical order based on one of the patterns shown in Table 7-1, beginning on page 193.

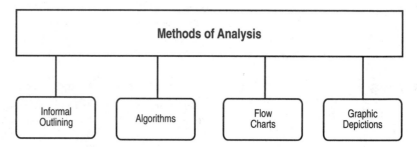

Let's look at several methods of analysis, including informal outlining, algorithms and flow charts, and graphic depictions.

Informal Outlining   Outlining is a form of analysis. It presupposes that you know your categories of information; you need only determine to which of the categories each piece of information belongs. Instead of working *inductively* from specific pieces of information to general statements, you work *deductively* from general categories to pieces of information.

The object of an outline is to indicate relationships between pieces of information by chunking related items together in their respective categories; and to indicate a sequence for the information that the writer can follow to create an organized text.

Let's consider how a member of the Ecological Materials Research Institute might prepare a paper on degradable plastic films to be presented at a symposium on degradable plastics. The writer recognizes he needs to use a problem/solution organizational pattern, since he is discussing a solution to problems in biodegradability. In thinking further about the topic, the writer recognizes that the problem he is discussing is twofold: there is a community problem and a technical problem. He also recognizes that the solution has evolved in two chronological phases: early and recent times. Having determined the major areas of concern, he sorts his information into the appropriate categories. He draws up the following informal outline. Notice how he chunks the information.

I. Degradable Plastic Films
    A. The Problem
        I. Community
            (a) Unmanageable dumping sites
            (b) Industrial refusal to recognize problem

2. Technical
    (a) Resistance to biological influences
    (b) Resistance to chemical influences
B. The Solution
  1. Early
    (a) Addition of cornstarch-background
    (b) Description of process
  2. New
    (a) Thermal degradability
    (b) Expanded uses

Algorithms and Flow Charts Algorithms and flow charts are both forms of analysis and both are comprised of a list of steps that problem solvers must take to achieve their purpose. Algorithms and flow charts usually refer to computer programming, but they can be used in any problem-solving endeavor, including composing a technical document. Both require that information be sequenced in a logical order according to what the reader must know/do in order to understand/do the next concept/step.

In preparing his paper, the member of the Ecological Materials Research Institute recognizes that he must explain the problem before providing the solution if his readers are to accept the solution. Furthermore, he knows he should present the information chronologically, narrating the story to show the evolution of the present solution.

An algorithm for the presentation on degradable plastic films might look like the one below. Consider the similarities in the results between this method and outlining.

ALGORITHM

1. List community and technical problems.

2. Explain early solution in relation to problems—which problems were solved, which were not.

3. Describe new solution: which problems were solved that weren't solved with early solution.

A flow chart might look like the one shown in Figure 7-5. Again, notice the similarities between its categories and those in the algorithm and the outline.

Graphic Depictions Some people prefer to picture their information. Figure 7-6 depicts one way to do this. Try to determine the major categories as you look at it.

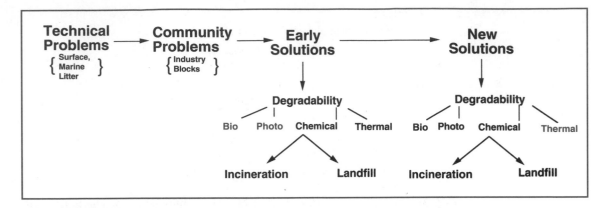

Figure 7-5 Flow Chart Organizing Information

Figure 7-6 Organizing Information Graphically

S T R A T E G Y
C H E C K L I S T

1. Organize your information analytically only if you already know the categories in which to chunk your information.

2. Use informal outlining, algorithms, or graphic depictions to organize information.

3. Determine if your information is relevant to your readers' needs.

4. Draft your information by following your organizational plan.

USING COMPUTERS

A number of software programs have been developed to help you organize your writing. Many of the programs include segments on audience analysis, outlining, and clustering for planning and drafting a document.

While these programs provide strategies for organizing, they do not evaluate your responses. If you respond inappropriately, if you fail to relate your organizational pattern to your readers and purpose, if you do not categorize your information into chunks that are logically related to your purpose, the computer will not warn you. You need to be careful that you don't fall into the trap of believing that the computer will take care of the planning for you. It won't.

CHAPTER SUMMARY

You need to consider your readers' needs as you organize your information. Your information should be focused around your purpose and situation, and it should be sequenced according to your audience's reading processes and behaviors.

The following strategies can help you organize your information:

- use preformatted documents
- refer to document genres
- synthesize information
- analyze information

Synthesizing information, which is based on inductive logic, involves chunking related pieces of information into categories, determining an organizing idea for each category, and a developing a main organizing idea for the entire text.

Analyzing information, which is based on deductive logic, includes outlining, creating algorithms and flow charts, and clustering.

Before drafting, you need to evaluate your information for relevance and sequence your categories.

When you draft, follow your organizational plan. Begin with an introduction which includes a forecast.

PROJECTS

As you work on the following projects, you may want to use the Audience Analysis Chart (page 97), the Decisions Chart (page 101), and/or the Purpose Chart (page 136).

Collaborative Projects: Short Term

1. Determining an organizational pattern

You work for the government of a state that is mainly farmland. Your supervisor has assigned you to a task force to study ways to use farm products for energy conservation. Your findings will be presented to the legislature in a proposal to encourage farmers and others in your state to produce these fuels.

a. Work in groups of five.

b. Study the information below.[1] Using collaborative debating, determine the organizational pattern in which the information should be presented in the proposal.

c. As a group, create an outline for the proposal.

- Twenty million acres of cropland could produce 100 million tons per year of biomass.
- Biomass is plant matter.
- The problem with growing plants for fuel is the large amount of land necessary.
- Biomass can be converted to liquid and gaseous fuels, such as methane and ethanol.
- Biofuels could potentially replace 30-90 percent of U.S. diesel fuel needs.
- Between twenty-five million and seventy-five million acres of farmland are idle.
- Three methods can be used to break plant matter down to gasses.
- Gasification is the most widely commercialized method.
- Gasification combines heat, steam, and air.
- Gasification used to power street lamps in earlier times.
- The potential market is large.
- The price of oil would have to rise to $40 per barrel for gasification to become profitable.
- Ethanol fuel was popular for a while in the 1980s when gas prices were high.
- Brazil has a much larger production of ethanol than the United States.
- Ethanol can be produced from corn and grain surpluses.
- If demand for ethanol grows, present surpluses will not meet the demand.
- Oils extracted from plants can be used to replace diesel fuel.
- Rapeseed, sunflower seed, and soybean oil can be used.

2. **Organizing data for a proposal: recycled paper.**

You are employed in a small firm. A group of employees, including you, would like the firm to use recycled paper. However, you are not certain that the owner of the company, who prefers to use traditional supplies and is always looking for a bargain, will approve your decision. The members of your group have gathered the following facts about recycled paper.[2]

[1]Data based on information in *Cool Energy*, published by the Union of Concerned Scientists, 1990.
[2]Data based on information from *Seventh Generation Environment Product Guide*, 1990.

Member 1

- The recycling loop consists of recycling and using recycled paper.
- Fewer mills make recycled paper than virgin paper.
- It's cheaper for some mills to make regular paper since they buy the lumber from the U.S. Forest Service, which sells its timber below market value.
- Recycled paper costs more than regular paper.
- In 1989 U.S. consumption of paper and paper products totaled eighty-five million tons.
- Consumption of paper products is expected to rise 50 to 100 percent over the next twenty years.

Member 2

- Paper comprises 40 percent of our landfills.
- The U.S. recycles 25 percent of its paper and paperboard (over eighty million tons).
- Recycling cuts down on trash dumped in landfills.
- Unbleached paper causes less pollution.
- Recycling rates in Denmark and Japan are 40 percent higher than in the United States.

Member 3

- Recycled paper that does not use bleach may be slightly gray, because bleaches which cause pollution are not used in the de-inking process.
- There are two types of recycled paper:

  a. pre-consumer—industrial scrap, left over from paper mills, that's never been used

  b. post-consumer—used by consumers and then recycled
- A ton of recycled paper saves 3.3 cubic yards of landfill space.
- Recycled paper differs from regular paper as follows:

  a. the fibers get shorter with each recycling

  b. papers that require strength can use both recycled and virgin fibers
- Bleached paper contains chlorine, dioxins, and toxins, such as furans and chloroforms, which cause pollution.
- Inks and dyes are removed from paper for recycling by two processes:

  a. bleaching the paper

  b. non-bleaching method

a. Work in trios. Assign each person in your trio a member number. Trio members are responsible for the information under their number.

b. You and the other members of the trio will collaborate to organize the information and prepare for drafting a memo to your employer proposing that the company purchase recycled paper. Use the worksheet in Appendix B to prepare for drafting the memo.

Option: Write the memo. Continue this project in Chapter 8, Persuading; Collaborative Projects: Long Term or see Chapter 14, Correspondence.

3. **Organizing data from a food and music survey.**

    Work in trios. You are a member of the entertainment committee for the Christmas party. In trying to determine the type of music and food to bring in, your subcommittee decided to conduct a survey. Each member surveyed twenty-five people. The results appear in Figure 7-7.

    In addition to the survey, you are aware of the following information:

    - a pot luck meal is the least expensive
    - deli is the most expensive
    - two large pizzas can feed eight people
    - the cost of two large pizzas is $9.00
    - a three-foot sub can feed 10 people
    - subs cost $9.00 per foot
    - the disc jockey on the pop radio station is a friend of one of the committee members and thinks she can get a discount on a pop band
    - member 1 hates heavy metal

    Assign each person in your trio a number. Trio members are responsible for the information under their number. Your subcommittee needs to ana-

| | Number of People | | | | | | | | |
| | Member 1 | | | 2 | | | 3 | | |
	Total	Male	Female	Total	Male	Female	Total	Male	Female
Music									
Jazz	3	2	1	2	1	1	0	0	0
Heavy Metal	4	2	2	6	2	4	2	1	1
Pop	14	2	12	11	2	9	17	9	8
Rhythm & Blues	2	1	1	3	3	0	4	2	2
Country Western	2	0	2	3	2	1	2	1	1
Food									
Pizza	16	8	8	11	6	5	10	5	5
Subs	5	3	2	11	5	6	10	5	5
Deli	2	2	0	2	1	1	5	2	3
Pot luck	2	2	0	1	1	0	0	0	0

Figure 7-7 Survey Responses

lyze the data so that it can decide which type of music and food to bring in. Then your subcommittee needs to make a report to the entire committee, which has to approve the decision. Use the worksheet in Appendix B to prepare to draft the subcommittee's report. Write the first paragraph of the report.

Option A: With your subcommittee, create two tables, one for music and one for food, illustrating the results of the survey. Create graphs to depict (a) the results of the music survey, (b) the results of the food survey. Create graphs to show the breakdown in gender between males and females in (a) the music survey, and (b) the food survey. (See Chapter 12.)

Option B: Make the presentation to your class. Continue this project in Chapter 8, Persuading; Collaborative Projects: Long Term, or see Chapter 21, Making Oral Presentations.

Collaborative Projects: Long Term

The following projects can be continued through Chapter 13. As you complete each chapter, you will find activities for these projects that are directly related to the chapter you are reading. In this chapter, you'll engage in activities related to organizing your information.

- **Communicating about a campus/community problem** continued from Chapter 4, Collaborative Projects: Long Term Exercise 1.
- **Proposing microcomputer laboratories** continued from Chapter 4, Collaborative Projects: Long Term Exercise 2, Projects 1, 2, or 3.
- **Gathering information for a feasibility report: diaper laundry service** continued from Chapter 6, Collaborative Projects: Short Term Exercise 2.
- **Gathering data for an information report: cleaning supplies** continued from Chapter 6, Collaborative Projects: Short Term Exercise 3.

Fill out steps 4 -7 of the worksheet in Appendix B.

- 4 Categorize your information
- 5 Relate ideas in sentences
- 6 Write organizing ideas
- 7 Write a main organizing idea

To be continued, Chapter 8, Persuading, Collaborative Projects: Long Term.

Individual Projects

1. The description in Figure 7-8 is written to provide consumers with information for selecting a low-priced CD player. Read the description. Determine whether the information is organized so that you can read fluently. Is there a

main organizing idea for the entire text, and an organizing idea for each paragraph? Place the main organizing idea in brackets. Underline the organizing ideas.

2. The description in Figure 7-8 is written to present consumers with information on four CD players. They can then select the one which fits their needs best. Rewrite it, using the worksheet in Appendix B, to recommend the Sony CD player above the others.

**Figure 7-8
Descriptions of
Four Brands of CD
Players**

Since Sony introduced the high-quality sound of compact disc players in 1983, dozens of audio companies have jumped on the bandwagon. More than 250 different models are on the market with prices ranging from $200 to about $8,000. Let's take a close look at four entry level models—Denon DCD-610, Kenwood DP-860, Sony CDP-550, and Yamaha CDX-510U—in the price range of $280-$330. You need to examine programming options, remote control capability, technical features, and pricing to determine the one best suited for your needs.

One of the most important features of a compact disc player is programming. All of the CD players can be programmed to play the tracks on a disc in any order, and to repeat a single song or a whole disc indefinitely. In addition, all of the players have shuffle play, which allows a player to randomly select the songs on a CD, a nice feature for providing variety to old CDs. However, unlike the others, Sony has a unique feature which allows the player to repeat segments of songs. The feature can come in handy for acquiring lyrics. This segment repeat option outweighs the fact that Yamaha's player has the capability to play up to 24 tracks while Sony's can only play 20.

The remote control capability is another important feature of a CD player. All four players have comparable features, including programming, direct track selection, fast forwarding, and eject. None, however, have volume control.

Although sound quality differs very little between models, there are technological differences that add to the reliability of a player. While all of the players use advanced three-beam lasers that are resistant to skipping, only the Sony uses a dual digital/analog converter system which provides a separate converter for each of its two channels. This dual system gives Sony the capability of deciphering the digital signals better than the single system used by the other models and can prolong the life of the player. While Sony, along with Kenwood, only has double oversampling (a system for filtering out distortion) and the other models have quadruple oversampling, there is no significant difference in sound quality. While each of the four models has a variety of other features, Sony is the only one with all of the following: a headphone jack with a volume control; a display indicating time elapsed, the track playing, and a calendar; and the capability to play 3 inch CD singles.

The four brands are priced within $50 of each other. Kenwood and Sony sell for $280, Denon $300, and Yamaha $329.

James McGrary

3. A member of the Quality Assurance (QA) division of a manufacturer evaluated an area in a plant and found problems relating to a regulation that requires operational personnel be aware of the plant's status at all times. In contrast, the evaluator noticed that information was not always transmitted from employees coming off a shift to those coming on. The QA evaluator needs to write up a report of her findings.

    The evaluator's supervisor is her internal audience. Her external audience includes the employees of the division she evaluated, their supervisor, and the director overseeing that division. Her primary audience is the vice president who receives all QA reports.

    The purpose of the document is to get problems corrected. The political situation is sensitive. The evaluator doesn't want to come down too hard on the department where the infraction has occurred; after all, she works for the same company.

    The evaluator made the following notes while observing the department:

Notes

Technical Specifications a.b.y. require that operational personnel should be cognizant of the status of plant systems and equipment under their control and should ensure that systems and equipment are controlled in a manner that supports safe and reliable operation.

Ten auxiliary operator turnovers were observed. The following deficiencies were noted:

- No one used a checklist or other formal means to guide the turnover process. A turnover checklist does exist, but was not used as a turnover guide. A more detailed checklist could provide a better guide for these turnovers.

- All exchanges of information at the turnovers were from operators' memory.

- There was no review of the round sheets at its turnovers.

- The turnovers were often interrupted by people in the area who were not involved in the turnovers.

- After one auxiliary operator had completed a shift turnover on the reactor building and rad waste, a second auxiliary operator on the same shift came in and said he would take the reactor building rounds. The second operator then left the auxiliary operator turnover room without receiving any turnover.

- Turnovers were often interrupted to discuss non-job-related topics and to review non-job-related reading material.

- As many as five auxiliary operators turned over simultaneously in the auxiliary operators' turnover room.

Using the worksheet in Appendix B, write the evaluator's finding.

4. Look at the software memo on page 182. Develop an outline based on the information and reflecting the reader's needs. See if you can also create a flow chart for the information.

5. The Director of Manufacturing of a 70-person manufacturing operation has received a complaint from a customer indicating that a plastic housing for a sealed beam light is not meeting quality specifications. The Production Engineering Supervisor began implementing statistical quality control (SQC) techniques about a year ago. The director has requested the supervisor respond to the customer, highlighting the overall performance of the process to produce these housings, and assuring the customer that, based on SQC data, the parts are not inferior and meet specifications. Assume the role of the supervisor and write the response. You need to submit it to the director for approval before sending it to the customer. The director wants the letter within twenty-four hours.

Exercise adapted from *Industrial Technology 313, Statistical Quality Control*
Hank Campbell, Industrial Technology, Illinois State University, Normal IL

Persuading

MUCH OF THE TECHNICAL WRITING you do is persuasive. You are often trying to influence your readers to do something or to think in a certain way.

You may want readers to follow certain procedures, to engage in certain activities, or to support a program. You may want readers to believe that one process or brand is preferable to another, that you have done a good job, or that your research results are successful.

Some of your documents will be overtly persuasive, such as proposals in which you attempt to influence readers to support a project or idea. In others you may be more covert, implying your needs. In a project report, you may attempt to persuade your supervisor to give you a raise by indicating you are doing a good job; in an annual report to a funding agency, you may attempt to influence the agency to continue supporting a project by indicating your successes in that project. Regardless of whether you write an overtly or implicitly persuasive document, your audience will probably be skeptical.

To persuade your audience to accept your message, you need to present an effective argument. An argument consists of a series of organizing ideas that help you achieve your purpose by presenting your point of view. These organizing ideas are expressed as claims you make about your topic.[1] In an effective argument, each claim is supported by evidence that is related to your point of view. In addition, the evidence relates to your readers' needs and values and to their perceptions of a logical explanation.

In this chapter you'll learn how to determine claims that your readers may accept and how to support such claims with sufficient and relevant evidence. You'll also learn how to organize your claims so that readers can follow your line of reasoning and arrive at your conclusions. Finally, you'll learn some strategies for countering your readers' claims.

RELATING CLAIMS TO READERS AND THEIR COMMUNITIES

Whether or not your argument is effective depends on the reader and the reader's community. Readers often determine the validity of a claim based on their prior knowledge and experience. During the Three Mile Island nuclear accident, local residents would not accept officials' claims that there was no risk to their health or safety. Their prior knowledge, which had been reinforced by the recent release of the Jack Lemmon-Jane Fonda movie, *The China Syndrome*, led them to expect that the release of steam from a nuclear utility would harm them.

Readers often gauge the validity of a claim by the values of their communities. If you want to persuade your instructor to raise your grade, you may claim that your answer is correct. Because teachers consider a correct answer a valid claim for raising a grade, you may achieve your purpose if you have evidence that you have answered correctly. However, if you claim that your grade should be raised because you spent a lot of time studying, you may not achieve your purpose, even if you have evidence to prove you studied all night. While teachers in an elementary and secondary school community may accept the amount of studying as a valid claim for raising a grade, most teachers in a college community do not. Of the two claims, the one that is valid—the correct answer—is the one based on the values that are shared by both the writer and the reader.

Arriving at a Valid Claim

Sometimes you know which claims readers will accept. Other times there may be a number of claims that you could make about your topic, but you're not sure whether or not your readers share your values or your interpretation of the facts

---

[1] The term "claims," and much of this discussion, has been adapted from Stephen Toulmin's concepts in *The Uses of Argument*. (Cambridge: University Press, 1958).

or policies. At other times, it's difficult to think of any claims that you may share with your readers. Two methods, brainstorming and collaborative debating, can help you arrive at a valid claim.

- Brainstorming

    In brainstorming, write down all of the claims you can make about your topic without judging any of them. Then, after you've gotten them all down, review them, and consider how readers will respond to them, based on readers' knowledge, biases, and community.

    Next, try to imagine what readers' claims about the topic are, and write them down. Do any of your claims correspond to the readers'? If so, then these claims are valid; they relate directly to readers' concerns. If not, see if you can think of a claim that you share with your readers.

- Collaborative Debating

    In collaborative debating, a small group of people gets together to discuss the possible claims that can be made to support a topic. They also consider rebuttals readers may make to these claims, and they discuss claims readers themselves may make. Through a process of elimination and negotiation, the group arrives at a consensus on which claims readers are most likely to accept.

■ Let's look at how a writer might select the claims for persuading an audience to accept a message. Dr. Albert Manville's purpose in writing a paper for the Alaska Wolf Management Team is to persuade his readers to stop using aerial hunting to kill wolves. If he were to brainstorm possible claims, he might come up with the following:

- Aerial hunting is unethical.
- Aerial hunting is unsportsmanlike.
- Aerial hunting kills too many wolves.
- Aerial hunting could result in placing wolves on the endangered species list.
- Aerial hunting could result in an imbalance in the region's ecology.
- Aerial hunting violates the federal Airborne Hunting Act.

He also considers his audience's claim:

- Aerial hunting rids the region of wolves who are depleting the moose and caribou populations.

In studying the two sets of claims, he realizes that none of his claims correspond to that of his audience. If his audience is going to accept his message, he needs to recognize their claim. He returns to brainstorming, but this time, he considers claims that address his audience's concerns. The following list is the result:

- Reduction of moose and caribou populations may be related to factors other than wolves.

- Studies indicating wolves deplete moose and caribou populations are flawed.

- Moose and caribou populations may be increased by strategies other than predator [wolf] control.

- Elimination of wolves may not increase moose and caribou populations.

- If wolves wind up on the endangered species list because of poor wolf management control, hunters will not be able to kill wolves at all in the future. ■

Types of Claims

Claims are based on three factors: facts, values, and policies.

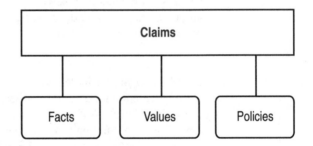

■ The first two claims Manville makes in his original brainstorming list—aerial hunting is unethical and unsportsmanlike—are based on values many people hold toward hunting. However, his audience does not share these values, and therefore will not accept the claim as valid. The next three claims—aerial hunting kills too many wolves, wolves may wind up on the endangered species list, and aerial hunting could result in an imbalance in the region's ecology—relate to the audience's attitudes. If the audience doesn't care how many wolves are killed, that wolves may be endangered, or that the region's ecology is not in balance, they will not be swayed by his claims. The final claim, that aerial hunting violates a federal act, is based on policy. However, because the audience does not interpret the policy in the same way as Manville, it does not consider the claim valid.

In Manville's revised list of claims, the first claim is still based on attitudes, but this time, the claim is related to something the audience cares about—hunting wolves. While the audience may not care about the ecological ramifications of wolves winding up on the endangered species list, they do care about the hunting ramifications.

The remaining claims on Manville's revised list are based on fact. If he hopes to persuade his audience to accept his message, he needs evidence to support these claims. ■

Defining a Claim

You can use definitions to support a claim, especially if readers are unfamiliar with concepts included in the claim. The way in which you define a term can help you persuade readers to accept your message. Definitions specify and clarify the aspects of a topic under discussion so that readers view a topic in the same way you do.

■ In addressing the Alaska Wolf Management Planning Team, Manville uses the following definition of his organization, "Defenders of Wildlife," to support his claim that he represents a group that is "neither a pro- nor an anti-hunting organization." He uses the definition to defuse some of his readers' hostility toward him by implying that while he may not be on their side, he is not against it.

[Defenders of Wildlife is] a group of old-fashioned Rocky Mountain types who like animals with teeth, not just the ones with cute cuddly faces. Of all the wildlife groups, Defenders of Wildlife has stayed closest to its original purpose: to protect species and habitat. ■

From "Inside the Environmental Groups," *Outside*, 1990.

(See Chapter 9 for more detailed information on writing definitions.)

Qualifying a Claim

A claim is a generalization about a particular situation, object, or process. However there are exceptions to most generalizations. When you make a claim, you don't want readers to argue that because of those exceptions, your claim isn't valid. Further, claims are often made about situations or conditions that haven't been proven to be true, but are *believed* to be true. To avoid implying that your generalization is true in all conditions or aspects, or that the situation or condition has been proven to be true, qualifiers are used to "hedge" the points in a claim. The news media uses the term "alleged," e.g., "the alleged murderer," as a hedge word to protect them from libel suits by a person who has not yet been proved guilty. The terms "appear," "possible," and "probable" are often used as hedge words.

■ As you read the following statement by Manville, notice how he uses qualifiers to modify two aspects of his claim.

The wolf, unfortunately, has been used in the past as the scapegoat for what may actually be a host of problems resulting in the apparent suppressed conditions of moose populations in Alaska.

Albert Manville, II, Ph.D. Defenders of Wildlife Presentation to Alaska
Wolf Management Planning Team, 1991.

Manville uses the auxiliary verb "may" to hedge on his claim that a host of problems are depleting the moose population, since research has not definitely determined the cause. Furthermore, he uses the term "apparent" to modify his audience's claim that the moose population is suppressed. The word "suppress" implies that something is holding it down. However, if Manville's theories are correct, the small population is natural. ■

STRATEGY CHECKLIST

1. Base your claims on facts, policies, and/or values which your readers and their communities share with you.
2. Recognize your readers' prior knowledge and experience in determining a claim.
3. Use brainstorming or collaborative debating to determine valid claims.
4. Define terminology to specify and clarify the aspects of a claim.
5. Use qualifiers to indicate that your claim is not absolute.

RELATING EVIDENCE TO READERS AND THEIR COMMUNITIES

To write an effective, persuasive document, each claim must be supported by evidence. And, like the claim, the evidence needs to relate to readers and their communities.

If readers are to accept a claim, the evidence must be related to their prior knowledge and experience as well as to their needs and values. The evidence must also be sufficient for them to generalize about the situation, object, or process, and they must be able to perceive the relationship between the evidence and the claim. In addition, the discussion must follow a logical line of reasoning.

■ If Manville cited a single study conducted during the eighteenth century in Illinois, which found that wolves were not the major cause of depleted moose and caribou populations, readers probably wouldn't accept his claim because they wouldn't have perceived the geographic location and time period as parallel to the situation in Alaska today. However, if Manville added more examples of studies, readers might begin to consider the evidence, because the number of studies cited would make it possible for them to make generalizations about conditions. If he then included examples of studies conducted in Alaska today, readers might accept the claim, because they could easily see that these studies paralleled the present situation. ■

As with claims, communities often determine what evidence is acceptable and what is not. Communities establish the guidelines for determining acceptable data, sources for data, procedures for conducting studies, and the statistical methodology for measuring results of studies. Expert readers are more willing to accept facts from journals of professional organizations than data from brochures distributed by profit-making businesses. In addition, each community recognizes certain people or organizations as authorities. Opinions expressed by these people or organizations are considered more seriously than are opinions expressed by others.

Furthermore, communities require that research be conducted under certain stringent conditions, and that certain statistical calculations be performed if results are to be considered valid. Most expert readers will not accept the results of a study in which the conditions specified by its community are not met.

■ Manville supports his claim that wolves may not be the cause of the depleted moose and caribou populations by citing a study conducted in Alaska by an expert in the field. Had he cited a study by someone unknown by the community, readers probably would not accept his claim. ■

Determining Effective Evidence

There are two ways to support your claims: you can use factual evidence, and/or emotional appeals.

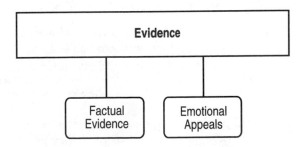

Factual Evidence   Evidence to support a claim can assume the form of data, opinions, or examples.

Data. Data can be historical, scientific, or statistical. You want to select those facts which support your point of view.

You can usually gather a great deal of data about a subject. Much of that data may not be relevant to your audience or your purpose. Select the data that supports your claim and that readers will accept.

■ To support his claim that the loss of the moose and caribou population is due to factors other than wolves, Manville cites such scientific factors as poor habitat conditions and depletion of soil nutrients. He also relates the loss to historical trends in wildlife population fluctuations that are unrelated to wolf population. In addition, he presents statistical evidence indicating that, while in Tanana Flats the moose population increased after 61 percent of the wolves were eliminated, the moose population in the Nelchina Basin failed to increase although all (100 percent) of the wolves were eliminated.

Manville's readers are familiar with the topography of the region, as well as with the incidents in Tanana Flats and the Nelchina Basin, and should accept the relationship between his claim and the data he uses to support it. By including statistics supporting the readers' view, and then countering with statistics that support his view, he preempts readers' claims related to these facts. Thus, by presenting data related to readers' needs, and with which they are familiar, Manville develops an argument that readers may accept. ■

Quantitative data should be directly related to the claim. Numerical data that support a claim can range from survey results to results of a research study. They can also include basic factual information, ranging from the cost of an object to the size of a parcel of land to the number of animals inhabiting an area.

■ Manville cites the percentage of wolves killed each year in Minnesota as a result of poaching to support his claim that the number of wolves killed in Alaska as a result of poaching is higher than that estimated.

Consistent with estimates of wolf poaching elsewhere (e.g., up to 25% of the wolves in Minnesota), poaching in Alaska could be far more extensive than the figure of approximately 10% cited in 1989. ■

Albert M. Manville, II, Ph. D., Defenders of Wildlife.
Presentation to Alaska Wolf Management Planning Team, 1991.

Opinions. Opinions are presented in the form of interpretations. Opinions may be quoted from experts or they may belong to the writer. Readers are more apt to accept opinions from experts, especially those in their own communities, and especially those whose point of view corresponds with their own. Readers need to be familiar with the person cited and consider that person a reliable source.

■ Manville cites a study by one expert, Ballard, to counter the claim of another expert, Gassaway.

> While the elimination of 61% of the wolves in the Tanana Flats in 1976 resulted in an increase in the moose population several years thereafter (Gassaway et al., 1983), elimination of all wolves in portions of the Nelchina Basin in the late 1970s resulted in no net increase in moose there since wolves turned out not to be the major predators of moose (Ballard et al., 1981).

These experts are respected by both Manville and his readers.

Manville also supports his claim that wolves may wind up on the endangered species list by citing the opinion of his organization that more wolves are dying than is being reported. However, his audience may or may not accept this claim, since his organization's viewpoints do not usually correlate with the viewpoint of his audience. ■

Examples. Examples that provide concrete evidence to support a claim are often used as supporting evidence in persuading both generalists and novices to accept a claim. For readers to accept an example as valid, they must be able to perceive the relationship between the example and the claim.

■ Manville supports his claim that reduction of moose and caribou populations is related to factors other than wolves by referring to examples of areas in which moose and caribou populations have rebounded despite the existence of healthy wolf populations.

As with data and opinions, you should select examples with which your audience is familiar. Manville uses the incident at Nelchina Basin, with which his readers are familiar, to support his claim that the elimination of wolves does not necessarily increase the population of moose and caribou. Figure 8-1 provides a diagram of Manville's argument. ■

Emotional Appeals Emotional appeals are made to readers' values and needs.

■ Manville's claims that aerial hunting is unethical and unsportsmanlike may be accepted by members of the National Humane Society or animal rights groups whose values correspond with his. However, the claims will not be accepted by those in his audience who perceive aerial hunting as a way of maintaining the wildlife populations of moose and caribou. ■

Readers of technical documents have a triple persona, one associated with their personal and social life, one with their corporate life, and one with their professional life. In attempting to persuade readers to act or think as you wish, you need to appeal to one or several of these aspects.

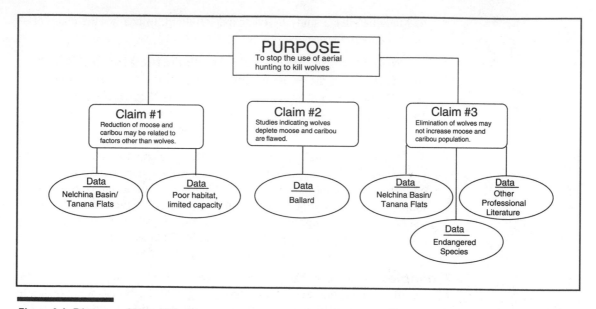

Figure 8-1 Diagram of Manville's Argument

Personal and social needs and values. A variety of personal and social needs appear to influence people. These include the needs of individuals as well as the needs of society at large. They range from physiological and emotional needs, to needs for ethical standards, to needs for authority.

- **Physical/Emotional Needs and Values**. Maslow (1954) developed a hierarchy of the needs of individuals within society. These include physiological needs, safety needs, the need to belong and be loved, the need for self-esteem, and the need for self-actualization. Readers are more willing to be persuaded by a discussion if it relates to these needs.

  The report from which the following excerpt was taken relates directly to a basic need—nourishment—and to a bias for eating low-fat products. One of the purposes of the report was to convince fast food chains and meat packaging companies to offer the new low-fat beef for consumer consumption. Notice how the writer tries to persuade restaurant owners by appealing to their need to attract customers:

These newly-developed lean ground beef products are tailored to meet the needs of the diet-conscious consumer. The products contain less fat and calories than traditional ground beef products. Upon introduction of these products to the retail market, consumers will have the opportunity to substitute the lean ground beef products for the higher-fat ground beef products currently available.

Dale L. Huffman & W. Russell Egbert
Advances in Lean Ground Beef Production, Bulletin,
606 Alabama Agricultural Experiment Station, 1990.

Because of people's need for self-esteem, readers are often influenced by the use of a courteous and concerned tone.

The purpose of the following letter is to persuade the owner of a franchise to correct a situation that was discovered during an evaluation of the operation. The writer uses a courteous, positive tone, opening the letter with an expression of thanks. Consider the list of factors used to create a tone of appreciation.

List of factors.
Use of term
"improve" rather
than "correct."

I would like to express my thanks for the time, courtesy, and cooperation you and your people extended to me during the service technician evaluation last week.

   The following summarizes my findings and the steps to be taken in order to improve the service department..."

- **Ethical Needs**. In addition to their physical and emotional needs, people also have ethical needs. They are more willing to accept a message if it reinforces their value system.[2] In the excerpt below, notice how the GreenCotton Company attempts to persuade people to purchase its products by relating the company's products to a clean and healthy environment, a value to which many people subscribe.

When dyes, bleaches and formaldehyde are eliminated [from processing cotton fabrics] so are the steps needed to apply them. Initially the manufacturers we work with were surprised to see how much water and energy they saved producing GreenCotton. Our untreated GreenCottons are every bit as soft and comfortable as their treated counterparts (sometimes more so!).
                              Environmental Product Guide, Seventh Generation, April, 1991.

- **Authoritative Needs**. People are also more apt to accept a claim if it is supported by an authority or a model they respect.[3] Teenagers are more willing to consider the ramifications of AIDS when Magic Johnson discusses it than when an unknown doctor does. Furthermore, people are persuaded by a promise of a reward or the threat of a punishment.

   Research reports usually relate the study under consideration to theories supported by authorities in the field. In addition, these reports usually relate a study's hypothesis to previous findings, which serve as models for the present one. In the following excerpt, notice the writer supports his disagreement with the majority opinion by citing research that agrees with his findings.

---

[2]This concept was developed by H. C. Kelman, 1961. "Processes of Opinion Change." *Public Opinion Quarterly*. 57-78.
[3]This concept was developed by J. R. R. French and B. Raven, 1968. "The Bases of Social Power" in *Group Dynamics: Research and Theory*. ed. by Cartwright & Zander. New York: Harper & Row.

Although it is a widely-held opinion that alkanes with molecular weights greater than 500 do not biodegrade, evidence exists to the contrary. In 1974, Haines and Alexander showed that normal alkanes with up to 44 carbons (the highest pure molecular weight alkanes available) were metabolized by heterogeneous microflora at rates comparable to the 12 to 18 carbon alkanes.

Regulations and procedures are often related to penalties. Readers know implicitly that if they do not follow a procedure, they may lose their job, or at least be reprimanded. In cases such as the one below, if they do not follow procedures, they could be hurt or even killed.

When you feel an earthquake, DUCK under a desk or sturdy table. Stay away from windows, bookcases, file cabinets, heavy mirrors, hanging plants and other heavy objects that could fall. Watch out for falling plaster or ceiling tiles. Stay under COVER until the shaking stops. HOLD onto the desk or table, if it moves, move with it.

State of California Governor's Office of Emergency Services

While procedures seem to be based on implicit threats of punishment, companies usually try to relate support of their products to a reward. In the following excerpt, notice how the writer, aware of the readers' desire to save the planet, promises a reward—conserving natural resources—for buying recycled paper

Buying recycled packaging is one of the most effective means consumers have to conserve our natural resources.

- **Organizational Needs**. When your purpose relates to a corporation or organization, you should relate your discussion to the needs of that organization. These needs include safety, profits, personnel, products, and self-actualization.

   The needs of various companies, organizations and institutions differ over a period of time as circumstances change. The prime concern at a nuclear power utility may be safety, whereas the prime concern at a plant that manufactures engines may be profitability. However, if the Office of Safety and Health Administration (OSHA) begins to investigate the manufacturing plant because of recurring accidents, safety may become a prime concern.

   In the following example, a writer tries to persuade clients to purchase a new computer chip by relating the new product's attributes to the clients' needs for a more accurate, sensitive, and flexible system for their own lines of digital receivers. As you read, notice how the writer develops a logical line of reasoning, leading from the problem to the solution.

Establishes the
client's problem.

The family of newly-released products focuses on a unique and challenging area of digital signal processing (DSP) for digital receivers. Because of limitations in the performance of microprogrammable-based DSP devices, the functions implemented by the chipset previously were often done utilizing analog circuit techniques with corresponding limitations in the accuracy, sensitivity and flexibility of those systems. In addition, the requirements to process coherently these signals often increases the hardware by a factor of two to four over single channel architectures for DSP...

Suggests solution
to client's problem.

To accomplish coherent processing of these signals, complex number arithmetic operations need to be performed. In the DSP family the devices provided are focused on offering solutions for this complex arithmetic. The latest introductions to the family provide a higher level of integration of functions necessary so that a two-chip set will implement the entire front end of a digital receiver system.

Professional needs. Professional discourse communities, such as those of biologists, oceanographers, or electrical engineers, usually have established standards by which they judge the validity of an argument. These standards usually determine the profession's needs. These needs revolve around the issues a profession is interested in addressing at a specific point in time. Each community determines an agenda, and establishes the relevant issues, the criteria for examining those issues, and the authorities it considers credible. Because they are not only members of an institution but also of a profession, readers are more likely to accept a claim if it meets the standards and needs of their professional field.

■ Just as the needs of institutions differ and change over time, so do the needs of a profession. While wildlife biologists may be interested in the ecological balance between wolves and moose and caribou in Alaska, wolf specialists may be more interested in determining the number of species of Alaskan wolves still in existence. The question of the number of species is irrelevant to the wildlife biologist. In trying to persuade members of the Alaska Wolf Management Planning Team to eliminate aerial hunting, notice how, in the following excerpt, Manville relates his discussion to the main issues recognized as relevant in the readers' field. He does not bring up the issue of the number of species.

I have watched this issue [of wolf control] closely and have been actively involved with it for nearly seven years. Hunters are pitted against nonhunters, consumptive users against nonconsumptive ones, sport hunters against subsistence users, aircraft gunners against ground trappers, and rural and urban residents against each other... ■

Determining Sufficient Evidence

It is difficult to determine the amount of data a reader needs to accept a claim. Sometimes a single example is sufficient; at other times a series of comparisons are necessary. Of course, it is more difficult to persuade a non-sympathetic or hostile reader than a sympathetic one. Hart (1990) found that at least two sets of data were necessary for every claim when writing for a hostile audience, and that most scientific communities do not accept claims based on a single sample or incident.

■ Manville uses three sets of data to support his claim that the elimination of wolves may not increase moose and caribou populations. He refers to the conflicting studies at Tanana Flats and Nelchina Basin. He also indicates that according to one expert, Ballard, predators other than wolves are depleting the moose population. Finally, he refers to another expert whose studies indicate that moose populations may rebound without any change in the wolf population. ■

STRATEGY CHECKLIST

1. Support your claims with factual evidence and/or emotional appeals.
2. Use data, opinions, and/or examples to support your claims factually.
3. Appeal to readers' personal and social needs, organizational needs, and/or professional needs to support your claims.
4. Select evidence related to your point of view.
5. Use opinions of persons who are respected and/or who are considered authorities by your readers and their communities.
6. Use at least two sets of evidence to support your claims when your audience is not sympathetic to your claims.

ORGANIZING TO FOLLOW A LOGICAL LINE OF REASONING

Once you have developed your argument, you need to present your claims and evidence to your readers so they can follow your line of reasoning and arrive at the same conclusions you do.

Adhering to the Given-New Contract

It's important to recognize the need to adhere to the given-new contract. To persuade readers to accept a message, you need to begin with evidence they already know and accept as valid before you introduce a new argument.

■ In the following example, consider how Manville moves his audience from familiar to unfamiliar data as he attempts to persuade his audience to accept his claim. Notice how he moves from what readers know to new knowledge.

Known data.

As you are certainly aware, the public is sharply divided on the issue of wolf control. Those who totally oppose it are largely residents of the larger cities in the state. On the other hand, the groups which strongly favor wolf control are generally urban sportsmen, outfitters, guides, and rural residents who hunt. This group generally views wolf control as necessary to increase populations of moose and/or caribou, so they in turn can replace wolves as the predator. While opinions expressed by both groups are vehemently defended as correct, the real truth of the matter is usually ignored, and political decisions often supersede those based on sound scientific principles and basic biology. Some major biological and political realities need to be included in your review of strategic wolf management. These include the following...

Bridge to new information.

New information.

The suppression of a prey population can be a result of one, several, many, or a combination of factors. These can include such variables as degraded habitat, changes in vegetative succession (seral stage) resulting in a lowering of the carrying capacity, harsh winters (in some cases, several back-to-back), overhunting (e.g., 19% of the moose population harvested in the Tanana Flats in 1973), disease, parasitism, predation by wolves and bears, and unknown variables. The wolf, unfortunately, has been used in the past as the scapegoat for what may actually be a host of problems resulting in the apparent suppressed condition of moose populations in Alaska.

Claim.

Manville begins with data the audience already knows, i.e., the opinions of the groups who are for and against wolf control. He goes on to indicate there are biological and political data his audience doesn't know, and he then presents this new data.

His claim is made at the end of the paragraph: the reduction of the moose and caribou population is not necessarily due to the wolf alone, but is probably related to other factors as well. By presenting the known data first, the writer helps his readers accept the new data and finally the claim. ■

Structuring the Text

Not only does the evidence need to be congruent with readers' prior knowledge, but readers must recognize the relationship between the evidence and the claim if they are to accept an argument. Therefore, you have to establish a logical relationship between your evidence and the claim it supports. Your information needs to be organized so readers can easily follow your argument and arrive at the same conclusions you do.

There are two ways to organize an effective argument, inductively and deductively.

Inductive In an inductive pattern you move from the evidence to the claim. This organizational pattern is often used for unsympathetic audiences or for audiences with no prior knowledge of a subject. Writers often use an inductive pattern to organize the problem section of a document, especially a proposal. The pattern lets you provide sufficient evidence for the reader to understand and accept your resolution to the problem (claim) when you finally present it.

In an inductive discussion, you begin by presenting the evidence. When you have provided sufficient information, you make an inductive leap by making a generalization in the form of a claim. The shorter the gap between the data and the claim, the easier it is for readers to make the leap also. By providing sufficient and relevant evidence, you can keep the size of the gap relatively small. However, if you fail to provide enough evidence, readers will fall into the gap and will not be able to accept your claim.

■ Manville's audience is unsympathetic to his purpose, and so he uses an inductive approach. Notice how, in the following excerpt, he leads the reader to accept his claim:

Evidence.
• History.
  Describes
  causes.

Describes
alternative
solutions.

Describes
contrasting
situation.
Claim.

Wildlife populations in Alaska (and elsewhere) have historically fluctuated naturally and probably will continue to do so well into the future, exhibiting peaks and troughs in population levels. The variables which cause these fluctuations—including changing successional patterns/habitat conditions, depletions of soil nutrients, variable weather and precipitation patterns, seasonal and yearly differences in available food, depletion of the habitat, presence of disease and parasitism, and other factors—are as yet not well understood.

Man, in the current form of the Alaska Department of Fish and Game (ADF&G), the Department of Wildlife Conservation (DWC) and private citizens, has attempted to alter these natural fluctuations through management strategies focused at maintaining artificially high levels of moose and caribou. These strategies have included habitat manipulation, adjustments in seasons and bag limits, and predator control, mostly targeted at the wolf.

Wolf control has often been looked upon as the "quick fix"—eliminate the wolves and the prey population will automatically rebound...Elimination of all wolves in portions of the Nelchina Basin in the late 1970s resulted in no net increase in moose there...The wolf unfortunately has been used in the past as the scapegoat for what may actually be a host of problems resulting in the apparent suppressed condition of moose populations in Alaska.

Manville provides readers with data related to the history of man's attempts to control wildlife populations to support his claim that controlling the population of wolves will not necessarily increase the population of moose.

If readers know that (1) wildlife populations fluctuate; (2) the variables which cause the fluctuations include changing successional patterns/habitat conditions, etc.; (3) man has attempted to alter these fluctuations; (4) one of these attempts has involved predator control; (5) predator control has been mainly targeted at the wolf; (6) it is believed that if wolves can be eliminated, their prey population will increase; and (7) incidents such as those in the Nelchina Basin don't support this theory, then they may accept the claim that the wolf is not the sole cause of a low moose population.

To arrive at this conclusion, readers need to make an "inductive leap" from the specific pieces of information (numbers 1 to 7) to the general claim that the wolf is not at fault. If the information about the Nelchina Basin incident had been omitted, readers might not have understood why wolves might not have been at fault. Thus, the information about the incident is a necessary link in the chain of supporting information. ■

If you do not provide each of the links necessary for the reader to follow your line of reasoning, you probably will not persuade the reader to accept your claim.

Deductive   A deductive pattern is the opposite of inductive: it moves from a general statement to the specific. In a deductive argument, the organizing idea, in the form of a claim, is placed at the beginning of a section or paragraph, and the evidence supporting it follows. Documents intended for readers who are already aware of a problem, or sympathetic to your claims, or knowledgeable about your topic, are often written in this organizational pattern. For instance, the solution section of a proposal is usually written deductively.

The following excerpt is from a brochure on renewable energy sources. The purpose of the pamphlet is to persuade the general public to lobby Congress to fund these resources. Consider the difference between the line of reasoning used in this excerpt and the previous excerpt.

Claim.

Evidence.

Renewable energy sources are forced to compete in a rigged market because our economic system fails to account for the environmental and social damage caused by energy technologies. These "external"costs—such as acid rain and air pollution caused by coal burning—are not included in the prices we pay for fossil fuels. This means that relatively harmful energy sources, like coal, are given an unfair market advantage over relatively benign sources, like wind power.

Union of Concerned Scientists. *Cool Energy*:
The Renewable Solution to Global Warming, Cambridge, MA 1990

The first sentence, the organizing idea, establishes the claim: our renewable energy sources can't compete in the present market. The remaining sentences in the paragraph support the claim by explaining the reason the sources can't compete: costs of acid rain and air pollution are not included when costs for non-renewable resources are totaled, and harmful energy sources are less expensive than renewable sources when environmental and social costs are not calculated.

STRATEGY CHECKLIST

1. Adhere to the given-new contract if readers' prior knowledge and experience is not congruent with your evidence.

2. Lead up to your claim if readers' attitudes are not sympathetic toward it, or if they are unfamiliar with the evidence.

3. Use an inductive pattern to help readers move from the evidence to the claim (in writing the problem section of a proposal) or when readers are unsympathetic to your claims or have little prior knowledge of the topic.

4. Provide as much detail as possible in an inductive pattern to keep readers from falling into the gap created by the inductive leap you make when moving from the specific evidence to the general statement of your claim.

5. Use a deductive pattern to provide readers with evidence after you present your claim (in the solution section of a proposal) or when readers are knowledgeable about the subject, are sympathetic to your claims, or aware of their needs.

COUNTERING READERS' PERCEPTIONS AND CLAIMS

In writing a persuasive document, you need to consider readers' responses to your claims. Writing a persuasive document is like holding a debate with your reader. You not only need to present claims that are relevant to your readers, but also you need to present claims and supporting data to rebut possible negative responses your reader may have to your initial argument. In the following excerpt from a paper to persuade readers to use degradable plastic bags for yard waste, notice how the writer preempts readers' questions and counterarguments.

Is composting better than landfill disposal? How difficult is it to implement composting programs? What is the best way to collect yard waste for composting? Vacuum trucks are expensive. Street debris gets into the waste. Paper isn't strong and breaks easily when wet. How expensive is a composting program?

Franklin Associates estimated that yard waste such as leaves and grass clippings constitute 20 weight percent of the waste in U.S. landfills. If yard waste were composted rather than landfilled, a significant savings in landfill space would be realized. Yard waste composting programs are relatively simple to implement and do not necessarily require a high level of technology to be functional and productive.

Collecting yard waste can be accomplished in several ways. One way is in bulk by vacuum trucks. Vacuum trucks are expensive, however. Other disadvantages include wind blowing the leaves which have been placed at the curb, and contamination from already present street debris. Paper composting bags are another method of collection. They will degrade in a compost pile, but are more vulnerable to losing strength when wet; thus, wet grass clippings and rain are a problem.

Degradable plastic composting bags for yard waste collection offer a more economic and facile means by which leaves and/or grass clippings can be transported to the compost site. Since these bags can be placed directly into the windrows without having to be emptied of their contents, an appreciable savings is gained through eliminating a previously labor-intensive step in the composting process.

Throughout these three paragraphs the writer appears to be responding to readers' questions.

When you write for a non-sympathetic audience, you need to use some special techniques.

Provide information to overcome readers' biases. These biases influence the way in which the reader responds to a message.

■ If you want to write a memo to your manager requesting a personal computer for your office, keep in mind your manager's biases toward money, time, and computers. ■

Anticipate and refute readers' objections. As in chess, the player who can anticipate and plan the furthest in advance has the advantage. A writer needs to determine in advance, as best possible, what the reader's questions and arguments will be.

■ In the memo for a personal computer, you'd want to include arguments which address your manager's biases in the areas of time and money. You might argue that the cost of a computer would be only slightly more than hiring the extra clerical help already approved for a certain upcoming project. In addition, you might point out that the computer would still be around after the completion of the project. You might also estimate that your own productivity would increase with a computer, based on the fact that you know how to type, but currently have to write everything in long-hand and wait for text processing to turn it around. ■

Provide readers with arguments against opposition from their supervisors, clients, etc. A writer needs to give readers ammunition to deal with other arguments they might encounter.

■ Your memo for a personal computer shouldn't stop with your manager's objections. You should also try to answer the arguments your manager might run into with *her* boss, who might object because she's afraid everyone else will want a computer, too. You might suggest that the PC would be available to others in your division when you're not using it. ■

Emphasize shared goals. Concentrate on those goals you and the reader have in common, rather than those on which you differ.

■ In the PC memo, indicate that both you and your supervisor are working toward an efficient staff that can turn out quality work quickly. ■

Maintain a positive point of view. You want to include information that makes your reader feel positive toward your topic and, if possible, exclude information which irritates your reader. If your readers are involved, praise their efforts.

■ In your memo for a PC, don't suggest that you can't do your job well without a computer; your supervisor might replace you with someone who can. Instead, suggest you can improve your performance with a PC. ■

Lead up to your point. You may not want to start off with your main point; it may be better to lead up to it. Present your case before presenting your request or solution. You may want to discuss a problem before recommending a solution, or you may want to discuss the positive aspects of your topic before presenting the negative ones. Don't ask your readers, "Do you want the good news first or the bad news ?" Give them the good news.

■ In the memo to your manager requesting a personal computer, you might begin by telling about the account you nearly lost yesterday because text processing was tied up and you couldn't get a report finished by the deadline. Then you might explain how much additional money it cost in overtime for one of the text processing people to stay and complete your report. Once you've established a problem that's costing the company money, you can present your idea for a PC, which will save money as well as keep the company from losing accounts. ■

Open with a concession. If you're writing a justification and you know you're in the wrong, admit it. Your admission effectively eliminates the reader's argument. Other types of concessions include admitting your biases or failures, or conceding that the reader's argument may be valid.

If management at the Three Mile Island nuclear plant had admitted that the problems at the plant were dangerous, there would have been far less criticism of the plant's management, and the public would have been fairly informed.

■ In your request for a computer, you might concede that funding for your division is tight, and your request isn't a priority one. You could then go on to explain how the PC could save money in the long run by eliminating the need to hire additional clerical help in the future, and by your increased productivity. ■

Choose your language and tone carefully. You need to consider the connotation of a word as well as the denotation. Denotation is the actual definition or meaning of a word.

Connotation is the emotional association people have with certain words and phrases. Connotation becomes especially important when dealing with a non-sympathetic audience. You want to avoid language that may have a high negative charge for your reader. Use words with positive—or at least neutral—connotations. Your tone should be conciliatory or objective.

■ In your memo for a PC, you wouldn't want to write, "I know you're cheap, so I've selected an inexpensive model." Instead, you might say, "I know you're concerned with how cost-effective such a purchase would be." ■

STRATEGY CHECKLIST

1. Maintain a positive point of view.
2. Preempt your readers' questions and arguments by responding to them in your document.
3. Provide information to overcome readers' biases and objections, and also arguments they may encounter.
4. Emphasize goals that you and your readers share.
5. Lead up to your main point if your reader may be unsympathetic to your topic.
6. Make concessions where possible. Admit biases or failures.
7. Choose your language and tone carefully.

CHAPTER SUMMARY

To accept a claim, readers must perceive evidence as true and valid. Readers accept claims as valid if they're congruent with their prior knowledge and experience. You can arrive at a valid claim by engaging in brainstorming and/or collaborative debating. A valid claim is one with which you and your readers share values and/or interpretations of facts and policies.

Ambiguous terms in a claim should be defined so that readers have the same interpretation as the writer. Claims should be qualified to avoid counterarguments over exceptions or questionable data.

When writing persuasive documents, base your claims on factual evidence, including data, opinions, and examples, and/or appeal to your readers' personal, social, organizational, and professional needs and values.

Personal and social needs include physiological needs, safety needs, the need to belong and be loved, the need for esteem, the need for self-actualization, and the need to reinforce value systems. Readers are also influenced by rewards or punishment, respect for authority, and models.

Organizational needs include safety, personnel, products, profits, and self-actualization. Professional needs are usually determined by each professional community. Hostile or non-sympathetic readers require more evidence than sympathetic readers to accept a claim.

Readers need to have arguments presented in such a way that they can follow the writer's line of reasoning. Readers perceive an argument as logical if they recognize the relationship between the claims and the purpose, and between the evidence and the claims.

You should use an inductive organizational pattern when writing for hostile or non-sympathetic readers, or when writing the problem section of a proposal. You may use a deductive organizational pattern when writing for sympathetic readers or readers who are aware of their needs.

PROJECTS

As you work on the following projects, you may want to use the Audience Analysis Chart (page 97), the Decisions Chart (page 101), and/or the Purpose Chart (page 136).

Collaborative Projects: Short Term

1. **Writing a persuasive letter: high tech entertainment**

    a. Work in trios. You and your roommates have been discussing the purchase of one of the latest high tech consumer products for your dorm room, apartment, or fraternity. Each of your respective parents has agreed to pay for half of your share of the cost. Select a product.

    b. You and your roommates need to write letters to your respective parents, requesting the money. To save time, you have decided to write a boilerplate[4] letter to be sent to each parent. You need to *persuade* them that your decision to purchase this product is a good one, and they should send the money. Use the techniques you learned in this chapter as you write.

    If you can't think of a product, you might select one of the following:

    - a CD player
    - a PC
    - a Walkman
    - a cellular phone
    - a wide-screen television
    - a camcorder

2. **Determining valid claims**

    A local hospital board has appointed you as a member of a group that will make presentations to various organizations in your community about health topics.

    a. Work in groups of five.

    b. As a group, select one of the following topics or a topic of your own choosing, related to community health:

    - get annual physical check-ups
    - engage in preventive medicine
    - have children checked for cholesterol levels
    - give blood
    - donate organs
    - immunize children against childhood illnesses
    - use approved car seats and seat belts for children
    - don't smoke
    - avoid chemical abuse

---

[4]A generic letter which you can personalize with a few simple changes in terms of the reader's and your own name, address, and individually-related data within the letter.

    c. Using collaborative debating, determine the claims you would make to persuade one of the following groups to follow your advice:

        • the Gray Panthers

        • the local high school senior class

        • the elementary school parent/teacher association

        • the Kiwanis Club

3. Persuading people to be safe

    a. Work in groups of five.

    b. We constantly hear about accidents that could have been averted, or at least been less severe, if proper precautions had been taken or safety procedures followed. As a group, select an area in which you would like to see more people "play it safe." Some areas you might consider are:

        • wearing safety belts in a car

        • wearing safety glasses and other safety accoutrements in an industrial workplace

        • changing light bulbs and wiring

        • lighting the briquettes in a grill

        • using a lawn mower, ladder, and tools such as saws, drills, lathes

        • lighting a wood fire in a fireplace or at a campfire

    Select one of the following audiences to approach to discuss safety precautions and procedures for your topic:

        • elementary school students

        • teenagers

        • members of your dorm, Greek society, or campus organization

        • members of a parent/teacher association

        • members of a professional organization, such as the Institute for Electrical and Electronic Engineers

        • members of a local union

        • members of the local chapter of the American Management Association

    d. Using collaborative debating, determine the claims and evidence you will use to persuade readers to accept your message. Determine an appropriate organizational pattern so readers can follow your line of reasoning.

    e. Write a one-page handout persuading people to follow safety precautions and procedures related to your topic. The handout could be distributed at the next meeting to the group you have selected.

Collaborative Projects: Long Term

The following projects can be continued through Chapter 13. As you complete each chapter, you will find activities for these projects that are directly related to the chapter you are reading. In this chapter, you'll engage in activities related to persuading your audience.

- **Communicating about a campus/community problem** continued from Chapter 4, Collaborative Projects: Long Term Exercise 1.

- **Proposing microcomputer laboratorie**s continued from Chapter 4, Collaborative Projects: Long Term Exercise 2, Projects 1, 2, or 3.

- **Gathering information for a feasibility report: diaper laundry service** continued from Chapter 6, Collaborative Projects: Short Term Exercise 2.

- **Gathering data for an information report: cleaning supplies** continued from Chapter 6, Collaborative Projects: Short Term Exercise 3.

- **Organizing data for a proposal: recycled paper** continued from Chapter 7, Collaborative Projects: Short Term Exercise 2.

- **Organizing data from a food and music survey** continued from Chapter 7, Collaborative Report: Short Term Exercise 3.

Based on the strategies you have learned in this chapter, review your worksheets.

a. Determine whether or not readers will accept your information as valid evidence. Do step 8 on the worksheet.

b. Determine whether or not readers will accept your organizing ideas as valid claims. Eliminate or revise those claims that are not valid.

c. Determine the appropriate organizational pattern. Do steps 8 and 9 on the worksheet.

To be continued, Chapter 9, Collaborative Projects: Long Term.

Individual Projects

1. Read the ad in Figure 8-2 placed in *USA Today* to inform people about biodegradable products and to persuade them to continue to purchase plastic products. As you read, think of the strategies discussed on pages 226-228.

   What strategies did the writer use for writing to readers who might not be sympathetic toward the subject? Why were they used? What other strategies that have been discussed in this chapter did the writer use?

2. You have been looking at one of the latest high tech consumer products and would like your parents to buy one for your home. Select a product and write a letter to persuade your parents to buy one. As you write, use the techniques you learned in this chapter.

Figure 8-2
An Informative
Document to
Persuade Con-
sumers to Contin-
ue Buying Plastic

DOING SOMETHING VS. DOING NOTHING
The Benefits of Using Biodegradable Plastics

The inevitable problem has finally caught up with our disposable society: America is inundated with garbage. Every day, we generate over 900 million pounds of it. That's more than 20,000 truckloads. And studies show that 30% of the volume is plastic, which can take hundreds of years to break down.

Recycling programs for glass, paper and aluminum are gaining promising momentum, but plastics continue to be a problem. Today less than 3% of the plastic garbage we dispose of is being recovered and much of the remaining may never be reused. It's ending up in our landfills, littering our landscape and endangering our wildlife. Particularly, fish and marine animals. Obviously, something must be done. And now, something is being done.

Biodegradable plastics: A step in the right direction

The National Corn Growers Association (NCGA) is assisting the companies listed below that are researching, developing and now producing and selling a special biodegradable plastic, made with a cornstarch additive.

Because of this additive, biodegradable plastic breaks down through the action of naturally occurring organisms, whether it's buried, floating in water, or lying on the ground. After decomposition, all that remains is carbon dioxide, water, and a soil-like humic material. None of which is harmful to the environment...

All degradable plastics aren't created equal

The degradable plastics being marketed today are either biodegradable or photodegradable. You should know that there are important differences between the two.

Photodegradable plastics require prolonged exposure to sunlight before they can begin to break down. Unfortunately, those conditions simply don't exist when the material is buried.

The cornstarch additive in biodegradable plastics, however, permits both biological and oxidative degradation. In other words, they begin to break down whether they're buried in a low-oxygen environment or exposed to water and sunlight above ground.

In addition, cornstarch-based biodegradables are recyclable. Plastic products and industrial scrap can be blended into new polyethylene resins for reprocessing.

For these resins, cornstarch-based biodegradable plastic is now used in garbage bags, disposable diapers, shopping bags and grocery bags...

Now, and in the future, you can feel good about using products made with cornstarch-based biodegradable plastic. Because when it comes to the environment, you're doing something versus doing nothing.

If you cannot think of a product, here is a list of possibilities.

- a CD player
- a PC
- a Walkman
- an ATV (all terrain vehicle)

- a cellular phone
- a wide-screen television
- a lawn mower

3. Write a letter requesting your instructor to extend the deadline for assignment 2. Use the strategies you learned in this chapter for writing persuasively.

4. Certain weather conditions can threaten life and limb. Depending on the weather in the region where you live, write a set of procedures you want to persuade people to follow to avoid getting hurt when the weather turns vicious.

    You might consider one of the following conditions:

    - a hurricane
    - a tornado
    - an avalanche
    - a snow storm
    - an earthquake
    - a drought

    - fog
    - a white out
    - an ice storm
    - winds over 40 m.p.h.
    - a severe thunderstorm
    - a dust storm

5. You have been asked by a committee of campus/community leaders to serve on a committee to develop plans for lengthening library hours. However, you don't want to serve on the committee. At the same time, you are pleased that they apparently respect your ability and integrity, and that they like you enough to want to work with you. Based on your own schedule, interests, and lifestyle, write a letter to the committee rejecting their request.

Rhetorical Techniques

THE DOCUMENTS YOU WRITE require a variety of techniques to present your information effectively to your readers. As you draft your text, especially for readers who are not members of your discourse community, you need to define terms with which your audience is not familiar. In research reports you need to describe the methodology, and often the apparatus, involved in experiments; while in evaluation reports you need to describe events; and in instruction manuals you need to describe objects or processes. In proposals you need to analyze problems as well as summarize information.

In this chapter you will learn techniques for defining, describing, analyzing, and summarizing.

DEFINING

Whenever you use a term or phrase with which your reader may not be familiar, you need to define it. Your definition depends on your reader's knowledge and attitude, and on your purpose and situation.

You may only need to provide a brief, parenthetical definition, or you may need to draft an extended definition. You may want to define an object or concept in terms of its appearance or properties, or in terms of its function. A definition may involve a description, an example, or an analogy.

Regardless of how you define a word or phrase, you don't want to use terms in your definition that your readers don't understand. You should adhere to the given-new contract in defining words or phrases, and relate the word to something with which your readers are already familiar.

Relating Definitions to Purpose and Audience

Depending on your purpose and audience, you can define words and concepts according to their history and etymology, properties and appearance, function, results, and process.

History/Etymology You can define a word historically by describing how it has been used over the years. You can also describe it etymologically by explaining the derivation of the word, e.g., the word "technology" derives from the Greek word "technologia," meaning "systematic treatment."

In the introduction of a report by the Environmental Defense Fund (EDF), intended to persuade Congress to fund projects that eliminate the continuing problem of lead poisoning, the writer provides an *etymological,* as well as an *historical*, definition of the word "lead."

Egyptians in the time of pharaohs used lead in ornaments and cosmetics. Chalices made of lead-silver alloys carried wine for the ancient Greeks and lead piping still carries rainwater from the roofs of medieval cathedrals. Indeed, the word "plumbing" is itself derived from the Latin word for lead, "plumbun" (as is its chemical symbol, Pb).

Environmental Defense Fund *Legacy of Lead: America's Continuing Epidemic of Childhood Lead Poisoning* (1990).

By defining the term both historically and etymologically, the writer helps readers in the role of decision makers perceive the extent, both geographically and chronologically, to which lead has been used.

Properties/Appearance Chemicals, minerals, mechanisms, diseases, microbes, etc., are often defined by describing their appearance and/or their properties.

In a paper that informs generalists interested in solid waste management about the degradability of plastics, a chemist defines "polymer" in terms of its chemical properties:

A polymer is a mixture of very long chain molecules. The repeat units or links in the chain are usually very simple units.

Defining a polymer in terms of its properties helps readers in the role of learners understand the aspects of the compound that allow it to become degradable.

Function Mechanisms, microbes, chemicals, etc. are often defined by their function.

In a booklet to farmers (generalists) advocating the use of a specific method of soil preservation, "field windbreaks" are defined in terms of the function they serve:

A field windbreak is a strip of vegetation planted in or adjacent to a field to reduce wind erosion, trap blowing snow, conserve moisture, and protect crops, orchards, and livestock from wind."

U.S. Department of Agriculture, Soil Conservation Service

Describing the term "field windbreak" by its function allows farmers in the role of users to consider how they would use the method.

Results Many processes, concepts, and products are defined in terms of their results or effects.

In another agricultural booklet, the writer defines "conservation tillage" in terms of the results of the method.

Conservation tillage is defined by SCS [Soil Conservation Service] as any tillage and planting system that maintains residue on at least 30 percent of the surface after planting to reduce soil erosion by water, or, where soil erosion by wind is the main concern, maintains at least 1,000 pounds of flat small-grain residue equivalent on the surface during the critical erosion period."

U. S. Department of Agriculture, Soil Conservation Service

Defining "conservation tillage" in terms of its results provides farmers in the role of decision makers with information for deciding whether or not to institute the method.

Process Mechanisms, laboratory apparatus, and concepts can also be defined in terms of their process, i.e., how they work.

In a booklet written to inform the general public about solid waste disposal, the term "photodegradable" is defined in terms of its process.

Photodegradable: A process whereby the sun's ultraviolet radiation attacks the link in the polymer chain of plastic. The breaking of this link causes the plastic chain to fragment into smaller pieces, losing its strength and ability to flex and stretch. As the photodegradable plastic is subjected to the effects of the natural environment, the material is flexed, stretched, and disintegrated into plastic dust.

Keep America Beautiful, Inc.

Defining the term photodegradation in terms of its process helps readers as learners understand how degradability occurs.

Brief and Extended Definitions

Depending on your readers' knowledge of your topic, you may only need a *brief* definition of a few words, or an *extended* definition of up to several pages.

Brief Definitions   You only need to provide a brief definition if your readers are in your discourse community, the context of a term or phrase provides sufficient clues, or the term is peripheral to your topic.

Brief definitions can be integrated into your text parenthetically, in appositional phrases, or as addenda.

Parenthetical.   The following definition of "reading disabilities" appears in the EDF booklet mentioned on page 236. The paragraph in which the definition appears discusses lead poisoning as a cause of an increase in children's reading disabilities.

Compared to the low lead classmates, the higher-lead group showed a 7.4 percent increase in school dropout rates and a 5.8 percent increase in reading disabilities (scoring two or more grade levels below that expected for the highest grade completed).

> Environmental Defense Fund, *Legacy of Lead: America's Continuing Epidemic of Childhood Lead Poisoning* (1990).

Readers don't have to know much about reading disabilities; the topic is peripheral to the writer's purpose. However, readers do need to know what constitutes a disability, so the writer tells them briefly in a parenthetical definition.

Appositional.   The following definition of "platelets" is excerpted from a newsletter used to persuade the general public to reduce its intake of polyunsaturated fats in foods.

These fatty acids reduce the "stickiness" of platelets, the blood cells that help to make blood clot.

> Excerpted from the September 1990 issue of the HARVARD HEART LETTER © 1990, President and Fellows of Harvard College

Because platelets are peripheral to the focus, readers don't need to know much about them; the definition is limited to an appositional phrase.

Addendum.   The purpose of the article in which the following definition of "arteries" appears is to provide readers with instruction for reading blood pressure tests.

The goal of this procedure is to evaluate the pressure within the arteries—the vessels that carry oxygenated blood away from the heart to the muscles, brain, and other organs.

> Excerpted from the September 1990 issue of the HARVARD HEART LETTER © 1990, President and Fellows of Harvard College

Because arteries are peripheral to the main topic, the reader does not need to know a great deal about them except how they relate to blood pressure tests. They are, therefore, defined briefly in an addendum to the sentence.

Extended Definitions A brief definition is not always sufficient for readers to understand the relationship of a word or phrase to the purpose of a document, especially if the object or concept is an integral part of the topic. Brief definitions can also be insufficient in discussing abstract concepts. Readers may need to relate abstractions to concrete items. In addition, because they have no prior knowledge of the object or concept being discussed, novices may need more detailed explanations than experts.

Let's look at how a writer defines the term "leachate" in a magazine article to persuade the general public to solve solid waste problems. The writer feels the reader needs more information than the appositional clause, "a contaminant produced when rain mixes with buried waste," to fully comprehend the relationship of the term "leachate" to the topic of waste disposal.

Apposition.
Extension.

> Leachate, a contaminant produced when rain mixes with buried waste, tends to be acidic because of newsprint dyes, car and flashlight batteries. It often contains heavy metals: lead, cadmium and zinc—all known carcinogens. Because garbage has an enormous water-holding capacity—30 percent of its own weight—landfill sites exposed to more than 20 inches of rain yearly often produce leachate.
>
> *Visions Magazine*, Oregon Graduate Institute of Science and Technology

This definition involves a description of the "properties" of leachate. The description provides specific, concrete details on the causes of its acidity (newsprint dyes and batteries), as well as the heavy metals involved (lead, cadmium, and zinc), and the percentage of water (thirty) and amount of rain (twenty inches) required.

Types of Definitions

Definitions can assume a variety of forms, including classification, description, comparison, exemplification, and visual representation.

Classification In this classical form of definition, a term is defined by classifying it and then differentiating it from other members of its class.

In the following excerpt from a magazine article written to inform the general public about problems with asbestos removal, the writer defines "asbestosis."

Class membership.
Differentiation.

> Asbestosis, a biological reaction to the inhalation of asbestos fibers, is a common form of lung disease in which the lungs become scarred, resulting in reduced lung function and creating a greater reliance on the heart.
>
> by Mike Buelow, *Tracks*, Spring, 1990. Reproduced by permission of Deere & Co.

Asbestosis is defined by classifying it as a lung disease, something with which most readers are already familiar. The writer then differentiates it from other lung diseases by describing its particular aspects, i.e., (a) biological reaction, (b) scarred lungs, (c) reduction in lung function, (d) additional reliance on the heart.

Description   Often, readers need to know what an object looks like or what its composition is to understand what it is.

In the following excerpt from an article informing the general public about water conservation, the term "groundwater" is described in terms of its location and size.

Under your house and below your factory buildings, beneath streets, farm fields, and the shopping mall is water—groundwater. However, there is a misconception that groundwater flows in underground caverns and rivers. Actually, groundwater flows through cracks and pores between soil and rock particles. The groundwater in this area is bountiful. If you could somehow pour all the water below the ground on top, you could cover the surface many feet deep.

by Loren Polak, *Tracks*, Spring 1990. Reproduced by permission of Deere & Co.
© 1990 Deere & Company. All rights reserved.

The description relates to images with which the general public is familiar—houses and streets. The writer uses everyday words, "cracks," "rocks," and "pores," rather than scientific terms.

Comparison   Readers from other discourse communities can often understand a new term or concept better if it is compared to something in their own discourse community. An engineer trying to define the term "noise" for an English teacher might compare it to misspellings in a text: it doesn't cause a lack of comprehension but it does interfere with the message.

In an article for the general public on the earth's future climate, James Trefil uses a comparison to define a "mathematical model."

In everyday usage, the term "model" refers to a simple representation of something real. When you buy your child a model car, you expect to get something that has the general shape of a car but that isn't exactly like a real car. The model won't have a motor, and may not have a trunk and doors that open.

by James Trefil, *Smithsonian* magazine, December 1990

Trefil compares a mathematical model, with which readers are not familiar, to a model car with which readers probably are familiar, in an attempt to help them perceive what a mathematical model is.

Exemplification   Often, readers need an example of a term to understand how it functions.

In the booklet to inform the general public about water conservation, the writer uses an example, based on mathematical principles with which readers are familiar, to explain the rating index for CFCs (chlorofluorocarbons).

The heating and ventilation industry has developed an index number to rate various CFCs for their theoretical ozone depletion capability. For example, refrigerant R-11 is assigned a value of 1.00 and all other refrigerants represent a percent of R-11. Refrigerant R-22 has an ozone depletion number of only 0.05.

In a presentation to a group of novices, a wildlife biologist defined himself as a "game warden who wears a necktie," and "wilderness" as "an area with only one television station and a considerable distance from the nearest bar."

Visualization Since many readers are visually-oriented and have difficulty understanding something without seeing it, a visual representation can help readers understand a new term or concept.

The excerpt below appears in a pamphlet on radiation for the general public. Read the definition of alpha, beta, and gamma radiation below, and as you read, try to perceive the differences among the three types of radiation. Then look at the diagram in Figure 9-1. Consider how much the visualization clarifies your definition of the three types of radiation.

Alpha radiation consists of positively charged particles and is emitted from naturally occurring elements such as uranium and radium as well as from man-made elements. Alpha radiation will just penetrate the surface of the skin; it can be stopped completely by a sheet of paper. However, the potential hazard that alpha-emitting materials present is due to the possibility of their being taken into the body by inhalation or along with food or water.

Beta radiation consists of electrons. It is more penetrating than alpha radiation and can pass through 1-2 centimetres of water or human flesh. A sheet of aluminum a few millimetres thick can stop beta radiation. Tritium, one of the materials present in fall-out from nuclear explosives tests, emits beta radiation.

Gamma radiation can be very penetrating. It can pass right through the human body but would be almost completely absorbed by one metre of concrete. Dense materials such as concrete and lead are often used to provide shielding against gamma radiation.

The graphic in Figure 9-1 helps readers understand the types and penetrating power of radiation by using familiar symbols and objects, such as a hand and a concrete slab.

**Figure 9-1
Defining Radiation
Visually**

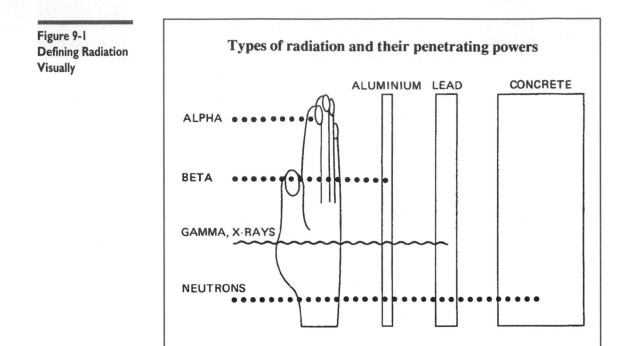

Types of radiation and their penetrating powers

International Atomic Energy Agency

**S T R A T E G Y
C H E C K L I S T**

1. Define words and phrases with which your readers may be unfamiliar.

2. Relate your definition to your purpose.

3. Follow the guidelines established in the given-new contract in defining terms.

4. Place short definitions (several words) of terms that are not an integral part of the topic in parentheses, in appositional phrases, or as addenda to a sentence.

5. Define terms peripheral to your topic briefly. Use parentheses or appositional phrases, or separate them by a dash from the main part of the sentence as addenda.

6. Use an extended definition, if the term is directly related to your topic, to help readers, especially those not in your discourse community, understand the relationship between a term and your topic.

7. Define a term by classifying it and then differentiating it from other members of its class.

8. Define words in terms of their etymology, history, properties, appearance, function, process, or results.

9. Use descriptions, analogies, examples, or visuals to define terms.
10. Don't use terms in a definition that your readers don't understand.
11. Define terms in documents for novices specifically and concretely when possible.

DESCRIBING

Almost every type of document involves a description in some form. Your résumé (see Chapter 15) provides a description of yourself; an accident report requires a description of the damage to your car; and a proposal often involves a description of a situation. The photograph in Figure 9-2 is a description of a form of soil conservation.

As you learned in the previous section, a description can serve as a definition; it provides a "picture" of a physical entity, process, or event for readers to see in their mind's eye.

Relating Descriptions to Purpose and Audience

As with definitions, the information you include in a description is related to your purpose and your readers.

Let's consider two readers. One, in the role of a decision maker, needs to determine whether or not to eat some cookies. The other, in the role of a user, wants to make some cookies. Both need descriptions of the ingredients. The reader who simply wants to eat the cookies needs a description of the ingredients in terms of nutrients, as well as the amount of each nutrient, to determine

**Figure 9-2
Describing a
Method of Conser-
vation**

Stripcropping
In stripcropping, strips of a row crop are alternated with soil-conserving strips of small grain or a cover crop, such as grass or a grass-legume mixture. The soil-conserving strips trap soil that erodes from the row-crop strips. The edges created by the alternating strips attract wildlife for feeding and nesting. Grasses and legumes in the cover strips provide food for many birds, such as pheasant.

U.S. Department of Agriculture, Soil Conservation Service

whether the cookie is nutritious, whether it is free of fat and cholesterol, and whether it is fattening. The reader who plans to make the cookies needs a description of the ingredients in terms of food stuffs, as well as the amount of each to determine how much to use.

To Eat		**To Make**	
Calcium	2 RDA	2 sticks	margarine
Fat	6 grams	3/4 cup	brown sugar
Calories	120	1 tsp	salt
Protein	1 gram	2	eggs

Organizing Descriptions

In writing descriptions, you need to provide readers with the large picture first, then the details. You need to provide a general statement at the beginning of a description, and then describe the parts of the product, process, or event. Related information should be kept together.

General Statement   By beginning with a general description of a physical entity, process, or event, you establish a frame in which readers perceive individual parts. It's like putting together the frame of a jigsaw puzzle. It's much easier to determine where the individual pieces go once the frame is constructed. If readers have a general idea of what a piece of machinery is or what a process does, then they can understand the relationship between the individual parts and the equipment, or between the various steps and the process. They can perceive how the pieces fit together to form the whole picture. This general statement serves as the organizing idea for the description section of a document.

Analysis   The body of a description should be divided according to the main parts of the product, process, or event.

As you draft a description, you need to keep related information together. Again, using the puzzle analogy, it is easier to recognize a barn or a pond when the pieces are clustered together than if they are scattered across the board.

The following excerpt comes from a brochure written to provide journalists covering nuclear utilities with sufficient information to understand how reactors work, and to persuade them that nuclear energy is a safe and viable alternative to energy production. In this description of a nuclear reactor, notice how the writer moves from the general statement to the specific parts.

General explanation.

A nuclear reactor is a complex machine designed to heat water, producing steam that runs a turbine that generates electricity. In a fossil fuel generating plant, the heat that boils water is produced by the burning of coal, oil or natural gas. In a nuclear generating plant the heat is produced by the fissioning of uranium atoms...

Breakdown of parts.

The heart of a reactor is the core that contains the fuel assembly where uranium fission occurs under controlled conditions. The fuel consists of uranium pellets of uranium oxide, each about four-tenths of an inch in diameter and about one-half inch long. In an LWR [Light Water Reactor] the core is immersed in ordinary water, which acts both as a moderator and as a medium to transfer the heat produced by fission to the turbines.

Reprinted with permission the U.S. Council for Energy Awareness

In this description the writer sequences the information as if it were a series of concentric circles, presenting first the outside structure, then moving inward to the core of the reactor.

The writer does not include many details, since readers only need to know general aspects concerning reactors, not the nuts and bolts; they won't be using reactors or repairing them. Because the writer's purpose is not only to inform readers how reactors work, but also to reassure them of the reactor's safety, the writer includes the prepositional phrase, "under controlled conditions," in the first sentence of the second paragraph to indicate the safe conditions under which the reactor operates.

Types of Descriptions

Most descriptions involve physical entities, such as mechanisms, land, the human body, etc.; processes or situations and events.

Physical Entities This information should be chunked according to the major parts of an entity. Then sequence the chunks from the top of an object to the bottom, or from left to right, from the outermost to the innermost part, or in the order of its operation.

The description in Figure 9-3 is excerpted from a technical manual to instruct readers in the use of a vehicle. The user needs to know how safe the vehicle is, and the kinds of materials it can transport. Notice how the writer moves from a general statement to an analysis of the parts.

The writer begins by providing a general description of the capacity of the AMT to carry cargo. All of the information relating to what can be carried in the cargo box, i.e., weight and size, is chunked together. The information is sequenced from back to front. By including specific dimensions of the cargo box, the writer provides readers with information they need for gauging the size and weight of materials the box can carry. The information concerning the vertical panel informs the user of the safety features.

Processes This category ranges from a description of the operation of a piece of machinery to procedures followed in discovering a new product. Descriptions of processes are usually dynamic—they indicate movement—whereas descriptions of objects are usually static.

The information should be chunked according to the major steps of the process, and then sequenced in the order in which they occur in the process.

General description.

Analysis of the parts:
 • cargo box

 • vertical panel
 • area in front

Cargo capacity for the AMT 600 is 272 kilograms (600 pounds) maximum, plus a 91 kilogram (200 pounds) operator. The cargo box area is 1092 mm long x 1226 mm wide x 279 mm high (43 x 48.25 x 11 inches). The vertical panel at the front of the box functions as a guard, to keep material from shifting forward against the operator. The area in front of the panel serves as a tool box/storage area.

Figure 9-3 Description of the Cargo Capacity of an All Material Transporter (AMT)

The following description informs readers about alternative methods of yard waste collection so they can determine which method their community should adopt. Readers in the role of learners need to know the number of persons required to engage in the process, and the amount of work it entails, to determine if it is feasible. Notice that the writer sequences the information in the order in which the activities are carried out.

List of components.
Step 1
Step 2

Step 3

Vacuum collection. The vacuum truck crew consisted of three workers, two rakers and a driver. The driver advanced slowly while the two rakers pushed leaves toward the vacuum hose. When the leaf box was full, the vacuum apparatus was detached and the truck driven to the compost site to deposit leaves.

Collection resumed when the truck returned and the vacuum was re-attached.

Biocycle, 419 State Ave., 2nd Floor
Emmaus, PA 18049 (215)967-4135

The subhead provides the summary statement for the description. The description begins with a list of the components—the workers, rakers, and driver—just as a recipe begins with a list of the ingredients.

The description has comparatively few details. It does not include such information as the time it took the driver to drive to the compost site and return, nor does it provide a weight or volume for the leaf box. Readers who are simply becoming aware of a method are not interested in this much detail. If they become sufficiently interested in the method to consider using it, they will need more specific information to make a final decision.

Situations and Events The need to write descriptions of situations and events has risen over the last decade. An increasing number of regulatory agencies are requiring industries to report problems which affect the environment, the well-being of citizens, or the economic status of business. These reports usually involve descriptions of events.

The information describing the event should be chunked according to the major incidents and sequenced chronologically. Assume an omniscient point of view, inserting explanations that include information learned later in the chronology, to help readers understand the significance of each incident.

The description in Figure 9-4 of an accident at a nuclear utility, Three Mile Island, is a classic description of a situation. The description is excerpted from a report that informed Congress and the American people of the events that occurred at the plant. Readers wanted to know the cause of the accident, who was at fault, and how it could have been prevented so they could create legislation to prevent similar problems in the future. As you read, notice that the description is sequenced chronologically

The writer helps the reader understand the meaning of each incident by including explanations that would not be necessary if the reader were an expert in that topic. The third paragraph explains the pressure reading, "pressure... [was] 100 psi above normal," and the final paragraph forecasts a problem, "Up to now the reactor system was responding normally..."

STRATEGY CHECKLIST

1. Relate your description to your purpose.

2. Begin a description with an overview of the object, process, or concept being described to provide your reader with a frame. The organizing idea should be part of this opening.

3. Forecast the major categories that you will discuss if your description is long.

4. Keep related chunks of information together.

5. Categorize information about a physical entity according to its major parts.

6. Sequence the major parts of a physical entity from top to bottom, left to right, outside to inside, or according to the sequence of operation.

7. Categorize information about a process according to the steps in the process.

8. Sequence the steps of a process in the order in which they occur.

9. Categorize information about an event according to the major incidents.

10. Sequence the chunks in a situation chronologically.

11. Describe occurrences from an omniscient point of view.

12. Use detail and language with which your reader is familiar.

Introduction.

Pump shutdown,
beginning of event.

Reactor shutdown
8 seconds into
the event.

Valves closed 9
seconds into
the event.

14 seconds into
the event.

Omniscient
viewpoint.
LOCA occurs.
Summary of events.

2 hours, 22 minutes
into the event.

A Loss of Coolant Accident (LOCA) occurred when a pilot-operated relief valve (PORV) was stuck open and valves on two emergency feedwater lines remained closed during a scram on the morning of March 1, 198X.

The accident began when a series of feedwater system pumps supplying water to X utility's steam generators tripped. The nuclear plant was operating at 97 percent power at the time. The first pump trip occurred at 35 seconds after 4:00 a.m. With no feedwater being added, the plant's safety system automatically shut down the steam turbine and the electric generator.

When the feedwater flow stopped, the temperature of the reactor coolant increased. The rapidly heating water expanded, and pressure inside the pressurizer built to 2,255 pounds per square inch, 100 psi more than normal. As designed the PORV atop the pressurizer then opened and steam and water began flowing out of the reactor coolant system through a drain pipe to a tank on the floor of the containment building. Pressure continued to rise, however, and 8 seconds after the first pump tripped, X's reactor scrammed: its control rods automatically dropped down into the reactor core.

Less than a second later, the heat generated by fission was essentially zero. But the decaying radioactive materials left from the fission process continued to heat the reactor's coolant water. When the pumps that normally supply the steam generator with water shut down, three emergency feedwater pumps automatically started. Fourteen seconds into the accident, an operator in X's control room noted the emergency feed pumps were running. However, he did not notice two lights that indicated a valve was closed on each of the two emergency feedwater lines and thus no water could reach the steam generators. One light was covered by a yellow maintenance tag. It could not be determined why the second light was missed.

With the reactor scrammed and the PORV open, pressure in the reactor coolant system fell. Up to now, the reactor system was responding normally to a turbine trip. At this point the PORV should have closed 13 seconds into the accident when pressure dropped to 2,205 psi. It did not. A light on the control room panel indicated that the electric power that opened the PORV had gone off, leading the operators to assume the valve had shut off. But the PORV was stuck open, and remained open for 2 hours and 22 minutes, draining needed coolant water—a Loss of Coolant Accident (LOCA) was in progress. In the first 100 minutes of the accident, some 32,000 gallons of water—over one-third of the entire capacity of the reactor coolant system—escaped through the PORV and out the reactor's let-down system.

Figure 9-4 Description of a Situation at a Nuclear Plant

ANALYZING

Most of the reports you write as a student involve analyzing. The major topic or concept is broken into categories and then into subcategories. The problem section of a proposal analyzes a problem. Research reports analyze previously-related studies, and evaluation reports analyze events or people.

Analyses are easy to write if you have used the organizing techniques discussed in Chapter 7. Your information is already chunked into categories and subcategories, and should be sequenced according to your purpose.

In the following report on the continuing problem of lead poisoning, the writer analyzes "chelation," the conventional treatment for lead poisoning. Chelation is a process in which a drug binds itself to lead in the body and makes the lead easy to excrete.

The writer develops the following outline, using the methods described in Chapter 7:

IV. Treatment
    A. Conventional Method
          1. Limitations
               a. neurologic impairment
               b. long term storage
               c. continued environment
          2. Options

Based on the outline, the writer drafts the following section on limitations. As you read, notice how the writer follows the sequence in the outline.

Forecast
I. Limitations

a. impairment

b. storage

c. environment

While chelation can substantially reduce lead levels, it has a number of significant limitations. The three major limitations of chelation are:

1) Chelation cannot repair neurologic impairment, but rather can only keep further damage to the nervous system from occurring. Children treated for lead poisoning are still likely to require special education and other cognitive or behavior-related therapy long after their initial treatment.

2) Chelation generally cannot reach lead that has found its way to long term storage sites in the hard body tissues (bone, teeth) or the brain and kidneys. As discussed above, lead can re-enter the soft tissues from bone at high levels. Chelation therapy does not prevent this. In fact, doctors have observed a "rebound" phenomenon in some patients, where blood lead levels rise after the cessation of chelation therapy.

3) Finally, chelation has little effect when the patient, as is often the case, returns to the same lead-contaminated environment in which the exposure occurred. Unless the course of exposure can be eliminated—for example, by removal of accessible lead paint, or by substitution of bottled drinking water for lead contaminated tap water—it is likely that the problem will recur.

Environmental Defense Fund, *Legacy of Lead: America's Continuing Epidemic of Childhood Lead Poisoning* (1990).

The forecast allows readers to accurately predict that they will read about three limitations of chelation treatment, and the numeration of each limitation helps readers recognize they are reading what they predicted.

STRATEGY CHECKLIST

1. Begin your analysis with an overview of your topic.
2. Include the major categories in a forecast.
3. Chunk your information according to your categories of information.
4. Sequence your categories from most to least important, or in the order in which the process you are analyzing occurs.
5. Use numbers or subheads to indicate the beginning of each category or subcategory.

SUMMARIZING

A large part of technical writing involves summarizing. Executive summaries are attached to the front of most long reports to provide executives with a thumbnail description of the report. Progress and periodic reports are actually summaries of work done over a period of time. Your supervisor may also ask you to summarize an article to save him time reading it.

Select information relevant to your purpose and synthesize it into a brief explanation. Use the strategies you learned in Chapter 7 for synthesizing.

The owner of a local restaurant needs to replace her manager, who has resigned. She writes a job description to advertise the position as well as to screen candidates. She bases the description on the responsibilities of the position, which she lists as follows:

- answers questions from customers
- answers questions from waitresses
- provides friendly atmosphere for customers
- seats customers
- greets customers
- arranges waitress and bus boy schedules
- evaluates staff performance
- offers advice to waitresses
- clears tables
- carries food to tables
- responds to customers complaints
- reports directly to the owner
- plans schedules for waitresses
- interviews and hires waitresses

By using synthesizing strategies, she concludes that the information falls into four categories: accountability, personnel management, customer relations, and waitressing. She then sequences the data in order of importance for hiring, so that if a candidate does not meet the major requirements, she will know at the beginning of the interview. She then drafts the following summary of the position to use for hiring a new manager. As you read, notice how she has categorized and sequenced the information.

PART-TIME MANAGER: Shirley's SteakHouse
Accountability: Reports directly to owner.
Major Responsibilities:
    Personnel management: Supervises waitresses and bus boys. Delegates responsibilities, plans and organizes schedules, hires staff, and evaluates staff performance.
    Customer Relations: Greets and seats customers, provides friendly atmosphere, responds to questions and complaints.
    Waitressing: Carries food to tables and cleans tables after customers are finished.
Requirements:
- Management capabilities to delegate work, interview and evaluate staff, plan schedules.
- Good interpersonal relations, friendly personality.
- Previous experience in food service industry.
- Previous management experience.

Using this summary, the manager will have no difficulty determining whether or not a candidate meets the requirements, and a candidate will have no difficulty understanding the duties and responsibilities of the position.

STRATEGY CHECKLIST

1. Synthesize information to determine major chunks of information.
2. Break up long lists into manageable chunks.
3. Use a subhead or opening sentence to provide a general overview of a chunk. The organizing idea of a chunk should be part of the subhead or overview.

CHAPTER SUMMARY

As you write, you use a variety of techniques to meet your readers' needs and achieve your readers' and your own purposes. These include defining, describing, analyzing, and summarizing.

Your purpose influences the type of definition you use in defining a term. You can define a term by its history/etymology, property/appearance, function, results, and process.

Definitions can be brief or extended. The length of a definition depends on the readers' previous knowledge and your purpose. Brief definitions can be integrated into the text in parentheses, appositional phrases, and addenda to sentences.

Definitions can be subjective or objective, and should adhere to the given-new contract.

The forms for definitions are classification, description, comparison, exemplification, and visual representation.

Descriptions are related to your readers and your purpose.

When you write a description, begin with a general statement. Keep related information together.

Descriptions involve physical entities, processes, and situations and events.

Use the organizing techniques in Chapter 7 to help you write analyses and summaries.

PROJECTS

As you work on the following projects, you may want to use the Audience Analysis Chart (page 97), the Decisions Chart (page 101), the Purpose Chart (page 136), and/or the Worksheet in Appendix B.

Collaborative Projects: Short Term

1. **Writing a definition: biodegradable trash bags**

   a. Work in pairs. One person should assume the role of a person on the marketing staff of a manufacturer of plastic bags in which cornstarch is used. The other person should assume the role of a manufacturer of plastic bags that do not use cornstarch.

   b. As a member of the marketing staff of a manufacturing company that produces plastic bags, your manager has asked you to write a letter to persuade purchasing agents to use your product for lining trash containers. Use the information from the article in Figure 9-5. Use the worksheet in Appendix B to organize your information. Then draft your letter.

   c. Give your partner a copy of the completed letter.

   d. Compare the two letters. Underline definitions. Place parentheses around descriptions. How are the definitions and descriptions related to each writer's purpose? What differences do you notice between the descriptions and definitions of each letter? Discuss why the differences exist.

2. **Writing a summary: solid waste**

   Your firm is interested in becoming more environmentally sound, and has established a committee to study the issue. You are a member of a subcommittee that has been formed to investigate ways to reduce solid waste. Each member of your subcommittee assumed responsibility for gathering

Companies misuse the word "degradable"

Many companies are using the word "degradable" on their boxes and in their advertising as a sales and promotional tool. They realize the general public is concerned with the solid waste problem and are capitalizing on these concerns by suggesting their bags are degradable and will help solve the problem. However, these are photodegradable plastics and it will take several hundred years to degrade when buried in a landfill.

Degradable plastics are materials which lose their physical properties faster than conventional plastics when exposed to a variety of environmental conditions, says Mike Erker, NCGA plastics assistant.

Various formulations have been developed to promote and accelerate the degradation of conventional plastic. This degradation may be initiated by living organisms (biodegradation) that are present in common landfills or through the action of the sun (photodegradation).

Researchers say the rate of these processes depends on a number of factors such as polymer (plastic) type, the concentration of active components in the polymer and product characteristics such as shape, thickness and color. Environmental conditions including temperature, moisture, biological activity and sun intensity also affect the speed of degradation.

Adding a cornstarch based master-batch to plastic entices microorganisms to feed on the plastic thus breaking apart the plastic product and rendering it biodegradable. Biodegradability is a valuable property because once plastic reaches a landfill it is covered daily and does not receive sunlight, Erker emphasizes.

Photodegradable plastics require several days to several months of exposure to direct sunlight in order to break down. "This variety of plastic will degrade if littered along roadways, but will not degrade quickly when placed in a landfill and buried," Erker says.

As quoted by Knight-Ridder News Service, Mike Levy, the Washington lobbyist for Mobil Chemical Co., manufacturers of Hefty degradable bags, concedes that the bags "are not an answer to landfill crowding or littering."

His statement, "degradability is just a marketing tool,"—especially as it relates to Mobil's "degradable" bags—truly demonstrates his company's nervousness because to Mobil, degradability does seem to be just a marketing tool, says Erker.

But Mobil sells them anyway, marketing the bags in a green box decorated with an eagle, pine tree and heavenly rays of sunlight designed to lure environmentally concerned shoppers. "Hefty Degradable Bags," the package says, "A step in our commitment to a better environment."

"We're talking out of both sides of our mouths because we want to sell bags," Levy said.

"Consumers must read the label and ask for cornstarch-based plastic that are biodegradable if they are concerned about plastics being buried in landfills," Erker says.

What is biodegradation?
Biodegradation is the degradation of a plastic product when exposed to the activity of natural organisms. Biodegradation of a plastic product is dependent on environmental conditions, including but not limited to temperature, moisture and number and type of organisms. Cornstarch based plastics are biodegradable.

What is photodegradation?
Photodegradation is the degradation of a plastic product when exposed to sunlight.

Degradable trash bags available today:
Contain cornstarch

EnviroGard	Biodegradable
BioBag	Biodegradable
Naturegrade-Plus	Biodegradable
President's Choice	Biodegradable

Does not contain cornstarch

Hefty Degradable	Photodegradable
Glad Degradable	Photodegradable
Good Sense	Photodegradable

Biodegradable trash bags at:
True Value (nationwide)
Ace Hardware (nationwide)
Kroger stores (KY)
Cubs Foods (IL)
FS farm supply (IL, IA, and WI)
Star Market (IL)
IGA (Midwest)
Some Jewel stores (Midwest)
National grocery stores

National Corngrowers Association, Copyright © 1990

Figure 9-5 An Extended Definition of "Degradable"

information about one of the following aspects of this topic: durable goods, nondurable goods, and containers and packaging. The results of your investigations follow:

- **Durable Goods.** Overall, durable goods are projected to increase in MSW generation, although not as a percentage of total generation (Table 31). The trends in generation of major appliances and furniture and furnishings are well established by production numbers, since lifetimes of up to 20 years are assumed. Generation of rubber tires, lead-acid batteries, and miscellaneous durables are projected based on historical trends, which are generally "flat" or exhibit low rates of growth.

    Substitution of relatively light materials like aluminum and plastics for heavier materials like steel has occurred in durables like appliances and furniture as well as other products. Also, cars have become smaller and tires have been made longer-wearing, which tends to reduce the rate of increase at which tires are generated. It was projected that these trends will continue.

- **Nondurable Goods.** As noted above, generation of nondurable goods has been increasing rapidly, and this trend is projected to continue (Table 32). Over 86 million tons of nondurable goods are projected to be generated in 2010, or over 34 percent of total generation.

    In 1988, paper products were over 77 percent of nondurables generated; it is projected that paper products will be over 81 percent of nondurables generated in 2010. These projections are based on trends developed in a study for the American Paper Institute (Reference 34). Books and magazines, office papers, and commercial printing are projected to increase their share of total generation more rapidly than other products. Newspapers, tissue papers, and other nonpackaging papers also are projected to increase, but not so rapidly.

    Based on historical trends, paper plates and cups were projected to show no increase in tonnage or percentage; plastic plates and cups were projected to show growth in tonnage, although not in percentage of total generation. (The plates and cups categories include hinged containers and other foodservice items, and it was assumed that there will be no widespread bans of disposable foodservice items.)

    Because of declining birth rates and processes that made individual diapers smaller and lighter, disposable diapers began to shown a decline in weight generated after 1985, and generation of disposable diapers was projected to remain rather "flat," which means that they decline as a percentage of total generation. (It was assumed that there will be no widespread bans of disposable diapers.)

    Clothing and footwear were projected to continue to experience the same slow growth exhibited in the past; these items thus will be a declining percentage of total generation.

    Finally, other miscellaneous nondurables, which include many items made of plastics, have been growing historically and the growth is projected to continue, making this category continue to increase slightly as a percentage of MSW generation.

- **Containers and Packaging.** As discussed earlier, historically containers and packaging have been the largest single category of MSW generation. This is projected to change, however, as nondurables are projected to exceed containers and packaging by the year 2000 (Table 33).

Tonnage of glass packaging generated has been in decline since the early 1980s as glass was displaced by lighter materials like aluminum and plastics. These trends were projected to continue; glass containers are projected to be a declining percentage of MSW generation (just over 3 percent of total generation in 2010).

Steel packaging generation has also been declining for much the same reasons as glass, and steel packaging is also projected to be a declining percentage of MSW generation (less than one percent of total generation in 2010).

Tonnage of aluminum packaging has been increasing steadily over the historical period, and this trend is projected to continue. Because of its light weight, however, aluminum stays at just over one percent of total generation in the projections.

Like other paper and paperboard products, overall generation of paper and paperboard packaging has been increasing rapidly. The increase is almost all in corrugated boxes, which are mainly used for shipping other products. Continued increases in generation of corrugated boxes are projected; tonnage of these boxes is projected to be 40 million tons in 2010, or 16 percent of total MSW generation. All paper and paperboard packaging is projected to be 20 percent of total generation in 2010.

Generation of other paper and paperboard packaging has not exhibited the same growth, generally due to displacement by plastic packaging. Thus generation of milk cartons, other folding cartons, and other paperboard packaging is projected to be almost "flat," while generation of paper bags and sacks is projected to decline, following historical trends.

Plastics packaging has exhibited rapid historical growth, and the trends are projected to continue. Soft drink bottles, milk bottles, other containers, bags and sacks, wraps, and other packaging are all projected to follow the increasing trends. Generation of all plastics packaging is projected to be 10 million tons in 2010, or 4 percent of total generation.[1]

    a. Work in trios. One member is responsible for the information related to durable goods, another is responsible for the information on nondurable goods, and the third is responsible for the information on containers and packaging.

    b. Write a one-page summary of your findings to present to the full committee. Each of you may want to summarize your own section and then put the three together, or you may want to combine your knowledge to write the entire summary. The summary should be no longer than 250 words.

    Option: With your subcommittee, create one or several graphs and/or tables to help readers understand your findings (See Chapter 11).

---

[1]Data from the U. S. Environmental Protection Agency. *Characterization of Municipal Solid Waste in the United States*, 1960-2000. March 1988. p.10, 14-16.

3. **Writing a description: food services**

    You are on the staff of a restaurant. The manager has decided to put together a handbook for new employees, such as bus boys, waitresses, food servers, hostesses, or cashiers. The purpose of the booklet is to help new employees understand how the restaurant operates. The manager has asked the current employees to write the booklet. Depending on the type of restaurant involved, the booklet should involve many of the following processes:

    - seating customers
    - serving customers
    - setting up tables
    - preparing the food
    - collecting customer's payment
    - dressing appropriately

    - taking orders
    - handing out menus
    - taking orders for drinks
    - checking on customers' needs
    - cleaning up tables
    - receiving tips

    a. Divide your class in half to form two large groups. Each group should select a restaurant to observe.

    b. Within each group, work in pairs. Each pair should select one of the above areas.

    c. Select a school cafeteria, fast food place, or other restaurant and observe the area for which you are responsible. Take detailed notes.

    d. Each pair should share their observations with each other, then collaborate to write a description of the process observed.

    e. Each group should write a summary to serve as an overview for the booklet on the operation of the restaurant.

Collaborative Projects: Long Term

The following projects can be continued through Chapter 13. As you complete each chapter, you will find activities for these projects which are directly related to the chapter you are reading. In this chapter, you'll engage in activities related to using rhetorical techniques.

- **Communicating about campus/community problems** continued from Chapter 4, Collaborative Projects: Long Term Exercise 1.
- **Proposing microcomputer laboratories** continued from Chapter 4, Collaborative Projects: Long Term Exercise 2, Projects 1, 2, or 3.
- **Gathering information for a feasibility report: diaper laundry service** continued from Chapter 6, Collaborative Projects: Short Term Exercise 2.
- **Gathering data for an information report: cleaning supplies** continued from Chapter 6, Collaborative Projects: Short Term Exercise 3.
- **Organizing data for a proposal: recycled paper** continued from Chapter 7, Collaborative Projects: Short Term Exercise 2.

- **Organizing data from a food and music survey** continued from Chapter 7, Collaborative Report: Short Term Exercise 3.

Using the strategies you have learned in this and the previous chapters, draft your document.

To be continued, Chapter 10, Revising, Collaborative Projects: Long Term.

Individual Projects

1. Your company has just received the letter shown in Figure 9-6 from a consulting firm. It is a response to a request for assistance in taking care of a toxic spill. Your manager has asked you to summarize the response for the V.P.

2. The counseling service division at your university has decided to publish a booklet to help sophomores select a major. You have been asked to write the section in the booklet for your field of specialization. The section should include a description of your major, your opinion of its difficulty, and the prerequisites needed to succeed in it. Since the division of counseling services does not have a lot of money to devote to this project, your description should not be longer than two typewritten pages. Draft a description.

3. Write an extended definition of the term "recyclable." The definition is to be included in a brochure that will be distributed to community residents. The purpose of the brochure is to persuade residents to participate in the community's recycling program, which has just initiated a curbside pick-up for recyclable solid waste.

4. Many universities have a parking problem. You have been selected to become part of an intercampus committee to discuss ways to alleviate parking problems. Prior to the meeting you have been asked to study the parking situation on your campus and to prepare a paper analyzing it.

   a. Observe the parking situation on your campus.

   b. Based on your observations, draft a brief paper summarizing the situation.

5. Signing up for classes can be a confusing process to many off-campus students. You have been asked to write a hand-out describing the process. Your hand-out will be included in a packet of enrollment materials sent to these students.

   a. Analyze the sign-up process.

   b. Draft a hand-out describing the process.

Figure 9-6
Proposal for
Assisting in Clean-
ing Up a Toxic
Spill

November 15, 1991

Re: Leaking Underground Storage Tank Compliance Incident   ESE   Environmental
Science &
Engineering, Inc.

Dear Mr. Smith,

This letter is prepared in response to our telephone conversation of November 8, 1991 and our November 9, 1991 meeting regarding the apparent petroleum release at the Anytown facility. Based on the information furnished to date, it appears that a release from at least one of the two diesel underground storage tanks (UST) removed from your facility has impacted the adjacent soils and potentially the local shallow groundwater. The Illinois Emergency Services and Disaster Agency was subsequently notified of the site status. The following paragraphs describe the typical steps involved in assessment and remediation of fuel releases in Illinois. As we discussed, we are prepared to assist you with all of these steps.

Remediation of soils which are in excess of the IEPA's objectives for benzene, toluene, ethylbenzene, and xylene (BTEX) and polynuclear aromatic compounds (PNAs) is required by the IEPA. In Illinois the IEPA generally promotes excavation and off-site disposal of contaminated soil which does not meet clean-up objectives. Although this method does not appear justified in many cases and does not permanently solve the environmental hazards, it is difficult to vary from this process. Corrective action activities associated with remediation of petroleum impacted soils (excavation and off-site disposal) do not require Agency approval of those activities in which the cost is less than $150,000. Corrective action may begin and continue up to the $150,000 without the need to submit and have approved a Corrective Action Plan.

During the tank removal operation, shallow groundwater was apparently encountered in both tank cavities. A slight product sheen was noted on the water surface and strong petroleum odors were reported in the soils displaced from the northern tank excavation. In such cases, the IEPA requires a subsurface investigation to better define the condition of the shallow groundwater aquifer surrounding the LUST. It would appear the groundwater quality near the northern tank cavity will require investigation. Based on laboratory analysis of the soil closure samples collected from the southern tank excavation, this area may or may not require a groundwater investigation.

If groundwater surrounding one or both of the tank cavities is found to be in excess of the IEPA's generic clean-up objective for groundwater, the Agency will require remediation of the shallow aquifer. Generally, when impacted groundwater is identified the IEPA requires an investigation which includes the installation of three to four (3-4) stainless steel monitoring wells and several soil borings. Unlike the soil remediation via excavation and off-site disposal, the IEPA must pre-approve each groundwater investigation step prior to its completion if State reimbursement for costs are to be sought. This pre-approval process involves the drafting and submittal to the IEPA of a Corrective Action Plan (CAP).

This CAP will detail the proposed subsurface investigation including the specific locations of proposed boring and monitoring wells, sample collection and analysis, aquifer testing, etc. The purpose of this phase of the investigation is to perimeterize the plume of contamination present in the local shallow groundwater, and to determine the characteristics of the aquifer. With the information obtained from the above-mentioned hydrogeological investigation, a plan for remediation of the petroleum impacted groundwater, which generally involves the installation of a groundwater pumping and treatment system, will be formulated for this site.

ESE can assist Anycorp in the required correspondence for submittal to IEPA including soil disposal permits, reimbursement application, 20-Day Certification Form, 45-Day Report Forms, Corrective Action Plans (CAP) and Corrective Action Report (CAR). We propose to perform these tasks for your review prior to submittal to IEPA. ESE will provide coordination of field investigations including sampling and laboratory analysis in accordance with IEPA requirements...

Revising

N O MATTER HOW MUCH PLANNING you do prior to drafting, you will still need to revise.

Look at the following excerpt from a memo to a manager from his supervisor. As you read, think about how you might revise.

TO: Robert Bauman
FROM: John Jones
RE: Management Requirements
DATE: September 24, 19xx

It has come to my attention that you need to set in motion an appropriately defined set of management activities and/or actions that will effectively result in timely development and proper implementation of a set of unmistakably interpretable obligatory rules of procedure and conduct.

If you revised this effectively, you should have arrived at something like *"Establish clear guidelines."* When Bob Bauman received the memo, written by an employee in a quality assurance division, he replaced the excess verbiage with a shorter version.

To reduce the message to its essence, you need to read the entire text. If you try to revise by changing only individual words or even phrases as you read, you'll never discover the meaning of the memo.

The etymological definition of the word *re-vision* is "to see again." The definition implies taking a fresh look at the entire text, such as the one above, by looking at the whole message. This concept differs from a limited view of revision as correcting punctuation, grammar, usage, and spelling, or rewriting legibly. Real revision, the kind of revision in which professional writers engage, involves far more.

In this chapter you'll learn that revising is a recursive process, occurring as you plan and draft, as well as between drafts. You'll also learn strategies for effectively engaging in the revision process. You will be introduced to a variety of techniques for evaluating and revising your text to meet your readers' needs and achieve your purposes. In addition, you will discover how to proceed through the three phases of revision: macro revision, micro revision, and proofreading.

THE REVISION PROCESS

The processes of composing are recursive: revising occurs during planning, during drafting, and between planning and drafting.

During planning, you may recognize that the organizational pattern you've selected isn't appropriate, and that you need to revise the pattern. Or, *during drafting*, you may realize that a paragraph needs to be moved, while *between planning and drafting* you may discover that a paragraph is not relevant to your purpose and should be deleted.

During drafting you may only read a page, paragraph, or sentence to determine whether or not to revise. However, *between planning and drafting* you need to read your entire text in order to consider your document from a global viewpoint. You cannot determine whether or not your information is organized to meet your reader's purpose unless you have a grasp of the entire document.

To revise effectively you need to engage in five steps: (1) read the text; (2) evaluate the text in terms of your readers, purposes, and situation; (3) determine the cause of reader problems, as indicated by the evaluation; (4) develop strategies to solve the problems; and (5) redraft the text.

Reading the Text

To determine whether or not you are communicating with your readers and achieving your purpose, you must assume two roles, that of a writer and that of a reader. When you read your text as a writer, you are engaging in internal revision; when you read it as a reader, you are engaging in external revision.

Only you can engage in internal revision because only you know what is in your head, what you want to say. You are the only one with that particular chunk of information in your memory. During internal revision, you must constantly ask yourself, "Am I saying what I want to say?" If not, you revise.

In external revision you "put yourself in your readers' shoes" and try to read the text as your readers will. During external revision, you ask "Am I communicating my message to my audience?" If you think the message may not be understood or accepted by your readers, you revise.

Evaluating the Text

Several methods exist for evaluating a text. These include self-evaluation, peer evaluation, checklists, reader response protocols, and usability tests.

Self-Evaluation Self-evaluation means reading your own text as if you were the reader. This method is good for a first go-round. It gives you an opportunity to reread your text and make obvious changes. However, self-evaluation is very limited. Even professional writers find it difficult to judge whether readers will interpret their texts as they intend. Writers have difficulty assessing their own writing accurately in terms of their readers because of the knowledge effect and the differences between their own frames of reference and those of their readers. Someone else needs to review what is written to determine whether or not the message is being communicated.

One strategy for improving your self-evaluations is to put a document away for several days, weeks, or, preferably, months. This reduces the knowledge effect. When a distance is created between writing a text and reading it, you forget the previously drafted work and some of the facts. For instance, you've probably had the experience of looking at a paper you'd written for a class the previous year and wondering how you could have turned it in, it seems so bad. It's only now, after you've forgotten what you were trying to say, that you realize how poorly you communicated your message.

While distancing can be helpful for evaluating your texts, it still doesn't give you insight into readers' frames of reference or biases. Another drawback to this method of evaluation is that you seldom have the opportunity to put a memo or report away for several weeks or months. Usually, such documents are due "yesterday."

In addition to revising a text yourself, you should have someone else review it. Editors do this for professional writers. You may be able to get similar help from your peers.

Peer Evaluation Peer evaluation means asking a peer to review and comment on your text. Even if that person isn't in the same organizational position as your actual reader, peer reviewers can still be helpful in determining if you have considered readers' processes and behaviors. If a peer reviewer argues with a point, fails to understand a discussion, description or instruction, or does not know the meaning of a word, then you need to consider these problems and determine whether or not your "real" readers will have similar ones.

While this method can be helpful, it has some problems. Peer reviewers may look for specific areas and overlook others. In addition, they tend to superimpose their own styles and ideas, even though yours may communicate your message appropriately. For example, a reviewer at a nuclear facility doesn't like sentences beginning with a preposition. Yet, there is nothing wrong with sentences which begin this way. They do not create a barrier to comprehension, nor do style guides consider them inappropriate. To overcome peer reviewers' prejudices, you and your reviewers need to develop a list of criteria for evaluating the effectiveness of your document.

Checklists   Because checklists contain objective criteria for determining readers' potential problems, they eliminate the intrusion of personal style preferences by peers and supervisors. You can use checklists for both self and peer evaluations.

Checklists can be designed to assess specific types of documents, such as proposals or executive summaries.

Checklists can also provide general guidelines for all types of documents by referring to such generic aspects as forecasts and signals for coherence.

Let's look at a draft of a letter written by a technician at a utility, and then at a generic checklist that a peer has filled out in relation to the technician's draft. As you read the letter in Figure 10-1, think about the various techniques related to organizing and presenting information to meet readers' needs, and determine if the writer has used these. If so, determine if they are used appropriately.

Now study the checklist in Figure 10-2, which was filled out by one of the writer's peers. Does your assessment agree with the evaluator's?

The peer reviewer has indicated that the writer has not organized the information well, nor provided the readers with a forecast to understand subsequent information.

While a checklist helps you concentrate on creating an organized and coherent document, it still doesn't help you perceive readers' biases or misconceptions. Reader response protocols help you get inside a reader's head.

Reader Response Protocols   These responses involve readers reading a text and commenting on their thoughts as they read. Their comments are either made aloud and recorded on a tape, or written in the margin of the document opposite the portion of the text they are reading.

Protocols are usually not done by the readers who will actually receive your message. Instead, they are often done by your office peers. However, their responses can indicate how your intended readers may read a text. The responses

**Figure 10-1
Letter to the
Nuclear Regula-
tory Commission**

Date: June 15, 19xx
Addressee   Waterfall Glen Power and Light Company
Subject: **Revised Implementation Schedule for Procedure
               Revisions for Detailed Control Room Design Review**

Dear Sir:

As part of the DCRDR (Detailed Control Room Design Review) Control Room Enhancement Program, the Authority is revising X plant's operating procedures to use consistent terminology abbreviations. The Authority provided a schedule for completing these revisions as part of Reference 1.

The schedule was divided into two phases. Phase I included installation of new control room labels, and updating the Annunciator Response and Emergency Operating procedures to reflect the labels. Phase II includes the revision of the remaining operating procedures to reflect the new labels.

Phase I has been completed. New labels have been installed and approximately 172 surveillance and operating procedures have been revised to reflect the new labels.

The Authority is scheduling completion of Phase II until April 30, 19xx. The additional time will permit the Authority to rewrite remaining procedures using a new Procedure Writers' Guide. This new schedule will have no significant affect on plant safety since the changes are editorial in nature and only normal plant procedures remain to be revised. The revised procedures will use consistent terminology in an improved, more human-factored format.

Should you have any questions regarding the changes proposed, please contact me.

Very truly yours,

J.C.B.

by your peers can give you general insights into readers' frames of reference, the biases they bring to a text, and the problems they have in understanding a document. If these peer readers have questions, need to reread passages, or indicate uncertainty, then the readers for whom the message is intended may have similar problems.

Reader response protocols are one of the best means of determining readers' potential problems because they indicate readers' specific problems in processing a text. By using a reader's response to a document, you can revise the exact aspect of a text that caused a problem.

■ Let's look at a report of a chemical technician's oversight in carrying out a sampling process, and the accompanying response by a peer (Figure 10-3). The document was drafted by an engineer responsible for writing reports of plant problems. As you read the document, the text of which is printed in the lefthand column, think about problems readers may have in understanding the text. Then read the righthand column to see whether or not the reader had a problem. ■

CHECKLIST

1. Is there a Main Organizing Idea (M.O.I.) for the section or document?　　Yes ( )　　No (x)
   (There is no statement explaining the purpose for the letter.)
   a. Is it the only M.O.I.?　　Yes ( )　　No (?)
      (There appear to be two main ideas—revising procedures
      and changing schedules.)
   b. Is the M.O.I. appropriate to the purpose?　　Yes ( )　　No (?)
      (Since there is no M.O.I. it's difficult to tell.)
   c. Is the M.O.I. strategically placed?　　Yes ( )　　No (?)
      (Since there is no M.O.I. it's difficult to tell.)

2. Is there a forecast?　　Yes ( )　　No (x)
   a. Does it contain the main organizing idea?　　Yes ( )　　No (?)
   b. Does it contain the sequence of categories to　　Yes ( )　　No (x)*
      be discussed, if this is appropriate?*
      (Since there are two phases, this might be appropriate.)

3. Are the categories related to the purpose?　　Yes ( )　　No (x)
   (The categories relate to each phase, not the
   reasons for the postponement.)

4. Does each paragraph comprise a separate category?
   Paragraph 1　　Yes (x)　　No ( )
   Paragraph 2　　Yes (x)　　No ( )
   Paragraph 3　　Yes (x)　　No ( )

5. Does each category have either an implicit or an
   explicit O.I. (and only one)?
   Category 1　　Yes (x)　　No ( )
   Category 2　　Yes (x)　　No ( )
   Category 3　　Yes (x)　　No ( )

6. Is the organizational pattern related to the purpose?　　Yes ( )　　No (x)
   (The pattern should be chronological, not analytical.)

7. Are the categories arranged in a logical sequence?　　Yes (x)　　No ( )
   (Since there are two phases, it would appear to be.)

8. Has the writer signaled to the reader the relationship
   between the O.I. of each category and the M.O.I.?
   Category 1　　Yes ( )　　No (x)
   Category 2　　Yes ( )　　No (x)
   Category 3　　Yes ( )　　No (x)

9. Is all the information relevant to the readers' purpose?　　Yes (x)　　No ( )

10. Is all the information in each category relevant to the O.I.
    of that category?
    Category 1　　Yes (x)　　No ( )
    Category 2　　Yes (x)　　No ( )
    Category 3　　Yes (x)　　No ( )

Figure 10-2 Assessment of DCRDR Letter Using an Evaluation Checklist

11. Is there sufficient information to meet the reader's needs?	Yes (x)	No ( )
12. Is there sufficient detail to support the O.I. of each category?		
Category 1	Yes (x)	No ( )
Category 2	Yes (x)	No ( )
Category 3	Yes (x)	No ( )
13. Do the assertions flow smoothly into each other?	Yes (x)	No ( )
14. Is the language appropriate for the audience/ purpose?	Yes (x)	No ( )
15. Is the tone appropriate for the audience/purpose?	Yes (x)	No ( )
16. Are the grammar, usage, spelling, punctuation, and mechanics correct? ("Affect" is misspelled. "Human-factored" is not an acceptable form of the term "human-factors.")	Yes (x)	No ( )
17. Do visual cues indicate chapters, sections, major categories?	Yes ( )	No (x)*
18. Are tables, graphs, etc. used to facilitate readers' understanding of the of numerical information?**	Yes ( )	No (x)
19. Are photos, drawings used to facilitate readers' understanding of objects, events, processes?**	Yes ( )	No (x)
20. Is the text legible?	Yes ( )	No (x)

If you have marked "X" next to "No" for any of these questions, you need to revise.

* You may or may not need to have a sequence in your forecast, depending on the content and your reader's needs.

Figure 10-2 Assessment of DCRDR Letter Using an Evaluation Checklist *continued*

When readers do a protocol, they are reacting to the content of a document, not editing it. They are trying to make meaning from a text, and they record their responses as they read.

To do a response protocol, readers need to engage in the following three steps.

1. Read a model protocol, such as the one in Figure 10-3. Note how the reader of the model protocol neither edits nor makes suggestions for corrections. The reader's response is simply a running interpretation of the text.

2. Read the text. As you read, try to interpret the piece. State what you think ALOUD as you read. You may stop after each sentence. However, you may find you need to stop even before a sentence has ended—at a phrase or even at a word that catches your attention, confuses you, causes you to think of something else, etc.

DIESEL GENERATOR COOLING WATER SAMPLE

Report	Response
SCOPE	
1. A nuclear equipment operator and two chemistry technicians were observed sampling the Unit 1 diesel generator jacket cooling water system.	Hmm, the writer seems to be observing three people sampling the cooling water system from Unit 1.
OBSERVATION	
2. The lead chemistry technician asked the NEO which diesel was 1 and which was 2.	One technician needed to know the difference between two diesel engines. Apparently there are two engines in Unit 1. Or are there two units, each with an engine? And what is an NEO? Was that mentioned before? I'll go back and look. Oh. It must be the acronym for Nuclear Equipment Operator.
3. The technician stated that it was difficult to identify the diesels on Unit 1.	Why couldn't the technician tell the difference between the units? Wasn't he trained well? He should have been. He's a lead technician. Apparently since it says diesels on Unit 1, there must be two engines on that unit.
4. While sampling the jacket cooling water, the technician partially filled the sample bottle three times, swirled the liquid and then dumped it into a five-gallon bucket.	The technician apparently filled the bottle with the cooling water, swirled it and dumped it into a bucket three times to do the sampling.
5. The cap to the sample bottle was not flushed on any of the samples.	The cap to the sample bottle didn't get rinsed. I guess he didn't have the cap on while he was doing those other things.
6. Proper technique is to flush the inside of the cap to minimize the intrusion of foreign substances that could be in the cap.	I guess the cap should have been on. Why? He keeps using the term "flush." It sounds more like he's rinsing the bottle than sampling water.

Figure 10-3 Reader Response Protocol

As you interpret the text, you may encounter any or all of the following barriers to comprehension:

- No clear, basic, overall problem addressed in the report.
- An unfamiliar or ambiguous phrase or word.
- Lack of logical sequence.
- Information conflicting with the reader's knowledge of the problem, or with previous statements by the writer.

- Insufficient data to understand the writer's claims, data that supports a different claim, or data that appears to be extraneous to the claim.

- An example that does not appear to relate to the topic being discussed.

- Reference to a previous topic which is difficult to locate or remember.

- Misspelled words, garbled language or jargon, confusing punctuation, and strange grammar. Although your job is not to edit, you will probably notice these errors and should comment on them. But you should do so as a reader, not an editor.

3. Discuss the interpretations aloud as they occur if they are being taped. Write them in the margin if they are being written.

Usability Tests   Usability tests determine readers' problems by observing readers' use of instructions or procedures, and by asking readers to record their problems as they use documents. These tests can be conducted for software documentation, and for other forms of instructions and procedures. Writers review the problems and determine strategies for improving their texts.

All documentation, instructions, and procedures should be tested before being published. But, because the procedure can be both costly and time-consuming, many firms do not test their texts. However, failure to test written material can be even more costly and time-consuming in the long run. As a result of failing to conduct usability tests, software firms may receive a large number of calls requesting assistance, thus requiring them either to hire additional employees to respond, or to purchase an 800 telephone number. In addition, consumers often stop purchasing products from companies whose poor user directions cause problems in using a product.

Determining the Cause of Readers' Problems

Once you have evaluated your text, you need to diagnose the cause of readers' problems. When readers fail to understand or accept a text, it is usually because writers fail to provide for readers' reading processes and behaviors, or their knowledge, attitudes, needs, and purposes. If readers are confused, it may be because unimportant information is emphasized and important information is subordinated; related information is not chunked together; or categories are not sequenced logically. You need to look at the reader's problems, as indicated by a protocol, and consider which of the following aspects may have caused the reader's problems:

- content
- organization
- coherence
- immediacy
- language
- sentence structure
- mechanical aspects

■ Figure 10-4 lists problems the reader had in understanding the document on the diesel generator, and the writer's diagnosis of the causes. Try to determine the cause of the problems, which are noted in the lefthand column. Then read the righthand column to learn the writer's diagnosis. ■

Developing Revision Strategies

Once the causes of readers' inability to understand and accept a text are diagnosed, you need to select one of the following strategies for successfully revising your document:

- **Reorganizing Information.** No matter how much time you devote to organizing your information during planning, you may find, as you read

Reader Difficulty	Cause of Difficulty
Sentence 2	
The reader isn't sure if the writer is discussing two engines in a single unit or two units, each with an engine.	Insufficient information for the reader to understand.
He doesn't know what an NEO is and has to go back to the previous sentence to see if it's mentioned. He doesn't find the acronym (abbreviation), but he concludes that the letters probably stand for the Nuclear Equipment Operator mentioned in sentence 1.	Related information is separated.
Sentence 3	
The reader doesn't understand why the technician can't tell the difference between the units. However, sentence 3 provides the answer to the reader's question about the number of engines and units; there are two engines on a single unit.	Insufficient information for the reader to understand. Related information is separated.
Sentence 5	
The reader becomes confused about the rinsing of the cap.	Insufficient information for the reader to understand.
Sentence 6	
The reader indicates he doesn't understand the term "flush."	Inappropriate language.
He is confused about exactly what the technician is doing—whether he's sampling or rinsing.	Unclear language.

Figure 10-4 Diagnosing Readers' Problems

your text, that the organizational pattern, focus, or sequence are inappropriate for your and your readers' purposes. Reorganize the information to reflect the needs.

- **Expanding Information.** The reader may not understand or accept your discussion because you have not provided enough detail. You may have omitted a step or a tool in your instructions, or you may have failed to include an explanation of a mechanism's use. Expand your information to provide sufficient supporting details for your reader to understand your text. You may need to gather additional information, or you may be able to insert information you've already gathered but haven't included.

- **Deleting Information.** You may have included irrelevant information the reader either already knows or doesn't need or want to know. You may also have repeated information in different parts of your text. Delete irrelevant or redundant information.

- **Emphasizing/Subordinating Information.** The reader may not have recognized a main point because it wasn't properly emphasized. Information can be emphasized by placing it first or last in a sentence, paragraph, section or document. It can be subordinated by placing it in a dependent clause or phrase instead of in a separate sentence, or by placing it in the middle of a paragraph, section, or document. Present the most important information, the information the readers need most, first.

- **Improving Coherence.** If readers have difficulty following your discussion, you need to signal the relationship between pieces of information. Insert transitions, repeat words, or restate ideas between organizing ideas and the main organizing idea of a text, and between assertions and the organizing idea in the same category.

- **Creating Immediacy.** Creating immediacy means flagging the reader's attention or including the reader in your document. Readers may not want to read your document, they may not be expecting it, or they may have a negative attitude toward the topic. If you are writing to persuade a reader, use the strategies discussed in Chapter 8, which get the reader on your side. If you are writing instructions, procedures, or regulations, involve the reader in the task. Use the first or second person.

- **Providing Visual Text.** Readers need cues to locate information. They need tables of contents, headings, and subheadings in long reports. Readers also need to be able to decipher the letters and words on a page easily. Type size should be easy to read, typeface should be simple, and plenty of white space should surround the type. Chunks of information should also be set off by white space so the reader can visually recognize a chunk of related information.

- **Providing Graphics.** Graphs and tables should present numerical data. Graphics should provide visual descriptions and definitions of complex verbal objects, processes, and concepts.

- **Improving Sentence Structure.** Sentences should not be too choppy, too long-winded, or too similar. Except for writing instructions, sentence structure should vary. It shouldn't be all complex or all simple. Revise your sentences so that they vary between short, medium, and long; simple, compound, and complex. Sentences should not "sound" awkward: awkward sentences usually indicate unclear thoughts, not incorrect grammar. If a sentence is awkward, rethink what you are trying to say.

- **Improving Vocabulary.** Make sure the language in the text is appropriate, neither above nor below the level of the reader. Be certain words are used correctly, and that a word doesn't have an inappropriate connotation. Check a dictionary or a thesaurus.

- **Correcting Grammar, Usage, Spelling, Punctuation, and Mechanics.** These proofreading activities polish your writing. Grammar, usage, spelling, punctuation, and mechanics, such as typographical errors, need to be checked. They serve as indicators of a writer's capability. Misspelled words, for instance, may indicate sloppiness, laziness, irresponsibility, and failure to attend to details.

The first eight functions are at least as important as the last three, which are the ones usually discussed in school. These functions address the real barriers to reader comprehension. If readers can't understand and accept a message, grammar, syntax, and punctuation become irrelevant.

■ In Figure 10-5, review the writer's diagnosis (column 2) of the reader's problems (column 1) with the report. As you do, consider which of the strategies listed above would solve the problem. Then read column 3 to learn the strategies the writer used. ■

Redrafting the Text

After determining the revision strategies that will solve your readers' problems, redraft your text.

■ Based on the strategies determined above, the writer revises the report. As you read the revised draft in Figure 10-6, compare it to the original and determine if the writer has solved the reader's problems.

Reader Difficulty	Cause of Difficulty	Strategies
Sentence 2		
The reader isn't sure if the writer is discussing two engines in a single unit or two units, each with an engine.	Insufficient information for the reader to understand.	*Reorganize information* so explanation in sentence 3 is chunked with sentence 1.
He doesn't know an NEO is and has to go back to the previous sentence to see if it's mentioned. He doesn't find the acronym (abbreviation), but he concludes that the letters probably stand for the Nuclear Equipment Operator mentioned in sentence 1.	Related information is separated.	*Reorganize information* so the acronym is next to the whole word it stands for.
Sentence 3		
The reader doesn't understand why the technician can't tell the difference between the units.	Insufficient information for the reader to understand.	*Expand information* to explain why engineers can't tell the difference between the two units.
However, sentence 3 provides answer to the reader's question about the number of engines and units; there are two engines on a single unit.	Related information is separated.	
Sentence 5		
The reader becomes confused about the rinsing of the cap.	Insufficient information for the reader to understand.	*Expand information* on why the cap needs to be rinsed.
Sentence 6		
The reader indicates he doesn't understand the term "flush."	Inappropriate language.	*Improve language* by clarifying "flush."
He is confused about what the technician is doing—whether he's sampling or rinsing.	Unclear language.	*Improve language* by clarifying "sampling" and "rinsing."

Figure 10-5 Strategies to Solve Readers' Problems

**Figure 10-6
The Revised Draft**

Defines NEO.

Expands information to explain why technician can't tell which diesel is which.

Provides transitions to indicate sequence of events. Uses terms "rinse," "swirl" appropriately. Eliminates term "flush." Explains about the cap. Expands information to include an explanation of proper techniques so reader can contrast with improper ones being observed.

DIESEL GENERATOR COOLING WATER SAMPLE

SCOPE

 A nuclear equipment operator (NEO) and two chemistry technicians were observed sampling the Unit 1 diesel generator jacket cooling water system.

OBSERVATION

1. The lead chemistry technician asked the NEO which diesel was 1A and which was 1B. The chemistry technician stated that it was difficult to identify the Unit 1 diesel because of a lack of clear, component labeling.

2. Before sampling each diesel generator jacket cooling water, the technician rinsed the sample bottle by partially filling the bottle with cooling water three times, swirling the cooling water and then dumping it into a five-gallon bucket.

 However, at no point did the technician rinse the sample bottle cap. Proper sampling techniques include rinsing the inside of the sample bottle cap to minimize the chance of contaminating the sample with foreign substances that could be in the cap.

 The writer has clarified many of the points that originally confused the reader. The table in Figure 10-7 can help you determine appropriate strategies for solving reader problems. ■

**STRATEGY
CHECKLIST**

1. Read the text to determine if you are saying what you want to say.
2. Evaluate the text using self- and peer evaluation to determine readers' problems.
3. Use a checklist to eliminate personal preferences.
4. Try to obtain a reader's protocol.
5. Conduct a usability test if you are writing instructions, procedures, or documentation.
6. Determine the cause of readers' problems.
7. Determine revision strategies based on readers' problems.

Readers' Problems	Potential Causes	Alternative Strategies
Don't understand why they're reading a document.	No M.O.I. No forecast.	Organize information. Include forecast, state M.O.I.
Confused about why they're reading a document or a paragraph.	More than one M.O.I. or O.I. Inappropriate M.O.I or O.I.	Reorganize information. State appropriate M.O.I. or O.I.
Don't understand how details relate to purpose.	No M.O.I. Categories not related to purpose. Relationships between M.O.I. and categories, between O.I.s and supporting information in a category not indicated.	Organize information. Improve coherence.
Don't understand how information relates to M.O.I., O.I., or previous paragraph.	No O.I. Categories not logically sequenced. Relationships not signaled.	Organize information. State O.I. Reorganize information. Improve coherence.
Don't understand focus of paragraph, section, or document, or don't understand relationship of information to focus.	Related information not chunked together.	Reorganize information.
Don't understand or accept description or argument.	Insufficient information. Important information not emphasized.	Expand information. Emphasize information.
Know information, don't want to know information, learned information previously in text.	Irrelevant or redundant information.	Delete information.
Biased against topic.	Induced negative attitude. Inappropriate tone. Prior knowledge and attitude.	Delete information. Improve language. Reorganize from analytic to synthetic pattern.
Don't read document or stop halfway through text.	Doesn't catch readers' attention.	Create immediacy.
Don't understand vocabulary.	Inappropriate language.	Improve language.
Confused about meaning.	Inappropriate language. Inappropriate or unclear sentence structure.	Improve language. Improve sentence structure.

Figure 10-7 Revision Strategies

continued

Readers' Problems	Potential Causes	Alternative Strategies
Fluency interrupted by grammatical, usage, spelling, punctuation, or mechanical deviations.	Errors in grammar, usage, spelling, punctuation, mechanics.	Correct errors.
Don't understand numerical descriptions.	Unclear numerical relationships.	Provide graphics. Develop numerical relationships.
Don't understand descriptions.	Unclear descriptions of objects, events, processes.	Provide graphics.
Difficulty predicting or locating information in long documents.	No cues indicating upcoming information.	Provide appropriate visual text.
Difficulty in reading fluently.	Typeface too small, text too dense, insufficient white space.	Provide appropriate visual text. Reformat.

Figure 10-7 Revision Strategies *continued*

REVISION PHASES

Because revision involves so many aspects of composing, ranging from content to typographical errors, it is almost impossible to be concerned with all aspects simultaneously. The best strategy for revising is to take it in three phases: macro revision, micro revision, and proofreading.

Macro Revision

In macro revision, you consider the content and organization of information on a global scale. You are concerned with reorganizing, expanding, deleting, emphasizing, and subordinating categories of information within an entire document, section, or chapter.

Your revision activities should begin with problems related to macro revision. Don't work on problems related to sentence structure until you are fairly satisfied that you have solved problems related to content and organization. You are wasting your time if you try to revise a confusing sentence contained in a paragraph that should be deleted because it contains irrelevant information.

Micro Revision

Once you have solved problems at the macro revision level, you can begin to work at the micro revision level. Micro revision deals with problems related to paragraphs, sentences, and words. It involves content and organization of information within paragraphs. It also involves readability.

This phase occurs in two steps. In the first step, you are concerned with content and organization; with reorganizing, expanding, deleting, emphasizing, and subordinating sentences and supporting details.

In the second step, you are concerned with a text's readability. You ascertain that relationships between ideas are indicated, and that the sentence structure and language are appropriate for the reader.

Proofreading

In this phase, you check to make certain all of the technical aspects of writing are correct. Proofreading should always be the final phase of revision.

Just as a car, aircraft, or boat manufacturer finishes a vehicle by polishing the chrome, checking that headliner seams are invisible, and touching up nicks in paint, a writer revises for grammar, usage, spelling, punctuation, and mechanical errors.

A RECURSIVE PROCESS

Because the process is recursive, you will find, as you engage in macro revision, that you are also doing some micro revision and proofreading as you correct misspelled words, awkward sentences, or inappropriate words. Alternatively, while you are engaged in micro revision, you may discover a need for adding visual text at the global level.

Let's look at how the executive director of the Heartland Water Resources Council revises a text that responds to a letter written by a local Sierra Club member. The Sierra Club member, who is knowledgeable in environmental issues but not an expert on water resources, opposes a project proposed by the U.S. Army Corps of Engineers. The director's purpose is to persuade the reader that the project is a viable solution for saving plant and animal life that is dying off because the Illinois River is filling up with silt.

As you read the letter in Figure 10-8, evaluate the text for problems the reader may have in understanding and accepting the message, then determine the causes of the problems. Finally, based on the macro revision strategies discussed in this chapter, consider how the writer might solve the reader's problems so she can achieve her purpose.

The director is satisfied that her explanations counter the claims Mr. Laszlo puts forth in his letter. However, the writer isn't sure she is communicating her message to her reader.

Phase I: Macro revision. In rereading her document to evaluate its effectiveness, the director concludes that, while she responds to each of Mr. Laszlo's objections in the same order in which he sequenced them in his letter to her, her

Figure 10-8
Director's Letter

Paragraph 1.

Paragraph 2.

Paragraph 3.

Paragraph 4.

Paragraph 5.

Paragraph 6.

Paragraph 7.

Paragraph 8.

Closing.

Dear Mr. Laszlo:

The monies and projects of the Environmental Management Program (EMP) as devised under the 1986 Water Resources Development Act can only be used in areas already being managed for fish and wildlife. Therefore, the suggestion that EMP funds be spent on acquisition of wetlands, greenbelts, and floodplains as well as siltation control, while laudable, cannot be accomplished through EMP.

EMP was not intended to completely address all impacts to the riverine system, but to identify some promising techniques to deal with impacts to fish and wildlife and to initiate a long term monitoring and analysis system that will contribute to our understanding of large flood plain rivers. This should help state and federal agencies better manage the uses of the river and its resources in the future.

Since the EMP must address impacts to fish and wildlife habitats, it was decided to focus efforts at Woodford County Conservation Area where sedimentation has destroyed much of the fish and wildlife habitat in Peoria Lake and the East River side channel. Shallow depths, along with wind and navigation generated wave action, promote resuspension of bottom sediment causing turbid water and a soft lake bottom where aquatic vegetation long absent from the lake cannot establish and survive. The silt plug in the East River has eliminated the flowing side channel and has increased the sedimentation rate to the waterfowl refuge area. All of these factors have collectively reduced fish and waterfowl populations.

This project will construct an extension of the peninsula to provide a protected, wave-free zone to promote re-establishment of aquatic plants now virtually non-existent in Peoria Lake. Establishment of aquatic plants will provide both a food source and resting area for fish and migratory waterfowl.

The removal of the silt plug from the East River side channel will re-establish a flowing side channel, one of the most productive aquatic habitats for fishery resources. This type of habitat has been the most impacted in the Illinois River by sedimentation. The placement of two rock substrate beds within the upper reach of the side channel for establishment of a mussel bed will address another habitat type that has been greatly impacted by sedimentation throughout the river system. Hydraulic considerations and rock gradation design features were studied to provide the best maintenance-free habitat and promote the colonization of mussels and other aquatic species.

These project objectives are supported by the results of intensive hydrologic research. We anticipate substantially improved management capabilities for fish and wildlife at Woodford County. In addition, we may learn techniques to help control sediment in Peoria Lake.

All levels of management of the EMP are concerned with the relatively high costs of these projects. Possible means of reducing these costs are being explored and hopefully future projects will reflect this effort. However, we do not believe that Illinois should stop participation in the EMP program and forfeit available federal funds to neighboring states.

If you have any further questions concerning this project, please contact Bill Donels, EMP Program Manager (217/782-3715).

Sincerely,

B. N.

Executive Director

responses seem haphazard; she does not appear to follow a logical line of reasoning in her discussion. She decides to reorganize the information, using a problem/solution organizational pattern and sequencing the categories within these two main sections from most to least important. By rearranging the information in this order, she emphasizes the positive aspects of the project. This sequence also subordinates the information on the project cost, to which the reader has objected. As the director makes these revisions, she realizes that she can delete paragraph 2; the reader already knows this information.

As you read the revised letter, shown in Figure 10-9, determine if the macro revisions the writer has made help the reader follow a logical line of reasoning understand and accept the message.

Phase II: Micro revision. After reorganizing the global aspects of the text, the writer is ready to look at the individual categories. Rereading each paragraph, she revises the first paragraph before going to the next. (See Figure 10-10.)

She *reorganizes* the last sentence in paragraph 1 by moving it to the beginning of the paragraph, and writes it to serve as the organizing idea for the problem category. Then she *deletes* the first and the second-to-last sentence, since the organizing idea now includes the same information. As she does so, she discovers that she needs to *reorganize* two categories. The last sentence in paragraph 2 and the first sentence in paragraph 3 appear related to each other as a category for research, so she creates a new category. (See the paragraph in parentheses in the following draft.)

She also *expands* information (see phrases in boldface type in the following draft) to provide the reader with additional background. After deciding she should be more specific in responding to her reader's charge that the project provides only short term relief, she *expands* paragraph 4. She is afraid that her reader might not understand that her reference to "learning techniques to control sedimentation" is directly related to providing a long-term solution.

Finally, she changes the emphasis in paragraph 5 by combining the two sentences into a single, complex sentence. By making the first sentence the independent clause, she *emphasizes* that funds cannot be used as the reader suggests, and by making the second sentence a dependent clause, she *subordinates* her reader's suggestion.

With her information organized to her satisfaction, she then reviews her message to determine whether she has clearly indicated the relationships among the ideas so the reader can follow her line of reasoning. She adds two transitions—"in addition" and "as a result" (see phrases in color)—to provide coherence between ideas.

She also examines the text to determine if she has caught her reader's attention. She decides to add the phrase "As we are both aware" (see type in color) to the beginning of the letter to create immediacy and indicate shared goals.

She then reviews her message for readability, checking sentence structure and language, and revises several sentences in various paragraphs (see bracketed areas for changes).

**Figure 10-9
Revision of the
Director's Letter**

Paragraph 1.
Problem section.
Main organizing
idea.

Paragraph 2.
Solution section.
Main organizing
idea.

Paragraph 3.
Ecological objec-
tives.

Paragraph 4.
Other objectives.

Paragraph 5.
Alternative use of
funds.

Paragraph 6.
Costs.

Dear Mr. Laszlo:

Since the EMP must address impacts to fish and wildlife habitats, it was decided to focus efforts at Woodford County Conservation Area where sedimentation has destroyed much of the fish and wildlife habitat in Peoria Lake and the East River side channel. Shallow depths, along with wind and navigation generated wave action, promote resuspension of bottom sediment causing turbid water and a soft lake bottom where aquatic vegetation long absent from the lake cannot establish and survive. The silt plug in the East River has eliminated the flowing side channel and has increased the sedimentation rate to the waterfowl refuge area. All of these factors have collectively reduced fish and waterfowl populations.

This project will construct an extension of the peninsula near the eastern shore of the river to provide a protected, wave-free zone to promote re-establishment of aquatic plants now virtually non-existent in Peoria Lake. Establishment of aquatic plants will provide both a food source and resting area for fish and migratory waterfowl.

The removal of the silt plug from the East River side channel will re-establish a flowing side channel, one of the most productive aquatic habitats for fishery resources. This type of habitat has been the most impacted in the Illinois River by sedimentation. The placement of two rock substrate beds within the upper reach of the side channel for establishment of a mussel bed will address another habitat type that has been greatly impacted by sedimentation throughout the river system. Hydraulic considerations and rock gradation design features were studied to provide the best maintenance-free habitat and promote the colonization of mussels and other aquatic species.

These project objectives are supported by the results of intensive hydrologic research. We anticipate substantially improved management capabilities for fish and wildlife at Woodford County. In addition, we may learn techniques to help control sediment in Peoria Lake.

The monies and projects of the Environmental Management Program (EMP) as devised under the 1986 Water Resources Development Act can only be used in areas already being managed for fish and wildlife. Therefore, the suggestion that EMP funds be spent on acquisition of wetlands, greenbelts, and floodplains as well as siltation control, while laudable, cannot be accomplished through EMP.

All levels of management of the EMP are concerned with the relatively high costs of these projects. Possible means of reducing these costs are being explored and hopefully future projects will reflect this effort. However, we do not believe that Illinois should stop participation in the EMP program and forfeit available federal funds to neighboring states...

Phase III: Proofreading. Finally, the writer proofreads for grammatical, usage, spelling, punctuation, and typographical errors.

The letter in Figure 10-10 is the final draft.

The director has engaged fully in the revision process. Figure 10-11 indicates the various strategies she used in each of these phases as she revised.

<table>
<tr><td>

Paragraph 1.
Problem.
Indicates shared goals.
Main organizing idea.
Claim 1. Evidence.
Paragraph 2.
Solution.
Organizing idea.
Claim 2, Claim 3.
Paragraph 3.
Claim 4, Claim 5.

Paragraph 4.
Evidence.

Paragraph 5.
Claim.

Paragraph 6. Appeals to
reader's self-esteem.
Claim based on policy.

Paragraph 7. Indicates
shared concerns.
Appeals to reader's values.

Paragraph 8.
Complimentary closing.

Signature.
Title.

</td><td>

Dear Mr. Laszlo:

As we are both aware, fish and waterfowl populations have been reduced in the Illinois River because of sedimentation. [Shallow depths, along with wind and navigation-generated wave action, have caused turbid water and a soft lake bottom which prevents aquatic vegetation from becoming established and surviving. **In addition**, it causes plugs in side channels which wipe out refuge areas for migrating birds.]

[By extending the peninsula **in the Goose Lake area** as the Army Corps of Engineers' report proposes,] we can provide a protected, wave-free zone in which aquatic plants, virtually non-existent now, [could be re-established. Their recovery will in turn provide a food source for fish and a resting place for migratory waterfowl.]

As the report also proposes, the removal of the silt plug from the East River side channel, which flows **between the peninsula and the shore**, [will provide habitats for channel catfish. Placing two rock substrate beds within the upper reach of this channel will also permit the establishment of mussel beds.]

(These project objectives are supported by the results of intensive hydrologic research. Hydraulic considerations and rock gradation design features have been studied to provide maintenance-free habitats and promote the colonization of mussels and other aquatic species.)

As a result of implementing these projects, we anticipate substantially improved management capabilities for fish and wildlife. While silt may gradually fill in over the next half century if nothing is done to control the situation, it is expected that during this period, local resources can maintain the reduced siltation level by using techniques for controlling sediment acquired through this project.

[While your suggestion that Environmental Management Program (EMP) funds be spent on acquisition of wetlands, greenbelts and floodplains is laudable, funds from EMP **as required** under the 1986 Water Resources Development Act can only be used in areas already being managed for fish and wildlife and, therefore, can not be used for acquisitions.]

[All EMP management levels] are concerned with the relatively high costs of these projects. Possible means of reducing these costs are being explored and hopefully future projects will reflect this effort. However, we do not believe that Illinois should stop participation in the EMP program and forfeit available federal funds to neighboring states.

If you have any further questions concerning this project, please contact the EMP Program Manager (217/782-3715).

Sincerely,

B. N.

Executive Director

</td></tr>
</table>

Figure 10-10 Final Revision of Director's Letter

* The color type, parentheses, and brackets have been placed in the letter to facilitate your seeing the changes made by the writer. Expanded information is indicated in color type, new categories are placed in parentheses, and sections in which micro revisions have been made are bracketed.

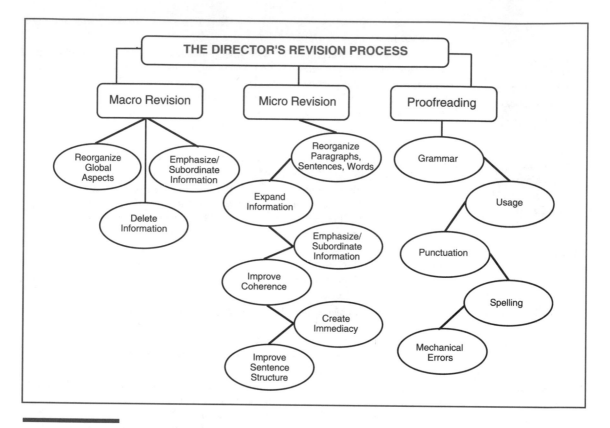

Figure 10-11 Director's Revision Process

STRATEGY
CHECKLIST

1. Determine (a) readers' problems, (b) causes of readers' problems, (c) strategies for solving readers' problems.

2. Engage in macro revision before engaging in other revision phases.

3. Revise your content before you revise your sentence structure and language.

4. Proofread your text after you have completed all other revision activities.

REVISING WITH COMPUTERS

Computers have facilitated revision by 1000 percent. Word processing programs make expanding, deleting, and reorganizing information quick and easy to accomplish. A simple keystroke allows you to add or delete text. Moving text around is a simple matter of highlighting the section to be moved, indicating the new location, and striking the return key. The "Find and Replace" function allows you to change words, spellings, and punctuation throughout a text at the press of a key.

Spell checkers and grammar checkers provide extra proofreading assistance. However, you need to use these checkers with caution. Some spell checkers only indicate a word is misspelled, leaving the correct spelling up to you. Others offer substitutions, but if you don't know the correct spelling, you might select an inappropriate alternative. In addition, spell checkers don't catch homonyms, such as "there" and "their." Nor do they catch typographical errors if the typo creates an actual word. For example, if you type "you" instead of "your," the spell checker won't catch the error because "you" is an actual word.

Grammar checkers are usually too simplistic, and often reflect either the programmer's biases or the most conservative grammar books. For example, several indicate that contractions are inappropriate. For most technical documents, they are perfectly all right.

CHAPTER SUMMARY

Revision is a recursive process. It can occur during planning and drafting, as well as between drafts. Revision that occurs during planning and drafting is internal revision, and can only be done by the writer. Much between-draft revision is external revision, and is concerned with communicating the message to a reader.

The revision process involves five steps: (1) reading the document; (2) evaluating the document in terms of the reader, the purpose, and the situation; (3) determining the cause of problems indicated by the evaluation; (4) developing strategies to solve the problems; and (5) redrafting the text.

There are five methods for evaluating texts: (1) self-evaluation, (2) peer evaluation, (3) checklists, (4) protocols, and (5) usability tests. Because of the knowledge effect, it is always helpful to have someone else evaluate your document. Checklists, protocols, and usability tests provide the most objective and reader-oriented forms of assessment.

Writers use the following revision strategies to solve readers' problems:

- expand information
- delete information
- emphasize/subordinate information
- reorganize information
- improve coherence
- create immediacy
- improve sentence structure
- provide visual text
- improve vocabulary
- correct grammar, usage, punctuation, spelling, and mechanical errors

Because revision entails so many aspects of composing, writers should engage in the process in three phases: macro revision, micro revision, and finally, proofreading.

PROJECTS

Collaborative Projects: Short Term

As you work on the following projects, you may want to use the Audience Analysis Chart (page 97), the Evaluation Checklist (Figure 10-2, page 264-265), and/or the Document Design Criteria Checklist in Appendix B.

1. **Revising your papers**

    Work in pairs.

    a. Select a paper you are writing for a class this term. If you do not have one, use a paper from a previous term.

    b. Swap papers. Do a reader response protocol on each other's paper. Then give them back.

    c. Review the reader's comments and determine the reader's problems.

    d. Based on the reader's problems, develop strategies for overcoming the problems.

    e. Revise the document.

    f. Swap papers again. Do another protocol, then return the papers.

    g. Review the reader's comments to determine if you solved the problems. If not, revise again.

2. **Revising for audience and purpose: insects at the super collider[1]**

    You are a member of the public information staff at the newly constructed Superconducting Super Collider in Texas. Over the next few months, a large number of people will be moving into the area from all over the country to work at the facility. One of the problems that these new residents will face is that of insect infestation in their homes and lawns. Many of the residents will be from other areas of the country, and will not be familiar with some of the bugs that will bother them. Several members of the staff, including yourself, have been selected to provide residents with the information they will need to deal with these pests.

    The information about the insects is included in the Environmental Impact Statement (EIS) on the Super Collider which appears below.

3.8.2.1 Insecticides

Insecticides may be needed to control fire ants and typical household pests such as cockroaches. Chemical insecticides labeled for use against fire ants include amidinohydrazone (Amdro, Combat, and Maxforce), fenoxycarb (Logic), avermectin-B1 (Affirm), acephate (Orthene), chlorpyrifos (Dursban), diazinon (Spectracide and AG500), isofenphos (Amaze and Oftanol), and pyrethrins (Accudose). Depending on the product and/or formulation, these insecticides are variously applied as baits, mound drenches, granules, dusts, injections, broadcasts, or liquid fumigants.

[1]If the super collider is ever re-instated, this is a scenario which might take place.

If high colony densities are causing a problem, baits are generally used. Baits are about 90% effective and can reduce population levels to a level that is tolerable and manageable by follow-up techniques (e.g., individual mound treatments) (Texas Department of Agriculture undated). Bait formulations typically consist of about 1% toxicant and are applied at a rate of 1.0-1.5 lb/acre. Control lasts up to one year if the bait is applied in the spring. Because the baits are sensitive to hydrolysis and the active ingredients are photosensitive, the toxicants do not persist in the environment. Of necessity, baits are slow-acting so that the poison can spread throughout the colony.

Aerial application would not be appropriate, as access sites are too small and the entire campus sites would not require treatment. Rather, only areas where people are present and the ants are causing a problem would need treatment. The "no control" option should be used in most areas.

If infestation is not severe, only individual mound treatment should be conducted. Although more labor intensive, this technique has the added advantage of not eliminating competing ant species that slow the spread of fire ants (Drees and Vinson 1988).

If applied correctly, boiling water is about 60% effective in controlling individual mounds. No effective biological agents have yet been found for fire ant control.

Regardless of the method used, the colony will not be eradicated if the queen survives. However, the number of workers may be temporarily decreased to tolerable levels. In the event that ants must be controlled, insecticides will probably need to be reapplied annually or semiannually.

Baits and chemicals used for fire ant control are commercially available to homeowners, and an applicator's license is not required to purchase the material.

However, commercial applicators must be licensed. Under 40 CFR 171.9, any federal employee using or supervising the use of restricted-use pesticides must be certified under the government agency plan for determining and attesting to their competency. Furthermore, federal employees must fulfill any additional requirements enumerated by the states under state plans in 40 CFR 171.7. In Texas, licensing and testing are administered by the District Offices of the state's Department of Agriculture.

A wide range of insecticides are used by licensed exterminators in the region for control of cockroaches and other household pests. Local exterminators indicate that typical insecticides used for cockroach control inside buildings and residences include Baygon, Dursban (chlorpyrifos), diazinon, DEMON (cypermethrin), boric acid, and pyrethrin. Most of the chemicals (active ingredients) are available in several formulations. The exact chemical and formulation used depends on the applicator and the nature of the area being treated. Similar insecticides are used to control cockroaches and other household pests throughout the United States.

The large number of insect pests attacking cotton and other agricultural crops leads to frequent applications of numerous insecticides from all classes, but principally organophosphates, synthetic pyrethroids, and carbamates. Applicators are licensed to ensure safe handling and use of insecticides and are required to adhere to label instructions. Texas Pesticide Regulations are contained in 4 Texas Administration Code—Chapter 7. Texas Pesticide Laws are given in Chapter 76.

3.8.2.2 Red Imported Fire Ants

The red imported fire ant is considered a major pest species throughout the Southeast. It has gained notoriety as an agricultural pest and for causing power outages, inflicting painful stings, and, more rarely, eliciting allergic responses. More importantly, it has gained notoriety because of the long-term "war" that has been waged to eradicate or control the fire ant since its introduction.

When their mound is disturbed, fire ants will attack and sting whatever is causing the disturbance. Alkaloids in the fire ant venom cause a burning sensation. Following the sting, a vesicle appears at the site of venom entry. The vesicle develops into a persistent pustule. If the pustule is broken, there is a risk of infection.

Reports variably state that 1-10% of the population is allergic. There is large variation in the severity of the allergic response. More serious complications can occur in individuals who receive numerous stings or in people who are highly allergic to the protein in the venom. Systemic reactions can include nausea, vomiting, dizziness, perspiration, and (in severe cases) anaphylactic shock. Since introduction of the red imported fire ant to Texas in 1957, only 19 (14 of these documented) anaphylactic deaths have been reported. Such deaths can be considered a rare event considering that more than 30% of people living in infested areas are probably stung each year (Rhoades et al. 1989).

    a. Work in trios.

    b. You will need to revise the text in the EIS for your readers. Collaborate on filling out an Audience Analysis Chart to determine the information your readers need.

    c. Use collaborative debating to decide how to present the information. You don't want to scare the readers so they won't move to the area, but you do want them to take the proper precautions.

    d. Write the handout.

Option: You may want to create a brochure, rather than a handout. If you do, you may want to include graphics. (See Chapters 12 and 13). Your whole team may work together on each aspect of the project, or each member may work on a separate aspect: text, graphics, layout.

3. **Revising for Usability**

    a. Work in groups of five. Each person should select one of the following tasks:

- Write instructions for holding a pencil.
- Write instructions for holding a cup.
- Write instructions for tying a shoelace.
- Write instructions for buttoning a button.
- Write instructions for zipping a zipper.

- Write instructions for putting on a sock.
- Write instructions for putting on a shoe.

(See Chapter 16 for writing instructions.)

b. Swap your instructions with someone in your group.

c. Try to follow the instructions. Do exactly what the instructions say. Don't second-guess the writer. Write down any problems you have next to the instruction with which you are having difficulty.

d. Return the instructions to the writer. Explain any difficulty you may have had.

e. Revise your instructions.

f. Swap your instructions with someone who has not yet tested them.

g. Repeat steps *c* through *f* until a user can carry out the instructions successfully.

Collaborative Projects: Long Term

The following projects can be continued through Chapter 13. As you complete each chapter, you will find activities for these projects that are directly related to the chapter you are reading. In this chapter, you'll engage in activities related to revising.

- **Communicating about a campus/community problem** continued from Chapter 4, Collaborative Projects: Long Term Exercise 1.
- **Proposing microcomputer laboratories** continued from Chapter 4, Collaborative Projects: Long Term Exercise 2, Projects 1, 2, or 3.
- **Gathering information for a feasibility report: diaper laundry service** continued from Chapter 6, Collaborative Projects: Short Term Exercise 2.
- **Gathering data for an information report: cleaning supplies** continued from Chapter 6, Collaborative Projects: Short Term Exercise 3.
- **Organizing data for a proposal: recycled paper** continued from Chapter 7, Collaborative Projects: Short Term Exercise 2.
- **Organizing data from a food and music survey** continued from Chapter 7, Collaborative Projects: Short Term Exercise 3.

You are now ready to revise your text.

a. Evaluate your document, using the checklist in Figure 10-2, pages 264-265.

b. Develop strategies to revise the document and solve problems indicated by the evaluation.

c. Revise the text by going through each of the four revision phases.

To be continued, Chapter 11, Readability, Collaborative Projects: Long Term.

Individual Projects

1. Engineers or technicians are sometimes asked to revise a report written by someone else. The following paragraph serves as an introduction to the "Problem" section of an evaluation report on the Lick Creek Watershed. The report is read not only by agriculture and conservation specialists, but also by county board members who have little knowledge of the field of soil and water conservation. In addition, the report may be read by news media reporters, and conservation groups such as the Sierra Club.

    a. Do a protocol of the paragraph.
    b. List the readers' problems.
    c. Determine the causes of readers' problems.
    d. Develop revision strategies to solve those problems.
    e. Revise the paragraph.

    PROBLEMS:

    Resource problems addressed during evaluation were those which have, or will in the future, result in the degradation of the soil and water resources. It was found that the majority of the problems were directly associated with erosion on cropland (e.g., productivity loss, soil degradation, and sedimentation). All alternative plan development was directed toward the cropland resource and the problems associated with erosion on these areas. Other resources such as woodland, grassland, and other land were not found to be major contributors to the problems, thus, they were addressed no further in the plan development. Opportunities for enhancement, especially to wildlife habitats and water quality, exist and will occur as a result of erosion control practices on cropland.

    Illinois Soil Conservation Service

2. Read the paragraph above. In the last sentence, the writer uses the term "opportunities for enhancement." What does this mean? Can you think of any strategies you have learned that would cause the writer to use this term? If you were to revise this paragraph, would you continue to use this term or would you change it? Explain the reason for your decision in terms of strategies that you have learned.

3. Writers often need to write about the same topic for different audiences, and in different genres. After the executive director of the Heartland Water Resources Council sent the letter on page 279, she received a response from Joseph Laszlo asking if she would like to publish her rebuttal on *Heartland Free-Net* (HFN), a free community electronic bulletin board serving central Illinois communities.

    Laszlo enclosed a copy of the note (Figure 10-12) stating his position, which he plans to place on the bulletin board.

    The director decided she would like to publish her rebuttal. However, she felt the format of her letter was inappropriate for the bulletin board. If you

**Figure 10-12
E-Mail Announce-
ment**

PEORIA LAKE ENHANCEMENT PROJECT (ISLAND PROPOSAL)

PROBLEM:

Peoria lake is filling up with silt, reducing plant and animal life in the lake and on its shores. The large among of silt suspended in the water prevents vegetation from growing (i.e., no light—no photosynthesis—no plant life—no animal life).

SOLUTION:

Increase biodiversity (widen range of plant and animal life) and improve habitat by:
1) Creating a 1.1 mile long peninsula in the Goose Lake area to serve as a wind and wave barrier, so that suspended silt settles out.
2) Establishing a flowing channel between the peninsula and shore to provide an exposed rock bed for fish, plant, etc.
3) Establishing a forested wetland management area, 168 acres, as part of the Woodford Conservation Area to enhance the habitat/refuge area for migrating birds.

COST:

$4.1 million plus approx. $20,000 annual maintenance cost

PROS:

1) Improved habitat for fish, bird, beast, etc.
2) Brings Federal dollars into area ($3.1 million).
3) Better duck hunting.

CONS:

1) Probably won't work.
• The channel will fill with silt over time and things will be back to the way they are now.
• Keeping the channel open will require intermittent dredging—which will destroy the established wildlife anyway.
• There is little evidence that the island barriers already in the Illinois River even come close to being as effective as the Corps proposal suggests.

Joseph Laszlo, Ph.D.

were to revise the letter for HFN, how would you do it? Refer to Laszlo's bulletin board note in Figure 10-12 as a model for the genre, and revise the director's letter for electronic publication.

4. The entire protocol, which begins on page 266, about a chemical technician's oversight in carrying out a sampling process was not printed. The remainder of the report is shown in Figure 10-13. Study it to determine the problems the reader has with this portion of the report.

    a. Make a list of the reader's problems.

    b. Next to each item in the list, write the revision strategy that would solve this problem.

5. Now that several weeks have elapsed since the beginning of this semester, reread the first paper you wrote for this course and, using the checklist in Figure 10-2, pages 264-265, revise it.

Text	Response
7. The chemistry technician drew approximately a 250 ml sample from the jacket cooling water heat exchanger outlet.	He drew a sample from the outlet. I guess he was just rinsing the sample bottle before.
8. In addition, a 250 ml sample of jacket cooling water taken at the pump suction was placed into the same bottle with the first sample.	He added another sample to the first sample.
9. The procedure called for a 250 ml sample from each sample point.	According to the procedure he was supposed to take a sample from each point...let's see that would be the suction and the...let me go back and see what the other point was...Okay. The heater exchange outlet.
10. This process was followed on both Unit 1 diesel generators.	He apparently sampled water in each generator in Unit 1. But why would he mix both samples together?
11. The contents of the 5-gallon bucket used to catch the spillage and rinse liquid was poured into the diesel generator room floor drain sumps.	He poured liquid into a floor drain. When did he use a bucket? I need to go back and look for it...Okay. He dumped the water he used to rinse the sampling bottle in it.
12. The chemistry technician stated the jacket water contained DALCO 1355, a corrosion inhibitor.	Apparently the water from the jacket contained a corrosion inhibitor in it. But that shouldn't matter since it was the rinsing liquid and not the jacket water which was in the bucket. Or did he use the jacket water?
13. Dumping this liquid into the floor drain system rapidly exhausts water processing resins because of its organic compounds.	Pouring the liquid down the drain apparently uses up resins for processing water.

CONCLUSIONS

14. Plant labeling practices necessary to facilitate identification of systems and components sometimes are not utilized.	Apparently labels aren't used. What does that refer to?
15. Good work practices necessary to ensure reliable and accurate chemistry data sometimes are not utilized.	Good work practices aren't used. What exactly does he mean? What kind of good practices?
16. Proper disposal of organic compounds may affect the water processing resins due to organic fouling.	The water processing resins could be affected by proper disposal...no improper disposal...of organic compounds. I guess that refers back to pouring the stuff down the drain.

Figure 10-13 Reader's Protocol

CHAPTER 11

Readability

289

WHEN READERS READ a text straight through and understand the writer's message easily, the text is *readable*. It is very hard to predict how readable a document will be, because readability depends on who is doing the reading. Readers who already know a lot about the topic being discussed have less difficulty reading a text than readers who are unfamiliar with it. If you know a lot about a topic, you can usually understand relatively complex texts, but if the text is complex and you are unfamiliar with the content, you may find yourself reading and rereading it, trying to understand the message.

When writers and editors try to determine how readable a text is, they are concerned with three major aspects: how well the parts of a text hold together (its coherence/cohesion); how the sentences are structured; and the kind of language used. Other aspects that affect readability include whether numerical information is interpreted, and how the text deals with such mechanical matters as grammar, usage, spelling, and punctuation.

In this chapter you'll learn how to create a coherent and cohesive document so that your reader can follow your discussion easily. In addition, you'll learn sentence structures that facilitate readers' fluency. You'll also learn to use language and to write about numerical concepts so your readers can easily comprehend your message. Finally, you'll learn how grammar, usage, punctuation, spelling, and mechanics add to a document's readability, and how to adapt these to different discourse communities.

COHERENCE/COHESION

The terms "coherence" and "cohesion" are derived from the Latin root word "cohere," meaning "to hold together as a mass of parts." Coherence relates to a message as a whole unit. All the techniques discussed in this chapter—coherence, sentences, language, and grammar—cohere. They are all related to the single concept of readability.

Cohesion refers to textual cues, such as forecasts, which serve as the glue to hold ideas together and help readers read a document fluently. Textual cues are often indicated by markers, such as subheads, which denote relationships between ideas in a text. These markers help readers follow ideas and understand a writer's logic.

Cohesive ties create a unified document in which all the parts are neatly tied together. As you read the forecast for this chapter, you predict the major parts of the chapter. Then, by relating the subheads in the chapter back to the forecast, you recognize you are reading the information you predicted.

Types of Coherence

There are three levels of coherence: global, interparagraph, and intraparagraph.

- **Global coherence** (between chapters and sections of a document). Each chapter or section relates to the table of contents and overview of a document, as well as to the chapter or section preceding and following it.

- **Interparagraph coherence** (between paragraphs). Each paragraph within a chapter or section relates to the main organizing idea, as well as to the paragraph preceding and following it.

- **Intraparagraph coherence** (within paragraphs). Each sentence within a paragraph relates to the organizing idea of that paragraph, as well as to the previous and succeeding sentence.

For the reader to perceive a single, unified message, you need to provide coherence at all three levels. The model in Figure 11-1 depicts the way in which the parts of a text are tied together so that it becomes a single coherent document. We'll discuss coherence at the global level in Chapter 20. In this chapter, we'll discuss intra- and interparagraph coherence.

Let's consider the excerpt from the Kemeny Report on the Three Mile Island nuclear accident to see how a writer achieves a coherent document. As you read the excerpt in Figure 11-2, notice how the writer ties all of the ideas in the text together.

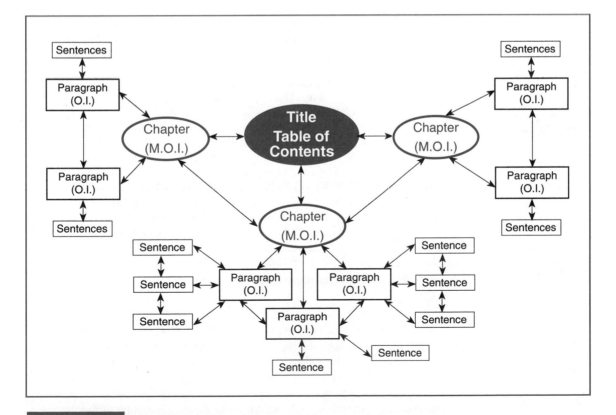

Figure 11-1 Model of a Coherent Document

Paragraph 1.
Main organizing
idea (M.O.I.).
Introduction.
Paragraph 2.
Organizing idea.
Events leading to
shutdown of safety
system.
Paragraph 3.
 Organizing idea
(O.I.). Events lead-
ing to reactor
scram.

Paragraph 4.
Organizing idea.
Feedwater valve
problem.

Paragraph 5.
Organizing idea.
LOCA occurs.

A Loss of Coolant Accident (LOCA) occurred when a pilot-operated relief valve (PORV) was stuck open and valves on two emergency feedwater lines remained closed during a scram on the morning of March 1, 1986.*

The accident began when a series of feedwater system pumps supplying water to X utility's steam generators tripped.* The nuclear plant was operating at 97 percent power at the time. The first pump trip occurred at 35 seconds after 4:00 a.m. With no feedwater being added, the plant's safety system automatically shut down the steam turbine and the electric generator.

When the feedwater flow stopped, the temperature of the reactor coolant increased. The rapidly heating water expanded, and pressure inside the pressurizer built to 2,255 pounds per square inch, 100 psi more than normal. As designed the PORV atop the pressurizer then opened and steam and water began flowing out of the reactor coolant system through a drain pipe to a tank on the floor of the containment building. Pressure continued to rise, however, and 8 seconds after the first pump tripped, X's reactor scrammed: its control rods automatically dropped down into the reactor core.

Less than a second later, the heat generated by fission was essentially zero. But the decaying radioactive materials left from the fission process continued to heat the reactor's coolant water. When the pumps that normally supply the steam generator with water shut down, three emergency feedwater pumps automatically started. Fourteen seconds into the accident, an operator in X's control room noted the emergency feed pumps were running. However, he did not notice two lights that indicated a valve was closed on each of the two emergency feedwater lines and thus no water could reach the steam generators. One light was covered by a yellow maintenance tag. It could not be determined why the second light was missed.

With the reactor scrammed and the PORV open, pressure in the reactor coolant system fell. Up to now, the reactor system was responding normally to a turbine trip. At this point the PORV should have closed 13 seconds into the accident when pressure dropped to 2,205 psi. It did not. A light on the control room panel indicated that the electric power that opened the PORV had gone off, leading the operators to assume the valve had shut off. But the PORV was stuck open, and remained open for 2 hours and 22 minutes, draining needed coolant water—a Loss of Coolant Accident (LOCA) was in progress. In the first 100 minutes of the accident, some 32,000 gallons of water—over one-third of the entire capacity of the reactor coolant system —escaped through the PORV and out the reactor's let-down system.
* Scram and trip are synonyms. When a system is scrammed or tripped, it is stopped.

Figure 11-2 A Coherent Document at the Interparagraph Level

The organizing idea in each paragraph is tied to the main organizing idea of the entire section through the repetition of words, i.e., accident, PORV, feedwater, and scram. By placing the main organizing idea, "a LOCA occurred," at the beginning of the description, the writer indicates the topic to be discussed and readers can easily predict what they will read. The writer achieves interparagraph coherence by tying each paragraph to the previous one chronologically. By using the term "began," and then delineating the progression of time with

The accident began when a series of [feedwater] system [pumps] supplying water to X (utility's) steam generators (tripped.) The (nuclear plant) was operating at 97 percent power at the time. The first [pump] (trip) occurred at 35 seconds after 4:00 a.m. With no [feedwater] being added, the (plant's) safety system automatically (shut down) the steam turbine and the electric generator.

Figure 11-3 A Coherent Text at the Intraparagraph Level

such phrases as "when the feedwater flow stopped," "less than a second later," and "up to now," the writer helps readers follow the narrative.

The ideas in each of the paragraphs are also tied together. If you examine the second paragraph in Figure 11-3, you'll see that the writer relates sentences two and three to the organizing idea by repeating the words "feedwater," "pump," and "trip," and by using the synonym "plant" for the word "utility."

By relating each organizing idea to the main organizing idea, and each sentence to the organizing idea in the paragraph in which it occurs, the writer provides the reader with topical coherence that helps the reader perceive a single, unified message.

Techniques for Achieving Coherence/Cohesion

The following techniques create coherent and cohesive documents:

- Chunking related information together.
- Using redundancy through repetition, restatement, or summarization of words, phrases, and concepts to indicate relationships between pieces of information as well as between categories of information.
- Using transitions to indicate types of relationships between ideas.
- Adhering to the given-new contract.
- Assuming omniscience.
- Locating the text in time, place, and sequence.

Chunking   Because chunking is inherent in the synthesizing method of organizing information, you should have little difficulty in achieving coherence in your document if you follow this method.

Notice that each of the paragraphs in the Kemeny excerpt contain chunks of related information.

Redundancy Redundancy includes repeating or restating key words or phrases from your forecast. These words build a bridge from one idea to another, from the main organizing idea to the organizing idea of each paragraph.

In the Kemeny excerpt, the words "accident," "feedwater," "valve," and/or "scram" from the main organizing idea are repeated in the organizing ideas of the paragraphs in the description section.

This technique may be counter to rules you have learned previously, in which you were instructed not to repeat words. In technical writing, repetition, especially between paragraphs, is an extremely useful strategy for achieving cohesion.

Repetition is also useful between sentences at the intraparagraph level. It's often better to repeat a noun than to use a pronoun, as the following example indicates.

Misleading   The motor turns the propeller. [It] is vibrating badly.

Clear   The motor turns the propeller. [The propeller] is vibrating badly.

In the first sentence, it is difficult to determine whether the pronoun "it" refers to the engine or the propeller. In the second sentence, it is clear that the vibrating part is the propeller.

Transitions Transitional words and phrases also build bridges. Transitions help readers recognize the types of relationships that exist between ideas. Following is a partial list of transitions for signaling relationships involving sequence or rank, cause/effect, possibility, and comparison/contrast.

Sequence/rank
first, second, third, etc., next, then, after, in addition, also, prior to, before

Cause/effect
consequently, therefore, because, due to

Possibility
if/then

Comparison/Contrast
however, but

In paragraphs four and five in the Kemeny report, the writer uses the transition "but" to indicate a contrast between what should have occurred and what actually did occur. In addition, in paragraph four, the writer uses the transition "thus" to indicate the effect of the closed valve.

Given-New Contract The given-new contract helps readers move fluently from one paragraph to the next, and from one sentence to the next, by indicating how the information in the new paragraph/sentence relates to the information in the previous (old) paragraph/sentence. By using redundancy and transitions, you signal readers that each paragraph relates to the previous one. The signal should appear in the first sentence of the new paragraph.

In the report, the writer relates paragraph three, the "new" paragraph, to paragraph two, the "given" paragraph. The writer indicates this relationship by summarizing the events from paragraph two in the phrase, "when the feedwater flow stopped," which appears at the beginning of paragraph three.

Omniscience Omniscience, meaning all-knowing, relates to the view from which a story is told. Thus, even though a series of events may be organized in chronological order, some events can be described out of order if it helps readers understand a situation.

In the fourth paragraph of the excerpt, the writer inserts an omniscient point of view, not only indicating the operator's failure to notice the two lights on the board, but also explaining why one light went unnoticed. This information is necessary to support the writer's later claim that the root cause of the accident was human error. By telling the reader what should have happened as well as what did happen, the writer helps the reader understand the relationship between the description section of the report and the solution section (global coherence).

By providing information that explains the cause of the problem, even though the cause was not determined until later in time, the writer helps readers understand the relationship between the problem and the solution.

Locating the Text If a text involves a situation that occurs over a period of time, or in different locations, or which involves several steps in a sequence, signal the reader that such a change is occurring.

In the excerpt, the term "began" in paragraph two not only signals readers they will be reading a chronological description, but also indicates that this is the first chunk in that description. Thus, readers are provided with a cue for situating the text in time. The use of the word "when" in the third paragraph places the paragraph in chronological sequence. The reader knows that the events being described in this paragraph follow those of the previous one. And the phrases, "less than a second later," and "up to this point," in the fourth and fifth paragraphs respectively, continue to provide readers with cues for locating the time of the text they are about to read.

Revising for Coherence/Cohesion

The writer of the description didn't plan all of the signals and ties included in the text before he began to draft it, nor did he make a conscious decision about including them as he drafted. However, once he was ready to consider whether or not he was communicating his message to his readers, he reviewed the text to make certain that signals were provided to facilitate readers' comprehension and fluency. Many of the signals were already in the text; he had included them subconsciously. Those that were not, he inserted. As noted earlier, creating a coherent and cohesive document is part of the revision process.

1. Signal readers that they are about to read the information they fore-cast they would read.
2. Indicate relationships between paragraphs and the main topic— between the organizing ideas of each paragraph and the main orga-nizing idea.
3. Indicate the relationship between each paragraph and the paragraph preceding it. Signal this relationship in the first sentence of each paragraph.
4. Indicate the relationship between the assertions in a paragraph and the organizing idea of that paragraph.
5. Indicate the relationship in each paragraph between each assertion and the one preceding it.
6. Use redundancy and restatement to indicate global, inter- and intra-paragraph relationships.
7. Use redundancy and nouns, rather than pronouns, to provide unam-biguous references to previous text.
8. Use transitional words to indicate the relationship between ideas, especially if there is a sequential, contrasting, causal, or contingent relationship.
9. Adhere to the given-new contract.
10. Provide omniscient information to fill up holes in a claim.
11. Provide cues for readers to situate themselves in the text.

SENTENCES

Until recently, the length of a sentence was considered a major criterion for determining readability. However, research now indicates that length alone is not a valid indicator of good or poor prose. A reader may have difficulty compre-hending a short sentence, but find a long one, such as the following, quite clear:

From these results it was concluded that the gas griddle was a better method of cooking ground beef patties than the convection oven, since ground beef patties cooked on the gas griddle were rated higher in beef flavor intensity and overall acceptability. (41 words)

The main reason readers have difficulty comprehending sentences is not their length, but their structure, as the following sentence from a report describ-ing a problem with a valve indicates:

During valve operability testing following an outage, the steam inlet isolation valve [(some-times referred to as the trip and throttle valve) to the turbine-driven auxiliary feedwater pump] would, [under certain test conditions,] stop or hesitate in a mid-position causing the motor to shut off on torque. (46 words)

The structure for this sentence is overly complicated. The subject of the sentence, "valve," is separated from its predicate, "would stop or hesitate" by fifteen words, and the auxiliary verb "would" is separated from the main verb of the predicate, "stop," by four words. The writer needs to present the information to readers in more readable form, as in the following revision:

Under certain conditions the steam inlet isolation [valve would stop or hesitate] in a mid-position, causing the motor to shut off on torque. The valve, sometimes referred to as the trip and throttle valve, operates the turbine-driven auxiliary feedwater pump. The problem occurred while testing the operability of the valve following an outage.

While the information remains the same, the sentence is restructured. Instead of a single sentence, there are three. While the subject, "valve," is separated from its predicate in the second sentence, the sentence is sufficiently short that the reader will have no difficulty understanding it.

Readability requires the following elements: length, structure, density, rhythm, emphasis, and variation. Figure 11-4 depicts the various elements you need to consider in writing readable sentences.

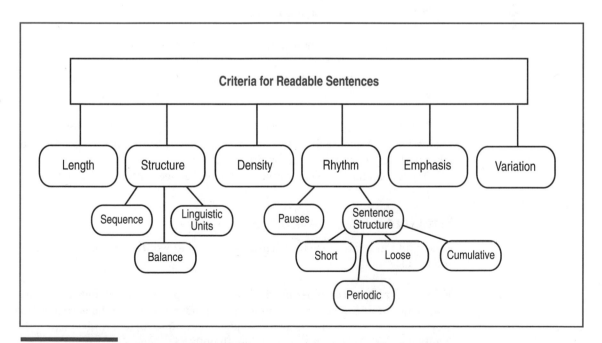

Figure 11-4 Elements of a Readable Sentence

Length

Sentences can be categorized by length as follows:

- Short (15 words or less)
- Medium (15-25 words)
- Long (over 25 words)

The average length for sentences in a technical document should fall into the medium category. The average length of the three sentences concerning the valve problem is eighteen words.

A good gauge for determining if a sentence is too long is to read it aloud. If you run out of breath before you reach the period, the sentence is too long.

Notice, we are discussing the *average* length, not the maximum. Often a sentence becomes long if it includes a list of items. Sometimes the list can be broken up, but often it can't. Nor is brevity always the "soul of wit." In technical documents you are often concerned with complex concepts. Because they may involve intricate relationships among ideas, you may have sentences that fall into the long category.

You may also wish to use a longer form of expression to achieve a more flowing syntactic rhythm. The following sentence from a follow-up letter is 28 words, but it is very clear, and the reader should be able to read it fluently.

[I would like to express] my thanks for the time, courtesy, and cooperation [you and your people extended to me] during the service technician evaluation last week.

If the writer were to shorten it, he might write it as follows:

Thank you for your time, courtesy, and cooperation during the service technician evaluation last week.

While the meaning remains the same, the sentence sounds more abrupt, giving the reader the sense that the writer is in a hurry. The original, longer version gives the reader a more gracious impression.

Try to maintain an average sentence length between fifteen and 25 words.

Structure

Readability in terms of syntactic structure relates to word order, structural units, and balance.

Word Order   The order in which words and phrases occur, rather than the complexity of a sentence, is a determinant of readability. When the normal word order of a sentence, clause, or phrase is changed or interrupted, readers have difficulty comprehending a text and reading fluently.

Just as readers predict the information and format of a text based on their knowledge for that content area and that type of document, readers predict the part of speech they will read, based on their prior grammatical knowledge. The various parts of speech in the English language follow a specific order, as do the parts of a sentence, i.e., an article precedes a noun (the ball), a predicate normally follows a subject (the child cries). A change in this sequence halts a reader's fluency. In English you predict the adjective will come before the noun (the blue ball), but in French you predict the adjective will follow the noun (la balle bleu). If you were to read an English text in which the adjective follows the noun, e.g., the ball blue, your fluency would be interrupted because the sequence is not aligned with your predictions.

Read the following sentence:

After plant shutdown, the shroud head bolts were found not to be fully tightened, allowing the shroud head to lift...

You may have paused for just a split second to re-align your grammatical predictions. After you read the predicate, "were found," you probably predicted you would read an adverbial or infinitive phrase next. Instead, you read a negative, which usually precedes rather than follows the main verb.

Structural Units Until recently, it was thought that readers understood simple sentences more easily than complex sentences. However, research indicates this is not necessarily true. Readers have no more difficulty reading complex sentences than simple sentences, and often can comprehend complex sentences better.

Simple   The termination of the program by the manager was caused by a reduction in funding.

Complex   The manager terminated the project because funding was reduced.

The complex sentence is much easier to understand, and it is shorter. It is also in the active voice, and it does not use "nominalizations" in which verbs are made into nouns by adding such suffixes as "tion," "ment," or "y," (termination, achievement, delivery).

It is not the type of sentence that makes a sentence readable or unreadable, but rather the way in which structural units are presented. Readers read words in structural units or phrases, such as noun or verb phrases (the blue ball, could have rolled quickly), prepositional phrases (under the table), and compound predicates (could be heard and seen). If these units are interrupted, readers' fluency is interrupted because they will not be reading what they predicted.

Look again at the sentence about the valve problem:

During valve operability testing following an outage, the steam inlet isolation valve [(sometimes referred to as the trip and throttle valve) to the turbine-driven auxiliary feedwater pump] would, [under certain test conditions,] stop or hesitate in a mid-position causing the motor to shut off on torque.

The complete subject is separated from the main verb by two phrases: first, a participial phrase, "sometimes referred to as the trip and throttle valve," and then a prepositional phrase, "to the turbine-driven auxiliary feedwater pump," which together total fifteen words. By the time readers get to the main verb they have forgotten what the subject is. In addition, the auxiliary verb, "would," is separated from the main compound verbs, "stop or hesitate," by a prepositional phrase.

To avoid interrupting the major constituents of a sentence, the ideas contained in the interrupting phrases need to be removed and placed elsewhere.

Under certain conditions during valve operability testing following an outage, the steam inlet isolation [valve would stop or hesitate] in a mid-position, causing the motor to shut off on torque. The valve (sometimes referred to as the trip and throttle valve) operates the turbine-driven auxiliary feedwater pump.

Balance   Finally, readers expect to read balanced sentences. You achieve balance by using parallel structure when you list words, phrases, or clauses, or compare elements.

The following sentence contains two causes for curtailing screening programs: discontinuation of funding, and dropping of data-collection efforts.

Screening programs [for lead poisoning] in many cities were curtailed in the early 1980s **[after the federal government discontinued funding for such program]** and **[after nationwide data-collection efforts were dropped.]**

By inserting the word "after" before the word "nationwide," you can recognize that the two concepts are contained in parallel clauses. If the two concepts were not parallel, you would have difficulty understanding the sentence. You can see this in the following version, in which the federal government's decision is contained in a clause, while the decrease in data collecting is placed in a phrase.

Screening programs [for lead poisoning] in many cities were curtailed in the early 1980s **[after the federal government discontinued funding for such programs]** and [dropping nationwide data-collection efforts.]

In this version the reader has difficulty comprehending the relationship between the gerund phrase ("dropping data-collection efforts") and the preceding concepts.

Density

Density means the amount of information contained in a sentence.

Read the following sentence from a report on the Upper Mississippi River System:

The improved water quality would stimulate the growth of submergent and emergent aquatic vegetation on the lee side of the island while deeper water on the windward side as a result of dredging would diversify the lake's aquatic habitat and provide valuable over-wintering areas for fish.

Even an expert would have difficulty absorbing all of the information packed into this sentence. While most people can remember a maximum of seven items, this sentence includes fourteen propositions:

- water quality needs to improve
- improved quality would stimulate growth
- growth would involve submergent vegetation
- growth would involve emergent vegetation
- growth would involve aquatic vegetation
- vegetation is on the lee side of the island
- water is deeper on the windward side
- water is deeper because of dredging
- deeper water would diversify habitats
- deeper water would provide overwintering areas
- overwintering areas are valuable
- overwintering areas are for fish
- water has quality and depth
- water quality stimulates while deeper water diversifies and provides

The sentence would be easier to read if it were split in two:

The improved water quality would stimulate the growth of submergent and emergent aquatic vegetation on the lee side of the island. In addition, water on the windward side, made deeper by dredging, would diversify the lake's aquatic habitat and provide valuable overwintering areas for fish.

Because it's easier to comprehend information that is presented in several small chunks rather than in a single, large chunk, you can facilitate readers' comprehension by reducing the number of propositions per sentence, as the writer does in the revised version of the sentence. The writer splits the sentence in two, thereby reducing the total number of propositions per sentence to seven.

Rhythm

You determine the rhythm of a text through your inner ear. If you read slowly, you "hear" the words as you read, and you recognize a certain "rhythm" as you move from phrase to phrase, clause to clause, sentence to sentence.

Some rhythms are staccato and choppy like the blows of a hammer; others flow smoothly and easily like a quiet river through a countryside; while still others flow slowly and smoothly like butterscotch sauce.

As you read the following paragraphs, try to recognize the different rhythms.

Purpose: social.
Audience: novices.
Rhythm: smooth, flowing.

1. The prevalence of lead in the environment, and the public health problem it poses, is almost entirely the result of human activity. Ice layers in Greenland, far from industrial centers, reveal a record of increasing lead use by humans, with certain phases (the industrial revolution, widespread use of leaded gasoline) clearly marked in the frozen strata. The increased exposure to lead in our society is so pronounced that the skeletons of modern humans contain 200 times more lead than those of their preindustrial ancestors.

Environmental Defense Fund. *Legacy of Lead: America's Continuing Epidemic of Childhood Lead Poisoning* 1990

Purpose: business.
Audience: experts.
Rhythm: staccato.

2. The maintenance reassembly activities were the cause of the piston/disc separation. The valve was last reassembled during the 1982 outage. The disc was not threaded completely onto the piston. A gap of approximately 0.25 inch was left between the top of the disc hub and the piston shoulder. The gap reduced the thread engagement length and eliminated the guiding between the disc and piston. The repair procedure and instruction manual required the disc/piston assembly to be "torque tight."

Source: Nuclear Regulatory Commission

Purpose: business.
Audience: novices.
Rhythm: varied.

3. All business and industry, including John Deere, depends heavily on government to establish appropriate environmental standards. The resources required to study and address global or even state-wide concerns are outside the capability of any single industry. However, business and industry supply many of the pieces of information which, together, provide the data base necessary for effective, scientifically based standards. Within its area of expertise, John Deere has been, and will continue to be active in providing information to regulators during the comment period in the regulation development process.

The audience for the report from which the first example comes is the United States Congress. The purpose of the document is to persuade members to support projects that eliminate the causes of lead poisoning, especially among children. Because of the social nature of the content, as well as the purpose, the writer's smooth, flowing style is appropriate.

In contrast to the first example, a more business-like style is appropriate for presenting the information in both the second and third examples. The second example is written for an expert audience. The purpose is to provide readers with information to determine whether a broken valve has been corrected. Because of the intricate subject, the text is written in a staccato rhythm.

The purpose of the third example is to persuade an audience of novices and generalists of the company's dedication to improving the environment. The writer wants the tone to be both serious and scientific. Because the audience is a general one, however, the rhythm is not as business-like as the second example. Instead, it falls between the other two examples.

You may want a document to flow smoothly and gracefully or you may want a more staccato, business-like rhythm, depending on your audience, purpose, and content. Regardless of which rhythm you select, technical documents should never be choppy or overflowing.

The rhythm in your prose is created by the pauses in your sentences. A string of short sentences creates a choppy rhythm, e.g., "The truck overturned. The truck had been carrying toxic chemicals. The chemicals spilled out. They oozed into the drain." An overflowing rhythm is often caused by long sentences containing many clauses and phrases: "The truck which carried a load of toxic chemicals to be recycled at a plant at the opposite side of the country overturned along the interstate which ran through the rich farmlands of the Midwest." However, long sentences that contain a series of phrases or clauses, especially participial (*ing*) phrases connected by coordinate conjunctions (*and, but, or, not*) or commas, can provide a flowing rhythm: "The truck overturned, spilling its contents and pouring its toxic chemicals along the highway, spreading a path of death."

Emphasis

Emphasis is related to the placement of a concept in a sentence. Concepts placed at the beginning or end of a sentence, paragraph, or section receive more emphasis than those placed in the middle. In addition, concepts placed in independent clauses receive more emphasis than concepts placed in dependent clauses or phrases.

The following four examples contain the same information, but each uses a different syntactic structure to emphasize different aspects of the content. The first version, the original sentence, is drawn from a presentation for generalists on degradable plastics.

Emphasizes first finding.

1. My own investigations of the behavior of LDPE in composted garbage [confirmed the early report] and added the observation that unsaturated cooking fat played a major part in the process by being selectively absorbed by the polymer and by subsequently generating peroxides by autoxidation.

Emphasizes both findings.

2. My own investigations of the behavior of LDPE in composted garbage [confirmed the early report.] In addition I observed that [unsaturated cooking fat played a major part] in the process by being selectively absorbed by the polymer and by subsequently generating peroxides by autoxidation.

| Emphasizes first finding, causes. | **3.** My own investigations of the behavior of LDPE in composted garbage [confirmed the early report] and added the observation that unsaturated cooking fat played a major part in the process. [The fat is selectively absorbed] by the polymer and subsequently generates peroxides by autoxidation. |

| Emphasizes both findings, causes. | **4.** My own investigations of the behavior of LDPE in composted garbage [confirmed the early report.] In addition I observed that [unsaturated cooking fat played a major part] in the process. [The fat is selectively absorbed] by the polymer and subsequently generates peroxides by autoxidation. |

Example 1 emphasizes the first of two findings—the confirmation of a previous report. Because the sentence is a long one, the information placed in the beginning of the sentence is emphasized, and the information in the second half of the sentence fades in the reader's memory. However, in example 2, by creating a separate sentence for each of the two findings, i.e., confirmation of the early report and the role of fat in the process, the writer emphasizes both results. In example 3, by creating one sentence to discuss the findings and another the causes, absorption of the fat and generation of peroxides, the writer emphasizes the causes as well as the first finding. Finally, in example 4, by creating a separate sentence for each finding and a sentence that discusses the causes, the writer emphasizes the causes as well as each of the findings.

Decisions concerning emphasis of information should depend on your purpose. Because the writer's purpose for the report shown above is to inform the audience about methods of obtaining degradable plastics, example 4 is probably the most appropriate, since it emphasizes the findings and the causes.

Variation

On the whole, technical documents should vary between long, short, and medium sentences; simple, compound, and complex sentences; and periodic, loose, and cumulative sentences. (See Appendix A, the Guidebook for additional information on types of sentences.) Placement of phrases and clauses should vary, with sentences occasionally beginning with a phrase or dependent clause rather than a subject or independent clause. The author of the following paragraph achieves this type of variation.

| Complex, long. (1)

Complex, medium length. (2)
Complex, medium length. (3)
Simple, medium length. (4) | (1) [At one time,] the general **consensus was** that children were exposed to lead-based paint primarily when they actually ate the flakes of the sweet-tasting product or chewed on readily accessible surfaces such as window sills.(2) [More recently, however,] **researchers have realized** that the primary exposure route starts with the transformation of lead paint into ordinary household dust. (3) **Children absorb** lead by playing in the dust that is contaminated with these fine particles of paint. (4) [Simply by behaving like children,] **they get** dust particles on their clothes and hands[,] and into their mouths. |

Environmental Defense Fund, *Legacy of Lead: America's Continuing Epidemic of Childhood Lead Poisoning* (1990).

The text flows. The sentences average 26 words, falling into the medium category. The sentence length, however, varies, with three sentences in the medium category and one in the long. Phrase and clause placement also varies. Two of the four sentences (one and four) begin with a modifying phrase, rather than a subject. In addition, the types of sentences are varied. Three of the four sentences are complex; the fourth is simple. All sentences follow a loose construction, giving the text a flowing rhythm. There are no interruptions in the natural word order, or in structural units. However, in sentence four the writer interrupts the flow of the two prepositional phrases by inserting a comma between them. The insertion of the punctuation mark causes readers to pause, thereby emphasizing the final phrase.

Revising Sentences

Unclear sentences can be revised by eliminating excess words; creating several sentences; restructuring the sentence; and rethinking the idea.

Eliminate excess words. Remove redundant words and irrelevant information if the sentence is too long.

Original	Although the event could not be duplicated, [it was most likely] caused by a failed relay in the control rod transfer logic and [by] improper step sequencing in the [control rod] transfer procedure.
Revision	Although the event could not be duplicated, the failure was probably caused by a failed relay in the control rod transfer logic and improper sequencing in the transfer procedure.

The sentence has been reduced from 33 to 29 words by eliminating the duplication of the preposition "by" and the noun "control rod." The double term "most likely" has been changed to the single term "probably," and the vague statement "it was" has been eliminated. As you can see, eliminating words does not always reduce the total wordage or density by much. However, it does tighten the sentence.

Create several sentences out of the original sentence if the sentence is too dense. Sentences can be split easily by separating clauses, phrases, or lists of items.

Original	The primary program elements focus on identifying and evaluating the deficiencies determining the proximate and root causes, and [specifying and implementing actions required to eliminate immediate and root causes of the deficiency.]

Revision   The primary program elements focus on identifying and evaluating the deficiencies determining the proximate and root causes. The program also focuses on specifying and implementing actions required to eliminate immediate
and root causes of the deficiency.

The sentence has been reduced from a single long sentence of 32 words to
two sentences, each less than twenty words. In addition, the number of propositions has been reduced from seventeen in a single sentence to nine in one sentence and twelve in another. While the number remains high, the reader, an
expert, has no difficulty comprehending, since many of the propositions are
chunked in dual sets and follow parallel construction, i.e., identifying and evaluating, proximate and root [causes], specifying and implementing, immediate and
root [causes].

Restructure the sentence.   If parts of speech, parts of a sentence, or
structural units are interrupted, not parallel, or out of normal order, rewrite the
sentence to correct the problem.

Original   After plant shutdown, the shroud head bolts were [found not to be] fully
tightened, allowing the shroud head to lift due to the increased differential
pressure at high core flow and [thereby leakage of reactor coolant] from the
core exit region inside the shroud to the vessel annulus region.

Revision   After plant shutdown, the shroud head bolts were not fully tightened, causing
an increase in differential pressure at high core flow which allowed the
shroud head to lift and the reactor coolant to leak from the core exit region
inside the shroud to the vessel annulus region.

The misplaced negative "not" is placed before the verb it modifies, "tightened," rather than the infinitive phrase, "to be." In addition, the effects of the
increased pressure, lifting of the shroud head, and leaking of the coolant, are
made parallel with "to lift" and "to leak." While this method creates a more
comprehensible text, it is still long and dense.

Rethink the idea.   Think about what you are trying to say. Is it really clear
in your own mind or do you need additional information or clarification? Try to
think of a new and better way to say what you want to say.

Original   There is an opportunity to minimize impact on management in the near
term and long term by establishing an organization and staffing plan to transition the organization from its present temporary state through the post
restart mode of unit 2 and back to the operational mode following restart
of unit 1.

Revision   Establish an organization and staffing plan to provide for the transition from a temporary to a permanent mode of operation. The plan should have a minimum impact on management.

The first sentence has been shortened to 22 words, and the reader should have no difficulty understanding the writer's message. In many cases, restating the sentence is the most effective method for reducing long, dense, and complicated or "awkward" sentences.

One of the best ways to engage in this form of syntactic revision is to work with a partner. The partner should have a pencil and paper and be prepared to take rapid notes. Explain to your partner what you are trying to say. As you talk, your partner should write down your explanation. After each explanation, check whether or not your partner has understood you. If not, try again. Keep trying until you finally explain your message satisfactorily to your partner. When you do, cross out the problem sentence and insert the explanation your partner has transcribed. You will probably find that you have replaced a single sentence with several, that you have eliminated all syntactic problems, and that your rhythm has improved.

STRATEGY CHECKLIST

1. Keep your text to an average 15 to 25 words per sentence.
2. Shorten your sentence if you run out of breath reading it aloud.
3. Do not interrupt the normal order of structural units or the parts of speech.
4. Do not interrupt structural units.
5. Vary the types of sentences (simple, complex, compound), sentence constructions (loose, periodic, cumulative), and placement of clauses and phrases.
6. Keep comparable items in a list, and coordinated grammatical units (nouns, verbs, infinitives, participles, prepositional phrases) parallel.
7. Keep elements you are comparing or contrasting parallel.
8. Try not to include more than seven propositions in a single sentence.
9. Relate the structure of a sentence to your audience, purpose, and content.
10. Emphasize aspects of your topic according to your purpose.
11. Keep your text from being choppy or from overflowing.
12. Use a staccato, business-like style when writing scientific prose.
13. Use a flowing style when writing about social concepts.
14. Revise sentences that are unclear, too dense, and too long by:
    - eliminating redundant and irrelevant words
    - reorganizing the structure
    - dividing the information into several sentences
    - rethinking the concept involved

LANGUAGE

Language involves the words you use. In this chapter we'll examine when to use technical terminology and how to avoid pompous verbiage as well as sexist language. In addition we'll examine the effects of the connotation of a word on the tone of a message.

Technical Terminology/Jargon

It has been said that one person's technical terminology is another person's jargon. When you write for readers in your discourse community, you can use technical terms that are recognized by the majority of persons in that community. However, readers outside your discourse community probably won't know most technical terms associated with your field, and will consider such terms as jargon. When writing for readers outside your discourse community, try not to use technical terminology. If you do use technical terms, define them.

Chemists working with enzymes are familiar with the term "hemicellulose" and, in fact, are expected to use the word to describe a certain type of fiber if they are writing for other specialists in their area. It is a common word for those in the chemist's discourse community. However, because the term would appear as jargon to those outside the community, a chemist might simply refer to the term as "one source of fiber in foods," if he were writing for novices.

The danger with using technical terms is that you need to be certain that all persons define the term in the same way. A document from the licensing division of a nuclear plant included the following sentence:

The original FSAR values were assumed to be nominal rather than minimum/maximum ratings.

During a workshop in technical writing, a consultant asked if the word "nominal" was a common term in the industry. She was assured it was. Yet, when she asked for a definition, the following four were proffered: "typical," "adequate," "standard," and "required." Each has a different connotation.

Not only is there the possibility that a technical term may not be uniformly defined by all members of an industry, but the term may actually only be company- or plant-specific. During the same workshop, the term "doghouse" was used. When asked to define the term, the engineers discovered their prior knowledge influenced their definition. Those who had previously worked in a boiler reactor had one definition, while those who had only worked in a pressure reactor had a different definition. The definition of the term was reactor-specific.

IBM publishes a dictionary of IBM-specific jargon with over 1400 words, including the following:

BiCapitalization—The practice of putting capital letters in the middle of words. This was done originally to refer to microcomputer software trademarks such as VisiCalc, Easy-

Writer, and FileCommand, but has since spread even to products totally unrelated to computing, and to many more than two capitals. The mainframe world, however, is still mostly devoid of BiCapitalization—In that environment the use of abbreviations is still the PMMR (Preferred Method of Reducing Readability).

Denotation/Connotation

The denotation of a word is its dictionary meaning. The connotation of a word relates to the emotional effect of the word on the reader. To an engineer, the word "catastrophic" in the phrase "catastrophic valve failure" simply means that a valve has broken. The broken valve may or may *not* result in dire consequences. However, to the general public the connotation of the term "catastrophic" implies that something has occurred which has dire consequences. While engineers who read a report in which the term "catastrophic" is used would understand the term as the writer had meant it, readers not in an engineering discourse community, such as reporters, would understand it differently. Had reporters gotten the report and published it, the public would probably have panicked, and, in fact, the entire area might have been evacuated before the public could have been convinced that no radioactivity had been released into the air.

When you are considering the impression a word may have on readers, you are often dealing with shades of meaning, controlled by a word's connotation. For example, the words "correct" and "revise" can be considered synonymous. Yet, they have two different connotations:

The procedure has been [corrected.]

The procedure has been [revised.]

The term "corrected" implies something is wrong. The term "revised" connotes something is changed. However, the change may not have been made because something was wrong, but because something needed to be updated or improved.

Connotations affect the tone of a message. They also imply additional meanings.

Tone Many of the words you use have emotional responses attached to them. The word "police" arouses good responses in those who are law-abiding and believe the police exist to protect them. However, during the Vietnam War, when protesters were beaten by police, the word "police" assumed a negative aspect.

Some words elicit immediate positive or negative responses. Words such as "mother" and "apple pie" are "purr" words; they arouse good feelings. On the other hand, words such as "dictator" and "war" are "snarl" words, eliciting negative responses.

Consider your impressions as you read the following evaluation report, which was written to the head of the division being evaluated.

Bob Smith's [failure] to recognize the valve problems as generic was caused by several factors:

(a) [Failure] of trending and/or root analysis programs to identify the significant failure rate of the valves because of procedural problems and inaccurate data. A program exists to initiate a root cause analysis if three component failures occur within 90 days. However, since the 3 previous valve failures occurred over a longer period, Smith did not initiate such a program...

(b) [Failure] of an operational critique performed in June, 1986, to identify a relationship between the 1986 failure of 2-HCV-23-31 and the 1982 failure of 3-HCV-23-37 or to recognize a relationship between 2-HCV-23- 31 and the other 1986 isolation inlet valve failure, 2-HCV-23-49.

(c) [Failure] of Smith to conduct an analysis of the valve failure. The operational critique recommended that the cause of the disk failure should be determined, but that analysis was never performed.

(d) [Failure] of the operational critique to provide appropriate division participation. The critique failed to initiate participation of the Division of Nuclear Engineering (DNE) into the identification and solution of the technical problem.

The organizing idea and all four subsections begin with the word "failure," an extremely harsh word. The writer is figuratively hitting the reader over the head with his division's "failure." Yet, most research in organizational effectiveness indicates that negative messages do not bring about change, the purpose of the report. The writer needs to rephrase the report without the constant use of the term "failure," as in this revision:

The valve problems [were not recognized] as generic because of several factors:

(a) Trending and/or root analysis programs [did not identify] the significant failure rate of the valves because of procedural problems and inaccurate data. A program exists to initiate a root cause analysis if three component failures occur within 90 days. However, since the three previous valve failures occurred over a longer period, Smith did not initiate such a program...

(b) An operational critique performed in June, 1986, [did not identify] a relationship between the 1986 failure of 2-HCV-23-31 and the 1982 failure of 3-HCV-23-37 nor did it recognize a relationship between 2-HCV-23- 31 and the other 1986 isolation inlet valve failure, 2-HCV-23-49.

(c) An analysis of the valve failure [was not conducted.] The operational critique recommended that the cause of the disk failure should be determined, but that analysis was never performed.

(d) The operational critique [did not provide] appropriate division participation. The critique did not initiate participation of the Division of Nuclear Engineering (DNE) into the identification and solution of the technical problem.

Implications Be extremely careful that you do not use words or phrases possessing connotations you do not wish to imply, especially negative connotations:

[Even if] the operators had used the D1 RHRSW pump, they could have instituted precautions by reinforcing the isolation barrier with multiple valves. However, they used the inlet isolation valve as a single isolation point, placing the system in the same configuration as that of the previous three valve failures with the inlet isolation valve, serving as a single barrier, closed and the RHRSW pump operational.

The term "even if" implies that the operators would have failed regardless of what they did. The passage leaves the reader with the impression that the operators were just plain dumb.

Simple/Excessive, Pompous Verbiage

While you may need to use polysyllabic, Latinate terms to describe specific entities and processes for readers in your discourse community, you should not use such terms throughout a document. For example, there is no reason to use the word "utilize" when the word "use" expresses the meaning just as well.

The following description was written for a patent application for Oatrim™, used in the preparation of low-fat ice cream. While the chemist uses technical terminology to describe the chemicals and processes involved, the language of the document is clear enough to be understood by a novice:

I have now discovered that water-soluble dietary fiber compositions can be recovered from milled products of oats such as oat bran and oat flour after enzymatic hydrolysis of these substrates with a-amylases. The resulting soluble dietary fiber compositions separated with the soluble hydrolyzate fractions are colorless and devoid of inherent undesirable flavors. The soluble fibers are, therefore, uniquely suitable for use in a variety of foods such as dairy products, dairy product substitutes, high-soluble fiber bakery products...

George E. Inglett, U.S. Department of Agriculture

In contrast, the following report, written for generalists involved with legislation affecting state rivers, contains technical terms to describe specific objects and processes, and a number of long, polysyllabic words.

Sedimentation of the waterways [impacts] aquatic habitats, [impedes] the operation of commercial and recreation traffic, [constricts] the [conveyance] channel, and transforms these water bodies into shallow water and wetlands. ...Various alternatives are available and can be implemented to [alleviate] the chronic sedimentation problems of this major river.

Unnecessarily long, complex, or seldom-used words and phrases have been bracketed in color. None of these words relates to specific technical or scientific aspects of the topic. Simple words can be used, as demonstrated in the following revision:

Sedimentation of the waterways [affects] aquatic habitats, [hinders] the operation of commercial and recreation traffic, [narrows] the transportation channel, and transforms these water bodies into shallow water and wetlands...Various alternatives are available and can be implemented to [help solve] the chronic sedimentation problems of this major river.

Many people mistakenly believe that "big" words impress readers. In fact, just the opposite occurs. Write to express your thoughts, not to impress your readers. You impress readers by writing so that they can easily comprehend your message, not by using unfamiliar, obscure, polysyllabic, or foreign words.

The following example illustrates how ridiculous writers can sound when they substitute long, foreign words for common ones.

DHV-43 was left in a partially open condition following its use as a drain path at 0430 on 09/18/90 due to [cognitive personnel error.]

If you're wondering what a "cognitive personnel error" is, you're not alone. It turns out to be "lack of operator knowledge." The phrase is not only excessive and pompous, but also totally incomprehensible.

However, every discourse community has its "buzz words," which the community adopts and which indicate that the writer is up-to-date on what is happening in that community. During the early 1990s, terms such as "ideology" and "deconstruction" became buzz words on college campuses as students and faculty discussed their viewpoints regarding the relationship between race, sex, gender, and the college curriculum.

Neutral/Gender and Other Biased Language

Gender-biased language, or language indicating a negative bias toward a group of people, is inappropriate under any circumstances. Further, by using such language you risk offending your readers and evoking a negative response to your message.

Sexist Language When you assign roles or characteristics to people on the basis of sex, you are using sexist language. The English language contains many words considered sexist. For example, the terms "*man*kind" and "chair*man*," which you may intend to denote both male and female genders, actually connote only the masculine gender.

Gender-specific language should only be used when you are certain of your reader's gender. When you are writing to readers whose gender you don't know, or when both men and women are your readers, your language should be gender-neutral.

Sexist language occurs in a variety of forms.

Global words. Many words used to indicate men and women in general have masculine markers, e.g. "*mankind.*" Substitute words such as "people" or "humanity" to avoid giving preference to one sex.

Compound words. Many words denoting occupational positions add the word "man," which creates a gender-specific definition, e.g., *mailman* or *congressman.* Replace the word "man" with the word "person," i.e., "mail person" or "Congressperson," or use a synonym, such as "mail carrier" or "representative," so that the word is neutral and does not favor a specific sex.

Pronoun Reference. Often writers use the masculine pronoun to refer to a generic antecedent, e.g., "The writer should avoid sexist language. *He* should be careful to refer to both sexes when referring to generic persons, such as doctors or lawyers."

Instead of the pronoun "he," use one of the following techniques to imply both sexes:

- Use the plural form of the noun, i.e., "writers," so that you can use the pronoun "they" rather than "he" or "she."

Sexist	A writer tries not antagonize [his] audience with sexist remarks.
Non-sexist	Writers try not to antagonize [their] audiences with sexist remarks.

- Rewrite the sentence in the passive voice.

Sexist	A writer antagonizes [his] audience when [he] makes a sexist remark.
Non-sexist	Audiences are antagonized by a writer [who] makes sexist remarks.

- Use the his/her form of the pronoun when nothing else works.

Sexist	A writer tries not to antagonize [his] audience with sexist allusions.

| Non-sexist | A writer tries not to antagonize [his/her] audience with sexist allusions. |
| Non-sexist | A writer tries not to antagonize [an] audience with sexist allusions. |

- Use a plural pronoun to refer to an indefinite pronoun, i.e., "everybody...they." Previously, words such as "everyone" and "anybody" took the singular form. However, to alleviate the problem of sexism, the grammar has changed and the plural form has become acceptable in all but very formal discourse:

| Sexist | Each uses the form most appropriate to [his] style. |
| Non-sexist | Each uses the form most appropriate to [their] style. |

Salutations. If the name of the person being addressed in a letter is unknown, the salutation often refers to a male recipient, e.g., "Dear Sir," "Gentlemen." To avoid offending a female addressee, use one of the following alternatives.

- Address the position rather than the person: "Dear Editor," "Dear Personnel Director."
- Refer to both sexes in the salutation: "Dear Sir or Ms."

Other Biased Terminology Just as you want to be careful not to offend either male or female readers, you also want to be careful not to offend other persons who have traits that have been considered negatively. For example, persons who cannot hear prefer to be designated as "hearing impaired" rather than deaf, and persons with AIDS do not want to be considered "victims."

| Use | Bob J., [who] has AIDS, |
| Avoid | Bob J., [a victim] of AIDS, |

STRATEGY CHECKLIST

1. Use technical terminology for readers in your discourse community. Be careful that the language is not industry- or plant-specific, and that all persons in the community define a term the same way.
2. Avoid using technical terminology for readers outside your discourse community.
3. Do not use excessive or pompous language. Don't use polysyllabic words if a one-syllable word is just as good.

continued

4. Consider the connotation of a word as well as its denotation.

5. Maintain a positive tone. Use words with a positive connotation rather than a negative one.

6. Avoid sexist and other biased language.

READABILITY FORMULAS

Over the years a number of formulas have been created to determine a document's readability. While these formulas provide guidelines, they should not be considered gospel: they are too limited. Most are concerned only with the syntactic and language elements of a text; they do not consider coherence. Furthermore, the only aspect of syntax considered is length: the formulas do not consider density, structure, rhythm, emphasis, or balance. And the only aspects of language which are considered are word length, number of syllables, and common acceptance. The formulas do not take into account the necessity for longer words because of the nature of technical terms, nor do they account for differences in readers' familiarity because of differences in discourse communities.

Two of the best known formulas are the Fog Index and the Dale/Chall Formula for Readability. Both of these are based on the syntax and language of a document.

The Fog Index[1]

The Fog Index, probably the best known of the readability formulas, indicates how much "fog" is in your writing. By calculating the number of words per sentence and syllables per word, you can determine the approximate grade level on which you are writing. As a means of comparison, the author of the index indicates that *Reader's Digest* is written at the eighth grade level, *Time* magazine at the tenth, and the *Atlantic Monthly* at the twelfth. Anything above the fourteenth grade level is probably not only too long, but also too dense.

To determine the Fog index in a document:

1. Take a sample of at least 100 words.

2. Determine the average number of words per sentence. Treat independent clauses as separate sentences.

3. Count the number of words with three or more syllables, omitting capitalized words and combinations of short words, such as "manpower."

4. Add the average number of words per sentence to the number of polysyllables.

5. Multiply the sum by 0.4

This formula is a guideline for determining the general readability of a text. If a text scores above grade fourteen, look it over to see if sentences can be

---

[1]The information in this subsection is adapted from *How to Take the Fog Out of Writing* by Robert Gunning, Chicago, IL: Dartnell Corporation, 1964.

shortened, irrelevant or redundant words eliminated, and polysyllabic words replaced by shorter ones.

The Dale/Chall Formula

Like the Fog Index, the Dale/Chall formula indicates the grade level at which a document is written, and is based on the average number of words per sentence. However, instead of the number of polysyllabic words, this formula is based on the number of words not included on the Dale list of common words, a list of approximately 3,000 words known by 80 percent of children in fourth grade. Because such words as "diseases," "immunization," and "protection" do not appear on the list, the formula is inappropriate for many technical documents. Despite this fact, the formula is used by a variety of agencies and institutions to determine readability.

STRATEGY CHECKLIST

1. Use readability formulas as guidelines in determining the readability of a document.

NUMERICAL DATA

Technical topics often involve statistical analyses, survey results, and other numerical data. Including numerical information in graphs and tables does not provide readers with sufficient information to understand a message. The relevant information should be extracted, summarized, and interpreted for the reader.

Relating Numerical Data to Purpose and Audience

The information you extract should be relevant to your purpose, and your discussion of it should relate to your readers' needs and numerical knowledge.

Figure 11-5 appears in a booklet published by the Alabama Agricultural Experimental Station for all interested persons, ranging from novices to experts. The chapter from which the information is drawn describes a research study to determine the amount (levels) of fat that consumers accept in a hamburger.

The writer interprets the results of the graph in terms of the purpose of the study. Rather than listing the results, leaving the reader with the task of interpreting them, he explains the results in terms of what is and isn't acceptable, i.e., ground beef patties at 20 percent fat level are the most acceptable.

Because the document on ground beef is written for the general reader, the statistical symbol (P<) is placed in parentheses. It isn't necessary for the novice reader to understand the symbol, but the expert reading the text receives assurance that the data is being analyzed statistically.

Consumer panelists gave the highest (P<0.05) ratings for overall acceptability to ground beef patties formulated to [20 percent] fat, followed by patties formulated to [15 percent fat.] An increase or decrease in the fat content of ground beef patties from a fat content of [20 percent] resulted in a decrease in overall acceptability of the products, figure 1.

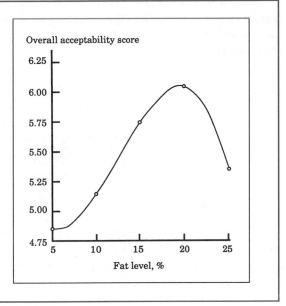

FIG. 1 Optimal overall acceptability of ground beef patties based on fat content, on a 10-0 scale (10 = like extremely, o = dislike extremely).

Alabama Department of Agriculture

Figure 11-5 An Explanation of Numerical Data

Organizing and Summarizing Numerical Data

In Figure 11-6, the writer describes the results of a study to determine which of two cooking methods creates the best-tasting hamburgers. The test is one of six used to develop low fat hamburgers for fast food restaurants. The report is published in a brochure for generalists.

. The table provides readers with the basic data if they want to examine it, but the most important aspects of it—in terms of the purpose of the report—are summarized in the first paragraph.

Numerical information is organized in the same way other kinds of information are organized—by chunking related information together. This writer perceives two major categories of related attributes:

- those measured by the panel of testers (juiciness, tenderness, beef flavor, acceptability)
- those measured by weight (cooking loss, moisture, fat, and protein)

He also recognizes that each of these categories is comprised of two subcategories:

- no difference between cooking methods
- differences between cooking methods

The categories are discussed in the same sequence in which they're later presented in the table.

Figure 11-6
A Synthesis of
Numerical Data

[No differences were found] (P.0.05) between ground beef patties when griddle-broiled compared to convection oven-broiled for the sensory attributes: juiciness (initial or sustained), tenderness, or texture, Table 6. Ground beef patties broiled using the gas griddle, however, [were rated higher] (P<0.05) by sensory panelists on beef flavor intensity and overall acceptability than those broiled using the convection oven, Table 6. [No differences (P>0.05) were found] between the two cooking methods for moisture fat, protein, Table 6. However, there was [a trend for patties cooked in the convection oven to have greater losses] from cooking than patties broiled on the gas griddle.

Table 6. Physical Compositional, and Sensory Differences Between Gas Griddle and Convection Oven Broiled Ground Beef Patties

Attributes	Griddle	Oven
Initial juiciness	5.6	4.9
Sustained juiciness	5.1	4.6
Tenderness	5.7	5.6
Texture	4.0	4.5
Beef flavor intensity	5.5	4.4
Overall acceptability	5.7	4.8
Cooking loss, pct.	26.7	29.3
Moisture, pct.	58.4	58.3
Fat, pct.	14.9	14.0
Protein	25.5	26.5

Juiciness (initial and sustained), tenderness, beef flavor intensity, and overall acceptability were rated on an 8-point hedonic scale (1 = extremely dry, extremely tough, extremely bland, extremely unacceptable, and 8 = extremely juicy, extremely tender, extremely intense, and extremely acceptable, respectively). Texture was rated on a 7-point hedonic scale (1 = more sandy, 4 = typical of ground beef, and 7 = more mushy).

Alabama Department of Agriculture

Indicating Numerical Relationships

Numbers by themselves are meaningless. For readers to understand what a number means, the number must indicate a relationship. In Figure 11-6, the numbers listed under the column for griddle-cooked patties are meaningless unless you know that they relate to a ten-point scale in which 1.0 is poor and 10.0 is good. Once you are aware of the scale, you can begin to understand the results. A score of 5.6 for juiciness indicates the patty is halfway between being extremely dry and extremely juicy. In addition, if you compare the score for convection-cooked patties with that for griddle-cooked patties, you recognize that the griddle-cooked patties are considered juicier. The writer discusses the statistical results in terms of these relationships, rather than the numerical scores.

Because readers need to understand numbers in terms of relationships, survey results are often presented as percentages, rather than absolute numbers. Readers won't perceive the significance of the fact that today's radial tires have an average life of 40,000 miles if they don't know that this is double the life of a tire fifteen years ago.

Approximating

Experts in a field may want to know the exact numerical data involved in a study, and persons putting something together or repairing something may need to know exact measurements. However, generalists and novices are not usually interested in data carried to several decimal places. Rather, they want to see numerical data expressed as simply as possible. Readers understand data fairly easily when the data is presented in terms of halves, thirds, quarters or tenths. They also grasp data more easily if it's expressed in terms of percentages that are divisible by 50, 25, 20, 10 and 5, rather than a number such as 17 percent.

When writing for generalists or novices, it is often easier to provide a general, rather than specific, numerical score so that readers can easily interpret the data. You can use one of several methods to derive approximate values:

- Round off a number to the nearest whole number, or to the nearest half, third, quarter, or tenth.

  About [three-fourths] of the waste-to-energy facilities in the U.S. are mass burned, where refuse is burned just as it is delivered to the plant, without processing or separation.

- Give a range of numbers.

  Fly ash tends to have higher concentrations of certain metals and certain organic materials and comprises approximately [10-25%] of total ash generated.

- Using the mean (average), median (the midpoint), or mode (the value occurring most, i.e., majority).

Qualifying Numerical Descriptions

The descriptions of the results of the two ground beef studies are specific. The writer refers to the exact numerical data, taking the percentages to the nearest tenth. However, numbers which are rounded off need to be preceded by a qualifier, such as "approximately" or "about."

Recycling can realistically reduce the amount of MSW (Metropolitan Solid Waste) by [approximately] 25% over time.

Figure 11-7
Interpreting
Numerical Data

GENERALIZATION OF MSW (Municipal Solid Waste) RECOVERY OF MATERIALS			
	Wt. Generated	Wt. Recovered	Percent Recovery
Paper & paperboard	71.8 million tons	18.4 million tons	25.6
Plastic	14.4 million tons	.2 million tons	1.1

Environmental Protection Agency, 1990

Qualifiers, such as "more than," "less than," "over, "under," and "nearly," can also be used when numerical data is approximated. By preceding numerical approximations with such qualifiers, writers can influence readers' perceptions of the meaning of numerical data.

Study the table in Figure 11-7 and then read the interpretations. Consider the different implications of each.

Interpretation 1   [Nearly] 26 percent of waste paper was recovered while [less than] 2 percent of plastic wastes were recovered.

Interpretation 2   [Less than] half of paper or plastic wastes were recovered.

Interpretation 3   [Approximately] 26 percent of waste paper and one percent of plastics were recovered.

The first interpretation implies that paper is a better packaging product than plastic in terms of recycling. However, the second interpretation implies that neither is very good. Finally, the third interpretation places no judgment on either product, allowing the reader to determine whether 26 and 2 percent, respectively, are good or poor ratings.

During the Vietnam War, correspondents would begin a story of the day's fatalities by stating, "Only x number of men were killed today," implying not only that the number was less than usual, but that this wasn't a very big number. Of course, if your son or brother happened to be one of those "few," it seemed very big. To avoid such problems, a casualty count was eliminated during the war with Iraq.

Whenever you interpret numerical data, be careful not to provide a false impression.

STRATEGY CHECKLIST

1. Extract relevant numerical data from tables, graphs, survey and research results. Present the data in the form of a summary. Interpret the data for the reader.
2. Determine relevant data according to your purpose.
3. Discuss the data in terms of your readers' needs and numerical knowledge.
4. Keep related data together.
5. Indicate the relationship between pieces of data. Use percentages rather than whole numbers.
6. When writing for generalists and novices, round off numbers.
7. Be careful not to imply a false interpretation in using qualifiers when rounding off numbers.

GRAMMAR, USAGE, PUNCTUATION, SPELLING, AND MECHANICS

Many organizations and industries have their own style guides. These provide such information as whether to put a comma before "and" in a series, i.e., "the man, the woman, and the dog," and what words to capitalize. If your company does not have its own guide, a large number of style guides are sold on the market. If you are writing in the behavioral sciences, use the style book of the American Psychological Association (APA). If you are in the physical sciences, use the style book issued by your particular academic society. Since each of these guides differs slightly from the others, many organizations that employ persons from a variety of disciplines use the *Chicago Manual of Style*, so that everyone is using the same guide and following the same rules.

One of the major reasons for using a style guide is to provide consistency in your writing. If you place a comma before "and" in a series, then do it throughout a text.

Appropriate Levels of Style and Usage

Each discourse community has special rules of grammar and usage for its texts, and many of the rules depend upon the community for which you're writing. In scientific texts, the introduction and conclusion of a report are written in the active voice, while the middle section is written in passive voice. To use active voice in the middle section is considered "ungrammatical." Yet, in English classes the passive voice is considered to be inappropriate. In the sciences, reviews of the literature are written in past tense, but in the humanities, reviews of the literature are written in present tense. Letters and memos may contain one-sentence paragraphs, but an article for a journal of the IEEE (Institute of Electrical and Electronic Engineers) may not.

To determine the rules to follow in your specific discourse community, refer to the guidelines published by your industry, company, or professional organization. In addition, as you are writing, read several documents in the same genre for the same audience to determine the text grammars used.

Computer Checkers

A variety of software products are on the market that provide writers with help in checking grammar, punctuation, and spelling. As with software that helps planning and revising, these should be used with caution. They provide red flags that indicate a problem may exist, but they are not programmed for your particular discourse community and audience, nor can they judge the context of a word or the rhythm of a sentence. Look at the items the programs flag, and then decide whether or not the problem exists for your particular discourse community and audience. You should also decide whether your phrasing and rhythm are appropriate for the context in which you are writing, including your discourse community, before making any changes recommended by these checkers.

Grammar checkers are not programmed for the text grammars of different discourse communities. In addition, they often contain rules that are no longer followed by modern technical writers. For example, many checkers flag contractions. However, contractions are acceptable in all texts but those categorized as formal academic discourse.

You cannot rely on spell checkers to catch all errors. For example, such errors as "cap" for "cup," or "caps" for "cap," go unnoticed since the errors are actual words. Some checkers suggest spellings for misspelled words. However, you need to know which of these is the correct spelling. You also need to be certain that the suggested words have the denotation and connotation that you intend. Your dictionary is still your best source for locating the proper spelling of a word you don't know.

Some programs also flag words that are repeated, and, if a thesaurus is available, may even suggest alternatives. Before accepting the alternative, make certain it has the same denotation and connotation. It's better to repeat a word than to use an incorrect one.

STRATEGY CHECKLIST

1. Proofread to make certain you have corrected all grammatical, usage, punctuation, spelling, and mechanical errors.

2. Use a style guide to determine correct grammar, usage, and punctuation.

3. Be consistent.

4. Correct grammatical errors by (a) rethinking your ideas, (b) checking usage, and (c) checking the text grammar of the genre and discourse community in which you are writing.

5. Use grammar checkers, spell checkers, and computerized thesauruses with caution.

CHAPTER SUMMARY

Readability is determined by coherence/cohesion, syntax, and language.

To create a coherent document, cohesive ties should be used. A document requires three types of coherence: global, interparagraph, and intraparagraph.

The readability of a sentence is determined by length, structure, density, rhythm, emphasis, and variation.

The readability of a word or phrase depends on the readers' familiarity with it. Pompous, polysyllabic words do not impress readers.

You can use a readability index that is based on sentence and word length, as well as word familiarity, to estimate the readability of a document.

Readability is also affected by the way in which you write about numbers. When you write about numerical data, summarize and interpret the data for the reader in terms of your purpose.

Your grammar, usage, punctuation, and spelling also contribute to or detract from the readability of a text. Use text grammars and technical terminology consistent with your discourse community. All grammar, usage, punctuation, and spelling errors should be corrected prior to submitting a final draft.

PROJECTS

Collaborative Projects: Short Term

As you work on the following projects, you may want to use the Audience Analysis Chart (page 97), the Evaluation Checklist (page 264-265), and/or the Document Design Criteria Checklist in Appendix B.

1. **Revising for readability**

   Work in trios.

   a. Select a paper you are writing for a class. If you are not writing a paper, use one you wrote during a previous term.

   b. Swap papers and check each other's papers for readability, including coherence/cohesion, syntax, language, numerical data.

   c. Swap papers with the member of the trio who has not read your paper, and proofread each other's papers for grammatical, usage, punctuation, spelling, and mechanical errors.

   d. Return your papers, revise for readability, and correct your errors.

2. **Using readability formulas**

   Work in pairs.

   a. Select a paper you are writing for a class. If you are not writing a paper, use one you wrote during a previous term.

   b. Swap papers and determine the readability of each other's paper using the Fog Index.

c. Return the papers and determine whether the grade indicated for your paper is appropriate for your audience. If not, revise.

3. **Writing about numbers**

Work in trios.

You are an employee of your local government. Your division is responsible for the county's landfills. Your supervisor has managed to get himself scheduled to make a presentation to the local Kiwanis Club to inform them of the problem of limited landfill space, and to persuade them to help reduce solid waste. He has asked you and another employee to help him write his presentation.

Your group plans to use the information in the table in Figure 11-8 to discuss what will happen by the year 2000 if people continue to discard waste the same as they have been.

Write a paragraph or two of no more than 250 words explaining the implications of this table in terms of your supervisor's purpose.

Option: With the other members of your group, design graphics to help listeners understand the table (See Chapter 12).

**Figure 11-8
Projections of
Municipal Solid
Waste**

PROJECTIONS OF CATEGORIES OF PRODUCTS GENERATED*
IN THE MUNICIPAL WASTE STREAM, 1995 TO 2010
(In millions of tons and percent of total generation)

Products	Millions of Tons			% of Total Generation		
	1995	2000	2010	1995	2000	2010
Durable Goods (Detail in Table 31)	28.6	31.3	35.7	14.3	14.5	14.3
Nondurable Goods (Detail in Table 32)	60.5	68.3	86.3	30.3	31.6	34.4
Containers and Packaging (Detail in Table 33)	61.9	65.7	75.8	31.0	30.4	30.2
Total Nonfood Product Wastes	150.9	165.4	197.8	75.5	76.6	78.9
Other Wastes						
Food Wastes	13.2	13.3	13.7	6.6	6.2	5.5
Yard Wastes	33.0	34.4	36.0	16.5	15.9	14.4
Miscellaneous Inorganic Wastes	2.7	2.9	3.1	1.4	1.3	1.2
Total Other Wastes	48.9	50.6	52.8	24.5	23.4	21.1
Total MSW Generated	199.8	216.0	250.6	100.0	100.0	100.0

* Generation before materials recovery or combustion.
Details may not add to totals due to rounding.

Source: Franklin Associates, Ltd.

U.S. Environmental Protection Agency

Collaborative Projects: Long Term

The following projects can be continued through Chapter 13. As you complete each chapter, you will find activities for these projects that are directly related to the chapter you are reading. In this chapter, you'll engage in activities related to readability.

- **Communicating about a campus/community problem** continued from Chapter 4, Collaborative Projects: Long Term Exercise 2.
- **Proposing microcomputer laboratories** continued from Chapter 4, Collaborative Projects: Long Term Exercise 1, Projects 1, 2, or 3.
- **Gathering information for a feasibility report: diaper laundry service** continued from Chapter 6, Collaborative Projects: Short Term Exercise 2.
- **Gathering data for an information report: cleaning supplies** continued from Chapter 6, Collaborative Projects: Short Term Exercise 3.
- **Organizing data for a proposal: Recycled paper** continued from Chapter 7, Collaborative Projects: Short Term Exercise 2.
- **Organizing data from a food and music survey** continued from Chapter 7, Collaborative Projects: Short Term Exercise 3.

Review your document to determine if you have facilitated your readers' comprehension and fluency in terms of coherence, language, syntax, discussion of numerical data, and grammar, usage, punctuation, spelling, and mechanical errors. If not, revise, based on the techniques you learned in this chapter.

To be continued, Chapter 12, Graphics, Collaborative Projects: Long Term.

Individual Projects

1. You are on the staff of the Fish and Wildlife Service of the U.S. Department of the Interior. One of your field supervisors has handed you the draft of a report written by one of the supervisor's field personnel (Figure 11-9). The report on the Peoria Lake Habitat Rehabilitation and Enhancement Project (HREP) is going to the U.S. Army Corps of Engineers, which will implement this project if the citizens in the area agree to it. While the primary reader is the Corps, interested citizens, reporters, etc., will also read it. Furthermore, since funds will have to be appropriated for it, congressional representatives will want to see it.

    You have already reviewed it once and sent it back for revisions related to content and organization. You are now ready to review it for coherence, syntax, language, and numerical discussions, as well as grammar, usage, punctuation, spelling, and mechanical errors. You need to get the report out quickly, and you don't have time to go back through the review process to have the writer revise it. Revise it wherever necessary.

**Figure 11-9
A Draft of a Letter
Proposing
Changes to Peoria
Lake**

Dear Colonel Greene:

The goal of the Upper Mississippi River System Environmental Management Program is to implement "numerous enhancement efforts...to preserve, protect, and restore habitat that is deteriorating due to natural and man-induced activities. The objective of these enhancement efforts is to recover some of the riparian habitat diversity that has been long due to construction of the navigation project and the effects of sedimention.

The Illinois River Basin served as an important fishery and migration path for waterfowl until the mid 1900's. An extensive system of backwater lakes, side channels and islands provided diverse aquatic and terrestrial habitats. The development of extensive levee systems, intensified agricultural practices, wetland fills and a system of locks and dams has greatly reduced fish and wildlife values along extensive reaches of the river. Water quality on the river has been improving in recent years, but contaminants from urban and agricultural developments eliminated some important invertebrate species, such as fingernail clams, in the early 1900's. These species may be returning to river in areas where habitats are suitable.

The Peoria Lake (HREP) is sponsored by the Illinois Department of Conservation (IDOC) and consists of four components. In order to improve fisheries habitat a plug in the East River channel would be removed to restore flows. Secondly, two small areas of rock substrate will be established in the newly opened channel to provide habitat for mussels. The third feature would consist of using dredged material to construct an island 1.3 miles in length that would serve to reduce wave induced turbidity in the State-owned portion of the pool known as Goose Lake. The fourth feature would consist of a three-celled forested moist soil unit in the Woodford County Conservation Area on the south side of Peoria Lake.

If special measures are not implemented the fish and wildlife values at Peoria Lake will continue to decline in the future. The East River will continue to silt in, as will the entire Goose Lake complex. Within 50 years upper Peoria Lake will consist of partially vegetated mudflakes. The silver maple forest on the eastern shore will continue to sporadically provide waterfowl habitat when flood conditions exist during migration seasons....

We look forward to working with your staff during the development of plans and specifications, and during construction of these worthwhile features. If you have any questions regarding this report please do not hesitate to contact Mr. J.K. of my staff.

Sincerely,

XX

Field Supervisor

Adapted from the U.S. Department of the Interior

2. Study the graph in Figure 11-10, which was included in a report on testing beef patties. The purpose of the test was to determine which patties tasted best when fat was removed. Write a paragraph to be inserted in a brochure for generalists, explaining the information in the graph.

3. Use a paper you have written recently for one of your classes. Review it in terms of coherence, language, syntax, discussions of numbers, and errors in grammar, usage, punctuation, spelling, and mechanics. Revise where necessary.

**Figure 11-10
A Graphic
Description of the
Results of a Study**

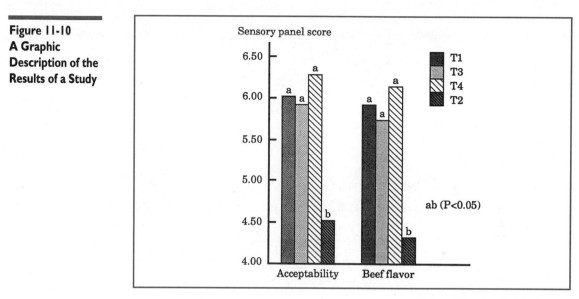

Alabama Department of Agriculture

4. You are a member of your fraternity/sorority/dorm council or other such group. You are going to have an end-of-semester pizza bash, and you have been asked to do a survey of local pizza parlors to determine which is the most cost-effective. Using the menus of the three most popular places in your area, compare the prices of various types of pizzas and determine which place offers you the best deal. Write a recommendation report to be presented at your group's next meeting.

Exercise adapted from Mathematics
Michael Sondgeroth, Illinois State University, Normal, IL

5. The technical writing faculty at your university is considering publishing a booklet for all students on how to avoid sexist language. To determine whether or not there is a need for such a booklet, they have asked all students in their classes to observe the use of sexist language on the campus and to report their findings.

   a. Keep a log or journal of all incidents for a one week-period. Each time you hear sexist language, make a note of the following:

      • what was said

      • whether it was uttered by a male or female

      • whether it was in or out of class

      • whether it was prejudiced toward men or women

   b. Write a memo to your instructor reporting on the results of your observations.

Exercise adapted from Sociology 341, Sociology of Sex Roles
Kathleen McKinney, Illinois State University, Normal, IL

Graphics

GRAPHICS INCLUDE A WIDE RANGE of visual images, ranging from pie charts and tables, to schematic diagrams, to line drawings and photographs. Graphics are an integral part of a text, and facilitate readers' thinking processes.

Including visual representations of verbal descriptions in a document enhances readers' understanding of a message. Readers tend to remember information better if they see it represented graphically. Furthermore, readers comprehend and remember concrete information better than abstract concepts.

The way you use graphics depends on your topic, your readers, and your purpose. The more complex or difficult your topic, the more you should use graphics to supplement, as well as complement, your verbal discussion, especially if you are writing for novices. However, even experts may need to have complex processes, mechanisms, and procedures complemented by diagrams, drawings, or illustrations. Visual presentations help readers see how phases, parts, or elements of an entity fit or work together. By presenting statistical concepts visually in tables, graphs, or charts, you help readers perceive relationships between various numerical values. In addition, if readers are not native-English speakers, graphics help ensure an understanding of your verbal description.

In this chapter we will examine the ways in which graphics are integrated with verbal text to facilitate readers' reading processes and behaviors. We will also look at the forms graphics assume and the conventions they follow. We will then consider three types of graphics: tables, graphs, and illustrations. Finally, we will study the criteria for determining effective graphics.

INTEGRATING GRAPHICS INTO TEXT

Graphics are an integral part of a technical document. You should determine the graphics your readers need as you engage in the composing processes. Graphics may be used to provide information in addition to that in your verbal text; indicate relationships between pieces of information; describe data visually; clarify, emphasize and/or summarize data; enhance readability; improve aesthetic quality; and increase reader interest.

Provide information. Sometimes, you may want to provide readers with additional, supporting details, though readers may not need to know this information to understand and accept your message. The additional data can be placed in a graphic instead of in the verbal text.

Consider the information presented in the graph in Figure 12-1. The graph appears in a report intended to persuade readers that plastic comprises a relatively small proportion of the waste problem. Recognizing that a verbal explanation may not be sufficiently convincing, the writer includes a graph that provides visual information relating plastic to other types of waste found in the municipal waste stream.

The nation's solid waste disposal problem is a growing concern. Today, about 80% of our waste is buried in landfills. This cannot continue. Many landfills have already been closed because they do not meet environmental standards. Many others are becoming full. And there are few communities willing to open new ones.

We need to reduce the amount of waste we put in these landfills. Many people single out plastic as the villain because until recently much of it was not biodegradable. But plastic comprises only a small percentage of the total volume by weight. Other waste products comprise a much larger percentage of the volume by weight. Paper and paperboard comprises more than four times the volume by weight that plastic does and yard waste comprises almost three times as much.

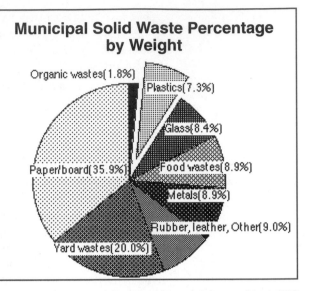

Environmental Protection Agency, March 1988

Figure 12-1 Information on Municipal Solid Waste (MSW)

Indicate relationships. Because a graphic depiction allows readers to perceive both the whole and the parts of an object or process simultaneously, graphics help readers visually perceive relationships among pieces of information as well as between pieces of information and an organizing idea.

The pie chart in Figure 12-1 indicates the proportion of plastic to each of the other types of waste. The graph also indicates the relationship of plastic to the overall municipal solid waste problem.

Describe data visually. Graphics help readers comprehend difficult concepts by presenting them visually as well as verbally. Consider how the graph in Figure 12-2, which provides a visual description of the verbal text and the table preceding it, shows the reader the difference between the weight and the volume of packaging discards after they have been placed in landfills.

Summarize information. When a great deal of detailed quantitative information is presented, a visual interpretation can offer an overview of it.

The graphs in Figure 12-2 summarize the details in the accompanying table.

Clarify information. A reader who has not seen a mechanism or followed a process being described may become confused in reading a verbal description. However, a graphic can depict a complex mechanism or process so that the reader not only can "see" the parts or steps, but can also "see" how they fit together.

Consider how the graphic in Figure 12-3, which is excerpted from a technical manual, clarifies the directions on the left.

Figure 2 shows the living area wastes. These wastes are 36 percent of total MSW by weight. Living area wastes also include about one-half of the packaging materials shown in Figure 1. About one-half of packaging is discarded at home, and about one-half at restaurants, other businesses, recreational areas, etc. Table 6 shows that packaging is about 44 percent of the living area wastes.

TABLE 6

·TRASH CAN AND LANDFILL VOLUME OF COMPONENTS OF MSW
ORIGINATING FROM LIVING AREAS OF HOUSES - 1986

	Discards (mil tons)	Weight % of Discards	Average Trash Can Density * (lb/cuyd)	Living Area Discards Volume in Trash Cans (mil cuyd)	Volume % of Living Area Discards Subtotal in Trash Cans	Average Landfill Density (lb/cuyd)	Living Area Discards Volume in Landfills (mil cuyd)	Volume % of Living Area Discards Subtotal in Landfills
PACKAGING								
Glass Containers	8.9	17.4	657	27.1	3.0	2,781	6.4	4.7
Steel Containers	1.6	3.1	204	15.7	1.8	561	5.7	4.2
Aluminum	1.0	2.0	53	37.8	4.2	317	6.3	4.6
Paper and Paperboard	6.8	13.3	44	307.1	34.5	777	17.5	12.9
Plastics	4.3	8.4	53	161.6	18.2	355	24.2	17.8
Other Misc. Packaging	0.1	0.2	200	1.0	0.1	1,000	0.2	0.1
Packaging Subtotal	22.7	44.4	83	550.3	61.8	753	60.3	44.4
NONPACKAGING								
Nondurable Paper	17.9	35.0	170	210.6	23.7	797	44.9	33.1
Nondurable Plastic	1.6	3.1	69	46.4	5.2	314	10.2	7.5
Other	8.9	17.4	215	82.7	9.3	877	20.3	15.0
Nonpackaging Subtotal	28.4	55.6	167	339.7	38.2	753	75.4	55.6
GRAND TOTAL	51	100	115	890	100	753	136	100

* Densities differ slightly from those in Table 5 because the product mix differs slightly.

Note: For more detail see Appendix Tables A-3 and A-4.

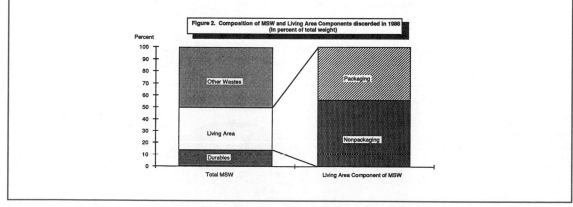

Figure 2. Composition of MSW and Living Area Components discarded in 1986 (in percent of total weight)

Estimates of the Volume of MSW and Selected Components in Trash Cans and Landfills, 1990. Prepared for the Council for Solid Waste Solutions. Franklin Associates, Ltd., Prairie Village, KS.

Figure 12-2 Visual Description of MSW Packaging Waste

Emphasize information. The use of color, shading, or position can emphasize a concept or mechanical part of a graphic.

The pie chart in Figure 12-4 appears in a brochure published for the general public. Notice how the chart emphasizes paper and paperboard as a major municipal waste problem by placing the slice at the top of the pie and shading it.

Enhance legibility. Graphics can be used to shorten lines, rows, and columns of type to create more readable text.

Notice how the diagram in Figure 12-3 shortens the line width so that readers' eyes do not have to travel across the width of the whole page.

Improve aesthetic quality. Graphics can improve the appearance of a page by breaking up the type. A cutout of a photograph can enhance the appearance of a page, as can the use of diagrams and graphs (Figure 12-1).

Increase interest. Graphics tend to catch readers' attention, drawing their eyes to a page. Readers will *see* and *read* a graphic before they read the accompanying text (Figure 12-1).

**Figure 12-3
Clarifying
Information**

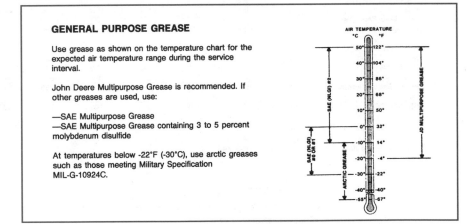

Reprinted by permission of Deere & Company © 1990 Deere & Company. All rights reserved.

**Figure 12-4
Emphasizing
Information**

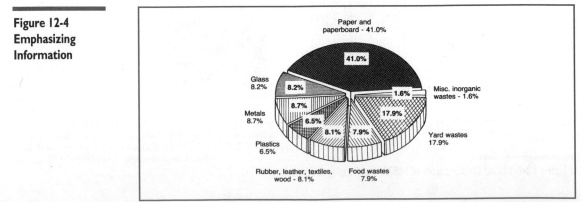

Keep America Beautiful, Inc.

FORMS, REPRESENTATIONS AND CONVENTIONS

Graphics can assume either symbolic or realistic forms. Tables, graphs, charts, and diagrams present information symbolically; line drawings and photographs present information realistically. Graphics can also provide visual representations of numerical, as well as verbal, concepts. Tables and graphs represent numerical concepts, while tables, charts, and illustrations represent verbal concepts.

When you include a graphic in a document, you need to provide readers with sufficient information to understand the graphic, and to perceive the relationship between the graphic and the verbal text. The following conventions are used to provide readers with information relating graphics to verbal text. You will want to consider these conventions in identifying visuals and in creating legible graphics.

Identifying Graphics

Titles. Graphics should be titled. The title should be brief but descriptive, and placed at the top of the graphic, as in Figure 12-5.

Labels. Various components in a graphic may need to be labeled. Labels can be placed within or next to the element to which they refer. If a label has to be placed at a distance from its referent, a line or arrow should be drawn between it and the referent so the reader is not confused.

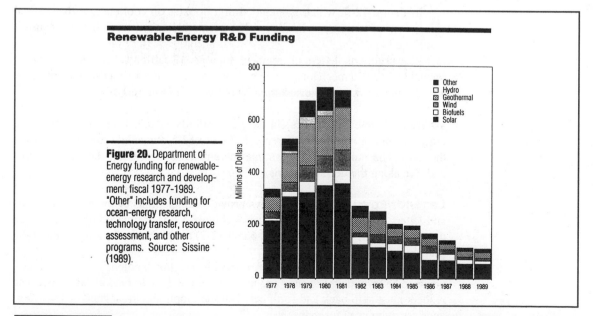

Renewable-Energy R&D Funding

Figure 20. Department of Energy funding for renewable-energy research and development, fiscal 1977-1989. "Other" includes funding for ocean-energy research, technology transfer, resource assessment, and other programs. Source: Sissine (1989).

Union of Concerned Scientists

Figure 12-5 Identifying Graphics by Title

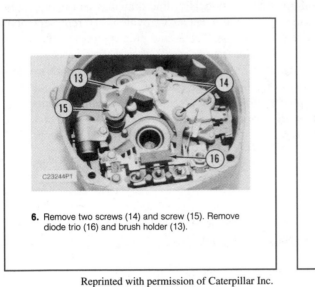

6. Remove two screws (14) and screw (15). Remove diode trio (16) and brush holder (13).

Reprinted with permission of Caterpillar Inc.

Figure 12-6 Labeling Graphics by Key

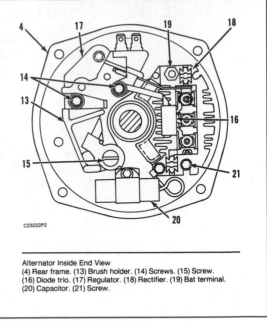

Alternator Inside End View
(4) Rear frame. (13) Brush holder. (14) Screws. (15) Screw.
(16) Diode trio. (17) Regulator. (18) Rectifier. (19) Bat terminal.
(20) Capacitor. (21) Screw.

Reprinted with permission of Caterpillar Inc.

Figure 12-7 Labeling Graphics in Text

In Figure 12-5, which is included in a booklet on cool energy, the vertical axis is labeled "millions of dollars." The horizontal axis is not labeled because it is self-explanatory.

In instructions, letters or numbers, known as "call-outs," are placed next to relevant parts of a mechanism. The call-outs are identified either in a nearby key, as in Figure 12-6, or in the accompanying text, as in Figure 12-7.

Values. Numerical values should be explicitly stated. Like labels, they are placed within or near their referents. In Figure 12-5, the values are marked along the axes. The numbers are in simple increments of ten along the vertical axis, and five along the horizontal axis.

Legends/keys. Legends and keys provide explanations of the lines, numbers, and letters of a graph or chart. Legends (Figure 12-5) are placed on the graphics themselves, while keys (Figures 12-6 and 12-7) are placed to the side or under the graphics.

If the various components of a graphic are not labeled, a legend or key needs to be provided. Legends and keys are not as easy to read as labels because readers must skip back and forth between the graph and the legend or key. Legends are often used for maps.

Most public health experts now agree that lead exhibits a "continuum of toxicity," where the smallest exposure can have a consequence somewhere in the body. This marks a radical departure from the approach to the problems only a few years ago....Prior to the mid 1960s blood lead levels of 60 micrograms of lead per deciliter of blood (ug/dL) or less were generally considered as not dangerous enough to require monitoring or treatment. This became an official standard in October 1970 when the U.S. Surgeon General issued a report defining 60 Ug/dL as a level of "undue lead absorption." But even within a year, further analysis prompted the Public Health Service to circulate a draft lowering that threshold for undue absorption by a third, to 40 ug/dL. Since then the threshold has been steadily revised downward (See Figure 2).

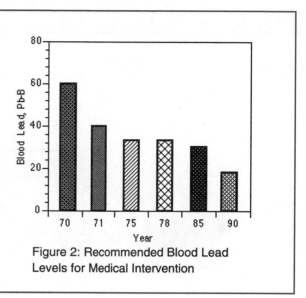

Figure 2: Recommended Blood Lead Levels for Medical Intervention

Figure 12-8 Referencing Graphics

Environmental Defense Fund, *Legacy of Lead: America's Continuing Epidemic of Childhood Lead Poisoning* (1990).

Captions. If a graphic is not specifically identified in a text, an explanation of the information should be included in a brief caption, placed beneath the graphic.

The caption in Figure 12-8 interprets the information contained in the bar graph. Captions should be kept short.

Table and figure numbers, text references. Graphics should be numbered sequentially. Figure 12-8 is listed as "Figure 2" in the document in which it appears. Tables are listed separately from figures. A graphic should be referenced in the verbal text by its table or figure number. Tables and figures should be listed with their accompanying page numbers in a "List of Tables and Figures" at the beginning of a document, following the Table of Contents (see Chapter 20). Notice how the writer follows the conventions in Figure 12-8.

Creating Legible Graphics

Graphics should be easy to read.

Type size. The size of the type used to identify graphics should be smaller than that used in the text. However, the size should be large enough for the reader to read without resorting to a magnifying glass. Eight-point is probably a good size. Typeface should be simple. Figure 12-9 uses legible typography.

Spacing. Ample white space should separate words and chunks of information. Notice how easily you can read Figure 12-9.

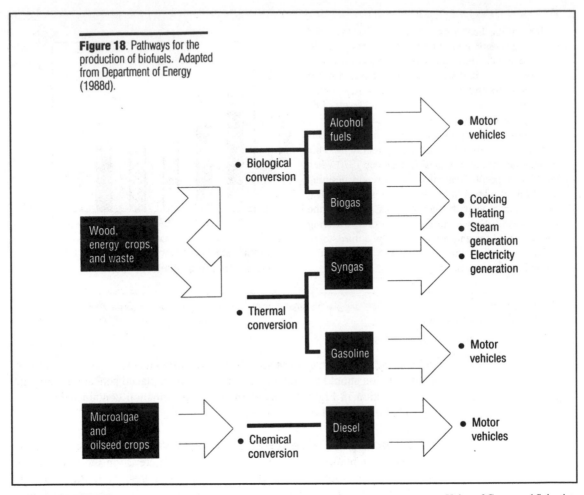

Figure 18. Pathways for the production of biofuels. Adapted from Department of Energy (1988d).

Union of Concerned Scientists

Figure 12-9 A Legible Diagram

TABLES

Tables help readers efficiently obtain specific numerical values for a variety of items. Tables include more information than texts, more precise information than graphs or charts, and can involve either numerical or verbal data.

Conventional Forms

Tables are often used when different readers need to examine different items, or when readers need to examine different aspects of an item at different times. Tables are comprised of rows (horizontal) and columns (vertical), with individual items listed in the far left column, and categories of information listed in columns to the right. This format permits readers to locate specific items in which they are interested by reading down the far left column. Once they locate an item, they can then read the row in the conventional left-to-right sequence to obtain information on that item.

The table in Figure 12-10 involves five research studies on hearing loss caused by loud music. The table includes information on the years the studies were conducted, the number of persons examined, and the number of points in hearing lost as a result of loud music. The table also provides a total for the latter two sets of data. The writer would probably need to produce at least a page of verbal text to discuss this much information. The table neatly condenses the information. Notice how the table adheres to the conventions.

By studying Figure 12-10, which appears in a professional journal article, readers (experts) can quickly perceive the results of each of the five studies on the hearing loss (Number Points—permanent threshold shifts) of persons listening to pop music.

Special Forms

Matrix A matrix is a special form of table (Figure 12-11). Readers use a matrix to determine if an item has a specific trait. Individual items are listed in the first column at the far left. Categories of traits are then listed in the columns to the right. A symbol, rather than a numerical value or verbal statement, is placed in the appropriate column if an item contains that trait. The matrix in Figure 12-11 allows readers to look up various chemicals, depending on their needs. If you want to determine whether or not aluminum dust is carcinogenic, notice how easily you can find the answer in the table in Figure 12-11.

**Figure 12-10
A Conventional
Table**

Table 1. Permanent hearing losses caused by pop music

Author	Year of Study	Number Examined	Number Points
Fluur	1967	13	0
Rintelmann & Borus	1968	42	2
Enertsen	1971	26	0
Catho & Hellman	1972	65	5
Chuden & Strauss	1974	14	1
Total		160	8 (5%)

A. Axelsson and F. Lindgren, "Pop Music and Hearing," *Ear and Hearing*, Vol. 2, No. 2, p. 65, © by Williams and Wilkins, 1981.

Figure 12-11
A Matrix As a Special Kind of Table

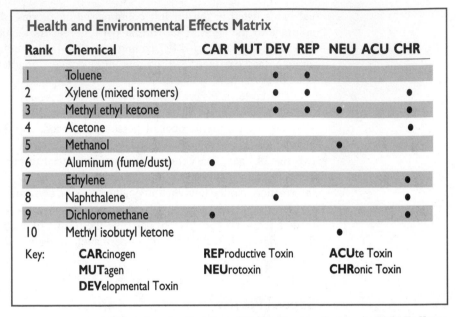

Health and Environmental Effects Matrix

Rank	Chemical	CAR	MUT	DEV	REP	NEU	ACU	CHR
1	Toluene			●	●			
2	Xylene (mixed isomers)			●	●			●
3	Methyl ethyl ketone			●	●	●		●
4	Acetone							●
5	Methanol					●		
6	Aluminum (fume/dust)	●						
7	Ethylene							●
8	Naphthalene			●				●
9	Dichloromethane	●						●
10	Methyl isobutyl ketone					●		

Key:
CARcinogen   **REP**roductive Toxin   **ACU**te Toxin
MUTagen   **NEU**rotoxin   **CHR**onic Toxin
DEVelopmental Toxin

Source: based on Paul Orum, Working Group on Community Right-To-Know
Ann Maest, Environmental Defense Fund/Environmental Information Exchange
David Allen, National Toxics Campaign Fund/Industrial Pollution Prevention Project
Ed Hopkins, Ohio Citizen Action

The matrix in Figure 12-11 is published in a booklet for the general public. The matrix indicates the health and environmental effects (listed by initials across the top) of various chemicals (listed down the far left column). People who live in communities through which chemicals are being shipped can scan the matrix to learn the effects of those chemicals if a spill should occur.

Budget   A budget is a special type of table. A budget involves expenses for a project and should include personnel; equipment; facilities; travel; office supplies; and operating costs, which include telephone, postage, duplicating and printing, computer services, etc. (For further information on budgets, see Chapter 17, page 499.)

Guidelines for Designing Tables

1. Format the table:
   a. Determine the categories for your rows and columns.
   b. Place the category label for each row in the first column, flush with the left margin.
   c. Place the category label for each column across the top row. Include the value, i.e., ft., $, %, etc.

Table 4

FLEXIBLE PACKAGING AS A PART OF MAJOR MATERIAL CATEGORIES IN MUNICIPAL SOLID WASTE DISCARDS, 1988

	Total Weight of Material in MSW		Weight of Flexible Packaging Material		Flexible Packaging as a % of Material Total
	(mil ton)	(bil lb)	(mil ton)	(bil lb)	
PAPER	53.4	107	3.224	6.4	6.0
PLASTICS	14.3	29	3.698	7.4	25.9
ALUMINUM	1.7	3	0.321	0.6	18.9
CELLOPHANE	0.05	0.1	0.048	0.1	100.0

Details may not add to totals due to rounding.
Derived from Tables 7, 8, 9, and 10.
Source: Franklin Associates, Ltd.

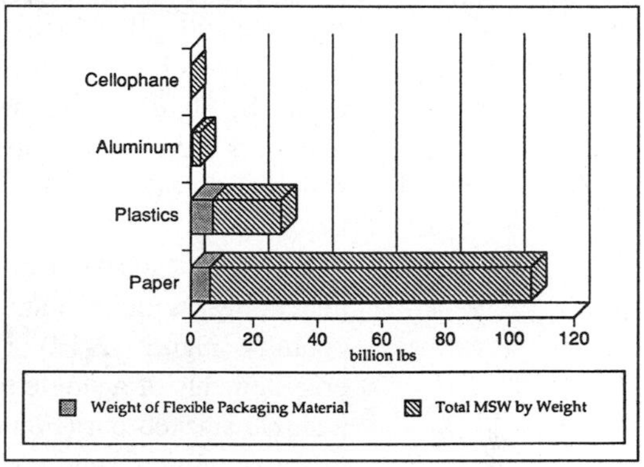

Franklin Associates, Ltd., for Flexible Packaging Association

Figure 12-12 Showing Statistical Relationships

d. Separate column labels from the data by a heavy horizontal line.

e. Place data in the appropriate boxes.

2. Identify the table:

   a. Title all tables. Give the table a descriptive title that reflects the topic. Place the title at the top of the table, in the center, in capital letters.

   b. Number tables sequentially. Use Arabic numerals. Capitalize the word "Table." The table number should appear in the text as follows: Table X. It should be centered above the title at the top of the table.

   c. Reference all tables in the text. Refer to tables by number.

3. Maintain legibility:

   a. Use a type size and font that is easily legible.

   b. Use bold type for row and column labels.

GRAPHS

Graphs help readers compare statistical information by indicating relationships between sets of quantitative data. Notice how easily you can "see" the data in the table in Figure 12-12 by looking at the accompanying graph.

There are numerous types of graphs, including bar, line, pie, and picture graphs.

Bar Graphs

Bar graphs (called histograms), which are used to compare related data, help readers perceive the quantitative relationships among various components of a topic. In Figure 12-12, readers can see the relationships between the various materials used for flexible packaging and their relative contribution to municipal solid waste (MSW).

Configurations Bar graphs can depict numerical data in a variety of forms, depending on the data and the purpose of the text. Bar graphs can be drawn horizontally (Figure 12-12) or vertically (Figure 12-13). If you need to compare several elements of a single topic you can use either multiple bars (Figure 12-15) or a single stacked bar (Figure 12-14). If you need to compare several components of several topics, then you can either stack multiple bars (Figure 12-12) or create sets of multiple bars (Figure 12-15). As you examine Figures 12-12 to 12-15, consider why document designers selected each configuration.

Types Bar graphs may be analytical or non-analytical, depending on your numerical data.

Analytical graphs. These graphs, in which the sum of the bars adds up to 100 percent, depict relationships between parts of a single entity. They help readers perceive quantitative differences within a single topic. These graphs can be presented either as a single stacked bar (Figure 12-14), or as multiple bars (Figure 12-15).

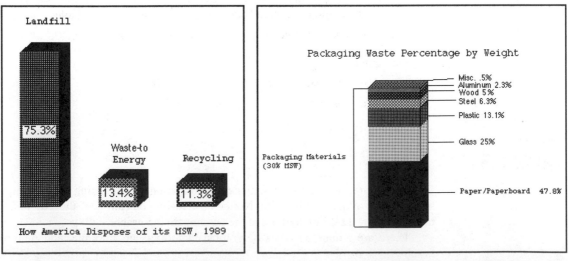

Keep America Beautiful

Adapted from Environmental Protection Agency

Figure 12-13 Vertical Bar Analytical Graph   **Figure 12-14 Single Stacked Bar Analytical Graph**

**Figure 12-15
Multiple Bar
Analytical Graph**

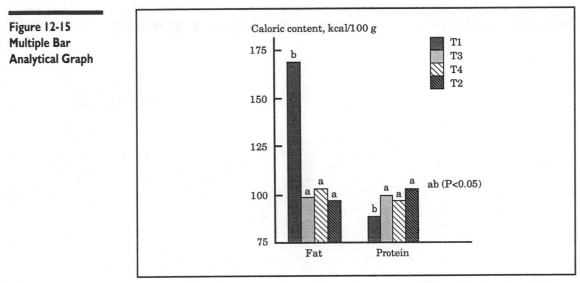

Alabama Department of Agriculture

Figure 12-13, included in a brochure for the general public, is a multiple-bar graph in which the sum of the bars equals 100 percent. From the graph, which depicts the relationship between three different techniques for municipal waste disposal, readers can see that most waste goes to a landfill and very little is recycled. The graph also allows the reader to "see" the huge discrepancy between the three forms of waste disposal presently being used.

The axes for both Figures 12-13 and 12-14 have been removed. The percentage of each portion of the bar for Figure 12-14 is printed to the right of it, while the percentage of each bar for Figure 12-13 is inserted on the bar itself. In comparing the two graphs, notice how much easier it is to perceive that the boxes are all part of a whole in Figure 12-14 than it is in Figure 12-13.

Non-analytical graphs. The sum of the bars in these graphs do not add up to 100 percent. These graphs help readers perceive statistical relationships among a variety of components in a topic. They indicate quantitative differences between several items in relation to single or multiple topics (Figures 12-12, 12-15).

Line Graphs

Line graphs help readers perceive statistical trends by presenting a continuous line indicating the up and/or down movement of a subject. In Figure 12-16, notice how easily you can perceive that paper recycling is increasing .

Line graphs are especially good for comparing several related topics and for depicting changes over time, as shown in Figure 12-16, which appeared in a study conducted for the American Paper Institute. The graph depicts the relationship between two aspects of paper recycling: the actual amount of paper recycled,

**Figure 12-16
Line Graph
Depicting Trends**

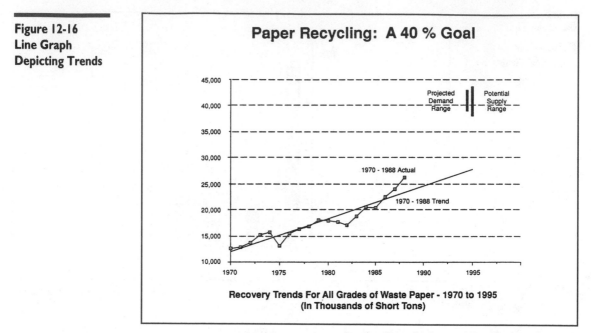

American Paper Institute and Franklin Associates, Ltd.

**Figure 12-17
Pie Chart Empha-
sizing Small
Amount of
Plastics in MSW**

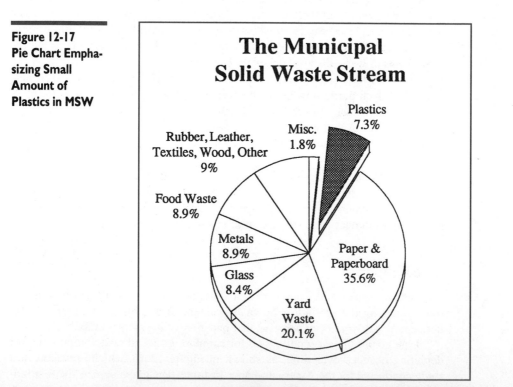

Council on Plastics and Packaging in the Environment

and the projected trend for recycling paper. Using the graph, the reader can perceive the relationships among these variables over a period of time. It would appear that the actual amount of paper being recycled is moving ahead of the projected trend.

Pie Graphs

Like analytical bar graphs, pie graphs (Figure 12-17) allow readers to see the statistical relationships between parts of a single entity. Pie graphs are usually preferred to analytical bar graphs, because readers do not automatically expect the percentages of a vertical bar to add up to 100 percent, whereas, based on their prior knowledge, they do expect the slices of a pie to equal 100 percent. Notice that it is easier to perceive the parts belong to the whole in Figure 12-17 than it is in Figures 12-13 or 14.

In Figure 12-17, the values and labels for the large slices are placed within the respective slices. However, since some of the slices are too small, the values and labels for these slices are placed outside.

Shading, coloring, and/or partially removing a slice from a pie chart emphasizes the information in that slice. The purpose of the article that includes Figure 12-17 is to persuade the public that plastic is not an environmental ogre. By pulling the plastic waste slice partially out of the pie and shading it, the document designer emphasizes the relatively small amount of waste comprised by plastics, especially when the size of that slice is compared to the slice representing paper and paperboard waste.

Picture Graphs

Picture graphs are bar, line, or pie graphs that are integrated with drawings directly related to the topic. In Figure 12-18, the single, partitioned bar graph is depicted as a horizontal line of trash dumped from a truck. These can be used to provide humor, aesthetics, or interest.

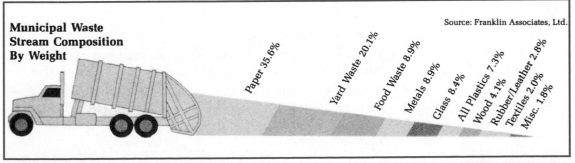

Franklin Associates, Ltd.

Figure 12-18 A Picture Graph of Elements in MSW

Guidelines for Designing Graphs

1. Determine appropriate graph:

    a. Use a line graph to indicate trends.

    b. Use a pie graph to indicate the parts of a whole.

    c. Use a bar graph to compare various elements of one or more topics that do not comprise a whole, and in which you do not need to indicate a trend.

2. Format the graph:

    a. Draw the axes of a line or bar graph so that the vertical axis is on the left side and the horizontal axis goes from left to right. The vertical axis should be shorter than the horizontal axis. The vertical axis should be between $2/3$ and $3/4$ the size of the horizontal axis.

    b. Indicate dependent variables (frequencies) on the vertical (Y, or ordinate) axis of a line or bar graph. Indicate independent variables (except scores, time lines) on the horizontal (X, or abscissa) axis of a graph.

    c. Indicate the zero point in all line or bar graphs. It should be located at the intersection of the two axes.

    d. Use convenient numerical units to identify positions along axes (i.e., 1, 2, 5, 10, 20, 25, 50, 100, or 1,000) in bar and line graphs. Units should be equidistant from each other.

    e. In dual and triple bar graphs and multiple line graphs, design each bar or line differently, using either different colors or shadings, line widths, or line types (continuous or dotted).

3. Identify the graph:

    a. Label each axis.

    b. Give each graph a title.

    c. If more than a single topic is graphed, provide a key to differentiate among the topics.

    d. Label each bar, bar stack, or pie slice.

    e. If the axes have been removed, or if a pie chart is used, indicate the quantitative value for each bar, partition, or slice.

    f. If labels or quantitative values are placed outside the bar, stack, or slice, use arrows or lines to relate the label to its respective stack or slice.

    g. Caption the graph.

h. Number graphs sequentially. Graphs are considered figures. The figure and number should be written as "Figure X," with the word "Figure" capitalized and placed at the bottom of the graph, flush left.

i. Refer to graphs as "Figure X" in the text. Reference graphs in the text.

j. Place the figure number and title under the graph and outside its frame. Place the figure number flush left with the frame, followed by a period. Then print the title.

k. Anything that is not considered a table is considered a figure.

l. The table number and title should be placed at the top and outside the frame of a table.

4. Emphasize important aspects:

a. Pull out, shade, and/or color a pie slice to emphasize it.

b. Use color, shading, or line thickness for emphasis.

5. Maintain integrity, clarity, elegance:

a. Provide a true representation of the numerical data.

b. Keep the graph simple, clear, and precise.

CHARTS

Whereas graphs present numerical relationships, charts indicate hierarchical, directional, and time relationships. Three of the most common types are organization charts, flow charts, and time lines.

Organization Charts

Organization charts show the relationship between different positions in an organization or project, as well as indicate lines of authority and responsibility (Figure 12-19). By looking at an organization chart, a reader not only learns the positions within an organization, but also the relationship of each position to the others. These charts help readers know who reports to whom.

Organization charts are comprised of boxes connected by lines in a pyramidal shape. Each box represents a position in an organization. Lines are drawn between boxes to indicate relationships.

Figure 12-19, which is included in proposal to a government agency, depicts the administration of a joint project between Caterpillar Inc. and a local community college (ICC). A reader from the funding agency can quickly scan the chart to determine how the program will be administered and who will be responsible for it.

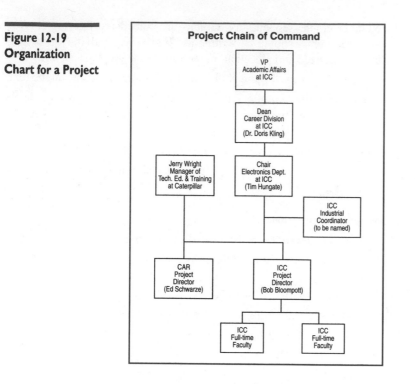

**Figure 12-19
Organization
Chart for a Project**

Flow Charts

Flow charts depict relationships between components of a project or phases of a process. They also indicate the sequence of these phases (Figure 12-20). Flow charts are comprised of boxes that are connected by lines/arrows indicating the relationship between the boxes and the direction of the flow. The first step or element of a project is placed either at the top of the page if the chart is vertical, or to the far left if it is horizontal. Figure 12-20 is a decision chart included in a report for community leaders (novices) and governmental agencies (generalists). It provides readers with an efficient description of the testing program used by the Environmental Protection Agency to determine whether standards for the disposal of toxic wastes have been met at disposal facilities. Readers can use the chart to understand how EPA makes its decisions. Notice how quickly you can see what happens if hazardous waste doesn't pass the TCLP test.

Flow charts are especially helpful in understanding large projects that involve a number of people, or even several different firms. They provide readers with an efficient description of each component, and show the relationship of the various components and phases to each other and to the overall project. Flow charts can also be used to show the steps in a complex procedure.

Computer programmers use flow charts to indicate the way in which the major parts of a program are connected. By studying Figure 12-21, included in document for clients (generalists), readers perceive the basic information users can obtain from a computer program. By following the lines, indicating data

**Figure 12-20
Flow Chart for
Disposing of Various Types of Toxic
Wastes**

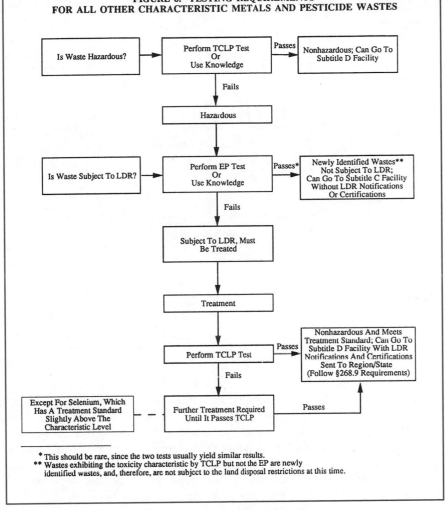

FIGURE 6. TESTING REQUIREMENTS
FOR ALL OTHER CHARACTERISTIC METALS AND PESTICIDE WASTES

U.S. Environmental Protection Agency

flow, from the rectangular box marked "Verify time card data entered," to the boxes at the far right of the page, you can easily perceive see that the user can verify four elements of the time card: labor data, employee data, job number, and phase data.

Time Lines

Time lines are often used in proposals to indicate the amount of time necessary to complete a project, and the time at which various components of a project begin and end (Figure 12-22). (See Chapter 20 for more information on time lines.)

**Figure 12-21
Flow Chart for
a Computer
Program**

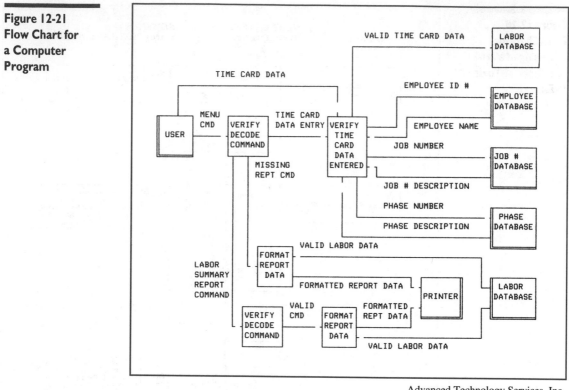

Advanced Technology Services, Inc.

The format for a time line is similar to that of a graph. The various elements of a project are listed in the far left (vertical axis) column, and time periods are shown on the horizontal axis at the bottom. In Figure 12-22, the elements are numbered, rather than placed on the vertical axis, and appear as captions below the time line. Time periods can be measured by minutes, hours, days, weeks, months, seasons, and years. A line is drawn from the beginning to the end of the time period required for each respective element.

By studying Figure 12-22, included in a proposal for experts in the field, readers can determine whether the program schedule appears feasible and will occur within the time frame of the funding.

Guidelines for Designing Charts

1. Determine the appropriate chart:

    a. Use an organizational chart to indicate positions in an organization, and relationships between those positions; lines of authority and responsibility; or management of an organization or project.

    b. Use a flow chart to show the relationship and sequence of various aspects of a project, or the steps in a program, or in complex instructions.

**Figure 12-22
Time Line for a
Research Study**

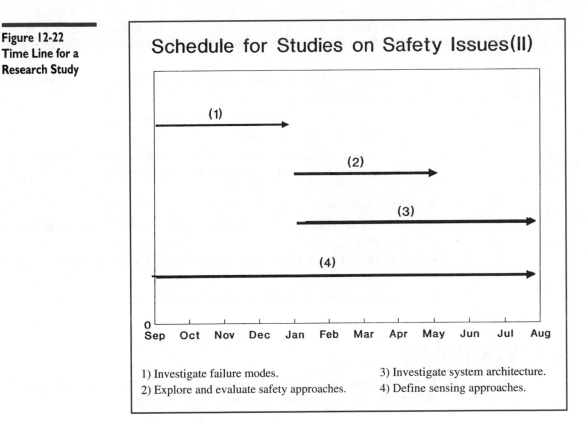

Schedule for Studies on Safety Issues(II)

(1)

(2)

(3)

(4)

0
Sep   Oct   Nov   Dec   Jan   Feb   Mar   Apr   May   Jun   Jul   Aug

1) Investigate failure modes.   3) Investigate system architecture.
2) Explore and evaluate safety approaches.   4) Define sensing approaches.

    c. Use a time line to provide an efficient description of the amount of time required to complete a project and the various components involved in the project.

2.   Format the chart:

    Determine the shape of the boxes in flow charts or organization charts. They should be simple—squares, circles, rectangles, ovals.

    The shapes for the boxes in a flow chart can reflect their functions:

- a diamond represents a decision
- a circle represents a connection
- an oval represents a terminal
- a rectangle represents a process
- a parallelogram represents input/output

3.   Identify the chart:

    a. Title all charts.

    b. Number all charts. Charts are considered figures.

    c. Place the figure number and title under the chart and outside its frame. Place the figure number flush left with the frame, followed by a period. Then print the title.

    d. Caption charts if readers need additional information to understand a message. Keep the caption brief.

    e. Provide a key to the chart if labels are not used.

    f. Label all parts of the chart.

    g. Refer to a chart in a text as "Figure X" with the word "Figure" capitalized and placed at the bottom of the chart, flush left.

4.   Maintain legibility:

    a. If text is included in a chart, use a type size and a font that is easy to read.

    b. Do not place boxes close to each other. Use plenty of white space between boxes.

    c. Keep verbal information in each box short.

ILLUSTRATIONS

Illustrations provide readers with pictures of parts involved in mechanisms, and the steps or phases involved in procedures, processes, and theories.

Photographs

Photographs convey realistic pictures of a mechanism or situation.

Figure 12-23, included in a technical manual for generalists, is a photograph that helps readers recognize and locate the battery support plate. Notice how the arrow superimposed on the photograph orients readers to the part of the photograph discussed in the text. By showing the entire battery, rather than just the plate, the designer of the document establishes the location of the plate in relation to its surroundings.

Drawings

Like photographs, drawings convey realistic pictures of objects. Drawings are used when it is impossible to obtain a photograph, when certain details cannot be shown by a photograph, when a combination of realistic and diagrammatic aspects are necessary, or when a photograph would include irrelevant items that would detract from the main one.

4. Remove battery support plate (A).

Figure 12-23 Photograph Illustrating Instructions for Removing a Battery Support Plate

5. Hold pulleys in rotated position with right hand.

6. Grasp belt in left hand pulling down and away while forcing belt into pulley with fingers of right hand.

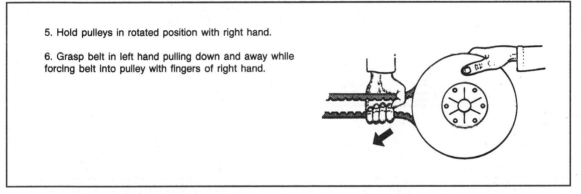

Figure 12-24 Drawing Illustrating Instructions

Figure 12-24, included in a technical manual for an All Material Transporter (AMT), depicts the positions of a person's hands when removing a drive belt from an AMT. A drawing is used because the photographer would have had to shoot from inside the engine to photograph the hands from that point of view.

Anyone who services the AMT can easily locate the various parts by referring to the diagram in Figure 12-25. Sometimes, in a drawing of a complex mechanism or procedure, it is impossible to show clearly all the details of a particular phase or part. To provide readers with these details, the particular aspect is redrawn larger and in more detail, and then placed above, below, or to the side of the smaller version. In Figure 12-25, the document designer "explodes" or "zooms in on" parts B and C to provide a more detailed view.

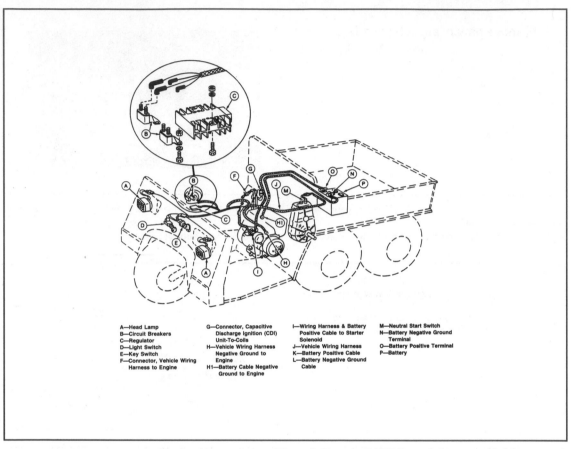

A—Head Lamp
B—Circuit Breakers
C—Regulator
D—Light Switch
E—Key Switch
F—Connector, Vehicle Wiring Harness to Engine

G—Connector, Capacitive Discharge Ignition (CDI) Unit-To-Coils
H—Vehicle Wiring Harness Negative Ground to Engine
H1—Battery Cable Negative Ground to Engine

I—Wiring Harness & Battery Positive Cable to Starter Solenoid
J—Vehicle Wiring Harness
K—Battery Positive Cable
L—Battery Negative Ground Cable

M—Neutral Start Switch
N—Battery Negative Ground Terminal
O—Battery Positive Terminal
P—Battery

Figure 12-25 Drawing Showing Components

FUEL

⚠ **CAUTION: Handle fuel carefully. If engine is hot or running, do not fill the fuel tank. Stop engine and allow to cool several minutes before filling fuel tank. Do not smoke while you fill the fuel tank or service the fuel system. Fill fuel tank only to bottom of filler neck.**

IMPORTANT: To avoid engine damage, DO NOT mix oil with gasoline.

Figure 12-26 Using Symbols in Instructions

Drawings often include symbols. As more and more organizations become multinational, it is increasingly important, especially in manuals, to provide visuals that can be recognized by everyone, regardless of their language or culture. Professional technical writers from around the world are working on establishing sets of internationally-accepted symbols.

The "no smoking" symbol is one that has become universally accepted. Figure 12-26, which is included in a technical manual for a vehicle shipped to countries around the world, uses the symbol to accompany a drawing cautioning users.

Diagrams

While drawings are usually static, showing the relationship among parts of an entity, diagrams are usually dynamic, and indicate the relationships between the functions of a mechanism, the phases of a process, or the steps in a procedure. Diagrams usually depict aspects that cannot be seen, such as the generation of nuclear power (Figure 12-27).

Conventional Forms Diagrams use symbols and lines to represent an object, process, or procedure. They can depict the object or process in a stable condition or in motion, as in Figure 12-27, which was included in a booklet on nuclear energy for journalists (novices). Readers can easily follow the process of creating nuclear energy in a pressurized water reactor by following the arrows in the diagram.

Special Forms: *Schematics.* Figure 12-28 shows a special type of diagram called a schematic. Schematics are highly technical and used only in documents for experts. An expert viewing the figure can easily comprehend the way in which the electrical system controls the braking system.

Maps. Maps are illustrations that inform and orient readers. Figure 12-29, included in a study on noise pollution for local government officials and citizens (generalists and novices), indicates the areas over which planes fly as they take off. If you were a resident in the area, you could easily determine whether or not you would be affected by the noise.

Guidelines for Designing Illustrations

1. Determine appropriate illustrations:
    a. Use photographs to portray realistic situations.
    b. Use drawings to depict specific details that cannot be shown in a photograph, or to indicate movement in a process or procedure.

Figure 12-27
Diagram of the
Process of Nuclear
Energy

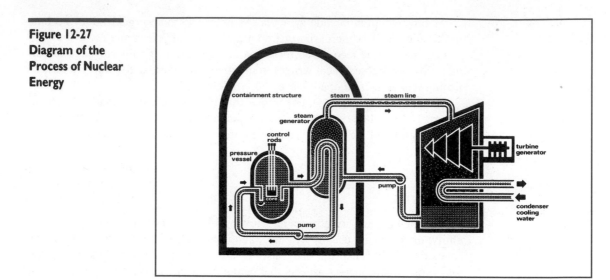

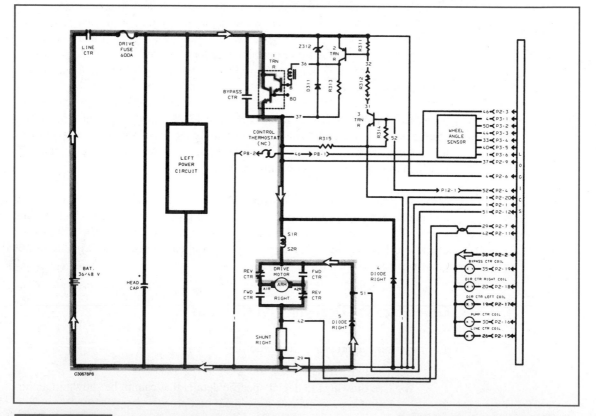

Figure 12-28 Schematic of an Electrical Braking Operation

**Figure 12-29
Map of Area
Affected by
Airport Noise**

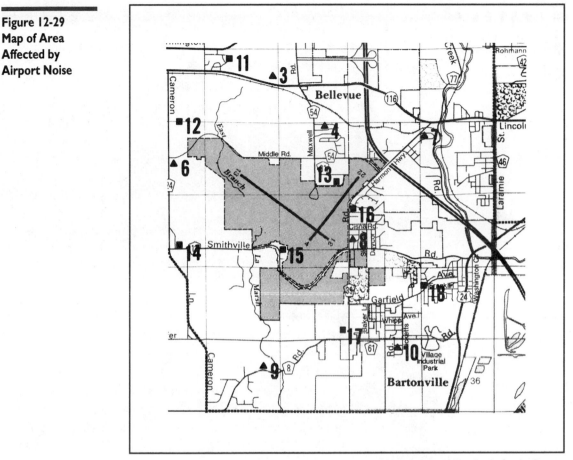

Crawford, Murphy & Tilly, Inc., 1991, F.A.R. Past 150 Noise Capability Study for the Greater
Peoria Regional Airport, Peoria, Illinois.

    c. Use diagrams to indicate the relationships between the functions of an object, phases of a process, or steps in a procedure to show the relationships among components of geographic areas.

2. Identify Illustrations:
    a. Title illustrations.
    b. Number all illustrations. Illustrations are considered figures.
    c. Describe the illustration briefly in a caption.
    d. Label illustrations when necessary.
    e. Refer to an illustration in a text as "Figure X" with the word "Figure" capitalized and placed at the bottom of the illustration, flush left.

STRATEGY CHECKLIST

1. To provide specific numerical data efficiently, use tables.
2. To compare quantitative aspects of one or more topics, use graphs.
   - Use a line graph to indicate trends.
   - Use a pie graph to indicate the parts of a whole.
   - Use a bar graph to compare various components of one or more topics which do not comprise a whole, and in which you do not need to indicate a trend.
3. To indicate hierarchical, directional, or time relationships, use charts.
   - To indicate positions in an organization, and the relationships between positions and lines of authority and responsibility, use an organizational chart.
   - To show the relationship, and also the sequence of various elements of a project, or the steps in a computer program, or in complex instructions, use a flow chart.
   - To provide an efficient description of the amount of time required to complete a project and the various components of a project, use a time line.
4. To depict a mechanism or object or a process, procedure, or theory, use illustrations.
   - Use photographs to portray realistic situations.
   - Use drawings to depict specific details that cannot be shown in a photograph, or to indicate movement in a process or procedure.
   - Use diagrams to indicate the relationships between the functions of an object, phases of a process, or steps in a procedure which cannot be seen.

EFFECTIVE GRAPHICS

Whether you create your own graphics or use ones already developed, you want to be sure that readers can read them easily. Readers should be able to read the lettering as well as "read" the graphic design itself.

General Criteria

Regardless of whether you are designing a table, graph, chart, or illustration, you can apply certain criteria to determine the effectiveness of your graphic. A good graphic is related to the text, focused, simple, precise, clear, and legible. Study Figure 12-30 to determine if it meets these criteria.

**Figure 12-30
An Effective
Diagram**

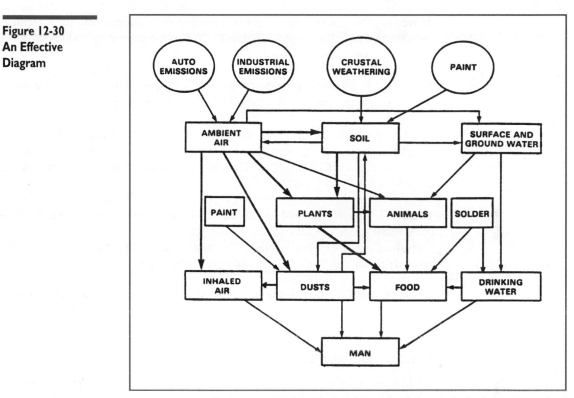

Environmental Defense Fund, *Legacy of Lead: America's Continuing Epidemic of
Childhood Lead Poisoning* (1990).

Related to Text. All graphics should be directly related to the written text.
Information that is not related to the text should be eliminated. If information is
included that is not specifically discussed, a caption should provide an explanation of the additional information and the relationship between it and the text.

As you discuss aspects of your text which are depicted visually, refer your
reader to the graphic. However, you do not need to call your reader's attention to
graphics used for aesthetic effects. Refer to the graphic in the verbal text either
as part of the sentence, e.g., "As you can see in Figure X," or parenthetically,
e.g., "The lead in our society is so pronounced that the skeletons of modern
humans contain 200 times more lead than those of their preindustrial ancestors
(Figure X)." Always refer to the graphic by its number.

The graphic in Figure 12-30 appears at the conclusion of a chapter on
sources and pathways of lead exposure. Each of the four sources at the top of the
graph are discussed in detail in the text.

Focused. The graphic should focus on that aspect of the text to which it is
related. The entire graphic in Figure 12-30 is focused on the path which lead follows from its source to a human being, the focus of the verbal text.

Simple. The graphic should not be crowded with information. Each element should stand out. The graph in Figure 12-30 is not cluttered by additional information, such as amounts of lead, etc.

Precise. Lines and points should be precise, shapes definite. Points of intersection on graphs and charts should be clearly designated. Photographs should be sharp.

The arrows in Figure 12-30 clearly indicate precise pathways. The geometric shapes depict the different elements, with the sources represented as circles and the receptacles as rectangles. The points of intersection are indicated by the arrows.

Clear. All graphics should be labeled clearly. In Figure 12-30, there is no difficulty understanding what each box stands for, or what the entire graphic represents, because of the clear labeling. The number of the figure and the caption are properly positioned at the bottom and outside the graph.

Legible. Written text within a graphic should be legible. The verbal aspects within the graphic in Figure 12-30 can be read easily.

Criteria for Numerical Data

As you learned in Chapter 11, numerical data can overwhelm a reader. To provide graphics that reflect a simplified presentation of numerical data, the following criteria, established by Tufte, should be applied to your design:

- Show the data without distortion.
- Focus the graphic on the data.
- Encourage the eye to compare different pieces of data.
- Relate the statistical description to the verbal descriptions in the text.
- Present many numbers in a small space.
- Make large data sets coherent.
- Reveal a broad overview as well as the specific points of the data.

As you look at the graph in Figure 12-31, determine whether it meets the criteria.

The graph appears to meet Tufte's criteria for effective graphics.

- **No distortion.** The Y axis begins at zero point. The X axis is marked in equal five-year segments from 1960 to 1995, and the Y axis is divided into equal segments of ten million tons.
- **Focused.** The reader's eye focuses on the graph.
- **Presents a comparison.** Two comparisons are presented: (1) change in materials recovery between 1960 and 1995; and (2) difference between high and low projections.

**Figure 12-31
Depicting Numerical Data Validly**

The range of projected recovery composting of materials in 1995 is shown in Table 35 and Figure 16... Recovered tonnage of materials made a significant increase between 1987 and 1988 (Figure 16). At the high end of the recovery projections, it is assumed that materials collection can be increased and materials markets can be found to sustain this growth. At the lower end of the recovery projections, recovery would grow in the more moderate pattern of previous years. Increasing recovery beyond the high end projections means that some fundamental changes will have to be made in the ways our wastes are managed.

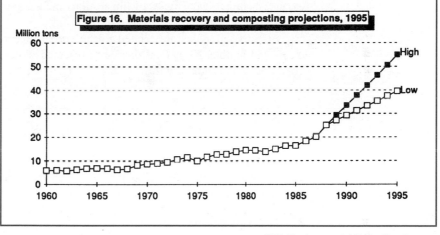

Figure 16. Materials recovery and composting projections, 1995

U.S. Environmental Protection Agency

- **Data related to text.** The graph is directly related to the text.

- **Data condensed.** The graph presents 35 years of data in a compact space of 3" x 5 ½".

- **Coherent.** The points along the graph create a single picture that indicates an increase in recovery tonnage from 1960 to 1995.

- **Broad Overview.** The graph presents an overview of the change in recovery over a 35-year period, while simultaneously depicting the difference in projected high and low recovery.

**C H A P T E R
S U M M A R Y**

The way in which you use a graphic depends on your topic, readers, and purpose. Graphics include a wide range of visual images, including, tables, charts, and illustrations. Graphics can assume symbolic, realistic, or representational forms.

Graphics are integrated with verbal text to provide information; indicate relationships among pieces of information; describe data visually; clarify, emphasize and summarize information; enhance legibility; improve aesthetic quality; and increase reader interest.

When designing graphics, conform to conventions in identifying visuals and designing legible graphics.

Tables help readers obtain specific numerical values for a variety of items. Special types of tables include matrices and budgets.

Graphs help readers compare statistical information. Major types of graphs include bar, line, pie, and picture.

Charts indicate hierarchical, directional, and time relationships. Major types of charts include organization and flow charts, and time lines.

Illustrations provide readers with realistic pictures of mechanisms, processes, and theories. Major types of illustrations include photographs, drawings, and diagrams.

Effective graphics are related to the verbal text, focused, simple, precise, clear, and legible. Numerical data are presented without distortion, condensed, coherent, and in summary form.

PROJECTS

Collaborative Projects: Short Term

As you work on the following projects, you may want to use the Audience Analysis Chart (page 97).

1. **Creating graphics**

    a. Work in groups of five. Each of you works for a company that manufactures one of the products listed below.
        *   Aluminum cans for food
        *   Bottles for soft drinks
        *   Aluminum cans for soft drinks
        *   Plastic milk bottles
        *   Paper milk cartons

    b. Each person should select one of the products listed. No product should be represented by more than one person.

    c. Various organizations have been complaining about the use of packaging and container materials that contribute to the problem of landfills. Each of your firms has decided to try to persuade the public that its product does not contribute much waste to MSW (municipal solid waste). Study the statistics shown in the table in Figure 12-32, which are related to the product your firm manufactures.

    d. Your firm plans to distribute a brochure to its customers. You have been asked to construct two graphs that show your product in the best possible light in relation to its contribution to MSW. The graphs will be included in the brochure.

    e. Compare your graphs. What are the differences? What are the similarities? What do you think accounts for the differences?

Containers and packaging amounted to 56.8 million tons of MSW in 1988 (nearly 32% of total MSW generated). While the weight of containers and packaging generated has increased steadily between 1960 and 1988, the percentage by weight has remained stable.

| | Generated MSW | | | | Recovery of Products | | | |
| | 1960 | | 1988 | | 1960 | | 1988 | |
	(Millions of$ons)	(%)	(Millions of tons)	(%)	(Millions of tons)	(%)	(Millions Of tons)	(%)
Durable Goods	9.4	10.7	24.9	13.9	0.4	4.3	1.9	7.5
Nondurable Goods	17.6	20.0	50.4	28.1	2.4	13.6	7.4	14.6
Containers & Packaging	27.3	31.1	56.8	31.6	3.1	11.4	13.8	24.3
Aluminum Packaging								
Beer, soft drink cans	0.1	0.1	1.4	0.8	0.0	0.0	0.8	55.0
Other cans	0.0	0.0	0.1	0.0	0.0	0.0	0.0	0.0
Total Alum. pkg.	*0.2*	*0.2*	*1.8*	*1.0*	*0.0*	*0.0*	*0.8*	*44.1*
Glass Packaging								
Beer, soft drink bottles	1.4	1.6	5.4	3.0	0.1	7.1	1.1	20.0
Total Glass pkg.	*6.2*	*7.1*	*11.4*	*6.3*	*0.1*	*1.6*	*1.5*	*13.3*
Plastic Packaging								
Milk bottles	0.0	0.0	0.4	0.2	0.0	0.0	0.0	0.5
Total plastic pkg.	*0.2*	*0.2*	*5.6*	*3.1*	*0.0*	*0.0*	*0.1*	*1.6*
Paper Packaging								
Milk cartons	0.0	0.0	0.5	0.3	0.0	0.0	0.0	0.0
Total paper pkg.	*14.0*	*15.9*	*32.9*	*18.3*	*3.0*	*21.4*	*11.0*	*33.5*
Other Wastes (Food, yard, inorganic)	33.5	38.2	47.5	26.5	0.0	0.0	0.5	0.5
Total MSW	*87.8*	*100.0*	*179.6*	*100.0*	*5.9*	*6.7*	*23.5*	*13.1*

Adapted from the U.S. Environmental Protection Agency, Characterization of Municipal Solid Waste in the United States: 1990 Update

Figure 12-32 Table Depicting Generation and Recovery of MSW

    f. Your firms have decided to work together to educate the public about the effects of packaging and containers on MSW. They will do this by distributing a brochure on the industry. Collaborate to construct two graphs to depict the packaging and container industry in relation to MSW in the best possible light. The graphs will be used in the brochure.

2. Creating graphics for a survey

    As a prelude to making a new shoe product, a shoe manufacturer requested a marketing research firm to conduct a survey of students' shoe preferences. You are on the staff of this marketing firm and have been assigned to the team for this research project.

    Your first task involved conducting a survey. The results of the survey of 250 students are tabulated in Figure 12-33.

    a. Work with two other people.

    b. Write a letter to the manufacturer reporting the findings of the survey.

**Figure 12-33
Results of a Cam-
pus Shoe Survey**

A. Activity Shoe Preference

Type of Shoe	No. of Students
tennis	36
basketball	74
running	93
walking	17
aerobic	18
leisure	11
cross country	1

B. Shoe Manufacturer

Manufacturer	No. of Students
Nike	53
Adidas	55
Reebok	57
British Knights	37
Keds	38

C. Cost

Price Range pd.	No. of Students
under $30	17
$30-50	16
$50-70	138
over $70	79

c. Determine the graphics your reader needs to understand and accept your message.

d. Design the graphics.

e. Integrate the graphics with the verbal text.

3. Determining graphics

You are working on the staff of a consulting firm hired by the Environmental Protection Agency to conduct a study on municipal solid waste. You have gathered all of your information and are in the process of writing your text. You have just completed the following paragraph:

Six major categories comprise the products found in MSW. Containers and packaging are the largest single product category generated in MSW by weight. Nondurable goods, such as newspapers and disposable food service items, are the second largest category. Yard wastes are approximately 18 percent and durable goods, such as furniture and tires, are 14 percent of total generation. Food wastes comprise only 7 percent.*

*Data adapted from Characterization of Municipal Solid Wastes in the United States: 1990 Update. Published by the Environmental Protection Agency, 1990. Washington, D.C.

You decide you should provide readers with graphs to help them understand your message. Your firm recently acquired a software package that draws graphs. After experimenting with the program, your team creates the graphs in Figure 12-34.

a. Work in groups of five.

b. Study the graphs.

c. Discuss whether or not they meet Tufte's criteria.

d. Arrive at a consensus on which graph should be used to accompany the text.

e. If you are not satisfied with one of those and have a graphics software program, generate your own.

Collaborative Projects: Long Term

The following projects can be continued through Chapter 13. As you complete each chapter, you will find activities for these projects that are directly related to the chapter you are reading. In this chapter, you'll engage in activities related to integrating graphics into your text.

- **Communicating about a campus/community problem** continued from Chapter 4, Collaborative Projects: Long Term Exercise 1.

- **Proposing microcomputer laboratories** continued from Chapter 4, Collaborative Projects: Exercise 2, Projects 1, 2, or 3.

- **Gathering information for a feasibility report: diaper laundry service** continued from Chapter 6, Collaborative Projects: Short Term Exercise 2.

- **Gathering data for an information report: cleaning supplies** continued from Chapter 6, Collaborative Projects: Short Term Exercise 3.

- **Organizing data for a proposal: recycled paper** continued from Chapter 7, Collaborative Projects: Short Term Exercise 2.

- **Organizing data from a food and music survey** continued from Chapter 7, Collaborative Projects: Short Term Exercise 3.

- **Revising for audience and purpose: insects at the super collider** continued from Chapter 10, Collaborative Projects: Short Term Exercise 2.

a. Determine the graphics your readers need to understand and accept your document.

b. Design the graphics.

To be continued, Chapter 13, Document Design, Collaborative Projects: Long Term.

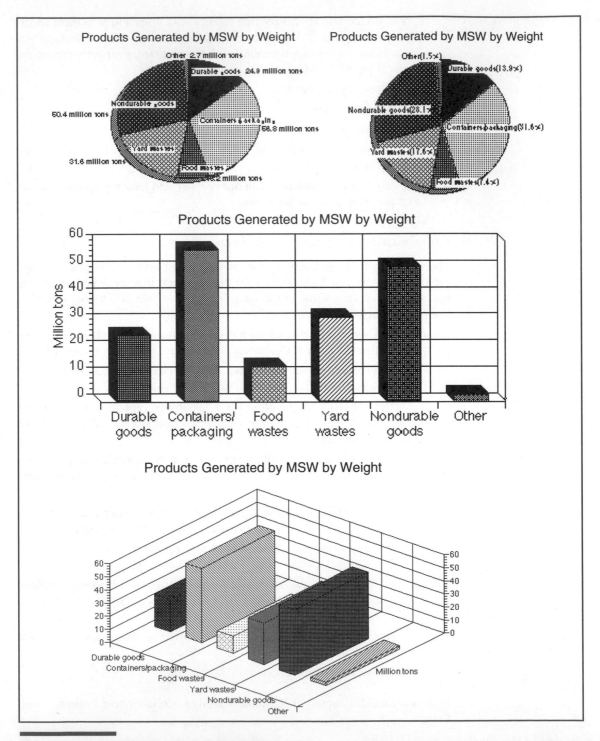

Figure 12-34 Graphs Depicting Major MSW Products

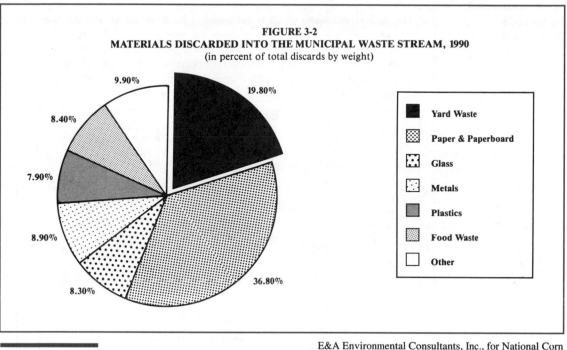

FIGURE 3-2
MATERIALS DISCARDED INTO THE MUNICIPAL WASTE STREAM, 1990
(in percent of total discards by weight)

Figure 12-35 Pie Chart of MSW Materials

E&A Environmental Consultants, Inc., for National Corn
Growers Association, © 1990.

Individual Projects

1. Study Figures 12-35 above and Figure 12-17 on page 342. Both of these charts are based on data from the U. S. Environmental Protection Agency. Figure 12-35 was published by an environmental consulting firm in a booklet for government agencies interested in composting yard waste. Figure 12-17 was published by the plastics industry in an article for the general public.

    a. Each graph emphasizes a different slice of the pie. Why do you think this has occurred?

    b. Figure 12-35 uses a legend, rather than labels, to identify the pie slices. Compare Figure 12-35 with Figure 12-17, which uses labeling. Do you prefer the labeling or the legend? Explain why in terms of readers' needs.

2. An excerpt from a study on aviation noise, which is sent to governmental agencies and interested citizens, appears in Figure 12-36.

    Using the criteria for effective graphics described in pages 356-358, determine whether or not Figure 12-36 is effective for the intended audience.

**Figure 12-36
Diagram of Different Sound Levels**

The A-weighted scale, which is measured in decibels (dBA), was developed to measure sound in a way that approximates the way it is heard by the human ear. This scale assigns more weight to frequencies that people hear more easily. It is recommended by the U.S. Environmental Protection Agency for describing environmental noise along with other Federal agencies. Common single-event sound levels in A-weighted scale are shown in Figure 1.

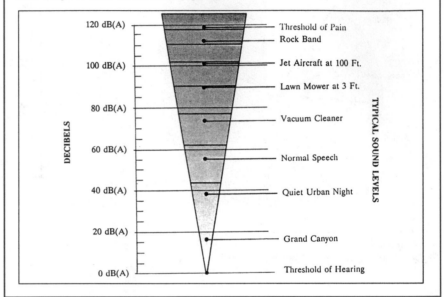

Crawford, Murphy & Tilly, Inc., 1991, F.A.R. Past 150 Noise Capability Study for the Greater Peoria Regional Airport, Peoria, Illinois.

3. Study the information below.

Component	Percent of All Refuse by Weight
Paper	51.70
Food Waste	7.23
Fine Material	6.99
Plastic	6.76
Glass	5.96
Ferrous Metal	5.15
Textiles	4.74
Garden Waste	3.66
Wood	3.00
Diapers	2.23
Non-Ferrous Metal	11.37
Ash-Rock-Dirt	.67
Rubber-Leather	.54

Data obtained from "Degradable Plastics and the Environment" symposium transcript, published by Solid Waste Management Solutions, Chicago, IL, 1988.

    a. Create a graph for a brochure to inform the general public about these categories of waste.

    b. Create a graph to persuade consumers that diapers do not cause much of a waste problem.

4. Write a set of instructions for consumers to construct, use, or repair something. Use appropriate graphics.

    If you cannot think of a topic, you might select one of the following:

Constructing	Using	Repairing
Setting a digital watch	A CD player	Putting in a new light bulb
Building a Lego vehicle	A VCR to record a program	Patching a screen
Checking on oil level in a car	A self-cleaning oven	Sewing on a button
Changing oil in a car	A microwave	Changing a flat tire
Training a dog to sit	A laser printer	Fixing a frayed wire

5. Select a document you have written, either for this class or for one of your other classes, that could be improved by using graphics. Determine the graphics you want to include, then design and integrate them into the text.

CHAPTER 13

Document
Design

WITH THE INTRODUCTION OF personal computers and the development of desktop publishing software, writers have the opportunity to control the design of a document. The visual aspects of a text are intrinsically related to the message, facilitating readers' processes and behaviors. Headings and subheadings help readers accurately predict what they will read as well as locate relevant information if they are skimming through a document. Legible typography facilitates the act of reading the words on a page. And listing helps readers perceive chunks of related data, while numerating the items helps readers perceive their sequence.

Writers make visual decisions at the following textual levels, according to Kostelnick:

Intratext   The lowest level of text, involving letters and words. This level involves selection of typeface, style, and size.

Intertext   The middle level of text, including sentences, paragraphs, and subsections. At this level, the writer determines sentence and paragraph lengths, headings and subheadings, boxes, and spacing.

Extratext   Graphic text. This level involves the designing of visual representations of verbal concepts, such as graphs, charts, photographs, line drawings, etc.

Supratext   Global aspects of a document. At this level, chapter headings, page breaks, etc., are selected.

Chapter 12 covered extratext. In this chapter we will discuss the other levels. Visual text concerns the way in which such elements as headings, bullets, and indentations provide readers with cues for filtering the contents of a message, as well as offer schematic representations of a text's organizational structure. The chapter then considers all four text levels as it examines the integration of visual text and graphics with verbal text in the formatting of a document.

VISUAL TEXT

Visual text provides readers with cues to facilitate their reading and thinking processes and behaviors. Readers can use visual cues to locate information while skimming or scanning a document. They can also use cues to accurately predict what they will read, and to align their reading with their predictions. In addition, through visual text readers can perceive the organizational pattern of a document and the sequence in which information will be presented. Furthermore, visual cues help readers perceive chunks of related information, as well as recognize important data.

Visual text also involves legibility, which relates to readers' ease in accessing the physical properties of a text. To read, a person must recognize the alphabetical letters on a page, and then perceive these symbols as parts of words, sentences, paragraphs, and textual chunks. The easier it is to recognize the letters and to perceive the beginnings and endings of words, sentences, paragraphs, or chunks of text, the more fluently readers can read a document.

Figure 13-1 is a page from an evaluation report. Consider how much you know about the contents of the page by looking at it without reading it.

Just by reading the chapter title, you know you are going to read about airport noise, and that this is the third chapter. Then, by scanning the headings and subheadings, you learn that this chapter includes some "General" information, as

Chapter Three

AVIATION NOISE

I. INTRODUCTION

A. General

The chapter "Aviation Noise" identifies the noise levels generated by aviation activity at the Greater Peoria Regional Airport. Predominantly, that noise is produced by aircraft arriving or departing from the Airport.

Before describing noise exposure at the Greater Peoria Regional Airport, it is necessary to give a brief presentation to readers in order that everyone understands some noise basics. The following discussion is a general examination of sound physics and noise measurements.

B. Background of Noise

General

Specific types of human activity may be difficult to enjoy due to the intrusiveness of certain levels of noise. An effective method for evaluating noise around airports should be utilized and an assessment made to determine the compatibility of existing, proposed or expanded airport use on the adjacent environs.

The human ear is very sensitive to changes in sound pressure fluctuations which are converted into auditory sensations. The two most important characteristics which must be known in order to evaluate a sound are its amplitude and frequency.

The amplitude is a measure of the strength of the pressure variations creating the sound. The hum of a bumble bee is of representative low amplitude, whereas the blast of a ship's horn is of a higher amplitude.

The frequency of sound refers to the number of times the pressure variations occur per second. The rumble of thunder is of low frequency, whereas a high pitched whistle from a flute is of high frequency.

A physical description of noise involves the measurement of sound pressure. These pressure measurements are expressed in decibels (dB). A sound pressure level of 0 dB represents the weakest sound that can be heard by the average human ear. At the other end of the human hearing range, some people may experience physical pain at a sound level threshold of approximately 120 dB. The minimum change in sound pressure that the human ear can detect is typically between 2.5 and 3.0 dB.

Unlike a linear scale, noise is added by the use of logarithms. For example, if two 60 dB sounds are coming from the same direction, the following relationship exists:

$$60 \text{ dB} + 60 \text{ dB} = 63 \text{ dB}$$

Methods of Measuring Noise

The topic of noise measurement has traditionally involved a rather confusing prolific number of units and indices. In

Crawford, Murphy & Tilly, Inc., 1991, F.A.R. Past 150 Noise Capability Study for the Greater Peoria Regional Airport, Peoria, Illinois.

Figure 13-1 Using Visual Text to Predict What You Will Read

well as specific information about "Background of Noise." As you continue, you find that if you read the section on "Background of Noise," you'll learn some "General" concepts as well as "Methods of Measuring Noise." You may have discovered some other things, too. Because the two words "amplitude" and "frequency" are underlined, you probably noticed them and concluded that the "General" discussion on "Background of Noise" would involve these two concepts. Furthermore, because the equation is on a line by itself, surrounded by white space, you probably also noticed it, and concluded mathematical equations are involved in determining "Background of Noise."

Now go back to Figure 13-1 and actually read it.

Because Figure 13-1 is quite legible, your eye should have easily spanned the short columns, and the headings and subheadings should have helped you easily recognize the relationships among the chunks of information. In addition, because the letters are not fancy and are fairly large, you should have been able to read the type easily.

You were able to identify the topics to be discussed, as well as read the text fluently, because of the document designer's use of visual text. There are two types of visual text, spatial markers and typographical markers.

- Spatial markers—titles, headings, subheadings, margins, indentations, centering, listings, numeration, bullets, spacing, and boxes.

- Typographical markers—capital and lower case letters, type styles, sizes, and faces, and columns.

These markers facilitate readers' reading processes and behaviors and enhance their fluency by signaling information, introducing information, indicating hierarchical order and sequence of information, indicating chunks of information, emphasizing information, and creating legible text.

Signaling Information

Visual markers help readers locate information. In long documents, readers who are skimming, scanning, or searching a document use markers to determine whether they want to pause to read a text.

Global markers may include the following:

- Numbered figures, tables of contents, lists of figures and tables, numbered pages.

- Document and chapter titles, and section and subsection headings and subheadings, to indicate chapters, sections, and subsections.

- Headers, usually placed in the top left corner of a page, to indicate chapter titles, and footers, usually placed in center at the bottom of a page, to indicate page numbers.

- Tabs or title pages to indicate new sections or chapters, especially in instructional manuals.

Local markers may involve headings and subheadings.

Introducing Text

A title introduces an entire document and/or a chapter. A heading introduces the major sections of a report. Subheadings introduce sections within a major section.

Titles, headings, and subheadings help readers accurately predict what they will read by indicating the main organizing idea for the information that follows.

In Figure 13-1, the heading "INTRODUCTION" indicates the overall content of the chapter. The subheadings, "General" and "Background of Noise," cue readers to the information contained in the chapter.

If documents are three or more pages, you should insert headings in your text to introduce major categories of information. You should also use headings in shorter documents if readers are expected to scan the text for pertinent information. See how easy it is to scan the information in Figure 13-2.

Headings or subheadings can introduce a chunk of information as short as a paragraph or as long as several pages. However, at least two lines of type should follow a heading or subheading. By introducing headings or subheadings every three or four paragraphs, you help readers predict as they read through a section.

Use the following conventions to compose headings and subheadings:

- **Truncate the main organizing idea.** A heading or subheading is often just a phrase, rather than a complete sentence, e.g., "Designing Documents," "Document Design," "The Design," "Designing Documents in Technical Communities."

  The heading in Figure 13-1, "Background of Noise," is a truncated form of the organizing idea, "Aviation noise provides a background of noise to those living near an airport."

- **Keep headings and subheadings short.** A heading or subheading should be shorter than the line of type it introduces. It is usually no more than four or five words.

- **Indicate action in subheadings for instructions and procedures.** Begin with a verb form, e.g., "Dusting the monitor," "To Dust the Monitor."

- **Do not separate a heading from its text.** If you reach the end of a page after a heading and do not have room for at least a line of text, move the heading to the next page.

- **Keep headings and subheadings grammatically parallel.** Headings should be grammatically parallel with other headings in a document. Subheadings should be grammatically parallel with other subheadings under the same heading.

  If one heading begins with a participle, all others should begin with a participle, e.g., "Dusting the Monitor," "Keeping the CPU Dust Free," "Cleaning the Disk Drives."

- **Use different type sizes to differentiate titles, headings, and subheadings.** A heading should be in a larger type size than the subheadings under it.

- **Use the same typeface for titles, headings, and subheadings in the same document.** If a heading is in Times Roman serif, don't use sans serif for a subheading.

**Figure 13-2
Introducing Text
with Headings and
Subheadings**

Critical Area Planting 342-B-1

CRITICAL AREA PLANTING
OIL WELL SALT DAMAGED AREAS
AND HIGH SODIUM AREAS

DEFINITION

Planting vegetation, such as trees, shrubs, vines, grasses, or
legumes, on highly erodible or critically eroding areas.

PURPOSE

To stabilize the soil, reduce a high concentration of salt (NaCl)
from oil well salt collection ponds or spills and high sodium
areas associated with natric soils.

CONDITIONS WHERE PRACTICE APPLIES

1. Salt concentration will vary significantly on any one site.
   Soil tests are recommended. Most values in the
   specifications are minimum. On areas where the salt
   concentration reaches the upper limits, increasing the
   amount of amendment may be beneficial.

2. Areas should be protected from excessive use. Plants
   selected should be the best available for the planned use.

3. Where erosion is not a problem, annual tillage may be
   performed to aid in the incorporation of organic material.

SPECIFICATIONS

1. Site Preparation

   a. Take soil samples and test to determine salt or sodium
      (Na) concentrations. Samples collected should be taken
      to represent the various zones of concentration of salts
      or sodium. Take samples of 6 inch increments to a
      minimum of 18 inches, ie., 0-6, 6-12, and 12-18.

   b. Where the salt or sodium concentrations exceed any of
      the following values in the top 12 inches, areas will be
      covered with a minimum of 6 inches of suitable rooting
      material.

      (1) Conductivity - 15 millimhos/cm. measured in
          saturated extract or a related measurement.

USDA-SVCS-Illinois March 1990

U. S. Department of Agriculture

Indicating Hierarchical Order and Sequence

By indicating the hierarchical order and sequence of information, readers can follow your discussion logically. Hierarchical order is indicated by headings and subheadings, typographical markers, margins and indentations, and numeration.

Headings and Subheadings Document titles supersede chapter titles, which in turn supersede section headings. Headings supersede subheadings. Notice how easily you can determine the hierarchical order in Figure 13-1.

Typographical Markers Capital and lower case letters used with headings and subheadings further demarcate levels of importance. Words can be written in all capital letters or in a combination of capital and lower case letters. Words in all capital letters are considered to be higher in an organizational hierarchy than those in capital and lower case. Notice, in Figure 13-1, that the subheadings are in capital and lower case lettering, while the heading is in all capital letters.

The following guidelines provide a basic model for designing levels among headings and subheadings. A large number of variations to this model exist, as you can see by examining the various examples in this chapter. The criterion for deciding whether or not a model is viable is whether or not the reader can determine the different levels. Regardless of the model, be consistent. Use the same model throughout a document. Figure 13-1, which has three levels of headings, follows these conventions.

Level 1   Use all capital letters. Place the heading on a separate line from the text. Either center the heading or place it flush left (against the left margin). Skip a line above and below the heading.

Level 2   Use capital and lower case letters. Place the subheading on a separate line from the text. Place the subheading flush left. Skip a line above and below the subheading.

Level 3   Use italicized letters. Use capital and lower case letters. Place the subheading on a separate line from the text and flush left. Skip a line above the subheading but not below it.

Level 4   Underline letters. Use capital and lower case letters. Place the subheading on the same line as the text and indent five spaces. Put a period after the subhead.

Type size may also indicate hierarchical order, as you can see in Figure 13-1. Chapter titles are usually in a larger type size than headings. While headings are seldom in a larger type size than subheadings, they appear larger when they are in all capital letters and the subheadings are in capital and lower case letters. Headings and subheadings are usually in a larger type size than the running text.

Margins and Indentations Margins establish the frame of a text on a page, i.e., top, bottom, right and left. Indentations place text inside margins. By indenting some text further inside a margin than other text, you can indicate hierarchical relationships between chunks of information.

The most important information is placed flush with (against) the left margin. The less important information is then indented to the right, as in Figure 13-3.

The deeper a left margin is indented into the main portion of a text, the lower in the sequence a chunk of information occurs. Readers automatically recognize an outline format, and comprehend the different levels of information.

Numeration You may use a numbering system with your headings and subheadings to indicate hierarchical and sequential order to make certain your readers know which level they're reading. A numeration system is often used for long texts with many levels (at least three) when you don't have access to such typographical markers as italics, underlining, and different type sizes.

Numbers preceding subheadings indicate the relationship between a chapter and its subsections, as well as the relationship between subsections. The number

Figure 13-3
Mapping a
Document's
Organization with
Visual Text

Inter-Office Memo

DATE: October 7, 19xx

TO: John Doe

FROM: Joan Smith

SUBJECT: Status Report - 3rd Quarter

<u>1.00 Corporate</u>

    1.01   Beaver Creek Road

        1.01.1. <u>Can Making Expansion</u> — Engineering work by consultant is underway, with drawings and bid package anticipated by November 1. Coordination meetings have been held with consultant and plant personnel to plan for maintaining traffic and ongoing operations during construction.

        1.01.2. <u>Press Room Dust and Environmental Control</u> - Approval of CPAR has been received and orders prepared for equipment. Deliveries of major dust collecting equipment are anticipated by February 1. Installation and change out of existing equipment is being coordinated with the Can Making Expansion Project to keep interferences to a minimum.

        1.01.3. <u>Compressed Air Expansion</u> - The building addition has been completed and the new compressor equipment is installed and running. Plant personnel have handled this project on a daily basis, with the Facilities group providing piping installation, equipment layouts, and inspection assistance....

1.01 in Figure 13-3 indicates that the section on "Beaver Creek Road" is part of the report on the corporate status. In addition, the numbering system indicates that the section on "Compressed Air Expansion" is the third of the various subsections under the Beaver Creek Road operation.

There are two major systems for numerating information, the numeric system and the outline system.

a. The *numeric system* (Figure 13-3) is comprised of:

- Chapter number(1; 2; 3...)
- Section number(1.1; 1.2; 1.3...)
- Subsection number(1.1.1; 1.1.2; 1.1.3...)

Example:
1. (Chapter 1)
    1.1 (Chapter 1, section 1)
        1.1.1 (Chapter 1, section 1, subsection 1)
        1.1.2 (Chapter 1, section 1, subsection 2)
    1.2 (Chapter 1, section 2)
2. (Chapter 2)
    2.1 (Chapter 2, section 1)
    2.2 (Chapter 2, section 2)
        2.1.1 (Chapter 2, section 2, subsection 1)

b. The *outline system* (Figure 13-4) is comprised of:

- Roman numerals (I, II, III...)
- Capital letters (A, B, C....),
- Arabic numerals (1, 2, 3....)
- Lower case letters (a, b, c...)

Example:
I. (Major category)
    A. (First subcategory)
        1. (Second level subcategory)
            a. (Third level subcategory)
            b. (Third level subcategory)
        2. (Second level subcategory)
    B. (Second subcategory)
        1. (Second level subcategory)
        2. (Second level subcategory)

**Figure 13-4
Using an Outline
System to Indicate
a Text's Organiza-
tion**

```
                              Cover and Green Crop Manure 390-2

   III. For Summer Cover, Green Manure, and Catch Crops

        A. Apply line fertilizer according to soil tests or crop needs.

        B. Prepare a suitable seedbed or no-till plant.

        C. Select one of the following seed mixtures. Mixtures are
           applicable to all plant suitability zones.

             Mixtures              Rates         Plant Suitability Zones
                                                     and Seeding Dates
             1.  Oats              50 lbs/ac.    (I) Early spring- June 1
                 Alfalfa          50 lbs/ac.    (II) Early spring-May 15
                 Sweetclover     10 lbs/ac.   (III) Early spring-May 15

             2.  Sudangrass       25 lbs/ac.     (I) May 1-July 1
                                                (II) May 1-June 25
                                               (III) May 1-June 15

             3. Catch crop in fall seed small grains

                a. Alfalfa      10 lbs/ac.
                                              (I) Feb. 15-March 15
                b. Red clover    8 lbs/ac    (II) Feb. 1-March 1
                                            (III) Feb. 1-March 1
                c. Sweetclover  10 lbs/ac.

        D. Incorporate crop into soil in late fall or the spring
           following seeding.

   II. Cover Crops for Orchards, Vineyards, and Nurseries

        A. Apply lines and fertilizer according to soil tests or crop
           needs...

                                        USDA-SCS-Illinois March 1990
```

U.S. Department of Agriculture

Numeration is also used to indicate the sequence of items in lists of information, and steps in procedures and instructions (Figure 13-5). Readers expect the first item to be either the most important or the first step in a sequence. Notice how the writer follows the convention in Figure 13-5.

Numeration replaces transitional phrases, such as "first," "next," "then," and "after completing X..." By replacing introductory phrases, such as "Before doing x...", numeration effectively eliminates dangling participles.

Sentence: After tightening the screw, you should turn the bookshelf around.
            OR
            After you tighten the screw, turn the bookshelf around.

Numeration: 1. Tighten the screw.
            2. Turn the bookshelf around.

**Figure 13-5
Listing Items in
Procedures**

If Evacuation Is Ordered

1. Check exit routes for damage and to make sure evacuation is possible.

   • Depending on timing of the earthquake, hospital corridors may be filled
     with food or medication carts. These can overturn and block access. So, an
     initial effort should be made to clear obstructions.

   • Expect to find doorways and exit routes blocked. Also, expect to find exit
     doors jammed.

   • Expect to find one or more of the fire stairs disabled. You may need to
     reroute evacuees.

   • Assign staff to help clear debris and open blocked doorways

2. Instruct ambulatory patients and staff to leave the building in an orderly manner to a
pre-established gathering place. Assist nonambulatory patients.

   • In past earthquakes, ambulatory patients were evacuated first,
     nonambulatory patients followed.

   • During evacuation it is important to make sure that medical records remain
     with the patient. This will help insure continuity of care and avoid
     unnecessary tests and other treatment.

   • DO NOT USE ELEVATORS.

   • Non ambulatory patients may take different exit routes than ambulatory
     patients to expedite evacuation.

3. Seek safety outside

   • Instruct those leaving your building to move into safe areas away from
     buildings and other potential hazards.

   • If your building has sustained heavy damage, preconceived interior
     locations for disaster response (e.g., triage, command center, etc.,) may be
     hazardous. Consider alternative locations based on your facility damage
     assessment.

California Governor's Office of Emergency Services

Numeration not only eliminates dangling participles, but also shortens sentences. Information in a list is usually reduced to the essentials readers need to know.

Although bullets do not indicate a specific sequence, readers usually infer that bulleted items are sequenced either numerically or in order of importance. Notice the writer uses bullets, rather than numeration, in Figure 13-6.

Indicating Chunks of Information

As you have learned, readers chunk information in order to comprehend a message. Chapters, sections and subsections, paragraphs, and lists can be chunked visually as well as verbally to facilitate readers' reading processes.

**Figure 13-6
Chunking
Information**

Other Forms of Ferrous Recycling

Other markets for ferrous metals recycling include tin, iron foundries, the production of ferro-alloys and the copper precipitation iron market. More information on these forms of recycling is contained in Keep America Beautiful, Inc.'s *Multi-Material Recycling Manual.*

Scrap Tires

Well over a billion discarded tires are stockpiled around the country. Tires are usually discarded at retail outlets that sell and mount new tires. They are sorted, and those found satisfactory are retreaded. The remainder, in some cases, are shipped to recyclers, frequently the seller of new tires. Where there are no markets, or acceptable disposal sites, the tires pile up.

Landfilling of whole tires is banned in many states because they cannot be compacted and often rise to the surface due to the resiliency of the rubber. Some landfills require that tires be shredded or split, while others discourage tire disposal by charging disposal fees of $1 or more per tire. Abandoned tires serve as a breeding ground for mosquitos and, when ignited, can take months to extinguish and produce oil-like liquids which can contaminate surrounding surface water and ground water.

The difficulty of disposal is directly related to the lack of markets for the scrap tires. Increased recycling promises a diminishing stockpile of discarded scrap tires, a savings of landfill space, and the assurance that a valuable material and potential energy source will not be wasted.

Following is a brief synopsis of how tires are being reused in the United States:

- **Reclaimed Rubber**
  Approximately 10 million tires are disposed of annually through a reclamation process involving grinding, shredding and pulverizing the tires. The material is formed into sheet rubber and sold in bales to producers of molded materials and semi-pneumatic tires.

- **Retreading**
  Retreading requires about 30% of the energy necessary to produce new tires and it provides nearly 80% of the mileage of a new tire. Almost 20 million truck tires and 17 million passenger tires are retreaded annually. The market for good quality used tires will always exist in the retreading industry.

- **Marine Use**
  Scrap tires are being used as artificial reefs, created by stringing tires together with non-corrosive cable and drilling holes in them so they will sink in the water.

- **Incineration**
  Tires added to mass burn units or shredded tires in a refuse-derived fuel system burn completely and are a good source of fuel. Additionally there are waste-to-energy plants that solely burn chipped tires, known as tire-derived fuel. Other plants burn whole tires to produce electricity.

Keep America Beautiful

Keep America Beautiful

Long lists should be divided into subcategories, creating manageable chunks to facilitate readers' thinking process. Use subheads to introduce each subcategory. Begin numerating each list under a new subhead with the number one. Consider how much longer the instructions in Figure 13-7 would appear if they were not subdivided.

You can visually indicate chunks of information by using a variety of techniques, including titles, headings, and subheadings; spacing; indentations; boxes; and lists.

- Titles, headings, and subheadings mark off large chunks of information. Capitalizing all the letters, or the first letter, of each main word in a heading or subheading further serves to create a visible frame around chunks of information. Readers can easily see that when they arrive at

**Figure 13-7
Subdividing Long
Lists**

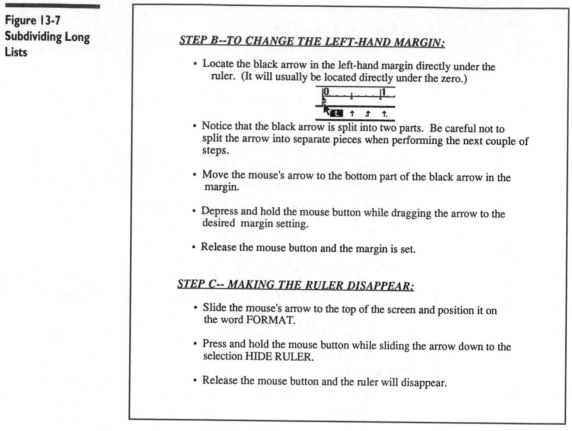

STEP B--TO CHANGE THE LEFT-HAND MARGIN:

- Locate the black arrow in the left-hand margin directly under the ruler. (It will usually be located directly under the zero.)

- Notice that the black arrow is split into two parts. Be careful not to split the arrow into separate pieces when performing the next couple of steps.

- Move the mouse's arrow to the bottom part of the black arrow in the margin.

- Depress and hold the mouse button while dragging the arrow to the desired margin setting.

- Release the mouse button and the margin is set.

STEP C-- MAKING THE RULER DISAPPEAR:

- Slide the mouse's arrow to the top of the screen and position it on the word FORMAT.

- Press and hold the mouse button while sliding the arrow down to the selection HIDE RULER.

- Release the mouse button and the ruler will disappear.

by Tim Curtin

the subhead, "Marine Use," in Figure 13-6, they will stop reading about retreading and begin reading about aquatic uses.

- Spacing separates a wide variety of textual aspects, including sections, columns, paragraphs, lines, words, letters, and graphics. It also separates text from the edge of the paper.

   In Figure 13-7, extra space separates the heading from the main text. Extra spacing also separates the items.

- Indenting helps separate lists of items (Figure 13-6), or steps in a process (Figure 13-7). Indenting also separates long quotes from a running text.

   There are two types of indentations—regular and hanging. Regular indentation occurs when only the first line of a paragraph is indented further to the right than the following lines. Hanging indentation occurs when the lines following the first line in a paragraph are indented further to the right than the first line or the heading or subheading, as in Figure 13-6. Hanging indentation is used to format lists and long quotes.

- Boxes mark off chunks of information. In Figure 13-8, a box is used to chunk the good aspects of sanitary landfills.

**Figure 13-8
Using Boxes**

VI.
SANITARY LANDFILLS

Modern sanitary landfills are the result of careful planning, engineering, monitoring and supervision. Federal, state and sometimes local regulations must be met and permits obtained from health and environmental officials. Additionally, the permitting process involves review of construction plans and safety features.

Modern landfill design features include liners, leachate collection and removal systems, methane gas controls, and environmental monitoring systems. In some cases, waste-water treatment plants, recycling facilities and energy production plants are also incorporated.

Sanitary landfills provide a final resting place for most solid and non-hazardous residential, commercial and industrial wastes. There are 6,000 solid waste landfills operating today. According to the U.S. Environmental Protection Agency, more than half of these will face closure within 10 years. Currently 75-80% of municipal solid waste is disposed of in landfills. While new technologies are gaining popularity, there will always be a need for the landfill, to dispose of nonrecyclables and ash from combustion.

Prior to the mid-1970's many landfills were basically open dumps and allowed to receive hazardous waste. In 1976 the Resource Conservation and Recovery Act (RCRA) was passed by Congress to protect human health and the environment from improper waste management practices. RCRA classified landfills and regulates what types of waste they may receive.

Site suitability is determined by a number of factors including careful analysis of the surface and subsurface geology, hydrogeology, the nature of adjacent environments, access routes and proximity to waste generation sources. It requires particularly careful analysis of ground water sources and flow direction, along with soil composition and site engineering. Only after a potential site passes the stringent legal, environmental and engineering criteria in all these areas can work begin in the elaborate preparation required for a modern sanitary landfill operation.

The bottom and sides of a landfill are lined with layers of compacted clay and/or thick plastic liners to insure that any liquid which enters the excavation is retained. Landfill sites are designed so that all rain or snow which falls on the site is retained, collected and treated before release. A network of drains collect the liquid (leachate) that has percolated through the wastes and directs it to recovery points where it is collected for treatment.

Within a typical landfill site, the area for waste disposal is divided into a series of individual cells. In daily disposal activities, only a small portion of the site (known as the "working face") is used, minimizing exposure to wind and rain. At the conclusion of each day's activities, a layer of earth (known as "daily cover") is spread across the compacted waste to minimize odor and prevent insect and vermin problems. Daily cover may consist of soil, foam material, or sheets of synthetic material. Each cell is filled and capped off with a layer of clay and earth and seeded with native grasses according to an approved closure plan.

Keep America Beautiful, Inc.

- Lists separate items visually from the main text, helping readers see the items both as separate entities and as parts of a single chunk. In Figure 13-6, readers can easily see that there are four uses for reusable tires, while they simultaneously recognize that the four items are all part of a single category—uses of reusable tires. By looking at Figure 13-6, readers know that "retreading" is one method of reusing tires.

Each item of information in a list, or each step in a process, should be placed on a separate line, as illustrated in Figure 13-7. Lines should be single-spaced within each item or step, but double spacing should be used between the various items or steps so that the beginning and ending of each chunk or step is visually demarcated for the reader.

Make all items or steps grammatically parallel. Notice that all items in Figure 13-7 begin with a verb.

**Figure 13-9
Emphasizing
Information
Visually**

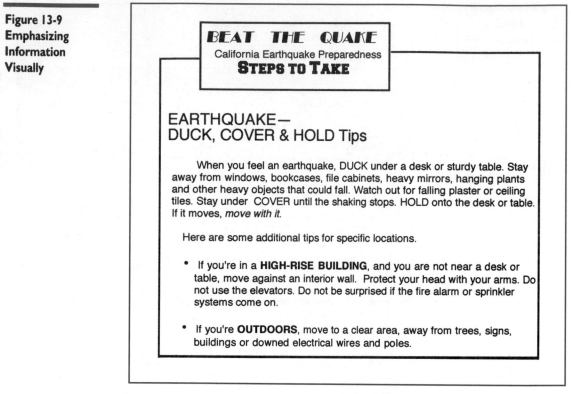

California Governor's Office of Emergency Services

Emphasizing and Subordinating Information

By visually emphasizing and subordinating information, you help readers comprehend your message. Emphasis and subordination may be indicated typographically as well as by spatial markers such as boxes.

- Typographical markers help emphasize concepts. Capitalizing an entire word, phrase, or sentence makes it stand out from the rest of the text (Figure 13-9).

  Type styles, which include **boldface**, <u>underline</u> and *italics*, also cause words, phrases, and sentences to stand out, thereby emphasizing information. The boldfaced type in Figure 13-6 enables readers skimming the document to locate specific information easily. The boldfaced type in Figure 13-9 emphasizes concepts readers need to remember.

- A box can either emphasize or subordinate information. Text which is boxed may include very important information the reader needs to know. However, a box may also include information that is subordinate to the reader's purpose. A concrete example of an abstract concept may be boxed if readers who are familiar with the concept do not need to read it. Supporting details that are not essential for readers' understanding and

**Figure 13-10
Boxing Subordi-
nate Information**

Tidd Test Program

During the 3-year operational phase, plant performance and operation will be carefully monitored and analyzed to verify technical performance of the plant and its components. Tests will establish the reliability, load following capability, and operating and maintenance costs.

Although the Tidd demonstration plant is not expected to be economical or operate at a high level of availability, the performance and test data collected during its operation are essential for confirming technical feasibility and economic viability. With this data, more accurate forecasts of the cost of electricity from a PFBC combined-cycle plant can be developed.

Acceptance Testing

Acceptance testing will demonstrate the technical feasibility of PFBC combined-cycle technology at the 70-MWe scale. Of primary interest are the

Parameters Continuously Monitored

During Tidd's operation, plant performance will be monitored continuously. Detailed analysis of the data collected will yield information about the deterioration of plant performance over time and will be used to ascertain preferred operating conditions and to improve future designs.

Major parameters being tracked are listed below:

1. Fluidization conditions (temperature, bed height)

2. Fluidized-bed air flow, gas flow, temperature distribution, excess air factor, bed ash removal rate, elutriation rate

3. Combustor mass flow, temperature, pressure, and heat transfer

4. Calcium-to-sulfur molar ratio

5. Gas turbine efficiency and output

6. Economizer mass flow, pressure, temperature, effectiveness, and heat transfer

7. Fly-ash mass flows, particle size distribution, and chemical analysis

8. Bed-ash mass flows, particle size distribution, and chemical analysis

9. Cyclone dust removal efficiency and pressure loss

10. Coal and sorbent mass flows, particle size distributions, and chemical analyses

11. Coal and sorbent feed system consumption of water and transport gas

12. Coal heat input

13. Auxiliary power needs for all PFBC-specific systems

14. Total combustor heat transfer efficiency.

Reprinted from "Clean Coal Technology, Topical Report No. 1," March, 1990, American Electric Power Service Corporation and U.S. Department of Energy.

acceptance of a message may be boxed, as in Figure 13-10, which appeared in a brochure to inform the general public about a new type of coal plant. You should be able to read the main text in Figure 13-10 without needing to read the text in the box. When you read the boxed text, notice that it is more detailed and technical. It provides for readers who are more knowledgeable about the subject, and who may be on the periphery of the technical community.

Because the information in a box is not part of the running text, readers have the option of reading the material or skipping it. By scanning the heading of a boxed text, readers can determine whether the information is extraneous to their needs and can be skipped, or whether it's important and should be read.

A box containing important information should be placed in a prominent position on a page so readers will not miss it, as shown in Figure 13-8. A box containing information that may be extraneous to readers' needs should be placed in a secondary position on a page, as shown in Figure 13-10.

Creating Legible Text

By creating legible text, you facilitate readers' fluency. Legibility can be improved by creating short chunks of text and by using typefaces that are easy to read.

Chunks By using subheadings to break long sections into several short ones, you help readers read a text more easily. Imagine how difficult it would be to read the text in Figure 13-11 if the heading and subheadings were removed, and the lists were written out as full sentences.

Typography Capital letters; type size, face, and style; line and column width; and justification all affect legibility.

Capital and lower case letters. Continuous Text Printed in Capital and Lower Case Letters or in all lower case letters IS EASIER TO READ THAN CONTINUOUS TEXT IN ALL CAPITAL LETTERS.

Type size. Type size relates to height and width of a letter. Type is measured in points; one point equals $1/72$ of an inch. Most typewriters and computers are set at 10 or 12 point, about $1/6$ of an inch.

Large type sizes for headings and subheadings help readers locate information as well as "see" various chunks of information. In addition, using different type sizes for titles, headings, and subheadings helps readers to perceive the various hierarchical levels of a text, as shown in Figure 13-12.

Columns. Small type printed in columns is easier to read than text printed in large type across an entire page. Pages can be divided into two or three columns of print.

Short columns, rather than a single line of print across a wide page, make it easier to read a text fluently. Readers' eyes don't have to span a long horizontal row of letters. Instead, they can take in several lines at a single glance. The large righthand margin in Figure 13-12 creates a short ($4^1/4$ inch) single column, which is not much larger than the $3^1/16$ inch columns in the double column page illustrated in Figure 13-11. Many writers who publish documents on $8^1/2$ x 11 inch paper are beginning to adopt the short, single-column layout shown in Figure 13-12.

**Figure 13-11
Easy-to-Read Text**

CHAPTER 9—COMPOST END USE

9.0 INTRODUCTION

A leaf and yard waste composting program is more than a solid waste management strategy. One is also entering the commodity market for soil amendments. Therefore, one of the first things to be done in planning a leaf and yard waste composting program is to determine local end uses for the compost. Adequate information on the potential users' requirements for quality and quantity, is a mandatory prerequisite for defining the composting method, equipment, and operations necessary to produce a compost meeting the user's demands. As long as a high quality, consistent, and stable compost is produced there should never be any difficulty securing more than adequate demand.

9.1 CHARACTERISTICS

Compost is a stable, soil-like material that is an excellent substitute for topsoil, peat moss, and mulches in horticultural and agricultural applications. Leaf and yard waste compost is considered to be an excellent soil conditioner or amendment. Typically, it is a very high quality compost with very low to not-measurable concentrations of heavy metals and toxic organic compounds. Leaf and yard waste compost requires post-processing in order to have the greatest value. Screening and shredding will remove clumps, twigs, branches, and inert contaminants, producing compost which has a very consistent quality and high value.

Compost's benefits to soil condition include:

- Improved soil aggregation
- Improved water infiltration
- Improved water retention
- Improved soil porosity
- Improved soil aeration
- Decreased soil crusting

In most cases, this compost has limited fertilizer value due to low nitrogen, phosphorus, and potassium content.

9.2 USER TYPES

Compost users can be grouped into four categories: commercial, residential, public agencies, and land reclamation. The specific user types are summarized as follows:

Commercial	Public Agencies
Landscape Contractors	Park Maintenance
Nurseries	Decorative Planting
Greenhouses	Curb Repair
Turf Farms	Backfilling
Topsoil Suppliers	Community Gardens
Soil Blenders	
Golf Courses	**Land Reclamation**
	Landfill Cover
Residential	Mined & Derelict Land
Garden, Lawn and Flower	Revegetation

Horticultural operations require large, continual supplies of high quality organic material. For landscapers, compost can be substituted for topsoil and peat moss in landscape construction and maintenance. Greenhouses and nurseries may substitute compost for peat moss in potting and planting soil mixes. Turf farms and golf courses may use compost to help establish new sod or as a fine-textured top dressing. Public agencies can use compost in a wide variety of ways, thereby reducing their purchases of soils and mulches. Compost can be used effectively as a final cover for landfills and for revegetation of strip-mined or derelict lands. If adequate amounts of free compost are available, public agencies may even expand their use of compost above previous levels that were constrained by budgetary limits.

9.3 DEMAND

In order to determine the potential demand for compost, one must gather the following information from potential users (Connecticut DEP, 1989):

- Specifications for organic materials
- Capacity to utilize compost (both seasonal and annual)
- Shipping and handling requirements
- Potential revenue from sales of compost

Another important factor is the price of competing materials such as topsoil, peat moss, or other compost products. Most potential users require high quality and consistency. Therefore, maintaining an adequate, stable demand for compost will be greatly dependent on having adequate supplies of a high quality compost at competitive prices. It also should be remembered that compost is in highest demand during the spring and early summer months.

Public education also plays a key role in increasing product demand. All potential markets need to be informed of the values and benefits of compost, stressing the value of adding organic matter as well as nutrients. The fact that the compost is a locally produced resource can be a secondary factor contributing to steady demand.

9.4 CONSTRAINTS

9.4.1 User Resistance to Change

In general, it is difficult to capture a market share for a new product, such as compost. If potential users are satisfied with certain existing materials, there is little incentive to change, unless there is a significant price difference. On the other hand, offering compost for free can make potential users suspect that there must be something wrong with it since it is being given away. Private sector users will be especially resistant to leaving their current supplier if it cannot be demonstrated that a large and stable supply will exist. It will require several years of work to overcome such resistance to change. Efforts can focus on providing accurate product information, user trials, cooperative extension and university testing programs, extensive public education, demonstration plots in visible areas, etc.

E&A Environmental Consultants for the National Corngrowers Association, © 1990

**Figure 13-12
A Legible
Document**

6. GOVERNMENT ACTION -- AND INACTION --
ON LEAD

*To date, Congress has responded to the lead prob-
lem on a number of fronts, enacting legislation to control
lead in paint, ambient air, drinking water and solid waste.
Initiatives to reduce lead in gasoline and lead-soldered
food cans have made headway in "de-leading" the nation
as a whole. But there have been only sporadic efforts to
control the most stubborn and significant source of
children's exposure to lead: house paint.*

The Lead-Based Paint Poisoning Prevention Act

The United States historically has been slow to respond to the health
threats posed by lead paint. While most European countries signed a treaty
banning the use of lead-based paint in the interior of buildings in 1921,[1] the
federal government took no action at all until 1971. That year, citing the
"epidemic proportions" of childhood lead poisoning in large cities, Congress enacted the first national lead abatement legislation. The Lead-Based
Paint Poisoning Prevention Act[2] sought to address three distinct aspects of
the lead-paint problem. Specifically, the Act set some limits on the use of lead
paint; created grants for lead-poisoning screening and treatment programs;
and required the submission of a report on abatement methods. The three
components have had notably different histories over the intervening two
decades.

Limits on the Use of Lead Paint: Contrary to general belief, the Act
did not ban the production of lead paint or even all of its uses in dwellings.
Rather, it merely prohibited the use of leaded paint on surfaces accessible to
children. The Act also authorized the Secretary of Health, Education, and
Welfare to issue regulations prohibiting the use of lead-based paint in Federal construction or rehabilitation of residential housing.

Recognizing the need to strengthen these provisions, Congress amended
the Act in 1973 to prohibit the use of leaded paint (defined as paint containing
0.5 percent lead by dry weight) in federally funded housing and extended the
prohibition to toys and other articles.[3] Not until 1977 was the use of lead paint
in housing actually banned -- and that ban was imposed by a regulation issued
by the Consumer Product Safety Administration rather than by statute.[4]

Grant Programs: The 1971 Act placed the administration of the grant
program with the Secretary of Health, Education, and Welfare. A decade

**Merely prohibiting the use of addi-
tional lead-based paint in dwellings
has proven to be an inadequate response to the problem.**

Environmental Defense Fund. *Legacy of Lead: America's Continuing
Epidemic of Childhood Lead Poisoning* (1990).

Justification. When a page is comprised of a single column, unjustified text,
text which has a ragged edge at the right margin (as in Figure 13-9), is easier to
read than justified text (text which is flush with the right margin as this paragraph is). The jagged edge helps readers recognize which line they have completed by giving them a visual image to remember.

However, when a page is comprised of two or more columns, text should
be justified, as in Figure 13-11. The justified text prevents the left column from
appearing to run into the one on the right. The right column should be justified
so that it is symmetrical with the left column.

Typeface. Typeface refers to the design of an alphabet. Some are fancy, such
as Zapf Chancery, while others are plain, such as Times Roman. Typefaces may
be serif or sans serif. It's easier for readers to read type with serifs; their eyes
move between the letters more easily. For examples of various typefaces, see
Fugure 13-13.

**Figure 13-13
Different
Typefaces**

> *Zapf Chancery is a fancy typeface.*
>
> Times Roman is a very readable typeface.
>
> Serif typefaces (such as Bookman) have "tails" on each letter.
>
> Sans Serif typefaces (such as Helvetica) do not have "tails."

Do not use fancy typefaces; readers often have difficulty determining the letters. Plain typefaces such as Times Roman, are much more readable. Figure 13-12 uses simple serif style for the body text, and a sans serif face (Helvetica) for the headlines.

STRATEGY CHECKLIST

1. Signal information at the global level.
2. Introduce text with titles, headings, and subheadings.
3. Indicate organization, hierarchical order, and sequence with headings and subheadings, typographical markers, numbering systems, indentations, and numeration.
4. Emphasize information by using type styles, boxes, and lists.
5. List information, instructions, and procedures.
6. Separate extremely important or extraneous information from the main text by using a box.
7. Use legible typography.
   - Use a type size which can be easily read. If you are using a typewriter or computer, use 10 or 12 point.
   - Use a typeface with serifs.
   - Do not use fancy typefaces.
   - Do not use all capital letters for continuous text.
   - Use unjustified right margins for single column pages; use justified columns for multi-column pages.
   - Keep line length within the reader's eye span. Do not exceed forty characters and spaces of type. Use columns if you are using a small type size or large paper. Two columns of approximately 3½ inches on a paper 8½ inches wide is a good width.

FORMATTING

As you have seen, both visual cues and graphics facilitate readers' comprehension and fluency. The way in which visual text and graphics are formatted can further facilitate readers' processes and behaviors. The format of a document refers to both the integration of visual and verbal text locally on a page and globally within an entire document.

Effects on Readers

Because a format involves both visual and verbal text, it affects readers' comprehension and fluency.

Effective formatting of a document enhances readers' memory, affects their attitudes, and facilitates their processes and behaviors.

Memory Because readers often visualize both the content and the appearance of a text, the format of a document can be as important in helping readers learn information as the verbal text.

Read the information in Figure 13-14.

**Figure 13-14
Helping Readers
Remember
Information
Visually**

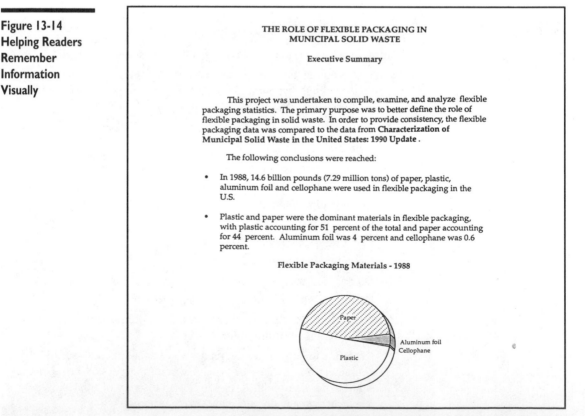

THE ROLE OF FLEXIBLE PACKAGING IN
MUNICIPAL SOLID WASTE

Executive Summary

This project was undertaken to compile, examine, and analyze flexible packaging statistics. The primary purpose was to better define the role of flexible packaging in solid waste. In order to provide consistency, the flexible packaging data was compared to the data from **Characterization of Municipal Solid Waste in the United States: 1990 Update** .

The following conclusions were reached:

- In 1988, 14.6 billion pounds (7.29 million tons) of paper, plastic, aluminum foil and cellophane were used in flexible packaging in the U.S.

- Plastic and paper were the dominant materials in flexible packaging, with plastic accounting for 51 percent of the total and paper accounting for 44 percent. Aluminum foil was 4 percent and cellophane was 0.6 percent.

Flexible Packaging Materials - 1988

Paper

Aluminum foil
Cellophane

Plastic

Prepared for Flexible Packaging Association by Franklin Associates, Ltd.

Without looking back at the information, what do you remember? You probably remember that the text is about packaging; that several materials are involved, including paper, plastic, and something else; that paper and plastic are used the most; and that the amount of materials is specified in the first bullet, though you may not remember the specific amounts. You probably remember these facts because they are stated in the headings, in bulleted items, and in a graph. As you think about the information, you probably pictured the page in Figure 13-14 in your "mind's eye."

A page of text becomes a chunk of information. Notice, in Figure 13-14, that there is enough space between the graph and the bottom of the page for several more lines of text. However, the designer leaves the bottom portion of the page empty and begins discussion of a new idea on the next page. Readers therefore see the new idea as a self-contained entity, rather than as part of the previous concept.

Just as a page indicates a chunk of information, a chapter or major section indicates a large chunk at the global level. For this reason, it is always good to begin a new chapter or section on a new page. The white space between the end of the previous chapter or section and the beginning of the new one visually separates the two chunks.

Attitudes Before you *read* a document, you *look* at it. If it is a long document, you may scan the table of contents or skim through the pages. Depending on the size of the paragraphs, the amount of white space surrounding the print, and the size of the type, you may consider the document "light" (lots of white space) as in Figure 13-14, or "dense" (packed with information) as in Figure 13-15. As you study the layouts in the two figures, consider the features we have already discussed that make Figure 13-14 easier to read.

These visual impressions may influence your attitude toward a message before you ever read the contents. Readers perceive dense text as containing more information or steps than does light, "airy" text containing the same information. A dense-looking message usually indicates that a good deal of time needs to be spent reading it. A long set of directions often indicates plenty of time should be set aside to engage in an activity. An airy text appears quick and easy to read, the steps easy to follow.

Processes and Behaviors In determining how to format a document at the global level, as well as at the level of a section or a page, you need to consider readers' reading processes and behaviors.

Consider readers' previous reading experiences. Just as readers need to relate new information to their previous knowledge, they look for formats which resemble those to which they are accustomed. Experts are accustomed to dense reports, such as the one shown in Figure 13-15. However, notice that even in Figure 13-15, the designer breaks up the text with headings.

On the other hand, if you are formatting a report for the general public, you should use the airier format illustrated by Figure 13-14. Novice readers often

EXECUTIVE SUMMARY

INTRODUCTION

The current intense debate over federal, state, and local waste management policies is taking place in an environment rich in speculation and opinion, but poor in factual evidence about the true contribution of various materials and products to the total quantity of municipal solid waste that must be managed.

The fact is that until now, there has been no comprehensive database available to policy-makers and the public that characterizes the volume of the various components of municipal solid waste (MSW). As a result, many estimates have been made and published--without any real scientific basis--that have had a profound impact on waste management policies. In the case of plastics, volume estimates reported in the news media have ranged from 30 percent to 70 percent of MSW.

This report presents the results of independent research which offers the first comprehensive, systematic characterization of the relative volumes of the components of MSW. The research was sponsored by the Council for Solid Waste Solutions. The report describes the development of an experimentally derived set of conversion factors which have enabled researchers to use an existing database that characterizes the weights of MSW components to determine the volume of those components in landfills. This research is important because, simply put, landfills do not close because they are overweight, they close because they have reached their volume capacity.

RESEARCH APPROACH

Franklin Associates, Ltd., prepares for the U.S. Environmental Protection Agency (EPA) a widely used database characterizing the weight of various materials and product categories in municipal solid waste. The challenge presented was to find conversion factors for each product category in MSW that would allow existing weight data to be converted into volume equivalents--expressed in cubic yards under landfill conditions.

Prepared for the Council of Solid Waste Solutions by Franklin Associates, Ltd.

need to be motivated to read a report. They need to be assured the information will not take too long to read, and they need to read the information in a familiar format.

The format to which the general public is most accustomed is that of a newspaper. In most newspapers the main story is in the upper right hand corner, story number two is located in the upper left hand corner, and story number three is located at the top of the page in the middle columns. Lesser stories are placed at the bottom of the page. In contrast, readers of *USA Today* have become accustomed to a format in which the main picture and story is placed in the middle of the page.

Newspapers, including *USA Today*, use a format which involves balance, contrast, and parallelism. If a two-column graphic is placed at the upper right of a page, it is balanced with a parallel, two-column visual—or at least headline— at the contrasting end (bottom) of the page. Most magazines follow a similar format, but with variations. Figure 13-16, which appeared in a brochure on clean coal, exemplifies this type of format.

**Figure 13-16
A Balanced Layout**

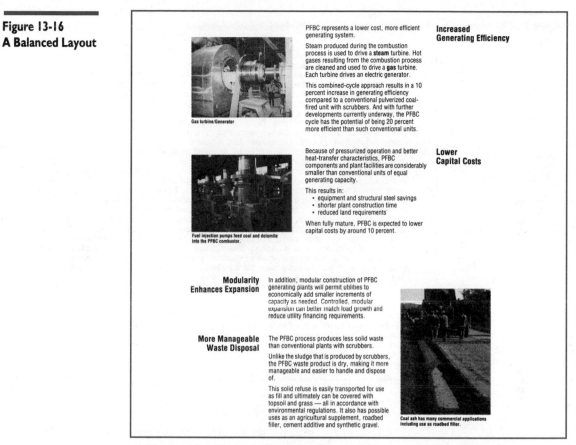

Reprinted from "Clean Coal Technology, Topical Report No. 1," March 1990, American Electric Power Service Corporation and U.S. Department of Energy.

Most of the time, unless you are a graphic designer or technical writer, you will not have the time, nor be expected, to create such visually elaborate documents. Most of the pages in your reports involving graphics will look like the one shown in Figure 13-14.

Designing Effective Formats

To design a document effectively, you need to use spatial and typographical markers, as well as graphics, appropriately.

Spatial and Typographical Markers Let's examine how a document designer uses visual text to create a legible, airy document out of a dense text. Look at Figure 13-7, and consider your attitude toward the text. It probably looks boring to you. Read the text and, as you go along, consider ways you could make the text more airy and look more interesting.

Figure 13-17
A Dense Text

PLASTIC PACKAGING

Since World War II, plastics have enjoyed impressive growth as a packaging material. A recent study by the industry indicates plastics will continue to grow in packaging applications, increasing in volume by 70 percent by the year 2000. There are many reasons for the growth in plastics packaging applications. For one thing, consumer demand for goods in plastic packaging is high. A recent study by Packaging Magazine showed 63 percent favored plastics for microwaveable packages.

There are countless packaging applications where plastics play a crucial part: large amounts are used for foods and personal care products (plastic films for liners and overwraps); bottles, jars, and containers in various molded shapes; pails, shrink wraps; bags and sacks.

There are several good reasons why plastics have enjoyed such a significant increase in packaging applications and will continue to do so. First of all they are break-resistant. Plastics are resistant to breakage which is important for a host of consumer products from food to personal care. And, unlike other packaging materials, plastics resist leakage. Plastic containers also have excellent insulation properties, keeping foods either hot or cold as desired. Furthermore, they can be molded into many convenient sizes and shapes, remaining rigid where the quality is important and flexible where that's a benefit. Finally, many plastics for packaging can be reused, including bottles and strong grocery sacks.

New Product News recently reported 1,000 new products were introduced to supermarkets and drugstores during just one month, including many plastic-packaged items that were microwaveable, disposable, and squeezeable. One new project, a squeezeable plastic package for frozen orange juice concentrate, makes it possible to make one serving at a time and helps extend consumer use for this product from a few days to several weeks.

Adapted from COPPE

Now look at and read the reformatted text in Figure 13-18. You should be able to read it more fluently. The designer creates the light appearance by shortening paragraphs, listing items, using hanging indentation, adding spaces between paragraphs, and inserting headings and subheadings.

The second heading includes five items. Each item is listed in a separate paragraph, thus adding white space between lines. In addition, each category is introduced by a subhead that emphasizes the item, just as bullets introduce each of the examples listed under the first subcategory. Each item is indented, thereby providing additional white space.

Visual Text. Using plenty of white space is the main determinant in creating airy text.

- Provide sufficient space in right and left margins for a reader to take notes, hold the paper, or bind the page in a book form. Left and right margins should be one or $1^1/_4$ inches. Top and bottom margins should be one inch.

- Use double spaces between paragraphs, and triple spaces between major sections, if the text is single-spaced, as in Figure 13-18. If the text is double-spaced, indent the first line of each paragraph, and use triple spaces between sections.

**Figure 13-18
Designing an Airy
Text**

PLASTIC PACKAGING

Since World War II, plastics have enjoyed impressive growth as a packaging material. A recent study by the industry indicates plastics will continue to grow in packaging applications, increasing in volume by 70 percent by the year 2000.

There are many reasons for the growth in plastics packaging applications. For one thing, consumer demand for goods in plastic packaging is high. A recent study by Packaging Magazine showed 63 percent favored plastics for microwaveable packages.

Applications

Plastics play a crucial role in countless packaging applications.

- Foods
- Personal care products (plastic films for liners and overwraps
- Bottles, jars, and containers in various molded shapes
- Pails
- Shrink wrap
- Bags and sacks.

Increased Usage

Plastics have enjoyed a significant increase in packaging applications and will continue to do so for the following reasons.

Break-resistance — Plastics are resistant to breakage which is important for a host of consumer products from food to personal care.

Leak resistance — Unlike other packaging materials, plastics resist leakage.

Heat/cold retention — Plastic containers have excellent insulation properties, keeping foods either hot or cold as desired.

Versatilility — Plastics can be molded into many convenient sizes and shapes, remaining rigid where the quality is important and flexible where that's a benefit.

Reusability — Many plastics for packaging can be reused, including bottles and strong grocery sacks.

According to *New Product News*, 1,000 new products were introduced to supermarkets and drugstores during just one month, including many plastic-packaged items that were microwaveable, disposable, and squeezeable. One new project, a squeezeable plastic package for frozen orange juice concentrate, makes it possible to make one serving at a time and helps extend consumer use for this product from a few days to several weeks.

Adapted from COPPE

- Keep paragraphs relatively short. Because you skip a line between paragraphs, short paragraphs allow more white space on a page. Notice how the final paragraph in Figure 13-17 becomes two paragraphs in Figure 13-18. However, if you are writing for experts in your own discourse community, determine whether short or long paragraphs are the accepted convention.

- List items when discussing several aspects of a concept, steps in a process, etc., as shown in Figure 13-18. Indent items in a list to open up a text, as in Figure 13-18.

Graphics As noted earlier in this chapter, you will want to place graphics in positions that not only indicate the importance of the graphic, but also add to the aesthetic effect of the page. In addition, you want to be careful that the graphic does not interfere with the legibility of a page.

- Place a graphic as close as possible to the text in which you discuss it.
- Place a graphic at the top of a page to emphasize it, or at the bottom of a page to subordinate it.
- Make a graphic large enough for the reader to easily discern each of the different elements within it.
- Place a graphic that contains a great deal of information on its own, separate page.

Figure 13-19 depicts another format for the information in Figure 13-12. A graphic has been added to Figure 13-18 to provide readers with additional information, as well as to emphasize the small amount of municipal solid waste that plastic comprises. The writer uses a two-column layout this time. Compare the format for Figure 13-19 with that of Figure 13-18, and determine its readability.

STRATEGY CHECKLIST

1. Create the impression of airiness in your text.
2. Provide sufficient spacing in margins and between sections, paragraphs, and lines.
3. Keep paragraphs relatively short.
4. Use indentation.
5. List items of information and steps in procedures and instructions.
6. Place text flush with the left margin.
7. Order text from most to least important by placing the most important information at the top of a page.
8. Balance text and graphics.
9. Study the formats of recent documents in the same genre and for the same audience for which you're writing to determine current formatting trends.

A RECURSIVE PROCESS

The visual aspects of a document need to be considered throughout the writing process. During the planning phase of a letter, you need to consider whether to create a formal or informal tone by formatting in block or semi-block style. During the drafting of a proposal for a new type of mechanism, you may realize you need to integrate a graphic to help readers understand a complex process. Finally, during the revising of a memo, you may recognize that your text is too dense to be legible; you need to add space to facilitate readers' fluency.

**Figure 13-19
Using Graphics in
a Layout**

PLASTIC PACKAGING

Since World War II, plastics have enjoyed impressive growth as a packaging material. A recent study by the industry indicates plastics will continue to grow in packaging applications, increasing in volume by 70 percent by the year 2000.

There are many reasons for the growth in plastics packaging applications. For one thing, consumer demand for goods in plastic packaging is high. A recent study by Packaging Magazine showed 63 percent favored plastics for microwaveable packages.

Plastic comprises only 2.4% of all flexible packaging discarded in Municipal Solid Waste

Applications

Plastics play a crucial role in countless packaging applications.

- Foods
- Personal care products (plastic films for liners and overwraps)
- Bottles, jars, and containers in various molded shapes
- Pails
- Shrink wrap
- Bags and sacks.

Increased Usage

Plastics have enjoyed a significant increase in packaging applications and will continue to do so for the following reasons.

Break resistance — Plastics are resistant to breakage which is important for a host of consumer products from food to personal care.

Leak resistance — Unlike other packaging materials, plastics resist leakage.

Heat/cold retention — Plastic containers have excellent insulation properties, keeping foods either hot or cold as desired.

Versatility — Plastics can be molded into many convenient sizes and shapes, remaining rigid where the quality is important and flexible where that's a benefit.

Reusability — Many plastics for packaging can be reused, including bottles and strong grocery sacks.

According to New Product News, 1,000 new products were introduced to supermarkets and drugstores during just one month, including many plastic-packaged items that were microwaveable, disposable, and squeezeable. One new project, a squeezeable plastic package for frozen orange juice concentrate, makes it possible to make one serving at a time and helps extend consumer use for this product from a few days to several weeks.

Reduced Waste

Plastic packaging can enhance the quality of food products. Laboratory analysis shows frozen produce to be more nutritious and bacteria-free. In addition, frozen vegetables in plastics packaging provide availability well beyond the growing season. Packaging for frozen food also reduces waste, since the produce is processed near where it is grown and where disposal of all leaves, stems, etc., is more efficient.

The National Science Foundation has reported a long-term trend to less packaging per pound of food, which can be traced to the increasing use of plastics. This helps reduce post-consumer packaging waste. For example, plastic, instead of waxed paper for bread, reduced packaging weight by 20 to 40 percent and resulted in a comparable reduction in waste volume. Newly developed plastic films have reduced weight a further 10 percent.

In its policy statement on "Single Service Materials, Resource Conservation and Public Health," the American Public Health Association stated that "Environmental health officials are in general agreement that single-use products contribute much to sanitation in food service facilities and that these public health benefits are greater than the possible disadvantages deriving from urban solid waste and litter."

Adapted from COPPE

DESIGNING WITH COMPUTERS

With today's computers, you can design your own documents with comparative ease. With most word processing programs, you can manipulate spatial markers, including line width, columns, spacing, margins, and bullets, and control typography by selecting typeface (font), size, and style.

Databases and spreadsheets (and some word processing programs) enable you to create tables. Many of these programs can also construct graphs from the tables, and even title and label the graphs.

Drawing programs give you the means to construct charts and diagrams. Other programs provide methods for formatting entire pages.

The writer for the article on plastic packaging created the pages in Figures 13-18 and 13-19 using a desktop publishing program. The graph was constructed using a graphics program.

Many programs provide global, as well as local, assistance for numbering pages, using headers and footers, and constructing tables of contents.

Scanners offer a method for copying graphics into a text. The scanners are capable of copying graphs, charts, and illustrations, including photographs. You need to be aware of copyright laws when you use published graphics.

CHAPTER SUMMARY

Document design involves visual decisions related to visual text, graphics, and formatting.

Visual text facilitates readers' reading and thinking processes and behaviors, as well as readers' fluency. It is used at the global as well as local (section and chapter) level.

Readers use visual text to locate information, predict what they will read, perceive the hierarchical order and sequence of a text, perceive chunks of related information, and recognize important information.

Visual text involves visual cues and legibility. Visual cues include spatial markers (titles, headings, and subheadings; margins and indentations; listing; numeration and bullets; boxes, and spacing) and typographical markers (capital and lower case letters; type size, face, and style; and columns).

Designing a text involves integrating visual text, graphics, and verbal text. In formatting a document, consider readers' reading and thinking processes and their previous reading experience.

Both visual cues and graphics are an integral part of a text and should be kept in mind throughout the writing process. Think about formatting during the planning, drafting, and revision processes.

Software programs provide writers with the capability of designing their own texts easily and professionally.

PROJECTS

Collaborative Projects: Short Term

As you work on the following projects, you may want to use the Audience Analysis Chart (page 97), the Decisions Chart (page 101), the Purpose Chart (page 136), Evaluation Checklist (page 264-265), and/or the Document Design Criteria Checklist in Appendix B.

1.   Revising for visual text

The following text is a U.S. Department of Agriculture handout on soil conservation. You are part of a team that has been assigned the task of designing the handout, which will be given to the general public.

Water is probably the natural resource we all know best. All of us have had firsthand experience with it in its many forms—rain, hail, snow, ice, steam, fog, dew.

Water covers nearly three-fourths of the earth; most is sea water, but sea water contains minerals and other substances, including those that make it salty, that are harmful to most land plants and animals. Still it is from the vast salty reservoirs, the seas and oceans, that most of our precipitation comes—no longer salty or mineral-laden. Water moves from clouds to land and back to the ocean in a never-ending cycle. This is the water or hydrologic cycle.

Every year about 80,000 cubic miles of water evaporates from oceans and about 15,000 cubic miles from land sources. Since the amounts of water evaporated and precipitated are almost the same, about 95,000 cubic miles of water are moving between earth and sky at all times.

Storms at sea return to the oceans much of the water evaporated from the oceans, so land areas get only about 24,000 cubic miles of water as precipitation. Precipitation on the land averages 26 inches a year, but it is not evenly distributed. Some places get less than 1 inch and others more than 400 inches.

The United States gets about 30 inches a year, or about 4,300 billion gallons a day. Total streamflow from surface and underground sources is about 8.5 inches a year, or about 1,200 billion gallons a day. This is the amount available for human use—homes, industry, irrigation, recreation.

The difference between precipitation and stream flow—21.5 inches a year, or 3,100 billion gallons a day—is the amount returned to the atmosphere as vapor. It is roughly 70 percent of the total water supply. It includes the water used by plants.

A person can exist on a gallon or so of water a day for drinking, cooking, and washing, though people seldom do or have to. In medieval times a person probably used no more than three to five gallons a day. In the nineteenth century, especially in western nations, a person used about 95 gallons a day. At present in the United States, a person uses about 1,500 gallons a day for needs and comforts, including recreation, cooling, food production, and industrial supply.

When water hits the ground, some soaks into the soil, and the rest runs off over the surface. The water that soaks into the soil sustains plant and animal life in the soil. Some seeps to underground reservoirs. Almost all of this water eventually enters the cycle once more.

People can alter the water cycle but little, so their primary supply of water is firmly fixed. But they can manage and conserve water as it becomes available—when it falls on the land. If they fail to do so, they lose the values that water has when used wisely.

Water management begins with soil management. Because our water supply comes to us as precipitation falling on the land, the fate of each drop of rain, each snowflake, each hailstone depends largely on where it falls—on the kind of soil and its cover.

A rainstorm or a heavy shower on bare soil loosens soil particles, and runoff—the water that does not soak into the soil—carries these particles away. This action, soil erosion by water, repeated many times ruins land for most uses. Erosion, furthermore, is the source of sediment that fills streams, pollutes water, kills aquatic life, and shortens the useful life of dams and reservoirs.

Falling rain erodes any raw-earth surface. Bare, plowed farmland, cleared areas going into housing developments, and highway fills and banks are especially vulnerable.

In cities and suburbs, where much of the land is paved or covered— streets, buildings, shopping centers, airport runways—rainwater runs off as much as 10 times faster than on unpaved land. Since the water cannot soak into the soil, it flows rapidly down storm drains or through sewer systems, contributing to floods and often carrying debris and other pollutants to streams.

Grass, trees, bushes, shrubs, and even weeds help break the force of raindrops and hold the soil in place. Where cultivated crops are grown, plowing and planting on contour terraces, and grassed waterways to carry surplus water from fields are some of the conservation measures that slow running water. Stubble mulching protects the soil when it has no growing cover. Small dams on upper tributaries in a watershed help control runoff and help solve problems of too much water one time and not enough another time.

Throughout the world the need for water continues to increase. Population growth brings demands for more water. Per capita use of water especially in industrialized countries, is increasing rapidly.

It is the obligation of the citizens of the world to return water to streams, lakes, and oceans as clean as possible and with the least waste.

U.S. Department of Agriculture

    a. Work in groups of three.

    b. Determine the layout, visual cues, and graphics. Then create the document.

    c. Your team may work together on the tasks for designing the handout or you may divide up the tasks, with one person writing the text, another developing the graphics, and a third doing the visual cues and layout.

    d. When the handout is completed, evaluate it, using the Document Design Criteria Checklist in Appendix B.

    e. Revise problem areas.

2. **Formatting a paper**

    a. Work in pairs.

    b. Select a paper you have written for this class or for one of your other classes.

    c. Format the document to facilitate your readers' processes and behaviors.

    d. Swap papers and assume the role of an editor. Use the Document Design Checklist and evaluate your partner's paper. Return the paper.

    e. Revise problem areas.

3. Developing a handout

The student/city council has asked you to be a member of a committee to design a handout for your campus/ community on procedures related to drinking and driving. The purpose of the handout is to provide students/citizens with specific procedures to follow if they or someone they are with is drinking.

    a. Work in groups of five.

    b. Analyze your audience, purpose, and situation.

    c. Gather your information.

    d. Determine the procedures.

    e. Determine the format for presenting the procedures.

    f. Write the procedures, then put them together in a handout.

    g. Duplicate five copies of the handout. Swap with another group. Do reader response protocols of each others' handouts.

    h. Return the handouts. Revise where necessary.

Collaborative Projects: Long Term

You have engaged in the composing processes as you have worked on the following projects. This chapter completes these projects. In this chapter, you'll engage in activities related to designing your documents.

- **Communicating about a campus/community problem** continued from Chapter 4, Collaborative Projects: Long Term Exercise 1.

- **Proposing microcomputer laboratories** continued from Chapter 4, Collaborative Projects: Long Term Exercise 2, Projects 1, 2, or 3.

- **Gathering information for a feasibility report: diaper laundry service** continued from Chapter 6, Collaborative Projects: Short Term Exercise 2.

- **Gathering data for an information report: cleaning supplies** continued from Chapter 6, Collaborative Projects: Short Term Exercise 3.

- **Organizing data for a proposal: recycled paper** continued from Chapter 7, Collaborative Projects: Short Term Exercise 2.

- **Organizing data from a food and music survey** continued from Chapter 7, Collaborative Projects: Short Term Exercise 3.

a. Determine the visual text needed by your readers at the intratext, intertext, and supratext levels to locate information quickly and easily, and to read your document fluently.

b. Design the layout for your document.

c. Integrate the graphics and visual text with your verbal text.

Option: To continue in Chapter 20, Document Locators and Supplements and/or Chapter 21, Oral Presentations.

Individual Projects

1. To see how well your visual memory works, follow these instructions.

   a. Read the information in Figures 13-20 and 13-21.

   b. Without turning back to the texts, answer the following questions. As you answer them, try to visualize the page and the position on the page where the information is located.

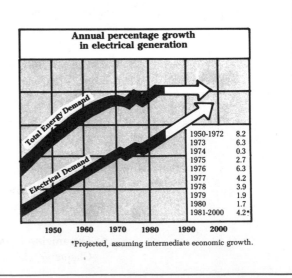

UNCERTAINTIES IN PLANNING
Demand Projections

Anticipating the amount of electricity that will be required in 5, 10, or 15 years is tricky. This is clearly illustrated by the experience of the last 15 years. Growth in total energy demand throughout the 1950s, '60s, and early '70s increased at a steady 10 percent per year. During this time the demand for electricity was growing at over 8 percent annually, and utilities planned new generating plants accordingly. Beginning in October 1973, a series of largely unforeseen events—an Arab oil embargo and a revolution in the Middle East producing gas lines and a sudden shift toward conservation—drastically revised America's energy forecasts. Today, experts project that electrical demand through the end of the century will grow at only half the rate experienced in the years prior to 1973.

Annual percentage growth in electrical generation

1950-1972	8.2
1973	6.3
1974	0.3
1975	2.7
1976	6.3
1977	4.2
1978	3.9
1979	1.9
1980	1.7
1981-2000	4.2*

*Projected, assuming intermediate economic growth.

U.S. Department of Energy

Figure 13-20 Visual Memory Check A

CHILDREN'S DISPOSABLE AND CLOTH DIAPERS

by Franklin Associates, Ltd.

Summary -   *All diapering options--cloth and disposable diapers--have some environmental and energy impacts. The results of this study illustrate the importance of considering the broad range of environmental and energy impacts of products rather than only one aspect.*

Introduction

Since the early 1970s, energy and environmental awareness among industry, government and the general public has grown significantly. As a result, understanding energy and environmental data regarding products, packages and materials has become important to decision and policy makers.

The purpose of this study is to thoroughly assess the energy and environmental impacts, through a product life cycle assessment, for children's disposable and cloth diaper systems. The analysis involves all steps in the life cycle of each diaper, including extraction of raw materials from the earth, processing these materials into usable components, manufacturing the product, distribution of the product, use and reuse of the product and final disposition of the product (whether recycled, incinerated, or landfilled).

The analysis not only includes the diapers themselves, but also the diaper packaging, plastic pants, detergents, etc. This report quantifies the energy and water requirements, solid wastes, atmospheric emissions and waterborne wastes generated by the three diaper systems.

Key Findings

The following conclusions were reached regarding the energy and environmental impacts for 1000 children's diaperings:

Energy	Home laundered cloth diapers consume more energy than disposable or commercially laundered cloth diapers.
Water Usage	Home or commercially laundered cloth diapers consume more water volume than disposable diapers.
Waterborne Wastes	Home or commercially laundered cloth diapers produce more waterborne wastes than disposable diapers.
Solid Waste	Disposable diapers produce more solid waste than home or commercially laundered cloth diapers.
Atmospheric Emissions	Home laundered cloth diapers produce more atmospheric emissions than disposable or commercially laundered diapers.

Franklin Associates, Ltd.

Figure 13-21 Visual Memory Check B

1. Which diaper does not have an environmental impact?

   (a) cloth   (b) disposable

2. Is demand for electricity growing?

3. What produces more solid waste, disposable diapers or home- or commercially-laundered cloth diapers?

4. Is demand for electricity between now and the end of the century growing at a rate (a) greater than, (b) about half, or (c) the same as it was for the years prior to 1973?

5. What unforeseen events occurred to change the demand for electricity?

c. You are probably able to answer the first four questions, or at least envision the area on the page in which the information occurs, because visual markers or graphics indicate the answers to each of the questions. Here are the answers.

   1. Figure 13-21: The statement is printed in italics at the top of the page. Even if you didn't remember the answer, you probably remembered where it was located.

   2. Figure 13-20: The graph visually depicts the growth. You probably got answered this question correctly.

   3. Figure 13-21: Solid waste is listed and set off in boldface type. It is also surrounded with white space. Even if you couldn't remember the answer to this question, you probably remembered where to find the answer.

   4. Figure 13-20: The graph shows that the rate of demand levels off. However, this trend is not as prominent visually as the growth from 1960, and so you may not have answered this question correctly, even though you could picture the graph.

      The question may have been more difficult to answer and to envision because it is embedded in the text, and visual markers do not emphasize it or place it in a special chunk.

2. Read the following text from a report on wetlands in the Peoria area. The report was written for government officials and interested citizens (generalists and novices).

Assessment of the existing resources available in the Upper Peoria Lake environs resulted in the following observations: permanent year round aquatic and side channel habitat is limited throughout the project area; emergent wetland habitat is minimally available at this location; waterfowl food production varies annually and in general is limited; and high turbidity values and suspended solids concentrations define Peoria lake's water quality status. Project goals that would address these conditions were developed for Upper Peoria Lake (Lower Peoria Lake was removed from consideration for rehabilitation and enhancement due to the extensive urban development along its western shore and the limited availability of public land). The project goals are the enhancement of aquatic and wetland habitats. The following objectives were determined to support the stated goals: (1) increase reliable food production and resting area for waterfowl; (2) increase diversity and areal extent of submergent and emergent vegetation for waterfowl, and (3) provide flowing side channel habitat. Multiple project sites, construction alternatives and design configurations have been considered for the purpose of realizing the stated project goals and objectives. Thorough analysis of all options resulted in the recommendation of the following design features: construction of a forested wetland management area: creation of a barrier island; and establishment of flowing side channel and rock substrate habitat.

Of several project sites with Upper Lake considered, the Goose Lake area was found to be the only location that met all of the minimal requirements for project site selection. These requirements included available foundations capable of supporting barrier island construction; State land ownership and management; minimal or no project-related impacts to the navigation channel; natural, flowing side channel development potential; and maximum environmental enhancement opportunities.

U.S. Army Corps of Engineers

    a. Determine the visual text your readers need to facilitate their reading processes and behaviors.

    b. Rewrite the text and insert the visual text.

    c. Evaluate your document using the Document Design Criteria Checklist in Appendix B.

    d. Revise problem areas.

3. Compare the format of an article in a professional journal in your major field of study with the format of an article on a topic in your field that is published in a popular magazine for the general public. What differences do you notice in the way visual text and graphics are used? Why do you believe these differences exist?

4. Select a paper you have written for this course. Based on the strategies you have learned in this chapter, format the paper, then evaluate it, using the Document Design Criteria Checklist in Appendix B. Revise problem areas.

5. The department in which you are majoring wants to provide freshmen and sophomores with information to help them decide whether or not to major in that field. The department has asked you to develop a handout to provide this information.

Engage in all of the steps of the composing processes, including analyzing your readers, purpose, and situation; gathering and organizing data; drafting and revising; selecting graphics; and designing visual text and format to develop the handout. You may want to use the worksheet in Appendix B as you plan.

The Documents

IN PART III YOU'LL APPLY THE PROCESSES and techniques you learned in Parts I and II to various types of technical documents. There are four major categories of technical documents: correspondence, instructions, proposals, and reports.[1] Each of these categories or genres[2] is comprised of a variety of subcategories or subgenres. For example, in correspondence, you may write a letter, a memo, a fax, or a computer bulletin board message.

EACH TYPE OF DOCUMENT FOLLOWS SPECIFIC conventions. You already know some of these. For example, the conventions of a letter involve the following:

1. a salutation, i.e., Dear Aunt Jean; Hi, Bob.

2. a complimentary closing, i.e,. Fondly, Love.

IF YOU WRITE A LETTER TO A FRIEND OR A thank-you note to a relative, you follow these conventions. When you write technical documents, such as proposals or reports, you need to follow the conventions related to them. You may already know some of these.

IN PART III YOU'LL LEARN THE CONVENTIONS and formats of the major technical documents you'll be expected to write in your professional and organizational communities, and you'll learn to adapt the processes and techniques you've already learned to these. □

---

[1] Adapted from Jimmie M. Killingsworth and Michael K. Gilbertson, 1986, "Rhetoric and Relevance in Technical Writing." *Journal of Technical Writing and Communication.* 287-97.

[2] Literary forms.

Correspondence

REGARDLESS OF THE FIELD you're in and the position you hold, most of the writing you do on the job involves short documents of one to three pages, and most of these are either letters or memoranda. A survey in the journal of the Society for Technical Communication (1991) found that letters and memos are produced more often than any other written document.

Letters are usually intended for external readers, memos for internal readers. While letters and memos have traditionally been transmitted by U.S. mail and office delivery personnel, an increasing amount of correspondence today is transmitted electronically.

This chapter begins with a general discussion of each of the types of correspondence—letters, memos, and electronic messages. You'll then study the conventions for writing two of the most common types of correspondence—requests and responses to requests. While you will learn the conventions related to these specific types, you will still need to make textual decisions related to the specific context in which you are writing and to the topic you are discussing. This chapter will help you apply some of the strategies you have learned in previous chapters to various types of correspondence so that you can create an effective document.

LETTERS

Results of the 1991 survey show that a major portion of technical information is transmitted in this form of correspondence.

Context

Readers Letters are sent to readers who are external to your organization. These readers may assume a variety of positions, roles, and fields, ranging from clients to consumers to representatives of regulatory agencies of state governments. Because these readers are external to your organization, you may need to provide them with background information, but keep technical detail to a minimum and limit your use of technical terminology. However, the amount of background and technical terminology you include depends on your reader's familiarity with your organization and topic. If you are writing to other experts in your field, or to people with whom you've been working for a long time, you probably won't need to include any background. You may also be able to use technical terms, even though your readers are external to your organization, since they will be familiar with them. On the other hand, if you are writing to readers for the first time, you'll need to be certain they have sufficient background to understand your message and that they are familiar with your terms.

All of your readers are busy. They want to know who is sending a letter to them and the topic being discussed. If they aren't interested in a topic, they may toss a letter into a waste basket. If they are managers, they may decide to give someone else the responsibility for fulfilling your request or instructions. However, if they feel they need to know about a topic, they will read the first paragraph. Often they won't read beyond the first paragraph unless they need to know additional information, or are interested in a subject. To provide for these reading behaviors, you need to include a subject line, below the inside address, that briefly states the topic of your correspondence. You also need to include the following information in the first paragraph:

- The purpose of the correspondence
- The discussion in a nutshell
- A list of the items you will be discussing in the proper sequence (optional)

Purpose Letters are written for many purposes. You may respond to a reader's request for information about a piece of machinery, you may request an interview for a job, or you may submit a progress report on a project. Or you may attempt to persuade readers to think in a certain way, as the Environmental Defense Fund (EDF) does in the letter sent to consumers of McDonald's (Figure 14-1). EDF wants to persuade readers that the fast food firm's decision to phase out polystyrene was appropriate, and to counter arguments against the decision made by other fast food businesses. As you read, notice how the writer attempts to achieve the purpose by following some of the strategies discussed in Chapter 8.

Subject line.
Introduction.
Presents purpose of
letter. Appeals to
readers' values.
Background.

Counters oppo-
nents' arguments.

Details included in
attachment if reader
wants to read more.
Most important
information summa-
rized here.
Indicates shared con-
cerns. Closes by
emphasizing value
readers share.

ENVIRONMENTAL DEFENSE FUND
1616 P Street, NW
Washington, DC 20036
(202) 387-3500

McDonald's Decision to Phase Out Polystyrene Foam

Due to your interest in solid waste issues, we wanted to provide you with details on the reasoning behind EDF's support of McDonald's decision to phase out poly-styrene foam packaging, and on the environmental benefits that will result because of that decision.

In the wake of McDonald's November 1st announcement, some proponents of polystyrene have mischaracterized, and some members of the press have misunder-stood, McDonald's decision to switch to a paper-based wrap. Some have even sought to frame McDonald's decision as a bad one for the environment, in particular, by misrepresenting the results of a recent study that compared polystyrene foam to paperboard boxes.

As the enclosed documents explain, McDonald's is switching initially to a paper-based wrap that is quite different from and preferable to containers. The docu-ments also describe, in detail, the environmental problems and costs associated with use of polystyrene foam, and the advantages and limitations of the new paper-based material...

After you review the enclosed information, I think you will see—despite the misinformation and misunderstandings surrounding this complex issue—that McDonald's switch, as recommended by EDF, is a clear victory for the environment.

Figure 14-1 Letter to McDonald's Patrons

Situation While letters are usually addressed to a single individual, peripheral readers may become involved. A primary reader may turn your letter over to an attorney, or a reporter may read a letter on your reader's desk. If you are discussing a project, the primary reader may circulate copies to others interested in the topic. In addition, you may send copies to other people who are involved with your topic.

Intermediary readers may also be involved. If you are informing a client or subcontractor about a problem, you may want your supervisor to have a copy of the letter so she is aware of your discussion. Often, supervisors want to review external correspondence before it is sent, to be certain the company is represented correctly. It is both courteous and good business to route a copy of all letters to your supervisor.

Style

Correspondence can be personal or impersonal, formal or informal, depending on your reader and your purpose.

Letters should be formal unless you know your reader well. If you feel comfortable addressing your reader on a first-name basis, and if you do not expect the document to be used legally, your correspondence can be personal and informal. You can talk directly to your reader, referring to yourself by using first person singular (I), and referring to your reader as "you." You can also inject personal notes, referring to the reader's family or health, or to an incident you two experienced together. If you don't know the reader on a first-name basis, or if the topic is important, you probably want to write a formal document.

When you write for your company, you may use first person plural (we), and you should use a formal style. In addition, don't make personal references even if you know the reader.

The writer of the EDF letter (Figure 14-1) assumes the combined persona of McDonald's and EDF by using "we." The writer also attempts to create a cooperative atmosphere between the reader and McDonald's by addressing the reader directly as "you."

Format

The conventions for the format of a letter are well established, and readers expect them to be followed. Most of these conventions permit readers to read a letter easily, and to locate information they need to know quickly. Letters should be single-spaced, with double spacing between paragraphs. They should be kept short. If possible, limit them to one page.

Business letters can be written in either block or semi-block form (Figure 14-2).

Block form presents the most formal appearance. All text is flush with the left margin; paragraphs in the body of the letter are not indented. In semi-block

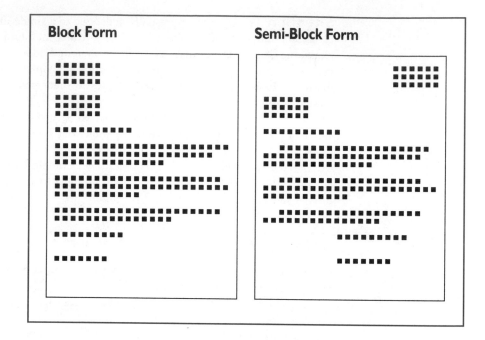

form, the heading, closing, and signature are placed about two-thirds of the way across the page, toward the right margin, and paragraphs are indented.

Regardless of the form, the body of the letter should be single-spaced with double spaces between paragraphs.

The conventional format for a business letter includes the heading, inside address, salutation, message, closing, and signature. A letter may also include a subject line, the typist's initials, a designation of an enclosure, and a distribution list (Figure 14-3).

Heading. Include your address and the date you send the letter. Do not include your address if you use a company letterhead.

Write out the entire word for street designations, e.g., "Street," "Avenue," "Boulevard," etc., as well as names of months, "February," "September," etc. Use the post office's designation for state names, e.g., "NY" for New York, "CA" for California, etc.

Inside Address. Include the name of the person to whom you are writing, the person's position and division, the name of the company, and the company's address.

Address the person by the appropriate designation, Mr., Ms., or Dr. If the person holds a special title, such as senator or professor, write out the full word; don't abbreviate.

If the recipient holds an administrative position, place a comma after the recipient's name, and add the recipient's position title, i.e., Manager, Director, Vice President. Don't abbreviate. Capitalize the title.

Heading.	Your street address City, State Zip Date
Inside address.	Name of addressee Address City, State Zip
Subject.	Topic
Salutation.	Dear ———:
Body of letter.	Xxxxxxxx xxxx xxxxx xxxxx xxx xxxxxxxx xxxxx xxxxx xxxxx xxx x xxxx x xxx x xxxxxxxx. Xxxxxxxx xxxx xxxxx xxxxx xxx xxxxxxx xxxxx xxxxx xxxxx xxx x xxxx x xxx x xxxxxxxx. Xxxxxxxx xxxx xxxxx xxxxx xxx xxxxxxx xxxxx xxxxx xxxxx xxx.
Closing.	Yours truly,
Signature.	Your name signed Your name typed
Typist's initials/ WP file number. Enclosures. Distribution.	GN/3405 Enc. cc: Rogers Smith

Figure 14-3 Conventional Format for a Business Letter

If you don't know the name of the person to whom you are writing, address the letter to the appropriate division, e.g., Personnel Department, Accounting Department.

Subject Line. This line has been adopted from the conventional memo format and should be used in formal letters only to specify the topic to be discussed. Readers scanning a letter use the line to determine whether or not they want to read the letter, and whether they want to read it immediately or at a later time.

If you include the line, use it to describe your topic. Your description serves as a title to help readers predict what they will read.

Salutation. Use the conventional "Dear." Use the person's first name only if you are on a first-name basis. Otherwise, use the person's last name preceded by the title. If you do not know the name of the person, use the person's position, e.g., Dear Personnel Director. Use Mr. or Ms. unless the person has a special title, such as "Dr.," "Honorable," or "Representative." Follow the salutation by a colon.

Message. Your message assumes a variety of forms, depending on your reader and your purpose.

Closing. Use the conventional closings, "Yours truly," or "Sincerely." "Yours truly" is considered more formal than "Sincerely." "Cordially" or "Best Wishes" implies more familiarity. Follow the closing with a comma.

Signature. Type your full name four lines below the closing. Between the closing and the typed signature sign your name in your own handwriting. If you are on a first-name basis with the recipient, sign just your first name. Never ask your secretary to sign for you. You insult your reader by not taking the time to sign personally.

Typist/word processing file number. If you have someone type the letter for you, include your initials in capital letters followed by your typist's initials in lower case letters. Separate the two sets of initials with a slash (/). If the letter is typed in a word processing file, include the word processing file number. This information allows both you and the reader to trace errors or problems if they arise.

Enclosures. If you are including reports, brochures, etc., list their titles here. Place each one on a separate line. This list enables the recipient to determine whether or not all of the materials are enclosed.

Distribution. List the names of all persons you want to receive a copy of the letter. The list informs the recipient of those who will also have the information. The list may include your supervisor and other persons in your organization who should know about the topic or are involved with it. In addition, it is always courteous to send a copy to anyone mentioned in a letter. The distribution list may also include people in the reader's division or organization whom you think should be informed about the topic. However, keep in mind that your list may not include all the people your reader thinks should have the information. In that case, the reader may send a copy of your letter to additional people.

Example

The letter shown in Figure 14-4, written by a manager at an airport, is written to a consultant engaged in a study of noise conditions at the writer's airport. While the reader is external to the writer's organization, he is involved with the writer's field. Therefore, the writer is able to use technical terminology and, because the reader and writer have been communicating about the project, the writer does not have to provide background information. Furthermore, because the reader is working closely with the writer and needs to know the specifics if he is to take them into consideration in finalizing the project, the writer provides specific details. Although he addresses the reader by first name in the salutation, he signs both his first and last name, since the letter can be used as a legal document. The letter is a request, and the writer uses his authority as the reader's employer to persuade the reader to make the suggested changes. As you read, notice how the writer adheres to the conventions.

**Figure 14-4
Letter of Request**

Letterhead.

Greater PEORIA AIRPORT AUTHORITY
Peoria, Illinois 61607-0147   /   PHONE (309) 697-8272

February 16, 19xx

Heading.
  Date.
  Inside address.

Mr. Bruce Jacobson
Crawford, Murphy & Tilly, Inc.
2750 W. Washington
Springfield, IL 62702

Comments on the Part 150 Noise Study

Dear Bruce:

Subject.
Salutation (by first name).
Introduction.
  Indicates purpose.
  Uses term "should" to
  indicate his authority.
Body.
  Lists details.
  Copies enclosed (see
  enclosure list).

    The Greater Peoria Airport Authority has received several comments following the February 7th P. A. C. meeting on the Part 150 Noise Study which should be considered in your final draft.

    Pastor Paul Tolo of St. John Lutheran Church commented on the noise and its effect on the church during weekend military exercises. A copy of his letter is enclosed.

    Furthermore, Col. Frank Rezac, Commander of the 182nd TASG had several comments regarding the MOA (pages 26-28). Col. Rezac believes the MOA extends up to 36,000 feet MSL, though 18,000 to 36,000 may be PCA (Positive Controlled Airspace). Secondly, the Peoria Guard unit apparently does not conduct "fighter interception training" as stated, though it does perform maneuvers in the LATN (Low Altitude Tactical Navigation) area, within the Howard MOA. Those altitudes can go as low as 100 feet and up to 7,000 MSL, not 10,000. Col. Rezac questioned whether this section was even relevant to the study, since it does not impact the immediate airport area. Finally, Col. Rezac stated the Guard unit presently has 20, not 24, OA-37s at Peoria.

Article enclosed
  (see enclosure list).
Conclusion.
  Assumes persona of
  organization.
Formal closing.
Signature (both first and
last name).
Typed signature, position.

    Finally, I have enclosed a copy of the newspaper article which appeared in the Peoria Journal Star the day after our meeting.

    We are looking forward to completing Phase I and moving into Phase II of this project. Thank you for your assistance in this matter. If you have any questions, please feel free to contact me at the Airport.

          Very truly yours,

          Tom Miller
          Tom Miller
          Tom Miller, Assistant Director

Typist's initials.
Distribution list.
  (Important for IAD and
  FAA to know about
  changes and schedule.)
Enclosure list.

TWM:rt
cc: T. Schaddel, IDA [Illinois Department of Aviation] J. Mork, FAA [Federal Aviation Administration]
Enc: Letter from Pastor Tolo
    Newspaper clipping

Computer Aids

The format illustrated in Figure 14-4 is used for all letters. You may want to use the mail merge and database features of your word processor to format correspondence automatically.

MEMOS

The survey in the journal of the Society for Technical Communicators indicates that readers receive much of their information from memos. Thus, members of an organization use memoranda to transmit a large percentage of their technical information to others in their organization.

Context

Readers Memos are written to internal readers. Such readers often don't need background information, because they are familiar with your institution's projects. In addition, they are familiar with jargon related to the organization, though they may not know the technical terminology of your field.

Your readers will be in a variety of positions, ranging from supervisors to peers to subordinates. They will also have many roles, ranging from decision makers to producers to users, from accountants to advertisers to attorneys, and they may be experts, generalists, or novices in your field.

Regardless of their role, readers of memos, like readers of letters, do not want to spend much time reading. Most recipients will look at the subject line of a memo to determine if the message is important enough to read. If they don't think so, it will wind up in a waste basket. Unless a memo contains complex information, most people only scan it.

Purpose Memos, like letters, are written for a variety of purposes. Often you write to respond to a request, or to request specific information or action from someone in relation to a project on which you are working.

The memo in Figure 14-5 is an informal one between two people who have been working together. Though they are both members of the same organization, they are in different divisions. The writer's purpose is twofold: to thank the reader for her support, and to persuade the reader to provide additional support. As you read, notice how the writer leads up to his request.

Situation While many memos are addressed to a single reader, you may also send a single memo to several readers, all of whom need to know the same information or respond to the same request. In addition, if you are informing a subordinate about a problem, you may send a copy of your correspondence to your supervisor, or some other person who should know about the problem and be aware of your action on it. You may simply want to keep your supervisor

**Figure 14-5
A Bread-and-
Butter Memo of
Appreciation**

Heading.

Subject line.
Purpose.
Statement of
appreciation.
Background so
reader under-
stands reason for
request.

Request.

METHODIST
MEDICAL CENTER OF ILLINOIS

221 Northeast Glen Oak
Peoria, Illinois 61636-0002

internal memo

DATE: October 15, 19xx
TO: Judy Hart, R.D.
FROM: Roger G. Monroe   *RM*
SUBJECT: Guardian Angel School

I can't begin to thank you enough for the fine support you provided for the partner-ship ceremony we had with the students and staff of Guardian Angel School. Thanks to your supply of ice cream, it was a big hit with the students and everyone.

In exploring additional ideas and activities for support of these 70 students, ranging in age from 8-18, it has been suggested we provide speakers. One suggestion was for dietitians to come and talk about nutrition as well as careers in the field. Guardian Angel has one class of females, 14-18, who are either single mothers or about to become single mothers. Perhaps a dietitian could talk to this class about the importance of nutrition in their circumstances, and another could talk to the other students.

If you and/or any of the dietitians in your department are interested, would you give me a call at 4989 and we'll discuss it further.

Methodist Medical Center of Illinois, Peoria, IL

informed of your decisions, or you may have a hidden agenda in which you want to cover yourself by notifying your supervisor of a potential problem. You should always send your supervisor a copy of any correspondence you send to your supervisor's peers or supervisors. You don't want your supervisor to think you're going over his head.

Style

Memos are often less formal than letters, especially if you are writing to a peer or subordinate. However, they become more formal when you address someone above you in the organizational hierarchy. Unlike letters, memos seldom contain any personal references; they get down to business immediately.

Format

Like letters, the conventions for the format of a memo have been established for a long time, and readers expect them to be followed. Memos should be single-spaced with double spaces between paragraphs, and kept to a single page if at all possible. Memos are usually written in block form (Figure 14-5).

The conventional format for a memorandum includes two sections: the heading and the body. The heading specifies the date, name of recipient(s), name of sender, and subject. The body presents the message. The memo in Figure 14-5 follows this format. Notice that the words "DATE," "TO," "FROM," and "SUBJECT" introduce each part of the heading. The words are written in all capital letters, followed by a colon, and the information following each is aligned.

Memos do not include a signature. Instead, senders sign either their initials, or their first or last names to the right of their name in the heading. Memos usually do not include a distribution list, since recipients are listed in the heading. The typist's initials and designation for enclosures used in letters may also be used at the end of a memo.

Computer Aids

The format in Figure 14-5 is used for all memos. You may want to use the mail merge and database features of your word processor for automatically formatting memos.

ELECTRONIC MESSAGES

Business and industry have greatly increased their use of electronic services, such as facsimile machines (fax) and computer networks, to transmit correspondence. Transmission is not only conducted within an organization, but between organizations, on an international as well as national basis.

Electronic messages are faster than either interoffice or the U.S. mails, allowing a person to receive a message in a matter of minutes rather than hours or days.

Fax Machines

A fax works like an old-fashioned wireless telegram, transmitting information across telephone lines to be printed at the other end.

Superior to the wireless, the fax copies a document in the same way a copy machine does, thus allowing you to add visual elements to your communications. This feature is especially helpful when you discuss machinery or a process, because you can include charts and diagrams with your discussion. Depending on how formal or personal you want your message to be, you can fax handwritten notes on personal memo pads, as well as reports on corporate stationery.

Transmission is almost simultaneous. You can be in Philadelphia at 1:55 p.m., sending a client in Atlanta a three-page letter describing a piece of machinery that he needs to discuss at a 2 p.m. meeting. The client will not only get it in time for the meeting, but also have time to pour a cup of coffee before going into the conference.

The fax in Figure 14-6, sent on a pre-established form by a mechanic at a general aviation facility, was a request for wiring diagrams for the 500 Shrike on which he was working. He had discovered a problem and needed a specific diagram to figure it out. Notice that the message is handwritten, sentences truncated.

In less than an hour, the diagram requested by the mechanic was faxed back to him. If the mechanic had relied on the mails, the repairs might have been delayed for a week.

Computer Networks

There are two types of computer networks: Wide Area Networks (WANs), which make it possible to transmit messages from one terminal to another anywhere in the world, and Local Area Networks (LANs), which limit transmissions to terminals within an organization.

Figure 14 6
A Fax Message

BYERLY AVIATION, INC.

GREATER PEORIA AIRPORT
PEORIA, IL 61607
(309) 697-5500

FACSIMILE TRANSMITTAL

DATE: 3-20-89
TO: . . . . Sandy
Attn:
FACSIMILE NUMBER: . . . .
FROM: . . . . BYERLY AVIATION, INC.
BY:
FASCIMILE NUMBER: . . . . (309) 697-2779
OUR REFERENCE: . . . .
TOTAL PAGES (Including Cover Page): . . . .
MESSAGE: Can you send a copy of Custon Kit 126 Papen work all Pages This is For a 500$ which has This Kit Installed

If you do not receive all the pages or if they are not legible, please FAX or Telephone Byerly Aviation, Inc. Immediately. Thank you.

GENERAL AVIATION

50 YEARS OF AVIATION SERVICE

WANs include networks such as "Source," for the general public; BITNET, which is largely for universities; and INTERNET, which is a global network connected with the Department of Defense and the National Science Foundation (NSF).

Messages are sent across these networks by modems connected to a PC or a mainframe computer. Each of these networks has a bulletin board service for posting messages, and an electronic mail service (e-mail) for sending messages to specific persons. Messages are received by calling up the bulletin board or "mailbox." Copies can be printed out.

LANs work the same way as WANs, except they're restricted to short, contiguous distances within a company. A LAN can connect the people in a single division, or everyone in a building, or even groups of buildings in a single company. Each person connected to a LAN has a code for a "mailbox." People can send messages by typing in the other person's code, and they can read messages by typing in their own code.

Style

Electronic transmission is affecting the style of the messages being transmitted. Whether the transmission is via fax or a WAN (LANs are not affected), cost is determined by the amount of time the transmission takes. The more words in a message, the longer it takes for the transmission to occur, and the higher the cost. The shorter the message, the more cost-effective it is. Thus brevity, conciseness, and clarity are keys for maintaining cost-effective documents. So documents are getting shorter. Even on LANs, people's documents are becoming more abbreviated and less formal.

This need for brevity is manifested in a number of ways. Abbreviations abound. Instead of the word "message" being spelled out, it may be abbreviated as "msg." Sentences are often truncated; such courtesies as the "dear" in salutations and the complimentary closings in letters may be eliminated.

In Figure 14-6, the first sentence is a shortened version of the following: "Can you send a copy of all of the pages included in the paper work for Custom Kit 126." In addition, the name of the aircraft, Shrike, has been abbreviated to "$." Since the reader is an expert, and familiar with these aspects of the topic, the truncated sentence and the abbreviations are appropriate; the reader understands the message.

According to Merv Rennich, Director of Marketing Support for Caterpillar Inc., e-mail is used extensively throughout the plants, and messages are much less formal than those sent in hard copy.

However, Ed Wahlstrand often uses a formal structure, such as the one in Figure 14-7, for his e-mail messages. Consider the differences in style between the Figures 14-6 and 14-7.

The key to making appropriate decisions, according to Wahlstrand, is the purpose of the message, and the situation in which it will be read. If it is a request or notice concerned with a single, simple topic that the receiver doesn't

Memo format.

John Deere Davenport Works
Training and Development Group

To: M. E. Addington
From: R. E. Wahlstrand
Date: 11 May 19XX
Subject: Organization issues

Continues conver-
sation. Jargon.
Complete
sentences.

Per our conversation, here are some issues which you may want to raise with your peers and at the same time be proactive.

These are issues which I believe are open (as a result of my one-on-one conversations with Mike's staff and general supervisors) and have potential for significant impact on our business and long-range plans. Dealing with these could also present an opportunity for those who have been trained in it to use their "decision focus" skills:

1. What products will Davenport Works manufacture into the 21st century? (This needs thorough definition.)
2. Should vehicle painting be part of our business?
3. What will the first level of management be supervising? If the answer is supervisors, what will their roles be? (Likewise, this needs thorough definition.)...

Figure 14-7 Formal Message on E-Mail

need to keep on file, an informal message is appropriate. However, if the request or information being transmitted is complicated, or if the receiver needs to print out and file the information, then a more formal message should be sent.

Harry Litchfield, Director of Training at Deere, suggests the level of formality is also related to the reader's position. If you're sending a transmission to someone at a higher level than you, or to someone you don't know, you probably want to send a more formal message.

Because the possibility exists that someone other than the recipient may see the information that's transmitted via fax or e-mail, documents shouldn't contain personal information embarrassing to the recipient, and, unless security precautions have been taken, they should not contain sensitive or confidential information.

Format

Electronic transmission has also altered the format of letters and memos. Because this is a relatively new form of correspondence, the conventions for it are not as well-established as those for traditional letters and memos, and you may find a variety of formats being used by organizations.

Fax Messages sent by fax require a cover sheet that includes the following information:

- Date
- Name of addressee
- Name of sender
- Fax number
- Name of person to contact in case message is garbled
- Total number of pages, including the cover page, so the receiver knows if all pages have been transmitted.

Most companies have created a form containing this information (see Figure 14-6). Personnel fill one out each time a fax is sent.

When the transmission is complete, a receipt is transmitted to the sender, indicating the message has been sent. The message itself can be included on the cover sheet if it is sufficiently brief, as in Figure 14-7, or it can be attached.

E-Mail Because e-mail is read directly off a monitor, the format needs to be one that facilitates reading from a screen.

Think of the screen as a page of text. A screen comprises a visual chunk of information just as a page does. The same strategies for writing legible text on paper apply to a screen. However, because it is more difficult to read text from a screen than from hard copy, legibility becomes extremely important.

Keep the following strategies in mind for text that will be read on a monitor:

- Messages should contain only one subject at a time. It's difficult for a reader to take in more than this when reading from a screen.
- Messages should be no longer than a single screen display, otherwise the reader may feel a need to print it.
- Paragraphs should be no longer than three or four lines on a screen so the screen does not appear full.
- Subheads or key words should be placed at the beginning of each paragraph to highlight the concept. Otherwise, all of the ideas appear to run together.
- Lines should be kept to a maximum of 60 characters and spaces so the reader can span an entire line easily.

The message in Figure 14-8 was transmitted from Peoria, Illinois, to Brazil via computer network. You should notice several differences between it and the memos in the previous sections of this chapter. The heading includes additional information related to computer codes. The heading also includes a reference line to indicate previous documents that are related to the topic being discussed. If readers don't remember the information in these documents or aren't familiar with the data, they can read the documents prior to reading the memo. Finally, the heading includes an attention line, which indicates the specific person in Brazil for whom the memo is intended.

The message deals with a single topic—966C differential noise. The writer adds spaces between paragraphs to make them easier to see. The writer not only keeps paragraphs short, but also lines. Notice that each line ends at the end of a linguistic unit, making the text easier to read (see Chapter 11).

**Figure 14-8
International
E-Mail Message**

```
Date: 08/16/xx
From:       SUTHERLAND           SUTHEMB -ADCCHOST
To:         RELAY BRAZIL PIRACICABA  28 CAT -NPEO
            TWEEDLOW             TWEEDLW -ADCCHOST
            ETHERIDGE            ETHERJR -ADCCHOST
SUBJECT:   966C DIFFERENTIAL NOISE
REF:       Luiz messages dated 7/04/xx & 7/16/xx
ATTN:      L.N., Product quality, Piracicaba
```
    After reviewing this problem with Aurora & East Peoria Engrg.
personnel, a number of facts have surfaced.

1. The IV9822/2V2006 Bevel Gear & Pinion Group
   used on the 966C has a 6.5 deg. spiral angle
   and low contact ratio.

2. A similar design with a 6.5 deg. spiral angle
   was confirmed "noisy" on the 950B. A change was made
   to 27.5 deg. spiral angle.

   Aurora has indicated that they do not have the manpower
   to come up with a similar design change for the 966C. However,
   they will review for approval a change designed by Brazil...

Lists each problem.

Short paragraphs.
Space to separate
paragraphs and
items in a list.

TYPES OF CORRESPONDENCE

Requests

Much of the correspondence you write is a request. There are three types of requests.

- **Action Requests.** You ask the reader to do something. You may request a job from a potential employer, or approval of a project from a supervisor.
- **Information Requests.** You ask the reader to provide data. You may request an update on a project from a subordinate, or data from a study by a colleague.
- **Complaints/Claims.** You specify a complaint or claim, and ask the reader to rectify it. You may request a refund for a defective product from a company, or a change in the design of a product from a division manager.

The Context: Readers, Purpose, and Situation
Requests are writer-initiated and persuasive. As a consumer, you may have written several letters of complaint about a product, and requested a refund or replacement. Sometimes you got what you wanted, but other times your request was rejected. In

writing as a member of a business organization, you want to do everything you can to persuade readers to fulfill your requests.

Your readers may be anxious to respond positively to your request, or they may regard your request negatively or even with hostility. You need to consider many of the strategies for writing for nonsympathetic readers.

Readers often read only the first paragraph of a request before deciding whether they plan to fulfill a request, and when they plan to fulfill it. If they do not plan to fulfill it, they are likely to relegate the message to a waste basket. If they do plan to respond, but not immediately, they may place it in their "to do" pile, where they will eventually get around to it. When they are finally ready to respond, they will probably scan the first paragraph again to refresh their memory, and then go on to study the remainder. Unless readers are highly motivated, they tend to place most requests on their "to do" pile. Therefore, if you want an immediate positive response, you need to motivate readers at the very beginning of your message.

Content, Organizational Pattern, and Sequence You usually need to provide background information or other supporting data so readers understand why the data or action is needed, or the claim or complaint is being made. If you expect readers will view a request skeptically or negatively, then begin with the background and lead up to the request. However, if you expect readers to approve your request, your background information should follow your first paragraph, which should specify the purpose of the message. In the remainder of the message you need to include enough information for readers to know exactly what you want them to do, think, or provide.

Readers of complaints or claim letters are apt to respond negatively, and often defensively. To reduce a reader's hostility or reluctance to fulfill a request based on a complaint or claim, you may want to follow an "It's good but..." organizational pattern. Begin on a positive note, with a discussion of shared goals. Then present the problem or the contrasting situation. Your request or complaint should be made toward the end of the message.

The memo in Figure 14-9 shows a request for both action and information. It was written by a programmer at ATS to his immediate supervisor and to two peers. The readers are all knowledgeable about computers, but not LANs. The writer's long-range purpose is to persuade readers to approve the idea of a Local Area Network for their division, and to determine the type of LAN they want. The writer's short-range purpose is to persuade readers to add suggestions for selecting a LAN to satisfy their own needs.

Notice the strategies the writer uses to persuade readers to read and approve his decision.

The information is contained on a single page. The memo presents a general explanation of the writer's need. Specific information is placed on succeeding pages, which readers can study when they have time. The first paragraph provides readers with sufficient information to determine whether or not to continue reading, to toss the message in a waste basket, to file it, send it on to a more appropriate reader, or set it aside to be studied later.

Conventional heading. Block form.	DATE: October 26, 19xx
	TO: Jim Brook, Extension 18
	Don Bowen, Extension 31
	Jody Howard, Extension 43
	FROM: Ron Rothstein, Extension 33
	SUBJECT: Local Area Network for Systems Engineering

DATE: October 26, 19xx
TO: Jim Brook, Extension 18
            Don Bowen, Extension 31
            Jody Howard, Extension 43
FROM: Ron Rothstein, Extension 33
SUBJECT: Local Area Network for Systems Engineering

Conventional heading. Block form.

While working on the project to locate a Job Tracking system for Systems Engineering, it became clear a Local Area Network (LAN) in the Powers building could be an effective solution to some of the applications being considered (e.g. time sheets, electronic mail, shared printers, shared VAX connection, and a shared Fax resource).

Introduction. Summarizes problem, solution. Indicates shared goals by using first person plural. Specifies purpose. Provides background so readers understand reason for request.

This memo is a first step at determining how we should evaluate, recommend, and select a local area network.

The PC local area network market consists of over 22 companies and products. These are listed on p. 2 under the headings of "LAN Products and Companies" along with the percentage market share under the heading "Installed Base Market Share by LAN Company." This information was used as the basis for selecting five (5) products for an in-depth evaluation. These include Netware (Novell), PC LAN (IBM), 3+ (3 Com), Starlan (AT&T), and DECnet PSA (DEC). Additional information on reasons for limiting our evaluation to these five (5) products is shown on page 3.

Request.

A requirements list is proposed on page 4 that will be used to determine which of the five network products best meets our needs. Requirements are identified as Mandatory (the selected product must have these feature) and Optional (nice-to-have features that will not eliminate a product if the feature is not supported).

Addresses readers directly.

Please review this information and consider adding requirements to the list. Who else would you like to have review the selection approach and requirements' list before we begin evaluating products?

Advanced Technology Systems

Figure 14-9 Conventional Memo Format

Style Two different styles can be applied to a letter of request: direct and indirect. In the direct style, the reader is *told* to do something, e.g., "I need the report by 5 p.m. Have it on my desk by then." In the indirect style, the reader is *asked* to do something, e.g., "Could you have the report on my desk by 5 p.m.?" An indirect style is considered less threatening and more polite. It is also used to indicate equal power between the reader and writer. While the direct style implies the writer has the power to make the request and the reader lacks power to refuse, the indirect style implies the reader has the power to reject the request.

A direct style is used when readers expect your request and/or will respond positively to it. Direct styles are usually used when you place orders for goods or services, or when a reader is a subordinate. An indirect style should be used whenever you write to a peer or superior. It should also be used when a reader does not know you, or knows you only on a very formal level.

There are several strategies for phrasing a request indirectly.[1]

- Asking a question, e.g., "Can you provide me with the results of your study?"

- Using a qualifier that hedges the request, e.g., "Can you possibly provide me with the results of your study?"

- Using the conditional verb form, e.g., "Could you send me the results of your study?"

- Apologizing for imposing on the reader, e.g., "I apologize for the inconvenience, but I would like to see a print-out of your results."

- Incurring a debt, e.g., "I would appreciate your sending me the results of your study."

- Depersonalizing the request by using the expletive "it" and passive voice, e.g., "It is important that the results of the study be sent to this laboratory."

It is possible to use several of these phrasings in the same sentence, e.g., "I would appreciate it if you could possibly send me the print-outs from the results of your study." By including several of these together, you increase the polite and nonthreatening tone of the request. You may want to use several phrases when you make a request that imposes on a reader's time, efforts, or resources; when readers may be nonsympathetic or even hostile to your request; when your readers are above you in the organizational hierarchy or hold the power for granting your request; or when readers are unknown to you.

The letter in Figure 14-10, written to the project manager of the proposed Superconducting Super Collider, is a claim that present plans for constructing the Super Collider do not provide sufficiently for sustaining fish and wildlife resources. Notice that the letter is in an "it's good but..." pattern, and follows many of the strategies for writing for a nonsympathetic audience.

Responses

You respond to a request with either a letter or a memo. Regardless of whether you're responding to a supervisor who has requested information about a project on which you are working, to a distributor who is requesting a part for a product, or to an organization requesting support for a project, your response should follow the conventional format.

- The purpose of the letter appears in the introduction.

- The details of the response comprise the body of the letter.

- Information for obtaining additional information, etc., if needed, is in the closing.

---

[1]Riley, Katherine. 1988. "Speech Act Theory and Degrees of Directness in Professional Writing." *The Technical Writing Teacher*, 1-30.

**Figure 14-10
Letter of Request**

Introduction.
Purpose of letter.

"It's good" section.
Positive beginning.
Notes shared
goals.

"But" section.
Background.

Complaint.

Request (indirect).
(Reiterates share
goal.)
Conventional
closing.

United States Department of the Interior
Office of the Secretary
Washington, D.C. 20240

ER 90/972

Mr. Thomas A. Bailleul
SSC-SEIS Project Manager
U.S. Department of Energy
Chicago Operations Office—EMD
9800 South Cass Avenue
Argonne, Illinois 60439

Dear Mr. Bailleul:

The Department of the Interior has reviewed the draft supplement to the final environmental impact statement for the Superconducting Super Collider, Ellis County, Texas, and has the following comments.

Commitment to Fish and Wildlife Resources

The Department of Energy (DOE) and Texas National Research Laboratory Commission (TNRLC) are to be commended for their efforts to ensure that fish and wildlife resources receive equal consideration during the planning and development phases of the Superconducting Super Collider project. The Department of the Interior commends the commitment in the draft supplement that all activities at this site will be consistent with the intent of E. O. 11990…

Migratory Waterfowl

During meetings between our Service, DOE, and TNRLC, the Service stressed the seasonal importance of water ponds with water level control capabilities for attracting and holding migratory waterfowl. Water level control would allow the ponds (or portions thereof) to be drained May through September to allow seeding of mudflats for forage production. The ponds can then be flooded the remainder of the year to attract waterfowl. Because of thermal loading of the cooling ponds, additional or multi-tiered ponds should be utilized. These should be physically separated by a dam and a closeable drain pipe.

While the separation of high temperature waters from the waterfowl portion of the pond appears to be indicated in Figure 4,5 (p. 4-16), there apparently is no inlet into the upper pond to allow for flooding if rainfall is inadequate…

[We would appreciate your making this capability available so that] the goal of "no net loss of wetlands" can be met and the ponds can function as they are designed to do…

We hope these comments will be helpful to you.

Jonathan P. Deason

Jonathan P. Deason, Director
Office of Environmental Affairs

U.S. Department of Interior

Responses may assume several forms, including acceptances, rejections, suggestions for further action, or requests for additional information.

Because letters of acceptance provide readers with good news, a straightforward approach, in which the acceptance is stated in the first paragraph, is the best organizational pattern. However, rejections, as well as suggestions or requests delaying an acceptance, almost always create a negative response in readers. To soften this response, rejections or delays are seldom stated in the first paragraph. Instead, the letter leads up to them by providing background information first.

This form of correspondence may serve as a legal document. Therefore, these letters and memos often assume a formal style even if the writer and recipient know each other well and are on a first-name basis.

The letter in Figure 14-11 is a response to a proposal for improving a waterway. The response is neither positive nor negative, but delays a decision until additional information can be obtained. The writer, whose agency must approve the proposed plans, has been working with the reader on the project. As you read the letter, consider whether the writer has followed the conventions for this type of correspondence and has used the appropriate strategies for this situation.

Because the writer does not meet the reader's expectations, the reader may be placed in a defensive position. Despite the closing paragraph's attempt to maintain a positive attitude, the reader may be left with a negative impression. Had the writer begun by discussing some positive aspects before commenting on his concerns, the reader may not have felt so defensive.

STRATEGY CHECKLIST

1. State the purpose of your correspondence in the introduction.

2. Begin and end on a positive note.

3. Be gracious.

4. In writing a request:

   a. Begin with the request if readers are expected to approve it.

   b. Begin with background information if readers may respond skeptically or negatively. Place the request at the conclusion of your correspondence.

5. Support your request with background information.

6. In responding to a request or writing a report:

   a. Provide the response or findings immediately if you are responding positively or communicating a positive message.

   b. Lead up to the response if you are responding negatively or skeptically.

7. Provide sufficient background or information for the reader to accept your rejection, suggestions, or request for further information if you are responding negatively to a request, or writing a report which communicates a negative message.

Formal heading.
Block form.

Addresses by first name.
Identifies purpose.
Indicates delay immediately.
Lists reasons.

Uses bullets.
Business-like, formal tone.
Uses technical terminology.

Ends on positive note.
Signs first name.

February 21, 19xx

Mr. Andy Bruzewicz
U.S. Army Engineer District
Clock Tower Building
Rock Island, IL 61204-2004

Dear Andy:

We have reviewed the Peoria Lake DPR outline that you submitted January 13, 19xx, and would offer the following concerns and comments.

At this time we are uncertain which source of water (well or river) would be most cost effective for the moist soil unit. We need to know well potential to determine if unit can be filled within optimum time of 5 to 10 days.

Size of spillway in levees need to be adequate to allow river flooding to take place without pressure on the levees.

- Can silt plug disposal be used as part of the island construction?

- Why does the cost of clam shell construction of the island continue to increase?

- We will need to assure concerns of sediment resuspension during construction.

We look forward to your continued cooperation on this project.

Sincerely,

Bill

William R. Donels

U. S. Department of Conservation

Figure 14-11 Letter of Rejection

CHAPTER SUMMARY

Much of the writing you do will be correspondence in one of three forms: letters, memoranda, and electronic messages.

Each of these forms follows a conventional style and format, which readers expect you to follow. However, the content, organizational pattern, and sequence depends on the reader, the purpose, and the situation. In many cases, electronic messages are less formal than traditional letters and memos.

Your correspondence will include requests and responses to requests. These follow the conventions of either a letter or memo. The organizational pattern depends in part on whether the reader will interpret your message positively, skeptically, or negatively.

- If readers are expected to respond negatively or skeptically, begin with background or information supporting your position or request.

- If readers are expected to respond positively, begin with your request or the results of your report.

PROJECTS

As you undertake a project, be sure to engage fully in all three processes—planning, drafting, revising. You may want to use the Audience Analysis (page 97), Purpose (page 136), and Decision charts (Page 101), as well as the Planning Worksheet in Appendix B, as you gather and organize your information. As you prepare to revise, you may also want to use the Evaluation Checklist (page 264-265), and/or the Document Design Criteria Checklist in Appendix B.

Collaborative Projects

1. Solid Waste

One of the major problems facing this country in the 1990s is solid waste disposal. The landfills where we dispose of our solid waste are filling up, and we are running out of room. City governments are responsible for solid waste disposal.

a. Create a committee of five members to determine how your college campus/community can reduce solid waste.

b. Each member should assume one of the following responsibilities:

(1) Write to the local city/county/state government requesting information about landfills and solid waste disposal policies and plans.

(2) Write to each of the following organizations requesting information on solid waste disposal.

- The Council of Solid Waste Solutions
  1159 Pittsford-Victor Rd.
  Pittsford, New York 14534

- American Society of Mechanical Engineers
  Solid Waste Processing Division
  345 East 47th Street
  New York, NY 10017

- Congress of the United States,
  Office of Technology Assessment
  Washington, D.C. 20000

(3) Write to the governments of the following places, which are engaged in effective ways to reduce solid waste, for information concerning their programs.

- State of Wisconsin
- Urbana, Illinois
- Woodbury, Minnesota
- Bristol, Connecticut
- Lincoln, Nebraska

(4) Write to the administrator of your college or the appropriate representative in your local/state government, requesting information about programs/plans for solid waste disposal.

(5) Write to the Environmental Protection Agency for the latest report on solid waste disposal.

- U.S. Environmental Protection Agency *or*
- Office of Solid Waste and Emergency Response
  Washington, D.C. 20460

c. Share your responses with each other.

Options:
a. Write a proposal for your college/ community leaders suggesting ways to improve the environment by reducing solid waste, based on the information your group receives from responses to these letters (See Chapter 17, Proposals, Collaborative Projects: Long Term.)

b. Make a presentation, summarizing the responses, to your class and recommending the top three suggestions for implementation. (See Chapter 21, Oral Presentations Collaborative Projects: Long Term.)

c. Write a summary of all the responses your group receives and distribute it to your classmates to persuade them to help reduce solid waste.

2. **Trees**

As part of a world-wide environmental effort, your class has decided to participate in a campus/community project to plant trees. You need to determine where on the campus or in the community the trees should be planted. As a class, select at least three alternative sites.

a. Determine a site on which you would like to plant the trees.

b. Work with others in your class who have selected the same site as you. As a group, draft a letter to the other students who have selected different sites, urging them to change their minds and select your site.

c. Make enough copies of your letter for each of the other groups. Deliver a copy of your letter to each group.

d. As a group, respond to each of the letters you receive. Depending on your reaction to each letter, you may wish to change your mind on the site for the trees, or you may still believe you have selected the best site.

3. **Recycling**

This exercise requires a computer-networked classroom.

You have been asked to work with university authorities to increase recycling of paper, plastic, glass, and aluminum waste on campus. You need to create a committee of five people to work on this project.

    a. Use your computer to communicate with other members of your class.

    b. Recruit four other persons in your class to work with you on the committee.

    c. Once you have a committee, determine a time for the first meeting. You need to select a time that is convenient for all members to meet.

    d. Select a place for the meeting that is convenient for all members.

    e. The committee should draft a memo to your instructor, indicating this assignment is completed.

Individual Projects

1. One of your parents' friends has a son, Joe, who is a high school senior and is considering attending your college. You remember Joe. He likes sports and isn't very motivated. He'd much rather watch a TV program than read a book. His parents have written the following letter to you.

   October, 19xx

   Dear _____:

       Your parents have been telling us about your college since Joe is considering going there. We'd like to ask you for your opinion. Do you think Joe would like it and, perhaps more importantly, do you think he can get in?

       Joe has a "C" average in school and made a composite score of about 975 in his SATs. We think he did better in the math section than the language part. He has no idea what he wants to major in though we're pushing for business.

       We'd really appreciate your writing us with your opinion. Thanks for whatever help you can give us.

   Fondly,

   The Rochmans

       Write a letter responding to these parents. You may or may not believe your college is appropriate for Joe.

2. Write a letter to a university requesting information on one of the following:
   a. a summer program
   b. a graduate program
   c. a study-abroad program

3. Write a memo to one of your instructors requesting a take-home or open-book exam instead of the traditional closed-book, in-class exam.

4. Write a letter to a local industry requesting information about the company's policy on an environmental issue.

5. Write a letter commending a company for its efforts in solving an environmental or social issue such as the following:

     a. Recycling the paper or plastic involved in its products.

     b. Toxic emissions from smokestacks.

     c. Day care for employee's children.

     d. Toxic waste.

     e. Oil/chemical spills.

     f. Maternity/paternity leave.

     g. Personal leave for taking care of sick parents/children.

     h. Access for disabled persons.

Additional Resources

Discussions and examples relating to follow-up bread-and-butter correspondence and correspondence containing proposals and reports can be found in the Instructor's Manual.

CHAPTER 15

Correspondence for Employment

BECAUSE GETTING A JOB is such an important aspect of life, this chapter is devoted solely to the particular forms of correspondence related to job hunting. These include the letter of application, the resumé, and the follow-up letter. A letter of application for employment is a type of persuasive correspondence. It is a subcategory of requests. A resumé serves as an "appendix" to a letter of application.

According to Chuck Williams, Manager of Professional and Technical Employment for Caterpillar Inc., the two most important documents you'll ever write are your resumé and a letter of application for a job. Your resumé and letter of application are documents in which you try to persuade prospective employers to hire you. To do so, you must persuade them you are both qualified for the position and better qualified than other applicants.

If you contact potential employers during a job fest sponsored by your college, you should have copies of your resumé available to leave with each company representative. If you apply for a position by letter, you will be expected to include a resumé with your correspondence. Whether or not you are invited to interview depends entirely on your letter and resumé.

Your potential employer's first impression of you is based on your correspondence. If your resumé looks sloppy, contains spelling errors, or fails to relate to the requirements of the position, the employer will probably conclude that you don't care if you get the position, you are not a careful worker, you do not look after details, and you are not focused.

Readers of a letter of application and resumé look for interest in the company, qualifications, quality of writing, and the appearance of correspondence.

READERS

Your letter of application and resumé may be read either by your potential employer, or by a member of a company's personnel department. In a small company, your application may be processed by the manager for whom you'll work. In a large firm, your application will probably be processed through a personnel/human resources department. Even if you write directly to a division manager, your documents in a large organization are usually sent to the personnel/human resources department for screening. However, if a manager knows you or is impressed by your credentials, she may tack a note onto your letter to indicate to the personnel department she wants to meet with you.

If you are writing directly to a potential employer, your letter of application may be more personal if you know something about the person to whom you're writing.

The frame of reference from which readers view your letter and resumé also depends on the type of company. Readers at a large, conservative company, such as Caterpillar Inc. or IBM, expect your documents to conform to the conventions. They are not impressed by purple paper, innovative typefaces and formats, and unique openings. On the other hand, innovative companies, such as Apple Computer, Inc., are interested in successful creative approaches. To determine your best approach, you need to research the companies to which you're applying. If you are writing to both types, develop separate letters and resumés for each.

Regardless of who is reading your documents, you need to keep in mind that the readers are busy, and that your application is probably one of a large number being reviewed. During the main hiring period each year, the personnel/human resources division of a large company may review as many as 3,000 applications per month. Most readers only scan a letter of application, searching for the most important information.

Letter 1

Addresses reader
by name.

Indicates she is
local to company.
Names company.
Names personal
contact.
Organizing idea.
Emphasizes experi-
ence in interna-
tional relations for
corporations since
she's applying to a
large corporation.

P.O. Box 281
Brentwood, TN 37024
January 17, 19xx

Bert Born
Professional Employment Representative
Caterpillar Inc.
100 NE Adams
Peoria, Illinois 61629

Dear Mr. Born:

As a native Illinoisan, I have long had an interest in Caterpillar Inc. and have followed your company regularly. I am currently in the process of relocating to Central
Illinois and am researching employment options in the Peoria area. Mr. Chub Dietz
suggested that I contact you for guidance in this regard.

My area of interest and expertise is international relations. During my seven years
in Tennessee I worked extensively with the Japanese in both America and Japan
developing my cross cultural, communication, and organizational skills. After
returning to the U.S., I was employed by The Japan Center of Tennessee as a
resource liaison and coordinator for Japanese and American corporations...

Figure 15-1 Resumés Written for Specific Readers (Letter 1)

PURPOSE

Your purpose is to persuade a company to hire you. Your reader's purpose is to
determine whether or not you have the qualifications for a position. Your letter
of application establishes your claims for a position; your resumé provides the
supporting evidence.

If you are applying for different positions at different organizations, you
should develop separate letters, because your claims need to relate to a company's
specific objectives as well as to the specific responsibilities of a position. As a
chemical engineer, you may apply for a position with a pharmaceutical company
to do research, or to a nuclear utility to work in its evaluation division. You need
to stress the aspects of your education and work experience that relate to each of
the specific positions. You can do this best by writing two separate letters.

Let's look at how Marlise Streitmatter alters two letters (Figures 15-1,
shown above, and 15-2, next page) to fit the needs of two different organizations
to which she is applying.

Letter 2

Addresses reader by name.

Indicates she is local to organization.

Organizing idea. Emphasizes intercultural experience in state organizations, since she's applying to that kind of organization.

P.O. Box 281
Brentwood, TN 37024
January 17, 19xx

Steven McClure, Director
Commerce and Community Affairs
620 East Adams
Springfield, Illinois 62701

Dear Mr. McClure:

In November, I plan to relocate from Tennessee to my home state of Illinois. I would very much like to use my organizational, communication, and cross-cultural skills to further bicultural understanding and economic development in Illinois.

My experience is unique in that I have worked extensively with the Japanese during the past seven years. I am presently employed at the Japan Center of Tennessee, an organization funded by the Tennessee Department of Economic and Community Development to promote mutual understanding between Japanese and Tennesseans.

As you may know, Tennessee has become a leader in attracting international investment, particularly Japanese investment...

Consultation services on writing provided to Streitmatter by Schatz & Schatz, 610 Mapletop Drive, Antioch, TN 37013 and PVT Enterprises, 2133 Chickering Lane, Nashville, TN 37215.

Figure 15-2 Resumés Written for Specific Readers (Letter 2)

SITUATION

Letters of application and the accompanying resumé are read in two types of situations. They may be read directly by the person who actually plans to employ you, or they may be read by a person in the personnel/human resources department of the firm to which you've applied. Let's look at each of these situations, and at the behaviors of the readers as they review the documents in these different situations.

Actual Employer If your letter and resumé are read directly by the person for whom you will eventually work, it will be read along with other applications, and it will be read in addition to the person's routine activities. Most employers don't have time to add this responsibility to their other ones. They want to find someone as quickly as possible, they want to find someone who has ideal qualifications, and they're skeptical. They know from previous experience that applicants who look good on paper and in interviews may not be as good as they appear once they're on the job. The employer who will supervise this person has a large stake in the hiring decision.

Your application and resumé go through several reviews.

- **Review 1.** The reader scans your letter of application to determine if you meet the criteria.

    If the reader considers these aspects satisfactory, your resumé will be examined for more specific information. If your qualifications are good, your letter will go in a pile with others for a possible future interview. However, if your letter is exceptional, if you not only meet but excel in important qualifications, the reader will probably place your letter on top, place an asterisk in a top corner, or in some other way indicate special interest in meeting with you.

- **Review 2.** The reader reviews the pile of acceptable applications more carefully, and also looks at the resumés this time. The reader does not want to interview many people, and uses the information on the resumé to narrow the list and possibly rank the candidates.

Personnel/Human Resources Department Members of the personnel/human resources department perceive your application as one of many they have read over the years. They probably have a stack of applications piled on their desks for the job for which you are applying and for other positions. Their task involves using job descriptions, which they have received from the potential employer, to screen candidates. Letters go through at least two reviews.

- **Review 1.** Like the potential employer, members of the personnel/human resources department scan the letters to determine if the writer's basic qualifications match those on the job description, and if the letter meets the criteria described on page 433. According to Williams, readers make up their minds within twenty seconds.

    If a letter meets their criteria, they place it in a pile which they return to later to read more carefully.

- **Review 2.** Readers now look at the resumés accompanying those letters of application they considered acceptable to determine an applicant's specific qualifications. At this point, readers weigh each applicant's qualifications against others in the pile to determine candidates for an interview. If the firm is in the same city as the applicants, a relatively large number of persons may be invited to come in. However, if the firm is out of town, the list is usually narrowed to only a few candidates. If your letter and resumé are exceptional, the personnel officer may indicate a special interest in bringing you in.

Once the personnel officer narrows the list, the letters and resumés are usually presented to the potential employer, who reviews and either approves or rejects the officer's decision before inviting applicants in.

THE RESUME

Write your resumé before writing a letter of application, since your letter is based on the information included in your resumé.

Readers who review a resumé expect certain information to be included, sequenced in a specific order, and following a basic format. You need to follow these conventions. Don't try to be original; you simply confuse readers who can't locate the information they want, and conclude you don't know how to write a good resumé.

Content, Organizational Pattern, and Sequence

Based on the information readers need to know, and your readers' and your own purpose, a resumé includes personal demographics, professional objectives, educational and work history, awards and honors, special activities and interests, and references. The information should be sequenced as it is listed here.

Personal demographics. Name, address, telephone number. Readers need this information to notify you that they are or are not interested in interviewing you. If you are living on campus but go home on vacations, or if you plan to move home after graduation, provide both your temporary (campus) and your permanent (home) address.

Permanent address	Temporary address
613 N. Allegheny Ave. Midland, PA 9xxxxx (412) 393-0987	413 Veterans Parkway #613 Normal, IL 6xxxx (309) 438-0965

Readers are also interested in learning whether you live near the company, or are familiar with the area in which it is located. Companies often prefer to hire local people. Furthermore, they need to determine whether you should receive travel expenses if they bring you in for an interview. And if they decide to hire you, they want to know whether they have to pay moving expenses.

Professional Objectives. Short and long term. In presenting your short term objectives, state the position for which you are applying and the level at which you expect to enter. If you have just graduated college, you are probably applying for an entry-level position. Apply for the broadest possible position rather than limiting yourself to a specific area, i.e., marketing rather than sales, electrical engineering rather than test operations. In expressing your long term goals, state the position you eventually hope to achieve when you pass the entry level. You can be specific here.

Short term:	Entry-level position in medical technology
Long term:	Position on staff for AIDS research

Your objectives should take no more than two lines.

Educational History. Name and address of college, major (include minor concentration if related to position), year graduated, degree received, and either cumulative overall average or cumulative average in your major if above a 2.8. Readers need to know if you have the necessary educational requirements for the position. Therefore, list your formal education chronologically, beginning with your most recent degree so that readers can quickly determine your highest level of education.

Include any additional educational experiences, such as study abroad and summer programs. Such experiences indicate that you are interested in acquiring information and willing to attend training programs to keep up in your field.

Work Experience. Name and address of place of employment, dates worked, position, and major responsibilities. Readers want to know where you have worked in the past, the positions you've held, and your responsibilities. Include the name and address of previous employers so readers can contact them for additional information. Include advancements you may have received; these indicate your employers have been impressed with your work.

Potential employers want to know your work experiences to determine your job knowledge. Employers recognize that you are more knowledgeable if you have worked in a field related to the one for which you are applying than if you have had only classroom experience. Most employers believe that classroom work gives you mostly theoretical knowledge. You don't really learn a field until you're working in it. If you have worked in the field for which you're applying, employers believe they won't have to spend as much time, effort, and money in training you for the position.

Potential employers also want to know your work experience to determine if you understand the responsibilities associated with a full-time job. If you have worked previously, even if it has not been in a related field, employers assume you understand such aspects of work as being on time, following a supervisor's orders, etc.

List the work experience you have had chronologically, beginning with your most recent job. If you have held jobs during the summers only, be specific when listing the dates.

Awards and honors. Awards and honors indicate that you excel in certain aspects. Employers are especially interested in these if they are directly related to the position, or if they indicate leadership qualities.

Special activities and interests. Clubs, campus organizations, community activities, and other groups, hobbies, family- and church-related activities, etc. Employers are not only interested in hiring someone who can do a job, they also want someone who can enhance their organization as well as the community in which the company is located. In addition, as more and more work requires collaborative efforts among employees and between employees and management, employers are looking for people who get along well with others. Your involvement in team sports; social activities such as campus organizations; school activities such as government or the newspaper; and community efforts indicate that you are an active citizen and interact well with people. Hobbies show that you are a well-rounded person. And you never know when your reader may have the same hobby and would enjoy hiring someone who shares an interest in model railroading, stamp collecting, or doing double crostics. Many employers also look for family-oriented applicants. If you engage in family activities, say so.

References. Do not include reference letters. If you are applying directly to an employer, or if your reader may know or hold in high regard one of your references, you will probably want to list your references. Include positions, place of employment, address or telephone number. If letters are on file with the university placement service, say so. However, if you are writing to the personnel/human resources department of an organization, simply indicate references are available on request (see Figure 15-3). Many employers prefer to speak with a reference rather than read a letter.

Always ask if you can use a person's name as a reference. *Never* list someone whom you have not asked.

Include three references who can, respectively, discuss your work experience, your educational qualifications, and your character. Since your readers are especially interested in your past work experience, include a reference from a job you have held, especially if you have worked in a related field. The higher up the hierarchical ladder your reference is, the more impressive the recommendation appears to the reader. Even if you have not had work experience in a related field, include a reference who can state that you are prompt and diligent. This assures the reader that you are a good worker.

Because your readers are also interested in your knowledge of the field, include a reference from your education. Your major professor, department chairperson, or the researcher for whom you have worked can provide a relevant recommendation, affirming your grasp of the subject.

The third reference may be someone with whom you have worked in a community or campus activity or in some other capacity. Select someone who can talk about you in relation to the position for which you are applying. A relative, neighbor, pastor, or rabbi is not able to do this.

Style

You do not need to use full sentences in your resumé. However, be certain your phrasing is parallel throughout. If you use a verb to begin the description of your work responsibilities for the first job listed on your resumé, then you need to begin your descriptions of other jobs similarly. For example:

- took notes for lead researcher
- aided in research studies

Visual Text

Visual markers are extremely important in a resumé. Readers need to scan a resumé quickly to locate specific information. Insert headings to indicate each of the major sections, and use white space to set off these sections as well as to make the page aesthetically pleasing. Use boldface, underlining, and/or italics to set off the important facts in a list, such as the place of employment and the name of your college.

No matter how good your resumé looks, you won't get the job unless you are qualified. However, no matter how qualified you are, you may not get the job if your resumé looks messy and contains inaccuracies and errors.

Format

Keep your resumé to a single page; your reader is busy. Readers prefer to see a resumé in the conventional format depicted in Figure 15-3.

Designing Resumés with Computers

By selecting legible fonts and using various type styles and sizes to emphasize sections and help readers locate information, computers allow you to create an attractive and efficient resumé. In addition, by using grammar and spell checks,

**Figure 15-3
Conventional Format for a Resumé**

	Name	
Present Address		Permanent Address
Street		Street
City, State, Zip code		City, State, Zip code
Phone		Phone

OBJECTIVE:

EDUCATION: Highest Degree. Major and minor. College.
(Date of graduation) Location (City and State). GPA or gradepoint average
                      for major.

EXPERIENCE   Most recent employer, your title
(Dates worked)   Brief description of duties performed

(Dates worked)   Other employers.

AWARDS AND HONORS

ACTIVITIES AND INTERESTS

MILITARY SERVICE

REFERENCES   Available on request

you get help in creating an error-free resumé. Finally, by printing your resumé on a laser printer, you can create a resumé that appears professional.

Models

The sequence of information in your resumé depends on whether you are just graduating from college or if you have been working professionally. If you are just graduating college, you probably have not had much professional experience. Therefore you will want to list your educational background before your employment record. If you have been working full-time at a related job, you probably want to list your employment record before your educational record. The resumés in Figures 15-4 and 15-5 provide models of each type of sequence.

College Student The resumé in Figure 15-4 was written by a college student. As you read, notice how he uses the conventional format.

Professional Applicant Now look at the resumé in Figure 15-5. As you read, consider the differences between this resumé and the one in Figure 15-4.

THE LETTER OF APPLICATION

Write your letter of application after you have written your resumé. Your letter of application is the first thing your reader sees. Your reader may never bother to look at your resumé if your letter isn't neat and without errors, and if it doesn't indicate you are qualified for the position and interested in the company to which you are applying.

Because your letter is probably one of many, begin with something that catches the reader's attention and sets you apart from the others. However, don't do anything outlandish unless you know your reader appreciates it.

Impress your readers with your knowledge of their company and your efforts to individualize the letter, not with cute language, egotistical statements, and clever designs. Just placing the company's name in the first paragraph indicates that you have taken the time to individualize a letter and will impress your reader. Black ink on white paper is appropriate.

Of course, if you are applying for a job that requires the clever use of language and design as well as imaginative approaches to a problem, you may want to be more creative. However, even in this case, function takes precedence over form. Readers still want all of the conventional information, and they need to find it where they expect it.

Rick Joseph Schwarzentraub
2834 Pleasant Hill Road
Peoria, IL 61614
(309) 697-4874

Career Goal

To obtain an entry level position in a financial field which will challenge my skills and permit me to grow into mid-management.

Education
1991

Illinois State University, Normal, IL
Bachelor of Arts and Sciences
Major: Finance and Law
Minor: Business Administration
GPA: 3.71 of a possible 4.00
Graduate Cum Laude with an honors degree in finance

Experience
1987-present

Schwarzentraub Refrigeration, East Peoria
Sales representative

1985-1987

Management: Shop foreman in charge of job delegation, invoice billing, and materials acquisition.

1980-1985

Mechanic in refrigeration

1979-1980

Edward D. Jones, Morton, IL
Intern. Duties included portfolio analysis, customer reviews, office management.

Scholastic Awards Illinois State University - Gold Key Honor Society, Phi Theta Kappa Honor Society, National Dean's list three semesters.

Interests

Member of Illinois Tae Kwon Do team and the Governor's Cup Tae Kwon Do Team. Enjoy golfing and other outdoor activities.

References

Mr. Cal Chandler	Dr. James Grundy	Dr. Carol Rosen
Edward D. Jones	Illinois State University	Writing, Assoc.
Morton, IL	Marketing Department	Peoria, IL
	Normal, IL	

Content, Organizational Pattern, and Sequence

The content depends on the position for which you are applying. Your readers want to know what experience and knowledge you have that qualify you for the specific position and make you the exceptional candidate.

In a letter of application, the conventions for presenting the information readers need are as follows.

Introductory paragraph. You need to catch readers' attention immediately by indicating that you are fulfilling their needs for a position. In this paragraph, establish the position for which you are applying, identify the company to which you're applying, and state your interest in a position at the company. If you're interested in working in a specific geographic area, add that information.

**Figure 15-5
Resumé of a
Professional in
Human Resource
Development**

Carl Daniels

1673 Chestnut Street   (215)838-5566
Plymouth Meeting, PA 19078

CAREER OBJECTIVE

Secure a middle management position in HUMAN RESOURCE DEVELOPMENT
for an organization dedicated to providing high quality, perfor-
mance-based training and development activities that produce
results.

EMPLOYMENT HISTORY

CURRENT   General Public Utilities Nuclear Corporation.
POSITION   Three Mile Island Training Center. P.O. Box
               480.Middletown, PA

               Educational Development Coordinator. Provide
               internal training and consultant services in
               program development and instructional process;
               design and conduct initial and continuing
               instructor training; conduct needs/job/task
               analyses; design training programs and procedures.

1984-85   Wilmington College. 320 DuPont Highway, New
               Castle, DE 19720

               Director, Graduate Conference Center. Managed
               rental of building facilities for training,
               conferences, seminars; conducted classes;
               supervised/ trained/ evaluated subordinates.

1974-85   Lancaster-Lebanon Intermediate Unit #13. 1110
               Enterprise Rd., East Petersburg, PA 17520.

               Lancaster County Coordinator for Adult Education.
               Administered the program for the county.
               Responsibilities included hiring, training,
               evaluating and counseling of staff; developed
               budgets; wrote proposals; developed/implemented
               training programs in response to needs analyses.

EDUCATION

1977   Master of Education. Adult Education. Temple
               University, Philadelphia, PA. GPA 4.0. Passed
               comprehensive examination with honors.

1972   Bachelor of Arts. Sociology. Millersville
               University. Millersville, PA.

AWARDS AND ACHIEVEMENTS

1985   Represented Pennsylvania at the 111 InterAmerican
               Conference on Community Education in Brazil.

1980   Cited by the Lebanon, PA, Jaycees as Outstanding
               Young Educator.

               Read, speak, and write fluent Spanish.

Main body. This section allows you to expand on your qualifications. Your
focus should center on the position for which you are applying. Discuss those
aspects in your resumé that are specifically related to your qualifications for the
position. This is your opportunity to expand on some of the items in your resumé
that you think readers will find particularly important, and that place you above
other applicants. For instance, if you are applying for a position as an X-ray

technician in a hospital and you worked your way through college assisting in the radiology department of your local hospital, you can emphasize that information by commenting that you were often invited to sit in on discussions between the radiologist and the technicians. In addition, you may want to comment on the most important or most interesting thing you learned while working there.

The information in this section should be sequenced from the most to least important. Begin with information indicating you are a qualified candidate for the specific position for which you are applying.

Closing paragraph. Use this paragraph to obtain a response by requesting an interview. Close by indicating you would like to meet with the company representatives. Do NOT ask them to contact you. That is presumptuous. If they're interested, they'll call. However, you can indicate that you will call to discuss the possibility of an interview. Then make sure you make that follow-up call ten to fourteen days after sending your letter.

If the company is in your home town and you plan to be there during a vacation period, suggest meeting with them at that time and indicate you will call to find out if it is convenient. Don't worry. If they're not interested, they'll tell you. If they are, you've relieved them from making a telephone call, and they will thank you.

Style   Use an impersonal and formal tone. Be gracious, since you are asking the reader to give you a job. Indicate enthusiasm, sincerity, care, and authority. You want to persuade the reader that you are interested in the position, you are serious about working in that position and for that company, you care whether or not you get the position, and you have the qualifications to do a good job. However, you don't want to appear conceited. Try not to refer to yourself constantly.

Modest: My educational background and experience have provided skills to help your company produce better equipment to meet today's needs.

Conceited: I feel that with my educational background and experience I may be able to help your company's goal of producing better equipment to meet today's needs.

Use technical terms only if they relate specifically to the position. The personnel officer probably won't have any familiarity with your field, and your potential supervisor may or may not be a member of your specific discourse community. Unless you know differently, consider your reader as a generalist .

Keep your sentences short. Your letter of application should not be more than a single page. Remember, regardless of which reader reviews your correspondence, both are busy.

Format

The letter of application follows a normal letter format. Use block form.

Address the person to whom you are writing by name in the salutation. If you don't know the name, find out what it is and how it is spelled. If you are writing to a division manager, the receptionist on a company's switchboard is often willing to give you the manager's name. If you're writing to the personnel division, ask your placement counselor for the name of the personnel manager, or look it up in the College Placement Counseling National Directory. Always make certain you have the name spelled correctly, and that you address the person by the correct title.

Models

The letter of application shown in Figure 15-6 is written by a college student about to graduate and the letter in Figure 15-7 is written by a person who has been working for a number of years in a professional position. As you read, notice how both writers observe the conventions for this type of correspondence.

Figure 15-6
Letter of Application by a College Student

Addresses reader.
Identifies position.
Identifies company.
Gives reason for
applying.
Organizing idea.
Discusses related
work experience.

Discusses related
educational experi-
ence.
Requests inter-
view.

203 Florence Ave.
Normal IL 61761
October 8, 19xx

Mr. Dallas Reynolds, Vice President
Personnel Department
State Farm Insurance Company
One State Farm Plaza
Bloomington, IL 61701

Dear Mr. Reynolds:

   I would like to submit my application for an underwriting position with State Farm Insurance Company. Donit Stewart, Deputy Regional Vice President of the Costa Mesa office encouraged me to apply. I will graduate May 31, 19xx, with a Bachelor of Science degree in English from Illinois State University.

   I have had experience working in the insurance industry (see resumé), serving as a data entry clerk in the Southwestern Regional Office and as an auditor for Workers' Compensation at the Illinois Regional Office. My current experience as a teleprocessing technician at Corporate Headquarters has given me a broad background in life, auto, and fire operations as well as a clear understanding of agents' and insureds' needs and expectations.

   In addition to my work experience I have taken insurance courses in CPCU, IIA and LOMA.

   I would like the opportunity to discuss my possible employment with you. I will contact you next week about setting up an interview. I look forward to speaking with you.

   Yours truly,

   Meleah Melton

**Figure 15-7
Letter of Application by a Professional Manager**

Addresses reader.
Identifies position.
Identifies company.
Qualifications:
 • work experience

 • philosophy

 • specialization

Requests interview.

January 24, 1991

Mr. Charles A. Williams
Personnel Department
100 N. E. Adams Street
Peoria, IL 61629-1490

Dear Mr. Williams:

   Do you need an experienced operations or administration manager at Caterpillar Engine Parts? If so, I would like to introduce myself to you.

   I believe my record of stable employment, coupled with a history of increasingly responsible manufacturing and general management positions and the accomplishments achieved in each assignment, speak clearly of what I can bring to your organization.

   The positions I've held over the last twenty years required the management of line, staff, and professional personnel as well as the mastery of a wide range of processes. I have consistently demonstrated that organizational and technical improvement coupled with a strong commitment to employee training and motivation results in products being manufactured to spec, on time, and at or below budget.

   My core disciplines are personnel management, quality control, budgeting, scheduling, maintenance, long range planning, cost reduction and tech-ops support. My approach has always been proactive; preventing problems whenever possible and implementing conclusive solutions as required.

   I am interested in having the opportunity to elaborate on my credentials in person and in learning about your requirements in detail.

   Yours truly,

   Allan B. Greene

THE FOLLOW-UP LETTER

If your letter of application and resumé interest a potential employer, you will be called and asked to come in for an interview. Immediately after your interview, you should write a thank-you note, not only as a courtesy, but also to inform the reader that you continue to be interested in the position, and to persuade the reader to hire you. These letters follow the same conventions as the "bread-and-butter" notes you send relatives to thank them for presents. They are short, beginning with your purpose—an expression of appreciation—followed by a brief discussion of the high points of the interview, and then a conclusion, which is usually a gracious closing.

   The letter in Figure 15-8 thanks the recipient for a job interview. Notice that the writer follows the conventions.

Inside address.
Date.
 	
Formal heading.
Block form.
 	
Formal salutation.
Expresses appreciation.
Begins with main organizing idea.
Brief discussion of position. Indicates continued interest.
Ends graciously.
Offer of more information.
Formal closing.
	406 N.E Peachtree Rd.
	Atlanta, GA 30333
	January 24, 1991
	
	Mr. Charles Williams, Manager
	College Relations
	Caterpillar Inc.
	100 NE Adams Street
	Peoria, IL 61629-4279
	
	Dear Mr. Williams:
	  I would like to thank you for your time, January 24, 19xx.
	  I found the information about the company fascinating and would very much like to become involved in your efforts to design machinery for today's needs. I am very interested in the position of design engineer in your research division.
	  If I can provide you with any additional information, please let me know. I hope I shall hear from you in the near future.
	  Yours truly,
	
	Betsy Randolph

Figure 15-8 Follow-Up Letter of Appreciation

STRATEGY CHECKLIST

A. General

1. Use conventional formats.

2. Research the company to determine:

   - Who will read your documents—your potential employer or a representative of the personnel department.

   - The size of the company—large or small.

   - The company's frame of reference—conservative or creative.

   - The name of the person to whom you are writing.

B. The Resumé

1. Select only information relevant to the position for which you're applying.

2. Include the following information:

   - personal demographics

   - professional objectives

   - educational history

   - work experience

   - awards and honors

continued

- special activities
- references

3. Organize your information according to convention.

4. Use the conventional format.

5. Make certain your items are parallel.

6. Keep your resumé to one page.

7. Use visual markers so readers can locate different categories of information easily, as well as see important information.

8. Use a computer and laser printer to print up your resumé.

C. The Letter of Application

1. Indicate the position for which you are applying, and explain how you learned about the position.

2. Include information indicating your qualifications for the position.

3. Personalize the letter by using the name of the person to whom you are writing in the salutation and the name of the company in the first paragraph.

4. Organize the information in the body of the letter from most to least important.

5. Focus the letter around the position for which you are applying.

6. Use an impersonal, formal tone, but indicate enthusiasm, sincerity, and authority.

7. Be gracious.

8. Do not use jargon.

9. Keep sentences fairly short.

10. Keep the letter to a single page.

11. Close the letter with an indication you would like to meet with the recipient.

D. Follow-Up Letter

1. Write a follow-up letter immediately after an interview.

2. Use the conventions of a follow-up thank-you note.

CHAPTER SUMMARY

Two of the most important documents you'll ever write are a letter of application and a resumé. These follow specific conventions involving both the content, style, and format. Readers expect to find certain information in a specific sequence in these documents.

Letters of application and resumés are read by two types of readers: potential employers and members of an organization's personnel department.

The main criteria readers use to review a letter of application and resumé include:

- Interest in the company.
- Relevance of qualifications to the position.
- Quality of writing.
- Overall appearance.

A follow-up letter should be written after an interview to thank the reader and to notify the reader of continued interest in the position.

PROJECTS

As you undertake a project, be sure to engage fully in all three processes—planning, drafting, revising. You may want to use the Audience Analysis (page 97), Purpose (page 136), and Decision charts (Page 101), as well as the Planning Worksheet in Appendix B, as you gather and organize your information. As you prepare to revise, you may also want to use the Evaluation Checklist (page 264-265), and/or the Document Design Criteria Checklist in Appendix B.

Collaborative Projects

1. **Applying for a job**
   a. Write a letter of application and resumé to a company for which you would like to work, in a position for which you expect to be qualified by the time you graduate.
   b. Swap applications with a partner.
   c. Assume the role of an employer.
   d. Read your partner's application. Based on the criteria for applications discussed in this chapter, write one of the following types of response letters to the applicant:
      - Come in for an interview.
      - You're okay if we can't find someone better.
      - We don't want you.

   In your letter, discuss the aspects of the application and resumé that influenced your response.

2. **Applying for a campus position**
   a. In the community newspaper you read a help-wanted ad for a college student to direct a project to help the homeless. You decide you would like to direct the project. Write a letter applying for the position. Make five copies.
   b. Meet in groups of five. Distribute your application to each member.
   c. Evaluate the applications, including your own. Select the application that is the best for the position.

d. Duplicate enough copies of the selected application for everyone in the class. Distribute a copy to each person in the class.

e. Evaluate the applications from each group, including your own. As a class, select the person to become the project director from among these applications.

3. Selecting an applicant

Work in groups of three. You are a partner in a law firm. You have received the letter in Figure 15-9. During a meeting with the other partners in your firm, you bring up the request for employment contained in the letter.

a. Based on your reaction to the letter and accompanying resumé, your group should decide whether or not to hire the applicant.

b. As a group, write a letter responding to the request. Indicate in your letter the aspects of the letter and resumé that made you decide to hire or not to hire the applicant.

**Figure 15-9
Letter of Application and Resumé from a College Student for Summer Employment**

112 West Mulberry Street
Apartment #12
Normal, IL 61761
March 7, 1990

Alderman Mr. Pat O'Connor
Attorney at Law
Berman and Lowe
5424 North Western Avenue
Chicago, IL 60625

Dear Alderman O'Connor:

My father, Harry Sweeney, informed me of an opening position you have this coming summer in your Community Legal Services Department. My exposure to the real work environment has helped me to learn the importance of performing routine duties such as drafting legal documents, gathering and analyzing facts relevant to legal research, and most importantly, working with the public in community service. I am very interested in seeking a position at Berman and Lowe, where I can best use my experience and Pre-Law education.

I am currently attending Illinois State University and majoring in political science and English with a specialization in legal studies. I feel confident that my legal experience and my education in relevant courses, such as Introduction to Paralegalism, Constitutional Law, American Judicial Process, and Law and Office Administration have broadened my knowledge and skills for my future occupation.

With my qualifications and work experience, I believe I am prepared to fill a position in your Community Legal Services Department. I look forward to further discussions of my qualifications.

Sincerely,

Meredith Sweeney

continued

Figure 15-9
Letter of Application and Resume from a College Student for Summer Employment
continued

Meredith L. Sweeney

Present Address
112 West Mulberry Street
Apartment #12
Normal, Illinois 61761
(309) 888-4699

Permanent Address
4529 North Sacramento Avenue
Chicago, Illinois 60625
(312) 539-7692

Objective	Seeking a summer position or an internship which will allow me to use my present skills and provide me with the opportunity for personal growth in Legal Studies.
Education	Attending **Illinois State University**, Normal, Illinois. Will receive a B.S. degree in May of 1991. Major: Political Science GPA: 3.7/4.0 Minor: English GPA: 3.7/4.0 **Relevant Coursework:** Introduction to Paralegalism; Judicial Process; Constitutional Law; Law and Office Administration; and Technical Writing.
Honors and Awards	Dean's List; top ten percent of college students, Political Science Honors Program, Golden Key National Honors Society, Phi Sigma Alpha Political Science Honors Society, Phi Sigma Kappa Little Sisters Honors, and nominated for Red Tassel/Mortar Board.
Employment Summer 1989	**Gino's East Pizzeria, Skokie, Illinois.** Worked as a hostess. Responsible for welcoming and seating customers, and balancing daily sales.
Summer 1989	**Terri's Clarendon Cleaners, Chicago, Illinois.** Worked as a cashier. Responsible for inventory control, merchandise display, and recording ticket sales.
Summer 1988-1989	**National Copper Paint Corporation, Chicago, Illinois.** Worked on the Board of Directors. Responsible for recording daily transactions and servicing customers over the phone.
1984-Present	**Clarendon Food Mart, Chicago, Illinois.** Employed as a cashier. Promoted to head cashier with other responsibilities including; handling keys, locking store, training new employees, weekly payroll, inventory control, merchandise display, and daily banking deposits.
Personal	**Activities:** Elected to the Academic Senate of Illinois State University; Active member of the Central Illinois Paralegal Association; Law Club; English Club; Irish Club; Zeta Tau Alpha National Sorority; Former member of Phi Sigma Kappa Little Sister Organization: Active member of Fundraising Committee and Rush Committee. **Interests:** Traveling, volleyball, football, and reading.
References	John Rutherford Clarendon Food Mart 4183 N. Clarendon Chicago, IL. 60613 (312) 281-7790
	Mary O' Toole National Copper Paint Corporation 4547 N. Kedzie Chicago, IL. 60625 (312) 539-7339

Individual Projects

1. Write a letter of application and resumé for either
    a. a summer job
    b. a job when you graduate

2. Look through the want ads in a newspaper. Select a job for which you would like to apply. Write a letter of application.

3. Determine a large company and a small company for which you would be qualified to work by the time you graduate. Write a letter of application to the owner of the small firm. Write a second letter to the personnel department of the large firm. Are there any differences between the two letters? If so, what are the differences, and why do the letters differ in these respects?

4. Select two companies in your field, one known for its creativity and one for its conservatism. Write a letter to each company applying for a position. Are there any differences between the two letters? If so, what are the differences, and why do the letters differ in these respects?

5. Write a letter applying for a job at a company located near your home or college. Then write a letter applying for a job at a company located at least 500 miles from your home or college. Are there any differences between the two letters? If so, what are the differences, and why do the letters differ in these respects?

Instructions, Procedures, and Documentation

AS AN ENGINEER, TECHNICIAN, OR PROFESSIONAL in a technical field, you are seldom required to write instructions, procedures, or documentation. These are usually written by technical writers. On the few occasions you will be asked to write these types of technical documents, they will probably be for internal readers. Your purpose will be to instruct employees in using job-related equipment, such as computers, assembly machinery, or drills; or to provide employees with procedures for carrying out duties or filling out forms. If you are a programmer, you may also write the documentation for your software programs.

Technical writers write instructions, procedures, and documentation for external as well as internal readers. Often, these materials provide clients and consumers with information in using and repairing a company's products.

The consequences of failing to meet readers' needs when writing these technical documents can range from a consumer's inability to assemble a child's toy for Christmas, to users returning computer programs they can't operate, to a nuclear accident. If instructions don't specify all of the steps, if documentation isn't sequenced appropriately, or if procedures don't include all essential information, readers may not be able to do what they want to do or they may do it incorrectly.

In this chapter you'll learn to write instructions, procedures, and documentation. You'll begin by learning to write individual sets of these materials. Then you'll study strategies for putting individual sets together to create manuals for large products, multifaceted activities, or entire software programs. You'll discover how users read instructions, procedures, documentation, and manuals, and you'll learn to apply the strategies you have acquired in previous chapters for determining the most effective organization, style, visual text, graphics, and format that will provide users with the information they need. While this chapter gives you the conventions for these materials, you will still need to create your text by making decisions related to your specific context and topic.

GENERAL CONTEXT AND CONVENTIONS

Context

Regardless of whether you're writing instructions, procedures, or documentation, the conventions and strategies you use will often be similar. In this section we'll discuss those aspects related to the context in which you write, and the decisions you make that are similar for all three types of documents. In the succeeding sections, we'll consider those aspects that are specific to each type.

Readers All of your readers are users. And, regardless of whether they are internal or external to your organization, regardless of their position in the organizational hierarchy, and regardless of their knowledge of their own field, they are novices in relation to your topic. If they were experts, they would not need your instructions. However, many may be experts in their own field. Duffy, et al., refer to these people as expert/novices. They are knowledgeable in their own fields, but not on the topic for which you are providing the instructions.

In many cases you are instructing readers who are generalists in a field, people who have acquired some theoretical—or even practical—knowledge of a topic, but not enough to do a task without following instructions.

If you need to instruct both generalists and novices, write for the novice. If generalists know the information, they can skip over it, but novices cannot fill in information that has been omitted.

Because readers are using your instructions or procedures to *do* something, they usually scan the information, searching for the verb that tells them what to do. They do not try to learn the information; instead, they place it in their short term memory, remembering it only long enough to accomplish their task. Therefore, you need to design your document so users can quickly and easily locate information. Using the imperative mood, beginning each line with an action verb, and providing numerous visual cues, can facilitate users' reading behaviors.

In the example shown in Figure 16-1, which instructs new employees in serving customers, the writer assumes that readers are novices. Though they may have eaten at restaurants, even in this particular one, they probably have never served in one. Notice how the subheads make it easy to locate each chunk of information, and how visuals complement the instructions.

Adhering to the given-new contract, the writer defines the term "pivot point" in terms of a point the readers have already learned (Host/Hostess/stand). Graphics are used to provide a visual complement to the verbal description.

Subheads divide the section into four major chunks: pivot points, wall table, guest check, and tray service, so readers can easily remember the four basic instructions.

**Figure 16-1
Instructions for
Novices**

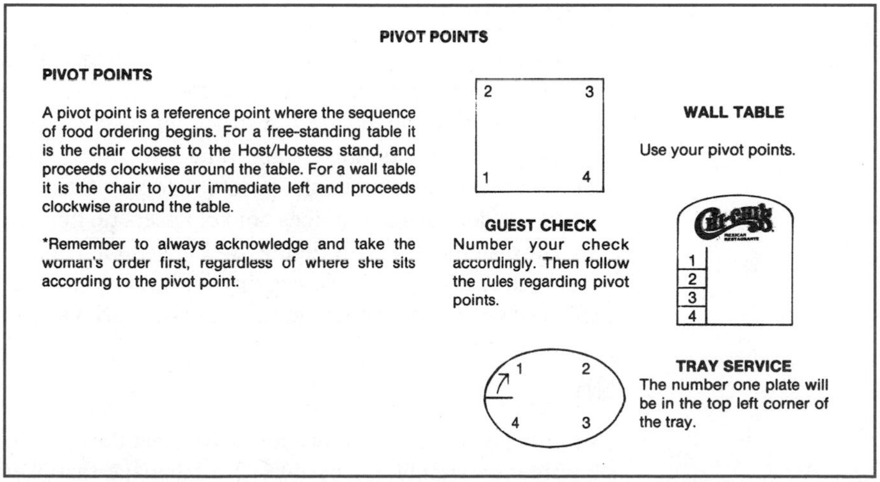

Purpose When you write instructions, procedures, and documentation, your purpose and your readers' coincide. Your purpose is to *help* readers *do* something. Your readers' purpose is to *do* it.

Situation Readers' attitudes toward your task differ. Some readers want to engage in the activities or follow the procedures, others are reluctant to do so. Their reluctance may originate from a dislike of the task, or from a belief that the task is superfluous, not part of their job, or too difficult, time-consuming, or expensive. Therefore, your instructions should be simple and "user-friendly."

Gathering Information

To gather the information that should be included, you need to perform the activity yourself, or watch someone else do it. You need to take meticulous notes, detailing everything you do from the very beginning to the very end. For instance, writers for technical manuals at John Deere literally build machines from scratch, taking detailed notes and photographs of everything they do.

Sometimes it is impossible to re-enact an activity. For example, if you are writing procedures for actions to take during an earthquake or if an emergency radiation leak occurs at a nuclear utility, you will not be able to actually simulate the situation. To be certain that all possibilities have been considered and are included in your document, you should engage in collaborative planning. By interviewing all of those persons who were involved or by conducting a brainstorming session with those people, you can obtain much of the necessary information. You may also want to search written sources to discover what others have done in these emergencies. And you probably want to conduct interviews with persons external to your organization who have written procedures for similar situations.

Point of View

The readers' point of view is that of a user. Keep this in mind at all times. The content you include and the sequence you use need to reflect the user's point of view.

In writing documentation, instructions need to be task-oriented rather than screen-oriented. Users don't want to be told what a program does. Rather, they want to know how to make the program do a task faster and easier than they can do it. If the documentation does not help users do this, they return the software.

The documentation in Figure 16-2 is included in a word processing program used by English majors. Rather than discussing how to set margins, the writer explains how to place long quotes in a text, a task users engage in frequently.

Style

Because readers need to concentrate on what they are doing, information should be written as simply as possible. Use language that everyone knows. Don't use technical terms. If you need to use them, define them.

Figure 16-2
**Instructions from
the User's Point
of View**

D. Long Quotes

When you include a quotation in your paper which is more than four typed lines, you need to set it off from the rest of your text by indenting it 10 spaces from the left margin, according to MLA guidelines. (Remember not to place quotation marks around it.)

1. Set the new margin BEFORE you type the quote.

2. Place the cursor where the quote should begin.

3. Press *Function Key #2* twice (2x).

4. Type the quotation. It will appear properly indented.

WARNING: Do NOT reform the quotes. If you need to reform your paper to double space it, etc., you will need to reform it paragraph by paragraph. If you use the quick reform command (^QQ^B), the quotes will no longer have the correct margins.

Keep sentences short. Don't place more than a single action in a sentence. Each action should comprise a single step, as in these instructions for preflighting an airplane.

PREFLIGHTING THE AIRPLANE

1. Walk slowly around the airplane, checking all external items.
    a. Check the propeller for security on the engine shaft.
    b. Check the propeller blade for nicks and scratches.
    c. Check the engine cowling for evidence of oil leaks.

Notice that while both items a and b are concerned with the propeller, the writer has placed them in two different steps, since they comprise two different aspects of the prop.

Effective instructions, procedures, and documentation follow specific text grammars, which you need to adapt (see Figure 16-2).

- Use the active voice. After all, the reader is acting out your instructions.
- Use the imperative mood.
- Begin each step with a verb.
- Keep sentences simple.

Format

Instructions, procedures, and documentation follow a conventional format which involves:

- Headings and subheadings to introduce sections/subsections respectively.
- Steps/items listed.

- Lists indented.
- Steps separated by extra spacing.
- Related steps chunked.
- Steps numerated.
- Items bulleted.
- Lines of text kept short.

In the instructions in Figure 16-3, the writer uses a conventional format to instruct users in stopping a snowblower.

Visual Text

Visual text is extremely important in writing these documents, since users often must leave the instructions to carry out various activities. Visual cues help users keep track of their steps, locate the step they're up to if they need to reread a direction, or find the next step if they have completed an activity.

To help readers quickly locate information, use subheads to introduce major chunks of activities. Numerate each subhead. To make subheads prominent, use such typographical markers as capital letters, boldface, and large type size. Hanging indentation helps readers quickly spot the number of the step they're working on if they have to leave the instructions. Surrounding chunks with white space visually sets off related steps.

Visual cues also help readers pace their activities by visually blocking chunks of work. If a project is a relatively long one, users often take breaks between the various chunks. Divide long lists of activities into short categories.

Because users are far more interested in *doing* an activity than in reading about it, visual cues can emphasize information that readers may miss in their hurry to work on a project. Typographical markers emphasize important information, such as warnings and things to remember.

The instructions in Figure 16-3 were written to help John Deere customers operate a snow blower. The writer assumes that users are novices. As you look at the instructions, consider how the visual cues help users.

The centered heading above the top margin makes it easy for users to locate the section. The subheads in all capital letters and boldface type facilitate readers' searching a section for specific information. The subheads also serve as organizing ideas for the information that follows.

The page includes two subsets of instructions for stopping: normal and emergency stops. The writer divides the instructions visually into three modules by boxing each set of steps around the major action to which the steps are related. Notice that a separate graphic is required for each.

In the first two modules, each step is numbered. In module three, the equipment's response to the operator's action is listed with bullets, rather than numbers, since these items do not need to follow a specific sequence.

The photographs contain call-out letters referenced in the text, making it easy for users to locate the specific mechanisms being discussed.

Chapter title.

Heading.
Boldface.
All caps.

Spacing.
Lists.
Hanging indenta-
tion.
Numeration.
Chunks.
Graphics (photos).
Call outs.

Boxes.

Short text lines.

OPERATING THE SNOWBLOWER

STOPPING SNOWBLOWER

Follow the steps below before parking, inspecting, adjusting, or unplugging snowblower:

1. Release traction drive lever (A) and auger drive lever (B): snow-blower will stop and auger/blower will stop.
2. Put speed selector lever (C) in NEUTRAL.

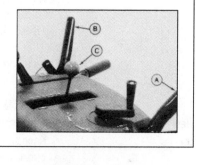

3. Push throttle lever (A) down to STOP.
4. Turn key (B) to OFF. Remove key.

Graphic symbol
(international
warning).
All caps.
Boldface.

Graphic (line
drawing).

EMERGENCY STOPPING

⚠ CAUTION: BE ALERT to possible emergencies while operating snowblower. BE READY to stop snowblower quickly.

Release both drive control levers:
• Snowblower will stop.
• Auger and blower will stop.
• Engine will continue running.

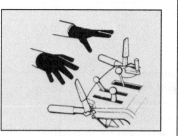

Figure 16-3 Using Visual Text and Graphics in Instructions

A visual symbol, a triangle with an exclamation inside, is used along with the word **CAUTION**, printed in all capital letters, to emphasize a warning. In addition, the entire warning is in boldfaced type. Furthermore, the two phrases the writer wants to persuade users to remember are in all capital letters: **BE ALERT** and **BE READY**.

Visual cues for procedures sometimes differ slightly from those for instructions. Because procedures often do not follow a specific sequence, you should use bullets rather than numeration, as in the following list of procedures for providing customer service.

Bullets.

SERVICE ASSISTANT GUIDELINES
• Anticipate needs of Guests in dining room.
• Keep an eye on detail and uniformity while maintaining a sense of urgency in clearing, cleaning and re-setting tables.
• Communicate table turns or other needs to fellow service assistants.
• Keep work area clean.
• Check every 15 minutes at H/H stand for restroom checks.

If readers are expected to engage in a specific number of procedures, however, you should use numbers to help readers remember rather than bullets. Numbers serve as a mnemonic device. The four procedures involved in serving chips at Chi-Chi's are numerated so readers remember to engage in all four as they go about their duties (Figure 16-4).

Graphics

The more you illustrate the tools, materials, and activities involved, the better chance your readers have of following your verbal instructions, procedures, or documentation correctly. People have different learning styles. Some people learn by reading an explanation of how to do an activity. Others need to see an

**Figure 16-4
Numerating as a
Mnemonic Device
in Procedures**

Table Chipping Guidelines

1. Warm welcome.

2. Present chips to the table.

3. Present hot or mild sauces.

4. Remove extra settings.

activity being performed. In addition, many employees speak English as a second language, and many consumers and clients may not speak English at all. However, they can usually follow graphics to learn what to do. Instructions for putting together "Legos," manufactured in Denmark, rely solely on graphics.

Furthermore, as you learned in Chapter 3, no matter how hard you try, you cannot get completely into readers' minds; you never know when a reader may misinterpret your words. Graphics help clarify verbal directions.

In Figure 16-5, the writer uses graphics almost exclusively to demonstrate the right and wrong way for waiters and waitresses to handle glasses, plates, and utensils. As you study the graphics, consider the reason the writer uses a graphic rather than a verbal description.

Many companies have stored graphics in computer databases so writers can call them up and use them in various documents. A large database of international symbols, such as the one in Figure 16-6, already exists. These symbols should be used if you write for an international audience.

**Figure 16-5
Instructions
Communicated
Graphically**

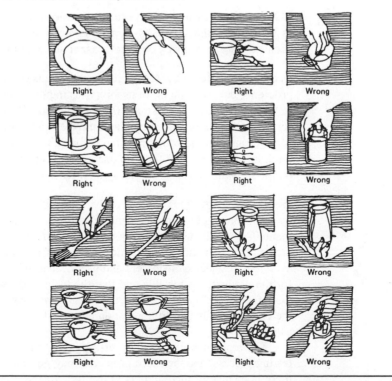

Handling: Glasses, Plates, Utensils

All service personnel should learn to handle dishes and utensils in a sanitary manner. Plates should be held by the bottom or at the edges; glasses by the handles or at the bottom; silverware by the handles.

Right   Wrong   Right   Wrong

Right   Wrong   Right   Wrong

Right   Wrong   Right   Wrong

Right   Wrong   Right   Wrong

Figure 16-6 Safety Symbols

Usability Tests

Once you have written your instructions, they should be field-tested to determine whether or not your readers can follow them and *do* what you are trying to instruct them to do.

A usability test is conducted to determine whether or not a manual or set of instructions fulfills its purpose, as determined by the end users. Usability testing can be conducted in a variety of ways. Redish and Schell suggest the following four types of usability testing:

User edits. Used for short sets of directions, users read the instructions aloud as they follow them, step-by-step. This technique is inappropriate for manuals, because it does not check on users' ease in locating information.

Protocol-aided revision. Readers read instructions aloud and comment on problems, etc., with them. In usability testing on software documentation, a tape is made of users' discussions, their keyboard actions, and the accompanying screens, and then reviewed to determine the exact problem.

Because people sometimes have difficulty in verbalizing their thoughts out loud, an alternative is to pair up users and record the conversation as they work on the software. Their discussions can often show problems with instructions.

Beta Testing. This term is derived directly from the early (1960s) product cycle of computer technology, and relates to the field-testing of a product and its manuals by actual users. The Dictionary of IBM Jargon defines it as testing "…a pre-release (potentially unreliable) version of a piece of software by making it available to selected customers and users…The A-test was a check that the idea for the product was feasible and that the technology was available and could be manufactured…The B-test was the most significant checkpoint; it showed that the engineering model (often just an advanced prototype) could run and perform as specified."

The problem with usability testing is that the writer has no control over the testing process, and therefore problems may go unreported. In addition, because there is no tape of the attempts to use the process, the writer may not be able to determine exactly what the cause of the problem is.

Laboratory-based usability testing. This technique combines the best aspects of the other three. It is conducted in a laboratory, but participants are given scenarios similar to those that users would encounter in the actual workplace. Both video and audiotape are used to record users' actions as well as their verbal descriptions. In addition, the appearance of the monitor is recorded. A data-logging program can also be used to record users' actions on the computer, the amount of time users take to fulfill a command, the number of times users move between the screen and the manual, and a list of keystrokes made by users. Finally, users respond to a questionnaire concerning their opinion of the program.

STRATEGY CHECKLIST

1. Consider readers as novices.
2. Write for the user's point of view.
3. Gather information by performing the activity yourself, whenever possible.
4. Place each step in a separate activity.
5. Write in small modules/chunks.
6. Use numeration to indicate that users must follow a specific sequence. Numeration is always used in instructions.
7. Use numeration if users must remember to engage in a certain number of procedures.
8. Use bullets if there are only a few procedures, and users either do not need to engage in them sequentially, or in all of them.
9. Use everyday language.
10. Define technical terms.
11. Keep sentences short.
12. Follow the appropriate grammar:
    - Active voice.
    - Imperative mood.
    - Begin each step with a verb.
13. Provide visual cues:
    - Create chapter organizers.
    - List all steps.
    - Numerate steps.
    - Use hanging indentation.
    - Use subheads to set off related steps.
    - Surround related steps with white space.
    - Use typographical markers to emphasize important information.
14. Use graphics to complement, as well as supplement, the instructions.
15. Use international symbols when writing for readers who may not read English.
16. Conduct usability tests to verify that readers can do an activity properly.

INSTRUCTIONS

Instructions range from putting together a child's mini wheel to setting a digital clock to repairing a 125-hp (horsepower) turbocharged combine. Because content, and organizational pattern and sequence differ between instructions and procedures, we will examine them separately.

Content

The content depends on readers' needs. Include all the information a reader needs to perform an activity. If you are teaching a first-time user to use a computer, you need to include information on how to turn on the machine. If you are teaching a novice to fly, you need to explain that the rudder pedals control the way the plane turns on the ground.

Do not take anything for granted. If you are providing instructions for checking the oil level in a car, begin by telling readers to open the hood.

Organizational Pattern and Sequence

The beginning of any set of instructions or chunk of instructions should indicate what the finished product looks like or will do so users have a frame in which to work just as the box cover of a jigsaw puzzle pictures the completed puzzle.

The following objective specifies what employees will be able to do after an RMS system is installed:

Objective: Employees will be able to receive mail, send messages, and update form letters on the RMS electronic mail (e-mail) system.

If tools, materials, ingredients, etc., are necessary for carrying out an activity, list them at the beginning of the document so users have them when they are needed.

The instructions with which you are probably most familiar are recipes in which ingredients are listed at the beginning. There is nothing more aggravating than discovering at step four that you need to add a cinnamon stick, and you don't have any in the house. While you may be able to borrow eggs from your neighbor, your neighbor may not have cinnamon sticks.

The writer places the key to the major parts of the brake system in an AMT at the very beginning of the instructions in Figure 16-7 so readers can easily locate the parts in the accompanying graphic as they're described in the verbal text.

Instructions should be written according to the sequence in which the user needs to work. Begin with the very first activity readers must do. Notice that the writer begins the instructions in Figure 16-8 with the opening of sidework and ends with the closing sidework.

**Figure 16-7
Listing Components in Instructions for a Mechanism**

BRAKE OPERATION

A—Lever	D—Adjusters	G—Balls	J—Self-adjuster Screw
B—Park Lock	E—Equalizer	H—Rotor	K—Disk
C—Cables	F—Stator	I—Self-adjuster Actuator	L—Pads

The brakes are a mechanical, disk and floating caliper, self-adjusting type. The hand lever is connected to the two brake caliper assemblies with separate cables. As the lever (A) pulls the cables (C), a rotor (H) in each caliper is turned. The turning rotor forces three balls (G) up ramps on the stator (F) which presses against one brake pad (L). That pad contacts the disk (K) and the caliper assembly moves causing the second pad to contact the disk.

The rotor also turns the self-adjuster actuator (I) which moves the self-adjuster screw (J) out. As the brake pads wear, the actuator rotates far enough to catch on a ratchet that prevents it from turning back

when the brakes are released. This then keeps the brake pads properly adjusted.

To adjust cable freeplay there are adjusters (D) at the lever end of both cables. And to allow for variations in cables and calipers, and to keep equal pressures on the brakes, there is an equalizer (E) that connects the lever to the cables.

The park lock (B) is a toothed cam that engages with a tooth on the brake lever. When the lever is pulled and the park lock is engaged, the lever is locked in the engaged position and the brakes stay engaged.

Instructions involving more than seven items should be divided into categories so the list doesn't appear too long, causing readers to approach it with a negative attitude. Shorter categories of instructions also help readers to remember activities more easily.

The instructions for servers at Chi-Chi's Restaurants in Figure 16-8 are divided chronologically into three categories: opening, on-going, and closing sidework.

Sometimes users need to select from among several steps. Provide readers with instructions for locating the particular set of steps they need. The writer of the following instructions provides users with two alternatives for billing guests in step 4.

Background.

Four credit cards are accepted at Chi-Chi's: Diners Club, American Express, Visa, and MasterCard. The forms for these cards are similar; therefore the greatest care must be taken to ensure that the proper form is used with each card.

1. Imprint charge slip and back of guest check.
2. Fill in the guest check number.
3. Fill in the amount of the check.

Alternative selections.

4. [a] If Visa or MasterCard is used, proceed to Envoy Instructions below.
   [b] If American Express or Diners Club, verify card with warning bulletins...

If readers need to select from more than two alternatives, the choices should be listed so that readers can easily identify them. Caterpillar Inc. presents these choices visually by using a decision box (Figure 16-9).

Category 1

Opening Sidework

1. Place lemon wedges, creamers and mints in a pan with a pan of ice underneath.
2. Make coffee (regular and decaffeinated) and iced tea.
3. Stock the Chajita stand with tongs, spoons, plates, lids, chip plates, clean wooden liners, and lemon wedges on ice....

Category 2

On-going Sidework

1. Keep all items in the service stations stocked i.e. lemons, creamers, full ice bins, fresh coffee (follow the rule: Empty a pot - Make a pot).
2. Keep the Chajita stand stocked.
3. Keep the server refrigerator stocked...

Category 3

Closing Sidework (Done after closing your station)

1. Thoroughly clean your assigned station. (Wipe all window sills and dust the blinds).
2. Clean the soda dispenser. Pull off the nozzles and soak them in soda water. Wipe all parts of the dispenser. Replace the nozzles after soaking them for 10 minutes...

Copyright © Chi-Chi's Inc.

Figure 16-8 Chunking Longs Lists of Instructions

**Figure 16-9
A Decision Box**

```
                            SUPPLIER ADDRESS INQUIRY
        Inquiring On Supplier Address
        ----------------------------


             *****
        7.   Press ENTER key.
             *****

             Response:  See table below.
```

IF YOU:	THEN:
Requested information for another supplier	The Supplier Address Inquiry screen (BAL) will be returned with existing data
Requested another supplier inquiry screen	The requested screen will be displayed
Requested an address that does not exist or entered an invalid address type	Space over the address type field to continue processing
Entered the same address type as previously requested	Processing will continue with the next address type

Reprinted with permission Caterpillar Inc.

Computerization

While most instructions, especially those for consumers, are included in printed instruction manuals, an increasing number of manuals for industrial machines and parts are being transferred to hypertext and stored in computers or on CD ROM.

Using computer terminals, users can locate the specific aspects of a mechanism in which they're interested from online menus (menus which appear on the monitor). By indicating the component, e.g., engine or fuel system, and then responding to onscreen questions, users can call up instructions for diagnosing or repairing problems. The information can be read online or printed out. Users can simply point to the battery of the AMT that appears on the monitor, and the information will appear.

Instructions that will be used on computers are written simultaneously for both hard-copy manuals and hypertext. For this reason, instructions are being written in small modules. An editor organizes the modules to be printed in a hard-copy manual which is sent to distributors and dealers who do not have computer capabilities. A computer programmer programs the modules and fits them into hypertext for users to call up at appropriate times.

STRATEGY CHECKLIST

1. Include ALL information necessary for readers to perform an activity.
2. Begin with the big picture, i.e., a description of what the mechanism looks like, or what the activity accomplishes.
3. List tools, materials, ingredients, etc., at the beginning.
4. Sequence the information according to the sequence in which the user does the activity.
5. Break down lists of more than seven steps into subcategories.
6. Instruct readers how to locate the steps they need if several alternatives are possible.

PROCEDURES

While instructions help readers *do* something, procedures provide readers with guidelines for *carrying out* activities. Procedures are usually written to ensure that employees or other persons conform to specific guidelines, such as those related to applying for vacation leave. When procedures are mandatory, they assume the form of regulations. Instructions are action-oriented, procedures are process-oriented.

The following procedures appear in the Illinois *Rules of the Road* booklet for student drivers. Notice that they are written using a point of view, style and format that is similar to instructions.

Items listed.
Short sentences.

Imperative mood.
Begins with verb.
Active voice.
One activity/item.
Hanging indenta-
tion.

In the event of an accident:

- Stop your vehicle in a safe place.

- Help an injured person if necessary or requested...

- Warn other drivers, using flares if possible.

- Ask all those involved for their names, addresses, phone numbers, driver's license number, and license plate number.

While some aspects of procedures are very similar to those of instructions, differences exist in content and organization.

Content

Procedures require more background than instructions. Not all readers want to follow the procedures you write. Provide background so they understand why they need to follow them. In the following set of procedures, the writer begins by explaining why a pilot needs to adhere to procedures for preflighting a plane.

Purpose.

Procedures.

...The airplane must be inspected before takeoff...The responsibility for it [the plane] rests solely with the pilot of a light aircraft. The line check [inspection] might not turn up *everything* that could possibly need repair or replacement, but unquestionably it prevents a good many potential accidents...
1. Walk slowly around the airplane, checking all external items...
2. Open the engine cowling...
3. Check fuel tanks and refuel as necessary...
4. Turn the propellers through two complete revolutions...
5. Enter the cockpit...
6. Start the engine and proceed through warmup and take-off procedures...

Like instructions, procedures may be complex, requiring a number of steps. Be certain you explain all aspects related to a procedure. Otherwise readers may not carry out the procedure properly.

In the following description, included in the Illinois *Rules of the Road* booklet, the writer recognizes that readers not only need to know when they can pass another car, but also when they can return to their original lane.

Passing
A driver should use caution when passing another vehicle. On a two-lane highway, the left lane should be clearly seen and free of oncoming traffic for a distance great enough to permit passing. Do not turn back into the right-hand lane until you can see the car you have just passed in your rearview mirror. You must return to your lane before you get within 200 feet of an oncoming vehicle.

Organizational Pattern and Sequence

The major difference between organizing and sequencing the steps in instructions and the processes in procedures is that many procedures don't need to be done in a specific sequence. For example, on page 468, item one needs to occur first; item five needs to precede item six, which needs to occur last; but items two through four can be done in any order. Procedures can be organized from most to least important/difficult, or even alphabetically, depending on how readers will use them.

**STRATEGY
CHECKLIST**

1. Follow the basic guidelines for point of view, style, and format given earlier for instructions when writing procedures.

2. Establish reasons for following procedures. Include background information at the beginning of a set of procedures.

3. Include all elements related to a procedure.

4. Organize procedures from most to least important/difficult, or in a logical pattern.

DOCUMENTATION

Documentation is a special form of instruction that accompanies computer software programs to help users use the programs. All of the strategies you learned for writing instructions and procedures apply to documentation. In addition, you need to be aware of some aspects specifically related to the context in which documentation is used. These aspects create a need for additional strategies.

Context

Readers and Purpose According to Carroll, users are easily overwhelmed, even though they are anxious to begin using a program. Some users are afraid to try anything on their own, while others move ahead of the instructions or out on their own. Many of these latter users decipher a program using the "guess-ability factor." Based on their prior knowledge of programs, they try to "guess" how this one works, resorting to documentation when they become stuck.

Most users purchase a program because of its major components. Users purchase a word processing program because of its capability in expediting the drafting and revising process. To use the program for these functions, writers need to learn only a few functions: insert, delete, move, print, and save. However, the program can perform many other functions, e.g., centering, spacing, paginating, italicizing, etc. In many cases users are not aware of these functions. Therefore, the documentation writer's purpose is not only to help users learn the functions to accomplish the tasks they want to do, but also to learn functions to accomplish tasks they may want to do, but aren't aware they can do.

Users refer to documentation when they first get a program so they can learn how to use it. Later, they refer to the documentation as they use the program to learn how to perform new functions, to review how to perform functions they've already learned, or to solve problems resulting from performing a function.

Documentation users fall into three groups, based on their purpose.

- Those who are completely unfamiliar with a program. This group wants to know how to perform the particular tasks for which the program was purchased.

- Those who are basically familiar with a program. This group wants to know how to perform particular functions while operating a program. These users are often in the middle of a task, with their fingers poised on a cursor key, unable to determine what to do. These users want immediate help.

- Those who want to learn what a program does. These users may be unfamiliar with a program, or they may have some familiarity with it. They are willing to play around with a program and try out the various options and bells and whistles included in it.

Focus, Organizational Pattern and Sequence

Because users need documentation for a variety of purposes in a variety of situations, different forms of documentation focus on different aspects of a program, and therefore use different patterns of organization.

Forms of Documentation

Documentation can be presented in three forms: tutorials, reference manuals, and online instructions.

Tutorials Tutorials help users *learn* to use a program. Documentation in a tutorial should be task-specific, providing users with an understanding of how to navigate through a program to obtain the information they want and to understand what the program can do.

To provide task-specific instructions, many tutorials simulate a problem that users may actually have, and then take users step by step through the program to solve that problem. Users can then transfer the strategies they have learned for solving a simulated problem to solving their own real problems.

The first few chapters of a tutorial provide users with the basic steps necessary to accomplish a task. Later chapters provide users with insight into additional functions a program can perform. Tutorials may be written as hard copy and compiled in a booklet or placed on a diskette.

The program Codebuster, from which Figure 16-10 is excerpted, provides architects with a quick means of determining building code regulations in various regions of the United States. Consider how the tutorial simulates a situation in which a user needs to determine regulations for a restaurant to be built in a midwestern state. Users can apply the steps they learn in the simulation to investigate codes for the types of buildings they design.

**Figure 16-10
A Task-Oriented
Tutorial**

1 - STARTING A NEW PROJECT

Introduction

The first ten lessons teach you how to use Codebuster to help you conduct code research on a project. They assume you're an architect who is using Codebuster for the first time. The project is a small seafood restaurant called *The Pier*. It's to be built of heavy timber construction and will be occupied by not more than 50 persons. You're in the process of designing it, and your assignment is to research exit requirements.

Objective

Your objective in this lesson is to select the appropriate building code and edition, as well as the relevant chapters for your restaurant. In doing so, you'll learn how to get around in Codebuster. Then you'll start a new project by creating project files for your restaurant.

Getting Started

To start Codebuster, assuming you've already installed it on your hard disk, at the DOS prompt type:

CB
<enter>

In a few moments the first introductory display will appear. Take time to record the serial number in the front of both your manuals. Then read the rest of the display.

Assuming you agree to the terms of the Codebuster License Agreement, strike:

<any key>

Architech, Inc., Chillicothe, IL

Tutorials should be divided into short lessons that take users no more than ten minutes to do.

Like all sets of instructions, the tasks to be learned in each lesson should be listed at the beginning. The list serves as an organizer, providing users with a frame in which to understand the various tasks, as well as with the opportunity to decide whether or not to engage in a lesson.

Tasks should be sequenced according to the order in which the user should perform them. If there is no specific order, then sequence the tasks from most-used to least-used, or from simplest to most difficult.

The major functions learned in each lesson should be summarized at the end.

A picture of each screen should accompany the text as it is discussed.

Reference Manuals Documentation in these manuals covers the basic structure of a program and the main commands needed to obtain information. It provides information about each function that can be performed, allowing the user to choose which function to use.

Documentation in reference manuals provides users with instructions so that they can perform a task immediately. Users already familiar with a program use documentation when they have forgotten how to obtain data or need to obtain new data or old data in a new form. They also use it for immediate help when they are in the middle of a program and don't know or have forgotten what to do.

A writer using a word processing program may forget how to change margins. The documentation in a reference manual, such as that shown in Figure 16-11, provides this information. Notice the difference between these instructions and those in Figure 16-10.

While the documentation in a tutorial (Figure 16-10) focuses on a specific task, the documentation in a reference manual focuses on specific functions, as in Figure 16-11, or on an entire screen, as in Figure 16-12. Notice that while both Figures 16-10 and 16-12 are concerned with the same aspect of Codebuster, the information and focus in each example is different.

In another example, let's look at how Michael Edwards, an archaeologist, used documentation. When Edwards completed a report on an archaeological site for the local township, he still needed to include a list of references for an appendix. To speed up this long, dreary task, he purchased a software program, Bibliography Generator. However, he had never used it. He needed to learn how to generate his list, but he didn't want to take time for a tutorial. Using the documentation in the reference manual, he got the program up and running within minutes, and in less than an hour had an alphabetized list of nineteen sources printed up. As you study the documentation in Figure 16-13, consider how the graphics helped Edwards move from screen to screen to accomplish his task quickly and easily.

When writing documentation related to a screen, discuss the various functions on the screen in the same way a user reads them—from left to right, top to bottom. Include a picture of the screen so users can identify the functions that they need before they have to look back at the screen.

Online Documentation Users prefer online documentation because it is immediately accessible. However, most programmers are not able to include complex instructions with their software.

At present, most online documentation relates to brief sets of instructions or lists of commands as illustrated in Figure 16-14, which aids a user in need of immediate help to deal with a specific screen and function.

Design strategies for online documentation are very similar to those used for writing electronic correspondence (See Chapter 14), and should adhere to the following three conventions:

- Each topic should stand alone as a separate module.
- Include one topic on the screen at a time.
- Keep sentences short. Be concise.

**Figure 16-11
Documentation
for a Function**

E. Changing Margins
1. Left and right margins
   a. Place the cursor where you want the new margin to begin.
   b. Type ^O.
   c. Type L for the left margin and M for the right margin.
   d. Press the *Escape (ESC)* key.
   e. Type ^B to reform the line(s) or paragraph or type ^QQ^B to reform the entire paper.

**Figure 16-12
Documentation
for a System**

6 - QUESTION MODE

If you're starting a new project, unless you inadvertently selected -*Select all*-, there's no need to select -*De-select all*-. Since you haven't answered any of the questions yet, they'll automatically be de-selected.

Composing the Question List

Selecting -*Proceed*- in the Question Menu prompts Codebuster to build the Question List and display it on your monitor.

During the current session, if you previously accessed the Question List, and the chapter set has not changed since that time, the Question List already resides in a file on disk. Therefore, Codebuster doesn't have to re-build it. It merely retrieves it and displays it on your monitor.

On the other hand, if it's your first time in the Question List with the current chapter set, Codebuster must build the Question List. To do this, it gathers questions that are pertinent to the statements in the chapter set. If your chapter set doesn't contain all chapters, Codebuster tailors the Question List to the chapter set.

E X A M P L E   Suppose none of the chapters in the chapter set have anything to do with structural steel. Even though -structural steel- is one of the questions available, Codebuster won't include it in the Question List. To include it would be wasting your time. When you established the chapter set, you effectively told Codebuster not to focus on chapters that involve structural steel.

Architech, Inc., Chillicothe, IL

HOW THE PROGRAM WORKS

After the title frame appears and the program is loaded, you will be requested to insert a File Disk (a blank formatted disk) into the disk drive. Before you can use a file, you must first select a file and then ask the program to open it.

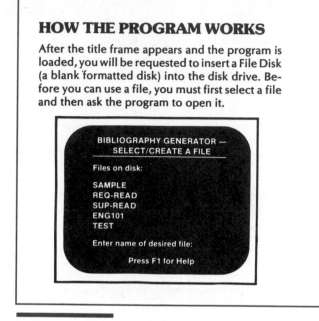

The Main Menu

After the file is opened, the Main Menu is displayed. While you are at the Main Menu, you can select the activity you want to do.

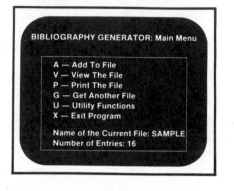

Six activities are listed on the menu. You select an activity by pressing its letter.

Educational Activities, Inc.

Figure 16-13 Graphics in Documentation

**Figure 16-14
Minimal Online
Help**

```
Disk: Disk 2 (Slot 6)              HELP              Escape: Main Menu
  _____
 |     Main Menu          |_____
 |  |                                                          |
 |  |   Help              |_____   |
 |  |                                                       |__|
 |  |   Ə-E:         This combination of keys lets you change  |
 |  |                between the "Insert" cursor (a blinking    |
 |  |                bar) and the "Strikeover" cursor (a         |
 |  |                blinking square).                          |
 |  |                                                          |
 |  |   Ə-H:         This combination of keys lets you print a  |
 |  |                "hard" copy of the contents of the screen. |
 |  |                                                          |
 |  |   Ə-Q:         This combination of keys lets you quit     |
 |  |                working with the file currently on the     |
 |  |                screen.  You can then choose from a list   |
 |__|                of all the files on the Desktop.          |
    |                                                          |
    |_____|

 Use arrows to see remainder of Help                    267K Avail.
```

S T R A T E G Y
C H E C K L I S T

1. Use the same strategies for writing documentation in reference manuals and for online instructions as you use for writing instructions and procedures.

2. Provide users with a simulation of a problem in a tutorial.

3. Divide tutorials into short lessons.

4. List tasks for each tutorial lesson at the beginning.

5. Sequence lessons in a tutorial from least to most difficult.

6. Sequence the segments in each lesson in a tutorial from a general overview, to a simulation of a specific problem, to an open-ended problem in which the user attempts to determine the solution.

7. End a tutorial with a brief summary of the lesson.

8. Follow the guidelines for sending electronic messages in Chapter 14 in writing online documentation.

9. Use the global guidelines in the following section for writing reference manuals for software.

MANUALS

The instructions, procedures, and documentation we have been discussing have involved simple products, activities, and programs. Instructions, procedures, and documentation for complex machinery, multifaceted activities, and intricate software programs involve multiple sets of instructions. Documentation for a word processing program involves instructions for inserting, deleting, and moving text, as well as centering, double spacing, changing fonts, styles and sizes, etc. Instructions for operating a 747 jet, procedures for operating a nuclear power plant, and documentation for an entire desktop publishing program involve hundreds of instructions, which are collated in manuals.

Manuals are usually written by a team of writers, all of whom must follow a single style and format to avoid confusing users. A manager, serving as editor, assigns different sections to each writer, and then puts all of the completed sections together. It is the manager's responsibility to see that all parts of the manual conform to a single style, so that it appears to be written by a single individual rather than a committee.

In this section we are going to look at instructions, procedures, and documentation from a global point of view. We are going to begin by considering how users read manuals, and then we'll examine how a manual is put together.

Reading Behaviors

Users seldom read an entire manual. Rather, they use a manual the same way they use a dictionary, as a reference. Users turn to a manual when they need to diagnose a problem, operate a piece of machinery, or repair a part. They read only that part of the manual specifically related to the task at hand.

Furthermore, people use a manual when they are busy. They may be ready to start doing something and need the manual to help them do it, or they may have had to stop doing something because of a problem, and they need the manual to help them solve the problem so they can return to their work. In either case, they need to locate information quickly and easily.

Global View

Determining an organizational pattern and sequence on a global level for the hundreds of sets of information that comprise a manual is perhaps the most difficult task involved in putting together a manual. It is also a crucial one, since users must be able to access the information they need quickly and easily. Readers also want all of the information in a single place; they do not want to have to look up half-a-dozen items to obtain the information they need to perform a single task.

Systems/Functional vs. Task-Oriented Organizational Pattern

Readers need manuals organized in either a systems/ functional or a task-oriented pattern, depending on the way in which they will use the information.

Manuals that use a systems/functional approach are organized according to the various systems comprising a mechanism; e.g., the engine, fuel system, electrical system, or power train, or the functions of a software program, such as editing, type face selections, or column guides.

Manuals that use a task-oriented approach are organized according to the user's use of a mechanism, e.g., accelerating, moving the seat, braking, steering, or lifting, or a software program, e.g., inserting, changing type size, creating columns.

Manuals traditionally follow a systems/functional approach. The problem with this approach is that often users must skip around from one section to another, and sometimes from one manual to another, to complete a single task.

For example, if the engine of an AMT is flooding, a technician must first look under the section on engine system diagnosis. If she determines that the problem is the fuel pump, she must turn to a different section to see a diagram of the fuel system of the AMT so that she knows where the pump is located. Then she needs to turn to still another section to learn how to remove the old pump. To order a new pump, she must go to a completely different manual, which lists parts.

A task-oriented manual places all of this information in a single section— engine flooding. The problem with this approach is that much of the information has to be repeated a number of times to provide for parts that are involved in multiple functions. The fuel pump is involved in six different engine problems: flooding, stopping when hot, surging, missing, losing power, and venting black exhaust.

Locators To provide users with easy access to the information they need, regardless of which organizational pattern a manual follows, document designers use various global locators.

Index. An index that is based on both a systems/functional approach and a task-oriented approach can provide users with quick access to the information they need. An index for the documentation to the word processing program discussed on pages 457 and 473 should include entries for "Long quotes" and "Margins," to provide users with access to the instructions on how to change margins to write a long quote.

Tables and flow charts. These graphics can help users locate information related to a task or system/ function regardless of a manual's organizational pattern. Notice how the flow chart shown in Figure 16-15 helps a task-oriented user locate information for solving a problem in an LED display, even though the manual follows a system organizational pattern. After reading the chart, the user knows to look up information on the battery, logics fuse, and/or key switch.

**Figure 16-15
Flow Chart for
Locating Informa-
tion About a
Mechanism
Globally**

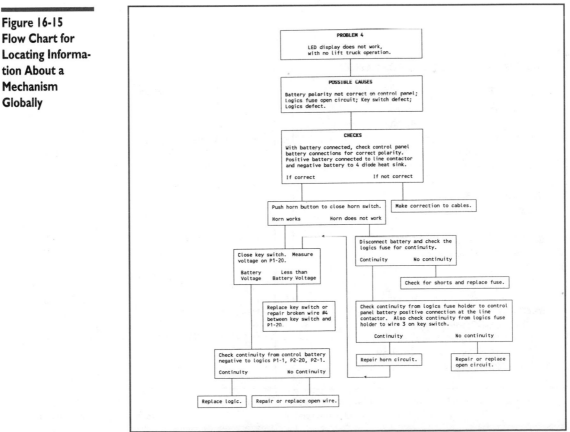

Reprinted with permission Caterpillar Inc.

Multiple tables of contents. Manuals often have tables of contents at the global, sectional, and chapter level.

- **A global table of contents.** A table of contents at this level indicates the major sections of a manual. In Figure 16-16, notice how a user with a problem concerning the headlamp can scan the information and locate the section (electrical—lighting and controls) containing the needed information.

- **A sectional table of contents.** At this level, a table of contents indicates major topics within a section. The reader can now locate the exact page dealing with headlamps in Figure 16-17.

- **Chapter organizers**. At this level, writers may use a variety of strategies to help readers predict what they will read as well as help them locate the specific information they need. These strategies include:

  - Listing the main topics at the beginning to indicate the subsections within a chapter.

Figure 16-16 Global Table of Contents

Figure 16-17 Sectional Table of Contents

**Figure 16-18
Chapter
Organizer**

COMPONENT LOCATION INFORMATION

This group contains component location drawings for the
following engine, fuel, and air components:
—KF82D/FZ340D Engine Internal Components
—Fuel and Air Components
—AMT 600 Fuel System
—Carburetor

Use the drawings when troubleshooting an engine
problem to help locate the components to be tested.

Reproduced by permission of Deere & Company, © 1990 Deere & Company. All rights reserved.

- Forecasting the major topics to be discussed in their appropriate
  sequence.

  The list of main topics shown in Figure 16-18 resembles the list of
  ingredients in a recipe.

Sequence The conventions that apply to small sets of instructions, procedures, and documentation also apply to large sections. To establish a frame in which readers can study the details, each section should begin with a description of the system to be discussed.

The various sets of instructions, procedures, and documentation should be sequenced in the order in which they occur. If this isn't appropriate, then they should be sequenced in order of importance, from least to most difficult, from general to specific, or from simple to most complex, depending on the way in which the reader uses the product, activity, or program.

Computerization Many organizations are investigating the use of hypertext to provide quick and easy access to information, regardless of whether readers need a systems/functional or task-oriented approach. Hypertext allows users to select the information they need from a series of menus.

Placing instructions in hypertext solves the problem of redundancy in functionally-organized manuals. Information on how to order a part, as well as how to remove an old part and install a new one, is programmed into the computer only once. Users can call up a specific problem and, at the touch of a single key, obtain information on removing the malfunctioning part. By pressing a second key, the user can obtain the necessary information for ordering a replacement part.

Because many people do not have computer capability, manuals continue to be printed in hard copy as well as being programmed for computers.

1. Use both a systems/functional and task-oriented approach if possible.
2. Begin each chapter, section and chunk with an overall description of the mechanism or process.
3. Use visual markers to help readers locate information quickly.
4. Provide a Table of Contents for the overall manual. If sections include many different aspects, provide a table of contents for each section.
5. Provide a Table of Figures so readers can quickly locate necessary diagrams and other graphics.
6. Provide an Index so users can find all of the information on each topic.
7. Include chapter organizers to serve as forecasts so readers know what information is and is not included in a chapter.

CHAPTER SUMMARY

If you are in a technical field, you may occasionally need to write instructions or procedures for an internal audience. If you are a technical writer, you may write instructions and procedures for an external audience, and if you are a technical writer or a programmer, you may need to write documentation.

Instructions help you perform an activity, procedures provide you with guidelines for carrying out an activity. Documentation is a special form of instruction for computer software programs.

The conventions for writing instructions also pertain to writing procedures and documentation.

Readers read instructions, procedures, and documentation to perform an activity.

Readers are usually novices in relation to your topic.

Instructions should always reflect the point of view of a user.

Be careful to include *all* necessary information.

Always present an overview of the mechanism or process before giving the steps. Include lists of needed tools, materials, or ingredients in the beginning.

Write instructions in the sequence in which users engage in an activity. In procedures, sequence information from most to least important/difficult.

Break long lists of steps into related chunks.

Use everyday language and keep sentences short.

Use the appropriate text grammars:

- active voice
- imperative mood

- begin each step with a verb
- simple sentences

List steps. Use hanging indentation. Numerate steps in sets of instructions. Bullet items in a list of procedures if readers don't have to follow a specific sequence or engage in a specific number of activities.

Use visual cues so readers can easily locate steps locally in a set of instructions, and globally in a manual.

Use visual markers to emphasize important information.

Use graphics to help users do the activity.

Validate your instructions by administering usability tests.

There are three types of documentation users:

- Those who are unfamiliar with a program and need to learn how to use it.
- Those who are familiar with a program but need to know how to perform a specific function while they are working.
- Those who want to know what a software program does.

Documentation assumes three forms:

- Tutorials. These provide simulations of problems that help users learn to use a program.
- Reference manuals. These provide instructions so users can do an activity and also understand the functions of a program.
- Online instructions. These are included in software programs and can be called up on the screen. They are usually limited to immediate help.

Manuals are written by many writers, each of whom follows specific guidelines.

Readers scan manuals to locate information they need. Manuals are organized in two ways:

- The systems/functional approach
- The task-oriented approach

Depending on the situation, users use manuals in both formats. Hypertext can provide users with a choice of either approach.

The conventions that apply to small sets of instructions apply to manuals.

It is especially important that global locators provide readers with cues for navigating large, cumbersome manuals.

PROJECTS

As you undertake a project, be sure to engage fully in all three processes—planning, drafting, revising. You may want to use the Audience Analysis (page 97), Purpose (page 136), and Decision charts (Page 101), as well as the Planning Worksheet in Appendix B, as you gather and organize your information. As you prepare to revise, you may also want to use the Evaluation Checklist (page 264-265), and/or the Document Design Criteria Checklist in Appendix B.

Collaborative Projects

1. **Developing a manual for a software program**

    Most documentation is not task-oriented, and users often have a great deal of difficulty using it. Your technical writing class has been asked to write a manual, using a task-oriented approach, for one of the software programs in your university computer labs. It is hoped that your manual will be easier to use than the one that came with the program.

    a. As a class, select the software program on which you want to work.

    b. Work in pairs.

    c. Each pair should select one or two functions to write up.

    d. Gather your information, then draft the documentation for your functions.

    e. Swap your documentation with another pair. Beta test the documentation and revise. Repeat this step until your documentation is perfect.

    f. As a class, determine the format, type size, font, and style for your documentation text, headings, subheadings, and graphics. Reformat your documentation to conform to the class guidelines.

    g. As a class, determine the sequence in which the different functions will be placed in the manual.

    Option: To continue in Chapter 20, Document Locators and Supplements.

2. **Developing a tutorial for a software program.**

    Your university computer lab has a variety of software programs, but it is often difficult to learn how to use them. Your technical writing class has been asked to write a tutorial for one of them, so that incoming freshmen can quickly learn to use them.

    a. As a class, select the software program with which you want to work.

    b. Work in groups of five.

    c. Each group should select several functions around which to build a lesson.

    d. Gather your information. Then decide on a simulation around which to center your lesson. Draft the lesson.

    e. Swap your tutorial with another group. Beta test the tutorial and revise. Repeat this step until your tutorial is perfect.

    f. As a class. determine the format, type size, font, and style for the text, headings, subheadings, and graphics for your tutorial. Reformat your tutorial to conform to the class guidelines.

    g. As a class, determine the sequence in which the different lessons will be placed in the manual.

Option: To continue in Chapter 20, Document Locators and Supplements.

3. **Writing instructions.** Select a common, everyday, portable appliance. Write instructions for using it so that everyone in your class can use it properly.

Bring the appliance, and any equipment or food necessary for using it, to class with you. Swap directions with a classmate. Follow the other person's directions. If you have difficulty, let them know. Try to determine why you had a problem. Revise the directions to overcome the user's problems.

Individual Projects

1. The procedures for handling glasses, plates and utensils on page 461 are communicated almost exclusively in graphics.

    a. Write verbal directions to replace one of the visual sets of instructions.

    b. Write verbal directions to accompany the graphics for the same set of instructions.

    c. Which of the three types of instructions—graphic, verbal, or graphic and verbal—do you think users prefer? Why? Which of the three types was the most difficult to draft? Why?

2. Select two places on campus. Write directions for going from place A to place B. Swap your directions with someone, and try to follow the directions. Discuss your experiences with each other at your next class. If you had a problem, discuss it. Try to determine why you had the problem. Revise the directions to overcome the user's problems.

3. Rewrite the set of procedures shown in Figure 16-19 so that users can follow them more easily.

4. High school students who take driver education learn a great deal about driving, but they still get into accidents. The local high school has asked you to help prepare these teenagers for driving. You have been asked to write and

design a leaflet containing the ten most important procedures for students to follow after they get their license, based on your own driving experiences. The leaflet may be passed out to the students.

5. One of the greatest problems we have today is an overflow of solid waste. We are running out of places to put it. Write and design a leaflet that suggests procedures students can follow to reduce their use of solid waste on campus. The leaflet may be passed out to students on campus.

**Figure 16-19
Procedures for
Popping Corn**

• *How to use.*

1. Remove the cover and butter/measuring cup from the popper base. Fill the butter/measuring cup level full with corn kernels, or use a standard ½ cup measuring cup to measure the corn. **Never** add more than ½ cup of corn to popping chamber at one time.

2. Pour corn into popping chamber of base. Place butter/measuring cup into top opening of cover. Place cover, with cup in place, onto base, positioning bottom edge of cover into groove of base. Do not put any other ingredients such as oil, shortening, butter, margarine or salt into popping chamber.

3. If buttered corn is desired, cut up to 2 tablespoons of butter or margarine at room temperature into 4 equal size pats. (Do not use refrigerated butter or margarine as it will not be completely melted during the short popping period.) Place pats of butter in a single layer in bottom of butter/measuring cup. Butter will melt during the popping period. Do not remove the butter/measuring cup from the popper cover until the popping cycle is completed and the popper has been unplugged. Do not pour the butter into the popping chamber.

If unbuttered popcorn is desired, place the empty butter/measuring cup into the top opening of the cover. Never pop corn without the butter/measuring cup in position on the cover.

4. Place a 4 quart or larger heat-proof bowl directly under the popping chute to collect the popcorn as it pops.

5. Plug cord into a 120 volt AC electrical outlet ONLY. Your corn popper has a short cord as a safety precaution. If you must use a longer cord set or an extension cord, be careful not to trip or become entangled with the cord! Arrange the longer cord so it will not drape over the countertop or table top where it can be pulled on by children or tripped over unintentionally. The electrical rating of the extension cord you use should be the same or more than the wattage of the corn popper (wattage is stamped on base of corn popper). Popcorn will be popped in about 3 minutes. After corn is popped, unplug cord from outlet to turn off.

6. If butter was melted in the butter/measuring cup, remove cup by the handle and pour over popcorn. Sprinkle popcorn with salt if desired. You may wish to use popcorn salt rather than regular table salt as it clings better to the corn.

Proposals

A PROPOSAL IS A SPECIAL FORM OF REQUEST in which you try to persuade readers to respond positively by approving your idea, or providing financial support or other assistance for your project. Proposals are concerned with three types of requests:

- Requests to provide services to readers. These are usually related to marketing a company or product. You may propose that you or your company assist a city in planning landscape waste collection or provide a firm with a tailor-made software package to meet its inventory needs.

- Requests to obtain permission from readers to make a change or improve a situation. You may propose a more environmentally-safe technique for painting engines that your company manufactures, or you may propose a method for preventing valve failures at the plant where you work. These proposals are usually internal or sent to a company's clients.

- Requests to obtain support and/or resources from readers for a project. These are often related to research and development, or to the design phase of a product. You may request funding from an external organization to study the use of bacteria to decompose solid waste, or you may request approval from your supervisor to undertake a new project.

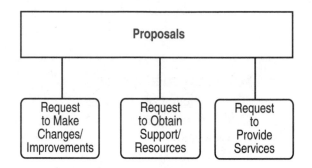

All proposals are persuasive. You are trying to persuade readers to support a project or idea. You may try to persuade management to institute a recycling program for office waste, or you may want to persuade an external agency to fund a research project on windshear, or you may try to persuade your company to develop a new and simpler door hinge for air compressors.

Proposals can assume a variety of forms. They can be as short as a page, or they can be several hundred pages. A proposal can be incorporated into correspondence, or it can stand as a separate document.

Until recently, proposal-writing was limited to senior-level personnel and technical writers. However, more and more companies are not only encouraging all employees, but also providing them with incentives, to propose ideas that may affect a company's products, efficiency, health, safety, productivity, and quality. In addition, as companies take a team approach to product development, personnel from the manufacturing, finance, marketing, and human resources divisions are becoming involved with projects from their inception and are, therefore, increasingly involved in developing proposals. Furthermore, small companies, especially not-for-profit organizations, often expect all professional personnel to help market a company by writing proposals that will sell the company's services or products.

Proposals concerned with small projects may be written by a single person. However, many proposals are the results of collaboration among a number of people. Sometimes a group of persons gets together to work on a project in which they're all interested. At other times, a supervisor may designate various people to work on a project. The proposal may be split into major sections, with each person writing a specific section; or a single person may be selected to write the entire proposal, based on information submitted by each of the other members of the project.

In writing a proposal, you need to invent a text that is relevant to your readers' needs. To do so, you need to use many of the strategies for persuasion discussed in Chapter 8.

In this chapter you will look at two types of proposals: preliminary and final. You will then consider the characteristics, processes, and behaviors of those who read and approve proposals as well as the purpose and situation in which proposals are written. You'll also examine the decisions involved in writing a proposal and learn the conventions which govern much of proposal writing.

TYPES OF PROPOSALS

Preliminary Proposals

A preliminary proposal contains a general overview of the proposed project or concept. A writer's purpose in presenting a preliminary proposal is to persuade readers that a project or concept should be supported. At this point, readers are not expected to contract for the project, but merely to indicate whether or not they are interested in pursuing the idea further. The reader who assumes the role of a decision maker must decide whether or not the project is worth considering. The reader is interested in the overall concept, not in the details.

The letter shown in Figure 17-1 is a preliminary proposal written to persuade the manufacturer of plastic bags to hire E & A Environmental Consultants to assist in research and development for a new type of bag. Notice that the letter is short and contains only a general discussion of the assistance the company could provide; specific details such as cost, personnel, etc., are not discussed.

If the letter persuades the reader and her company that E & A can assist them, then E & A will need to write a final proposal, specifying how it would assist in demonstrating that the bag is degradable and does not impact the environment unfavorably.

Final Proposals

The purpose of a final proposal is to persuade readers to support a particular project or idea. A final proposal is more detailed than a preliminary one, and contains all the information readers need to know to make a decision.

The proposal you see in Figure 17-2 was initiated and written by Nick Weaver, a transmission supervisor with Indiana and Michigan Electric Company. Weaver wrote the proposal to persuade management to adopt a different type of cable for wiring tower airplane warning lights.

Even though the proposal in Figure 17-2 is short, it is a final proposal. Readers were able to make a decision based on the information included in it. The plant not only adopted the proposed idea but gave Weaver a monetary bonus for it.

Figure 17-1
A Preliminary
Proposal to Provide
a Service

Salutation.
  Personal.
Introduction.
  Explains purpose of
  letter.
  Impersonal. Formal.
Problem.

  Appeals to readers'
  needs.

Solution.
  Main organizing idea.
  Provides evidence
  reader will consider
  valid.

E & A Environmental Consultants, Inc.

95 Washington Street
Suite 218
Canton, Massachusetts 02021
(617) 575-9099
Fax: (617) 575-8915

Offices Located in:
Minnesota
North Carolina
Washington

Dear Marcia:

  It was a pleasure to speak with you about biodegradable bags and yard waste composting. As I mentioned, E & A is interested in assisting your company in its efforts to develop degradable plastic bags that can be successfully integrated into municipal yard waste composting programs.

  As you know, there is a lot of politics and publicity surrounding degradable plastics. The agenda vis-a-vis marketing a degradable plastics suitable for municipal yard waste composting includes demonstrating that plastics will decompose within 6 to 8 months in a yard waste compost operation; that the final product shows no plastic residue that detracts from its marketability; and that degradation or decomposition does not generate measurable environmental impacts.

  E & A's active involvement in the start-up, training and monitoring of numerous municipal yard waste composting programs can provide expertise to meet this agenda. E & A has extensive experience in experimental design, operation, and analysis of innovative composting processes...

CLAIM
↓
EVIDENCE

CLAIM
↓
EVIDENCE

E & A Environmental Consultants, Inc. 1990

READERS

Proposals may be written to persons either internal or external to your organization. They may be written to persons who are experts in your field, or to people who have almost no knowledge of it. Regardless of readers' organizational characteristics, they assume the role of decision makers, deciding whether or not to grant your request.

  While a proposal may be read by a single reader, several readers are usually involved. A proposal to an external organization, such as the National Science Foundation (NSF), requesting funding for a research grant, is read by several experts in the writer's field and also by generalists who are administrators. In addition to a primary reader, a proposal written for a project by a member of a corporation's research and development division is usually read by intermediary readers, i.e., the writer's supervisor and division director, or possibly a committee consisting of representatives from various internal divisions, such as marketing and business.

  Proposals may also be read by peripheral readers. If the proposal is related to a government project, or if the writer is applying for a government grant, the proposal is considered in the public domain (available to the public), and may be

**Figure 17-2
Final Proposal to
Make a Change**

Idea No. <u>G.O. T&D #289</u>

Innovations

Idea Form   Date Submitted <u>July 22, 19xx</u>

Name <u>N.E. Weaver</u>   Title <u>Line Crew Supv.-N.E. Ext. 511-1209</u>

Dept. <u>T&D Transmission</u>   Location <u>Jackson Road</u>

Supervisor's Name <u>J.P. Grantham</u>

Categories (check one)   ☒ Procedures/Methods (plant, office, field)   ☐ Safety

☐ Improve design of tools/equipment   ☐ Improve customer service/relations

My Idea IS: Cost Cutting Procedures for Tower Light Projects
                     (Airplane Warning Lights on 765 KV Towers)

<u>Problem:</u>   Using standard conduit to wire FAA warning lights on
transmission towers is a strenuous task and very time consuming.
Standard conduit is not flexible, and the wiring path up the tower
contains turns that require bends in the cable. Besides being
difficult to work with, rigid conduit takes excessive time to
install. A crew of five will spend two days to wire just one 765
kV tower.

<u>Solution:</u> My solution to the wiring and conduit problem is to switch
to Signa/Clad MC cable, wire installed inside a metal jacket.
Sometimes called "conduit on a reel," it is listed in the National
Fire Protection Association Electric Code Book, Article No. 334
(See attached pamphlet, wire sample, and pictures.)

Applied to towers, Signa/Clad cable is drawn from the reel with
very little effort and self-straightens as it is pulled up the
tower.   The cable also bends easily when positioned for clamping
to the tower structure (Photo 2). In addition, Signa/Clad
accommodates the use of a running ground to drain static buildup.
This feature eliminates the requirement of rigid conduit to be
grounded at each section as it is assembled.

We monitored three 765kV Signa/Clad tower light installations for
one year.   The project consisted of three 765 kV towers on which
airplane warning lights were installed.   The towers were all over
200 feet. The year-end report showed an absence of wear on the
Signa/Clad wire at the three installations.   Also, we found the
wiring free of damage from lightning or water.

We found evidence that one of the monitored towers was struck by
lightning, but the Signa/Clad cable appeared undamaged. However,
on the same installation, a piece of standard rigid conduit showed
burn marks at the breaker panel.

Because of the ease in manipulating and laying the cable, a tower
crew can complete a 765kV tower warning light installation in one
day, eliminating 50% of the labor, or eight hours.   This saves
$800 for a crew of five for just one installation.   This savings
more than offsets the additional cost of the material. (Material
cost for installing 220 feet of Signa/Clad for a 765kV tower is
approximately $600 compared with $375 for rigid conduit.)   Using
Signa/Clad, the net savings per tower is $575 [$800-($600-$375)].
We have 12 towers at I&M that require FAA lighting.   Signa/Clad is
also usable for other applications, such as microwave towers or
other obstruction lighting.

read by the news media or other interested parties. In addition, proposals can be considered legal documents to which attorneys may gain access if a suit is involved.

Your reader's attitude may be skeptical or even hostile. If a reader's funding is tight, if a reader doesn't recognize the problem for which you're proposing a solution, or if a reader prefers a solution different from yours, then the reader may be biased against your document from the very beginning. Even if readers accept your argument, they will want to be certain that your project is a viable solution to a problem; that you and your staff are capable of executing the project successfully; that the materials, equipment, and facilities are appropriate; and that the amount of funding is cost-effective.

Readers usually read preliminary proposals quickly, often merely scanning the text. They are interested in understanding the main idea and are not looking for details. Readers spend more time with a final proposal. If your final proposal is one of many, readers may scan it initially to determine whether your document meets their basic criteria, i.e., that it is concerned with an important problem, the solution is viable, personnel are qualified, and the funding request is within acceptable limits. If it does meet the criteria, readers place the proposal in a pile to be reread and studied further. During the second reading, readers look more carefully at details, as well as consider your idea in relation to others.

PURPOSE AND SITUATION

Your primary purpose in writing a proposal is to persuade readers to fulfill your request.

Types of Situations

Proposals can be either solicited (reader-initiated) or unsolicited (writer-initiated).

Solicited Proposals Companies and government agencies often subcontract work to other companies or individuals when they do not have the necessary expertise within their own organization. To locate companies or persons with the necessary expertise, and to determine which can perform the necessary task best for the least amount of money, firms may publish a Request for a Proposal (RFP). The RFP is a document inviting interested parties to write a proposal and bid on the contractor's work. Proposals written in response to an RFP are "solicited" proposals. For example, as cities throughout the country decide to recycle municipal waste, they often publish RFPs for waste pick-up and disposal. Often an RFP will include guidelines for the content, organizational pattern and sequence, and format for a proposal. Companies submit proposals to do a job based on RFP guidelines.

Many funding agencies also publish RFPs for projects that they are interested in supporting. Often these agencies not only publish information about the types of projects they will fund, but also provide guidelines for writing the proposals. The following guidelines are used for submitting proposals for grants in science and engineering. As you read, consider the kinds of information a writer needs to include.

The proposal should present the (1) objectives and scientific or educational significance of the proposed work; (2) suitability of the methods to be employed; (3) qualifications of the investigator and the grantee organization; (4) effect of the activity on the infrastructure of science, engineering, and education in these areas; and (5) amount of funding required...The proposal must contain a 200-word summary of the proposed activity suitable for publication. This summary should not be an abstract of the proposal but rather a self-contained description of the activity that would result if the proposal is funded by NSF. The summary should include a statement of objectives...The main body of the proposal should be a detailed statement of the work to be undertaken and should include: objectives for the period of the proposed work and expected significance; relation to longer-term goals of the investigator's project; and relation to the present state of knowledge in the field, to work in progress by the investigator under other support, and to work in progress elsewhere. The statement should outline the general plan of work...Brevity will assist reviewers and Foundation staff in dealing effectively with proposals. The project description may not exceed 15 single-spaced pages...

National Science Foundation

When you write a proposal in response to reader-initiated invitations, you must follow the guidelines closely. While you know that readers want to receive your proposal, you need to keep in mind that readers' attitudes will be skeptical. They want to know if you can achieve the goals, and if you can do so better than your competition.

Unsolicited Proposals There will be many times when you have an idea for a product or a project. You need to get both permission and funding to carry it out. To obtain your supervisor's permission or the funding, you need to write a proposal explaining your plans.

You may also find, especially if you are in a small firm, that to market your organization, you are expected to write proposals for providing services to potential clients. The proposal from E & A was written from such a purpose.

GATHERING INFORMATION

Regardless of whether you write solicited or unsolicited proposals, the content is related to your work, and you will know much of the information you need to include. However, you may need additional information.

Many professional communities require that you support proposals, especially proposals related to research projects, with evidence from previous research. You may need to check written sources, not only researching books and journals, but also investigating patents and searching catalogs to determine if similar products or solutions are being or have been developed or tested. In addition, to support your statement of needs, you may need to undertake field research, administering a survey, conducting a focus group, or interviewing persons familiar with the subject. You may even conduct field tests or pilot projects to demonstrate that your solution is viable on a small scale.

In the proposal in Figure 17-2, Weaver field-tested the new type of cable before he recommended to his supervisor that the cable be used throughout the system. He uses the results of this test as evidence to support his recommendation (see paragraph 5). He further supports his recommendation with photographs taken during the field test. In addition, because the reader's organizational community accepts supplier's pamphlets as valid evidence, Weaver includes in an appendix several that discuss the Signa/Clad wire.

TEXTUAL DECISIONS

General Conventions

Content, Organizational Pattern, and Sequence The content, organizational pattern, and sequence for solicited proposals are usually specified in an RFP or in special guidelines. If they are not provided, use the conventions for unsolicited proposals.

The content for unsolicited proposals depends on readers' knowledge of a topic, what readers need to know to make a decision concerning your request, and factual evidence or emotional appeals related to the needs and values of readers and their communities.

In each type of proposal, you request some type of change. For readers to agree to your proposal, they must want to make a change. Therefore, you not only need to persuade readers that your proposed project or idea is a good one, but you also need to convince them that a change is necessary. Thus, you must develop two arguments, one indicating that a problem exists with the situation, process, or mechanism you are trying to change, and a second argument indicating that your proposed change provides a solution to the problem. Because you must convince readers of a need to change in order to persuade them to approve your request, the organizational pattern of a proposal follows the problem/solution pattern. You need to prepare readers to accept your solution by leading up to it. Therefore, begin with the problem section. Both the E & A letter in Figure 17-1 and Weaver's proposal for a flexible wire in Figure 17-2 follow this pattern.

Your claims in both the problem and solution sections need to be built on evidence that relates to readers' personal/ social, organizational, or professional needs. In addition, the evidence must be accepted as valid by the readers' communities. The content, therefore, must be relevant to the needs, values, and attitudes of your readers and their communities.

If readers are not knowledgeable about a problem, you may need to provide a great deal of information in the problem section in order to prepare them for your proposed solution. Adhere to the given-new contract and sequence the information in an inductive pattern, providing the evidence first, and leading up to the claim. Ned Weaver provides specific evidence (a crew of five took two days to wire one tower) to support the claim that using standard conduit is difficult to work with and takes excessive time.

If readers are aware of a problem, you only need to allude to it briefly in the problem section. In the proposal in Figure 17-1, E & A does not provide an in-depth discussion of the problem of municipal yard waste, or of degradability, because the reader is aware of it. E & A simply lists the major problems Patterson's company faces in marketing its product.

Once you have persuaded readers of the need for change, they are ready for a solution. Thus, the solution section of a proposal is sequenced deductively. Make your claim and then provide the evidence to support it. In the proposal in Figure 17-2, Weaver discusses the advantages for flexible wiring in terms of cost effectiveness and durability because these aspects appeal to the readers' needs and values.

Point of View and Focus In a proposal, regardless of whether it is preliminary or final, you need to keep the reader's point of view in mind as it relates to personal/social, organizational, or professional needs. A proposal must focus around readers' needs, as we saw in the proposals in Figures 17-1 and 17-2.

Style Most proposals, especially those written in response to an RFP, maintain a formal tone. The language should be precise. Technical terms that are familiar to readers should be used. The E & A proposal uses the phrase "measurable environmental impact," while the wiring proposal discusses a "765kV tower."

Visual Text and Format Preliminary proposals are often written as letters. If they are comparatively long, three to five pages, use headings and subheadings to indicate the problem and solution sections. This helps readers perceive the major categories easily.

Global locators should be used for long proposals, in addition to headings and subheadings for major sections and subsections. Objectives should be numbered and listed, using hanging indentation.

To determine the specific format for your proposals, study previous proposals for the technical community in which you are writing. You can observe the use of visual text in the proposal in Figure 17-2.

Graphics Graphics should be used to help readers understand and accept a writer's claim. Weaver includes a diagram and photographs to complement information included in the proposal in Figure 17-2.

Global Supplements Final proposals that are not part of a memorandum or letter often need to be accompanied by a variety of other documents. A cover letter or letter of transmittal should accompany all proposals. (If a proposal is solicited, a transmittal form may be included in the guidelines.) Always place a title page at the beginning of the proposal. If the proposal is over five pages, include an executive summary or abstract, as well as tables of contents, figures, and tables. You may also want to include appendices to provide readers with additional information that supports your discussion. Other supplements to a proposal may include a glossary and a list of references. (See Chapter 20 for a discussion of global locators and document supplements.)

Preliminary Proposal

Information contained in a preliminary proposal is general. Provide a summary of the problem and the proposed solution. You may wish to provide a few examples or references to support your claims. However, you don't need to go into detail in describing your proposed solution. Rather, you need to emphasize how it will solve the problem you've delineated. Because, in a preliminary proposal, you are trying to catch the reader's interest, you need to emphasize the problem, relating it to the values of the reader's community and supporting your interpretation of it with evidence the reader's community considers valid.

The writer for the proposal in Figure 17-1 uses a deductive organizational pattern in the solution section. Supporting evidence is provided after making the claim that the company can provide expertise to solve the problem.

Final Proposal

Final proposals, especially those which may be used as the basis of a contract or may serve as a legal document, should be detailed and specific. A final proposal usually includes three sections: a discussion of the problem; a listing of short- and long-range objectives; and a discussion of the solution.

The sections and subsections of a proposal parallel the aspects of a project on which readers base their decisions. They have become conventionalized, and readers expect them to be organized and sequenced accordingly.

■ As we study the different sections and subsections of a project proposal, we will examine how the writers adhere to the conventions in the proposal shown in Figure 17-3, which is a request to the National Science Foundation for financial support to study head injuries. The proposal is read by experts in the writers' professional community, bioengineering. ■

Figure 17-3 Final Project Proposal

<table>
<tr><td colspan="2">

COVER SHEET FOR PROPOSALS TO THE NATIONAL SCIENCE FOUNDATION
</td></tr>
<tr><td colspan="2">

For consideration by NSF Organizational Unit
Bioengineering, Aiding the Disabled
</td></tr>
<tr><td colspan="2">

Submitting Institution Code 001312800
</td></tr>
<tr><td colspan="2">

Name of Submitting Org. to which Award should be Made
The Regents of the University of California, Berkeley Campus
</td></tr>
<tr><td colspan="2">

Address of Organization (Include Zip Code)
Sponsored Projects Office, University of CA, Berkeley, CA 94720
</td></tr>
<tr><td colspan="2">

Title of Proposed Project
Rational Head Injury Criteria
</td></tr>
<tr><td>

Requested Amount
$543,857
</td><td>

Proposed Duration (Mos.)
36
</td></tr>
<tr><td colspan="2">

Desired Starting Date
1/1/91
</td></tr>
<tr><td colspan="2">

PI/PD Dept./Administering Unit
Mechanical Engineering/ORS
</td></tr>
<tr><td>

PI/PD Organization
University of California
</td><td>

PI/PD Phone
(415) 642-3739
</td></tr>
<tr><td>

PI/PD/Name/Title
Goldsmith, Werner, Prof. Emeritus
</td><td>

Highest Degree/Year
Ph.D. "49
</td></tr>
<tr><td>

Additional PI/PD
Sackman, Jerome, L., Prof.
</td><td>

Eng., Sc.D. "59
</td></tr>
</table>

Purpose.
Summary of
problem.

SUMMARY

This proposal is concerned with the development of a new concept of head injury criteria. In the past, either peak acceleration or a norm of the acceleration history has been mandated as the dosage for both skull fracture and brain damage regardless of position on the head or type of insult employed. Per se, acceleration is not a direct measure of damage and hence is not the best variable for designating the threshold of trauma. The present investigation in intended to construct a more rational head injury criterion based on deformation limits; separate damage levels will be specified for the brain and skull...

The results of the two phases of the proposed study will not only lead to a better understanding of critical levels of loading and damage processes, but should also materially assist in the design of better protective equipment.

ii

continued

Problem section.
Background.
Appeals to interests of professional community.
Chronological description of research on subject.

EVIDENCE

INTRODUCTION

Head injury, one of the most important occurrences in modern society, can be divided into non-penetrating and penetrating categories...

A critical aspect of head injury investigations is a knowledge of what degree of damage will be caused by various load levels and durations...

The first formal head injury criterion was proposed [in the early 1970s] by Gurdjian, Lissner, and Patrick [14] based on six experiments involving the drop of cadavers down elevator shafts from various heights, with the forehead striking a metal surface in the A/P direction. The dividing line between tolerable and unacceptable or fatal injury was specified as the observed onset of linear skull fracture with the dominant hypothesis that the major correlate with head damage is the translational acceleration. The results were correlated with... experiments involving concussion...and subsequently with long-duration tolerance information from volunteers. In addition, the tests were modeled by a sample linear spring-mass system implying rigid-body behavior of the head. All this resulted in the establishment of the Wayne State Tolerance Curve (WSTC) that plots effective head acceleration against impact duration and provides a demarcation between acceptable and major damage regions...The Wayne State Tolerance Curve either in its original form or as a derivative has up to the present formed the basis of human tolerance limitation in research investigations, industry-wide performance criteria for safety equipment and corresponding government regulations.

Analytical description of types of scales.
Uses technical terminology.
Describes scale problems in analytical sequence.
Cites articles in professional journals accepted by community as evidence.
Appeals to professional community's beliefs.
Main organizing idea for problem section.

The quantification of the WSTC resulted in the Severity Index (SI), obtained from an acceleration history of the mass center of the head used in a particular impact test. A further revision of WSTC replaced the SI with the Head Injury Criterion (HIC) which attempts to account for the initial jerk (acceleration rate) whose upper limit was also set first at 1000 and subsequently at 1500, and which is currently in use for certain dummy experiments...

The HIC has been subjected to severe criticism on various grounds, such as involving failure to correlate with head injury data and with the WSTC [15, 22, 23]. In contrast, federal motor vehicle helmet standards specify in professional permissible peak levels of acceleration and their journals accepted maximum duration for acceptance of the device [24]. Furthermore, a substantial segment of the head injury research community has adopted the position that rotational rather than translational effects are the principal cause of brain damage [25-29]; an effort has also been made to combine these two kinematic effects [30]. Thus, **it is evident that the current head injury criteria are generally regarded as unsatisfactory and severely limited**...

CLAIM

Figure 17-3 Final Project Proposal ***continued***

Discussion of previous field research as evidence to support claim that present study is feasible. Data accepted by professional community as evidence.

Short term objectives. Meet NSF criteria for advances in field. Long term objectives. Meet NSF criteria for improved technology & solution of societal problems. Short term objectives. Meet NSF criteria for advances in field. Long term objectives. Meet NSF criteria for improving engineering research.

Overview of solution section. Forecast. Presents study in terms professional research community accepts as valid. Relates to problem section. Relates directly to objective section.

Previous [Related] Investigations...

An initial experiment was conceived and executed to ascertain whether it is possible to rupture a small, sealed, fluid-filled tube immersed in a surrounding liquid contained in a closed vessel by means of an impact with an energy of the order of 60J...A second series of specimens consisted of...human blood vessels...In these tests, rupture of any vessel was not achieved at striker velocities of 22 m/s corresponding to the peak pressures shown in Figure 3, which were about two an a half times greater than those produced previously.

OBJECTIVES

The proposed investigation constitutes a fundamental research program whose purpose is the acquisition of NSF criteria for experimental data and the development of advances in field theories and solution algorithms that will lead to new national head injury criteria. These are essential for a better understanding of the critical universal problem of cranial trauma and its prevention. The work is intended to replace current standards that do not discriminate between skull fracture and brain damage...

This approach will produce new and pertinent data that includes the delineation of the mechanical loading parameters producing failure of blood vessels and, separately, fracture of the sandwich shell constituting the skull. The procedure will also be able to differentiate the effects of location and character of the applied load. The resulting information will form the basis of a rational head injury criterion that can be used both for regulatory purposes and for the construction of protective devices. The work can lead to important extensions involving failure components of the cardiac and human structural systems...

PROPOSED RESEARCH

The proposed investigation consists of two parts: (1) A study of the external loading to be imposed on a model closely simulating a human head containing cerebral vessels or their replications that will produce vessel rupture. This will involve a correlation of the loading conditions and impact configurations with the vessel failure response. The objective of this phase is to provide a more meaningful and accurate criterion of brain damage due to non-penetrating impact than is currently utilized. (2) An examination of the relationship between the fracture response of a humanoid skull and the force histories due to a non-penetrating impact which will characterize the damage to this component. This aspect will accomplish one of the major objectives of this study by distinguishing the trauma experienced separately by the major components of the head.

continued

Figure 17-3 Final Project Proposal *continued*

Methods conform to those required by readers' community. Methods are described in detail for experts.

Methods

The experimental blood vessel failure study will involve the construction of a more representative container with acoustic and elastic/brittle properties closer to the prototype and a more accurate geometric configuration of the human skull than employed in the preliminary investigation. Instead of the soft Tygon material employed there, a stiffer material more closely resembling the acoustic impedance of skull bone will be utilized; PMMA might be a suitable candidate. A spherical shell is the closest approximation of a human head that also admits of a relatively simple analytical representation and hence should be considered a subsequent physical model of the cranium. It will be very advantageous to secure a commercially available shell of appropriate properties as the manufacture of a special series of such shells will be both time-consuming and costly.

CLAIM

↓

EVIDENCE

Procedure I.

The craniums will be emplaced on an artificial neck from a Hybrid III dummy which will be secured at its base...

Personnel. Qualifications meet requirements of readers' discourse community.

Personnel

Dr. Werner Goldsmith has over 40 years of experience in the conduct of experiments involving dynamic phenomena, including a major concentration on the biomechanics of head and neck injuries...

The Co-Principal Investigator has more than 30 years of research experience in solid and structural mechanics...

A substantial portion of the program will be performed by students in both departments using this topic for their graduate research projects...

CLAIM
(Implicit Goldsmith is qualified)

↓

EVIDENCE

Equipment and facilities. These meet the criteria of the readers' community.

Equipment and Facilities

The laboratory of the proposed faculty investigators is uniquely equipped to conduct experiments involving dynamic loading on structures. There exists a substantial amount of space in two separate rooms for the conduct of impact experiments, including projection facilities permitting impact speeds ranging from about 1 m/s to 2 km/s. This includes pendulum and drop mechanisms as well as a variety of fast and power guns. A substantial amount of polymeric material has been accumulated in conjunction with previous studies that will assist in the use or else in the acquisition of substances to be used as prototypes of head tissues..

CLAIM

↓

EVIDENCE

Time line specifies duration of project.

Schedule

The funding requested will permit the investigators to accomplish the objectives of the proposal within the 3 year time frame.

Bibliography

Goldsmith, W. "The Physical Processes Producing Head Injury." In: W.F. Caveness and A. E. Walker (eds.) *Head Injury Conference Proceedings.* J.P. Lippincott, Philadelphia, 350-382, 1966.

Goldsmith, W. "Biomechanics of Head Injury." In: Y.C. Fung, N. Perrone and M. Anliker (eds.) *Biomechanics: Its Foundations and Objectives.* Prentice-Hall, Englewood Cliffs, N.J., 585-634, 1971...

Figure 17-3 Final Project Proposal *continued*

B7720.XLS 7/5/90

UCB Eng-7720
Goldsmith/Sackman

Budget - Year 1
(January 1, 1991 - December 31, 1991)

	Amt/time	Rate/mo*	Subtotal	Total
Personnel				
Principal Investigator W. Goldsmith Research Engr. Recalled	1 mo. @ 100%	$10,756	$11,294	
Co-Principal Investigator J. Sackman Professor VIII	1 mo. summer	$11,131	$11,688	
2 Grad. Student Researchers VI	9 acad. yr. mos. @ 50%	$2,989	$26,901	
	3 summer mos. @ 100%	$2,989	$17,934	
1 Grad. Student Researcher III	9 acad. yr. mos. @ 50%	$2,362	$11,160	
	3 summer mos. @ 100%	$2,362	$7,440	
1 Engineering Aid step 1	9 acad. yr. mos. @ 25%	$1,567	$3,702	
	3 summer mos. @ 25%	$1,567	$1,234	
TOTAL PERSONNEL				$91,353
Employee Benefits				
Principal Investigator		3.80%	$429	
Co-Principal Investigator		10.00%	$1,169	
Eng. Aid, Graduate Students (acad. yr.)		2.14%	$894	
Eng. Aid, Graduate Students (summer)		3.80%	$1,011	
Grad.Student Health Insurance**		2.24%	$1,421	
TOTAL EMPLOYEE BENEFITS				$4,924
TOTAL PERSONNEL & BENEFITS				$96,277
Other Direct Costs				
Materials; plastics, molds, strain gages, crystals, photographic supplies			$4,000	
Shop time @ $32/hr.			$9,600	
Computer; CRAY XM-P @ $432/hr.			$4,000	
Clerical services			$6,833	
TOTAL OTHER DIRECT COSTS				$24,433
TOTAL DIRECT COSTS				$120,710
Indirect Costs: 49% of Modified Total Direct Costs base of $119,289				$58,452
TOTAL AMOUNT - YEAR 1				**$179,162**

*Current (1/1/90) salary rates shown. Co-Principal Investigator's monthly rate includes a mandatory 4.03% retirement benefit allowance. Subtotals include a projected cost of living increase of 5.0% effective each January 1 thereafter.
**Graduate Student Health Insurance is applied to total GSR salaries and is not part of indirect cost base.

36

Adapted from a proposal by Professor Emeritus Werner Goldsmith, and Professor Jerome L. Sackman, University of California, Berkeley

Figure 17-3 Final Project Proposal *concluded*

Problem Depending on the discourse community for which you are writing, this section may be referred to as the "Statement of Need," "Problem Statement," "Rationale," "Background," or "Justification." Your description of the problem should include specific facts, and you should cite both written and field references. In this section you should also include previous and present attempts to solve the problem, and an explanation of why these have not succeeded. Readers can then understand why your attempt is important or better than others.

This section may follow either a chronological or an analytical organizational pattern. You may want to discuss the history of the problem (chronological), or you may want to discuss the various aspects of the problem (analytical). Or you may want to combine the two approaches by looking at the various aspects of the problem over a period of time. The pattern you follow should depend on the information your readers need to have to accept your claim. Sequence the information so that readers can easily follow your line of reasoning.

■ The project proposal in Figure 17-3 follows a modified analytical organizational pattern. It is categorized according to the various scales used to measure head injuries. These are discussed in the chronological sequence in which they were developed. The introduction then moves from a description of the specific scales to a general description of the problems related to existing scales. Consider how the writers lead up to the main organizing idea rather than presenting it at the beginning. ■

Objectives The objectives of a proposal should be specified so readers recognize the purpose of a document, and understand the relationship between the purpose and the solution being proposed.

State the objectives in terms of the needs and values of the reader's community. Make certain the objectives are directly related to the problems discussed. Both long- and short-range goals should be included. Objectives should be sequenced from most to least important.

■ The objectives in the head injury study are presented in paragraph form and relate directly to the values NSF has established for determining whether or not a proposal should be approved. These include (1) "the likelihood the research will lead to new discoveries or advances within the field," (2) "the likelihood the research can...serve as the basis for new or improved technology or assist in the solution of societal problems," and (3) the potential to improve "the quality, distribution or effectiveness of the nation's scientific and engineering research." As you read the objectives, notice the relationship between them and NSF's values. ■

When you do not need to follow guidelines from an RFP, as the head injury proposal in Figure 17-3 does, objectives should be listed to facilitate readers' perception of the purpose of your proposal. A list is also easy for readers to locate if they want to compare your results, conclusions, or recommendations with your original objectives. Use a truncated sentence, beginning with an infinitive phrase (to + verb) to show action, since objectives are action-oriented. Each objective should be no longer than a single sentence.

The following short- and long-term objectives are listed in a proposal to the Toxic Substances Research Program. This proposal is a request for funds to study the removal of toxic compounds from Advanced Integrated Ponding Systems (AIPS). Notice how the writer follows the conventions for writing objectives.

Relates to problem.

Short-Term Objectives

1. To investigate the removal rate of the compounds PERC, DBCM, bromoform, trichloroethylene (TCE) and 1,1,1-trichloroethane (TCA) by AIPS anoxic sludge...
2. To determine the maximum concentration of the compounds that can be treated successfully,
3. To investigate the intermediate steps of degradation and their relative rates,
4. To investigate the rate of volatilization of the compounds in the AIPS at Hollister.

Relates to values and needs of readers' community.

Long-Term Objective

To estimate the importance of the biotic, abiotic, and volatilization paths of toxic compound removal by the AIPS.

Professor William J. Oswald, University of California, Berkeley

Solution This section is the heart of a final proposal. You must persuade readers that you have the knowledge and qualifications, and an effective plan for succeeding in carrying out the solution. Readers will judge the potential effectiveness of your solution, as well as your qualifications, by determining whether your solution is valid according to the criteria used by their organizational and professional communities. Readers will consider whether or not you will conduct the project appropriately; whether qualified persons will be working on the project; whether you will be using appropriate materials and equipment; whether the time frame is acceptable; and whether the project is cost-effective, based on the results, time, and cost.

To provide readers with this information, you need to include evidence that is accepted as valid by the readers' organizational and professional communities. You should provide a discussion of the method and procedures to be used; a list of the personnel to be involved, and their qualifications for doing the work; a description of the materials and equipment required for undertaking the project; a timeline delineating the amount of time required for completing the project; and a budget specifying the amount of funding required. Each of these aspects should be discussed in separate subsections.

The solution section should begin with an overview, so that readers understand the information which follows. Therefore, the solution usually follows a deductive pattern, with the claim presented at the beginning of the section.

■ Notice that the overview to the solution section of the head injury proposal (Figure 17-3) provides a frame for understanding the remainder of the information and relates directly to the problem and objectives. Readers have a general idea of the solution and understand how it relates to the objectives of the study. ■

Methods and procedures. The section on methods and procedures describes how the research will be conducted, the product developed, or the project administered. The information should be sequenced according to the order in which the methods and procedures are used, so that it is clear to readers how they relate to each other and to the whole project. To persuade readers to accept a proposal, the writer must indicate that these methods meet the criteria established by the readers' community.

■ The methods and procedures section of the head injury proposal in Figure 17-3 is extremely detailed. They must persuade their readers that they will conduct a valid study according to criteria established by the scientific community. Thus, the writers must provide sufficient evidence to support their claim that the methods they plan to use are appropriate. ■

Personnel. The personnel section includes a list of personnel who will be responsible for administering the project, conducting the research, or developing the product discussed in the proposal. As with the previous section, the qualifications must meet the criteria established as valid by the readers' community.

This section should be organized according to the positions involved, and sequenced from the most to the least important position.

■ The personnel section of the head injury study discusses the facets of the researchers' careers that are directly related to the study. Notice that by indicating the investigator is experienced and knowledgeable, the qualifications meet the criteria for valid evidence that are required by the readers' community. ■

Materials and equipment. This section not only describes the materials and equipment to be used, but also indicates how they will be obtained and the condition they are in. Once again, if the proposal is to be accepted, materials and equipment must meet the criteria established by the readers. The information can be sequenced in a variety of ways, including alphabetical or from most to least important.

■ The equipment and facilities section in the head injuries study indicates that much of the necessary equipment is already available. Readers consider this availability a plus since funding will not be necessary for this aspect of the proposed project. Again, notice that the description of the equipment and facilities is tailored to meet the criteria for evidence required by the readers' community. ■

Time Line. Many projects require a relatively long period of time to complete. Readers need to know how long it will take. In addition, if the project is comprised of several phases, readers need to know the length of each phase, how the phases interact, and whether they overlap, to determine how personnel, materials, etc., will be used. The writer must indicate that the time frame is within the period established by the reader. A time line, which can be described in text and/or in a graph, provides this information.

■ While the head injury proposal provides only a brief statement indicating the project will take three years, it meets the readers' need. ■

A more elaborate time line is presented in outline form in a proposal to NASA for conducting research related to the effects of high turbulence on convective heat transfer. Notice that while the time frame is the same as that for the head injury study—three years—the writer provides specific information, dividing the time line by year, phase, and major task.

First Year Tasks: Phase I
1. Design, fabricate, and assemble the test plate and probe-traversing equipment.
2. Qualify the test plate for energy balance and baseline data running in HMT-2 with low turbulence.
3. Install the test plate in the free-jet facility and begin the free jet tests, varying free stream velocity and turbulence independently. Map out the operating surface shown in Figs. 8 and 9.
4. Take turbulence data at each flow condition, using the existing triple-wire system, to establish the flow characteristics.

Second Year Tasks: Phases I and II
1. Complete the free jet tests.
2. Install new blower speed controls on HMT-2.
3. Rerun the baseline tests on the test plate in HMT-2 with low turbulence.
4. Design and install "combustion chamber simulators."
5. Rerun St vs. RE *2 tests with high turbulence, measured with the existing triple-wire system.
6. Begin adapting STAN5 for high turbulence.
7. Coordinate with, NASA, Calspan and Oxford Programs

Third Year Tasks: Phase II and III
1. Conduct studies of high turbulence effects in conjunction with curvature and discrete hole injection.
2. Execute water-tunnel studies on wall-layer interactions.
3. Complete STAN5 modeling data set.

Professor Robert J. Moffatt, Stanford University

**Figure 17-4
Graphic Depiction
of a Time Line**

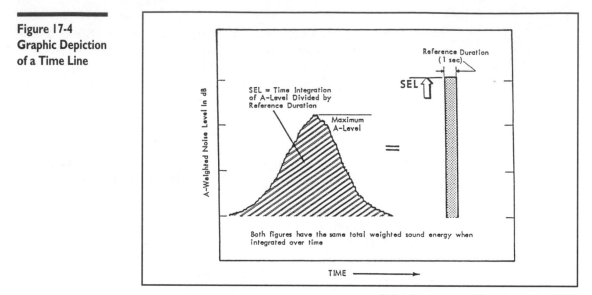

E & A Environmental Consultants, Inc., 1990

E & A consultants include a visual time line, shown in Figure 17-4, in their proposal for a municipal yard waste compost program. Consider how the differences between this visual time line and the previous outline and paragraph description affect readers' fluency and comprehension.

The graphic form allows readers to see overlapping phases as well as perceive the whole project, while the outline form provides readers with more detail concerning the various aspects of the phases.

Budget.   Readers need to be certain a budget is within the range established by the reader in reader-initiated documents, or within reasonable limits for writer-initiated documents. The budget should be placed on a separate page of a proposal. Rows contain the major categories, and columns contain the time periods.

■ The budget in the head injury study is for the first year of operation. Notice how the writer uses visual text so readers can quickly and easily determine the funds which will be required for each item. ■

1. Support your argument with information from previous written and field research.
2. Follow a problem/solution pattern.
3. Begin with the solution section in reader-initiated proposals.
4. Begin with the problem section in writer-initiated proposals.
5. Use a chronological and/or analytical organizational pattern for the problem section.
6. Place your main organizing idea either at the beginning or the end of the problem section.

7. Sequence your information as follows:
   • Problem
   • Objectives
   • Solution
8. Discuss both long- and short-term objectives in the objective section.
9. Begin the solution section with an overview.
10. Include the following subsections in the solution section in the sequence listed:
    • Methods and procedures   • Time Line
    • Personnel   • Budget
    • Materials and equipment
11. Use a formal style.
12. Use global locators for long proposals.

CHAPTER SUMMARY

Proposals are a form of request in which you try to persuade readers to support a project or idea, to approve a suggestion for change, or to accept your services or product.

Proposals request readers' support in terms of money, time, personnel, etc., for a project. There are two types of proposals, preliminary and final. Final proposals contain more specific details than preliminary proposals.

Proposals can be written individually or collaboratively. They may be read by a single reader or by multiple readers, all of whom assume the role of decision makers. Proposals can be either reader- or writer-initiated. Reader-initiated proposals usually include guidelines for the content, organization, and format. Readers read proposals skeptically, and sometimes with hostility. They read preliminary proposals quickly, but study final proposals.

The content for a proposal depends on the readers' knowledge of the topic, as well as the readers' personal/social, organizational, or professional needs. Proposals follow a problem/solution organizational pattern, and usually include the following information in the sequence provided:
1. Problem
2. Objectives
3. Solution

The solution section of a project proposal contains the following:
1. Overview   4. Equipment and materials
2. Procedures and methods   5. Time line
3. Personnel   6. Cost

The specific format for a proposal varies according to the community for which it is being written. Writers should study previous proposals in the community for which they are writing to determine the specific format to use.

PROJECTS

As you undertake a project, be sure to engage fully in all three processes—planning, drafting, revising. You may want to use the Audience Analysis (page 97), Purpose (page 136), and Decision charts (Page 101), as well as the Planning Worksheet in Appendix B, as you gather and organize your information. As you prepare to revise, you may also want to use the Evaluation Checklist (page 264-265), and/or the Document Design Criteria Checklist in Appendix B.

Collaborative Projects: Short Term

1. **Preliminary proposal for community assistance**

    a. Work in groups of five.

    b. Consider various community agencies or local charitable organizations that you might be able to help by collecting donations, distributing information or goods, or providing services.

    c. Decide on an agency or organization and determine the type of help you wish to offer.

    d. Write an unsolicited preliminary proposal to the agency or organization offering to help.

2. **Final proposal for a community brochure**

    A community agency or charitable organization has published the following Request for Proposal:

    Purpose: In order to persuade the public to become involved, provide a brochure informing the public about one of our programs .

    Guidelines: The proposal should provide (1) objectives including purpose and audience to which brochure will be sent; (2) description of brochure; (3) personnel involved; (4) time line; (5) cost.

    a. Work in groups of three.

    b. Write a final proposal to the agency or organization offering to develop, design, write, and print a brochure for the agency.

    To complete this assignment, you need to determine the purpose, focus, and audience for the brochure. You also need to determine the graphics, visual text, and layout, as well as the paper and number of brochures, in order to determine a budget.

    Option: To continue in Chapter 20, Document Locators and Supplements, Collaborative Projects: Long Term.

3. **Evaluating proposals**

Many agencies that provide grants for projects use the following criteria to determine whether or not to fund a proposal.

  a. Technical soundness of the proposed approach, including:

  - qualifications of proposer
  - applicable methodology
  - appropriate facilities and equipment
  - reasonable budget

  b. Potential for proposed project:

  - to lead to new discoveries or fundamental advances within a field
    OR
  - to serve as the basis for new or improved technology
    OR
  - to assist in the solution of societal problems
    OR
  - to contribute to better understanding or improvement of the quality, distribution, or effectiveness of the nation's scientific and engineering research, education, and manpower base.

You are a member of a committee of five persons whose task is to determine whether or not to fund various unsolicited proposals submitted to your organization.

  a. Read the proposal in Figure 17-3 and, using the above criteria, determine whether or not you would fund it.

  b. Work with the other four persons to arrive at a consensus on whether or not to fund the proposal.

  c. As a group, write a memorandum to the director of the agency to recommend or reject the proposal. Explain your decision in terms of the criteria listed above.

Collaborative Projects: Long Term

1. **Solid waste** continued from Chapter 14, Collaborative Projects: Short Term Exercise 1.

  a. Share your responses with the people in your group.

  b. Select those suggestions you believe your campus/community should implement.

  c. Write a preliminary proposal to the person(s) on your campus or in your community who would be responsible for implementing the recommendations.

Option: to continue Chapter 20, Document Locators and Supplements, Collaborative Projects: Long Term.

Individual Projects

1. Each of us has qualifications that could be used to help others or to help improve our environment. Think of a quality you possess that could help others or improve some aspect of your local area. You may be good at art-work, or at repairing household items, or at gardening. Write to someone on your campus or in your community, proposing that your services be used. You may want to write to the director of public information at your local museum, proposing that you create posters for an upcoming exhibit; or you may want to contact your city's human services department, proposing that you repair appliances for senior citizens; or you may want to contact your local park district, proposing that you develop a garden in a specific area of the city.

2. Study the outline for the time line on page 503. Create a graph to display the information. Compare the outline and graph. If you were writing the proposal, would you use an outline, a graph, or both, to give the reader information about the time for the project? Why?

3. Many aspects of your major field remain unknown. Your department has established a program to support majors who wish to engage in research in their fields. Your department chairperson has issued the following guidelines for grants for this program. These guidelines are patterned after guidelines for writing research grants for the National Science Foundation.

The Department considers proposals for support in most areas of the discipline. The proposal should present (1) objectives and scientific or educational significance of the proposed work; (2) suitability of the methods to be employed; (3) qualifications of the investigator; (4) effect of the activity on the infrastructure of the area explored; (5) facilities, materials, and equipment required; (6) amount of time required; (7) cost of the project; and (8) amount of credit hours for independent study requested.

THE PROPOSAL

To facilitate processing, proposals should be stapled in the upper left-hand corner, but otherwise unbound with pages numbered at the bottom and a 1-inch margin at the top. The original signed copy should be printed only on one side of each sheet. Additional copies may be printed on both sides. Review of proposals is facilitated when the contents are assembled in the following standard sequence: information about cover page, table of contents, project summary, project description, bibliography, biographical sketch of researcher, budget. Information concerning material, equipment and facilities required should be included in appendices.

Cover Page

The cover page should include the following information:
1. Name of person submitting proposal
2. Title of proposed project
3. Requested amount of credit hours
4. Requested amount of funds
5. Proposed duration
6. Desired starting date

The title of the proposed project should be brief, scientifically valid, intelligible to a scientifically literate reader, and suitable for use in the public press.

Grants may be awarded for a period of up to 4 1/2 months.

Table of Contents

The table of contents should show the location of each section of the proposal as well as major subdivisions of the project descriptions, such as the methods and procedures to be used.

Project Summary

The proposal must contain a 200-word summary of the proposed activity suitable for publication. This summary should not be an abstract of the proposal but rather a self-contained description of the activity that would result if the proposal is funded by the department. The summary should include a statement of objectives, methods to be employed, and the significance of the proposed activity to the advancement of knowledge. It should be informative to other persons working in the same or related fields and, insofar as possible, understandable to a scientifically literate reader.

Project Description

The main body of the proposal should be a detailed statement of the work to be undertaken and should include (1) objectives and expected significance and (2) relation to present state of knowledge in the field. The statement should outline the general plan of work, including the broad design of activities to be undertaken and an adequate description of experimental methods and procedures.

Bibliography

A bibliography of pertinent literature is required. Citations should be complete.

Biographical Sketches

Vitae should include list of coursework related to research, papers written related to present research, other research conducted.

Budget

Include materials, equipment, facility rental; travel; other direct services, including office supplies, publication costs, consultant services, computer services.

Facilities and Equipment

Equipment to be purchased, modified, or constructed should be described in sufficient detail to allow comparison of its capabilities to the needs of the proposed activities. Whenever possible, the proposal should specify the manufacturer and model number.

EVALUATION

All proposals are reviewed carefully by a faculty member specializing in the area in which the research has been prepared as well as by a graduate student who is involved in that area in addition to the Department Chairperson.

Proposals are evaluated according to the following criteria:

1. Research performance competence—the capability of the investigator, the technical soundness of the proposed approach, and the adequacy of the institutional resources available.
2. Intrinsic merit of the research—the likelihood that the research will lead to new discoveries or fundamental advances within its field.
3. Utility or relevance of the research—the likelihood that the research can contribute to the achievement of a goal that is extrinsic or in addition to that of the research field itself, and thereby serve as the basis for new or improved technology or assist in the solution of societal problems.
4. Effect of the research on the infrastructure of science, engineering, and technology—the potential of the proposed research to contribute to better understanding or improvement of the quality, distribution, or effectiveness of the department's scientific, and engineering research, education, and manpower base.

Write a final proposal to your department chairperson for support for a research project under this program.

4. Your technical writing instructor continually tries to improve this course. The instructor has requested that students who have ideas for improving the course should propose them in a memo. Write a memo to your instructor proposing an improvement for the course.

5. Because of tight budgets, organizations are always looking for ways to be more cost-effective. In addition, organizations try to improve the quality of their services and products to increase the demand for them. Organizations also try to improve the working conditions for their employees. Consider something at the university or at your workplace that might be improved or made more cost-effective. Write a memo to the director of the appropriate division proposing the improvements.

Interim
Reports

I N ADDITION TO WRITING a great deal of correspondence, you will find yourself writing many reports. You will often be asked to provide supervisors, peers, clients, and supporting and regulatory agencies with information about the progress of a project or the ongoing business of an organization. This information is usually written up as an interim report which readers use to determine whether a project or business will meet its goals.

Some reports, like progress or periodic reports, are written on a daily, weekly, monthly, quarterly, or annual basis. Others, like trip reports or minutes of meetings, are written sporadically. Some will be long, but many will be only a few pages. For some you will need to write complete narrations and descriptions, while for others you will simply fill in the blanks on a preformatted form. You may write an annual report for a regulatory agency, summarizing the problems you've corrected and discussing those still requiring action. You may keep a lab journal of each day's work on an experiment. Or you may write a trip report of a seminar you attended, summarizing the information you gained for improving the quality of your company's services.

GENERAL CONTEXT AND CONVENTIONS

While interim reports assume a wide variety of forms, many of the conventions and strategies for writing each kind are similar. In this general section, we will discuss those aspects that relate to all reports. We will then study specific conventions for each of the three most common types of interim reports—progress, project, and trip reports.

Context

Readers   Your readers may be internal or external to your organization. However, even if an external organization is your primary reader, your reports will probably still need to be approved by an internal supervisor.

Your readers are usually knowledgeable about your topic. They may be closely associated with either your organization or your project. They are anxious to learn whether programs are succeeding, and if progress is being made toward organizational goals. They also want to know what problems may prevent a project or organization from reaching its goals so that they can correct the problems before it is too late, or they can change their schedules to meet yours. They often skim through a report, pausing to study it only when the information indicates a deviation from projected plans.

While readers may be closely involved with a project or an organization, they are probably generalists, rather than experts, in a subject.

Purpose and Situation   Your main purpose in writing an interim report is to inform readers about the present status of a specific project, division in an organization, or activity as it relates to the goals of a program or organization. You may also have a secondary purpose—to persuade readers to view your project or activity in a certain way. You may want to indicate your success to date in running a project or your need for additional personnel.

A reader's purpose in reading these reports is to make decisions concerning the future course for a project or division, including a project's continuation, funding, manpower, and interface with other divisions.

Most interim reports are reader-initiated. Readers request these reports so they can keep up-to-date. They need to know what has been done, what is presently happening, and what still needs to be done to meet projected goals, especially in terms of deadlines and budget. Readers may also be interested in health and safety, quality, and personnel.

Annual reports, which indicate an organization's progress toward its goals, are initiated by stockholders or boards of directors. These groups use these documents to determine whether or not to continue investing in a company, and whether the present management should continue in their positions. The following excerpt, from an annual report by Genentech, Inc., is reader-initiated. Notice how the writer uses the report not only to provide readers with information about the product, but also to persuade stockholders to continue investing in the company by appealing to stockholders' needs—increasing profits.

Relates to stockholders needs.

Protropin: Growth hormone inadequacy. Protropin, human growth hormone, approved in 1985, was the first product to be developed, manufactured and marketed by a biotechnology company. It has grown steadily in sales and maintains a three-quarters market share.

The product is used to treat children who do not produce enough of their own growth hormone and therefore have growth hormone inadequacy, a serious disease. Treatment typically starts at childhood and continues through puberty. ...Studies are underway to determine the benefits of continuing to treat adults with growth hormone deficiency.

Genentech, Inc.

Occasionally, interim reports are writer-initiated. You might initiate a report when it is important for a supervisor or client to know your progress on a project or problems you've encountered. You may also write a report to cover yourself by placing in writing the details of a situation or your opinion, so that if there are future problems, you can support your claim.

At Advanced Technology Services (ATS), Jody Howard writes a report delineating the architecture she plans to use to meet her clients' computer needs. She initiates the report to determine whether or not she has covered all her client's requirements, so that she does not have to make expensive modifications once she begins to develop the computer system. If her readers are not satisfied with her plan, they can notify her now, and she can make the changes while still in the planning stage.

Purpose.

The purpose of this document is to describe the functional requirements of the Armour Guard Machine Control System. This specification outlines the system architecture, operational characteristics, and performance requirements that were established through discussions. This document shall serve as the foundation for implementing the control and information management of the Armour Guard Machine...

Howard's readers, like most readers of a progress report, scan the document to make certain that everything is satisfactory. If they notice a problem, they usually pause to study it, either immediately or later when they have time.

You may find that your report becomes part of another report. If your supervisor is required to submit a periodic report to her supervisor, she may integrate your periodic report with hers. In turn that report may be integrated with reports from other projects within the division, and submitted as a division report to the vice president. Figure 18-1 shows how your report becomes integrated into other reports as it goes up the hierarchical ladder.

Figure 18-1
The Report Chain

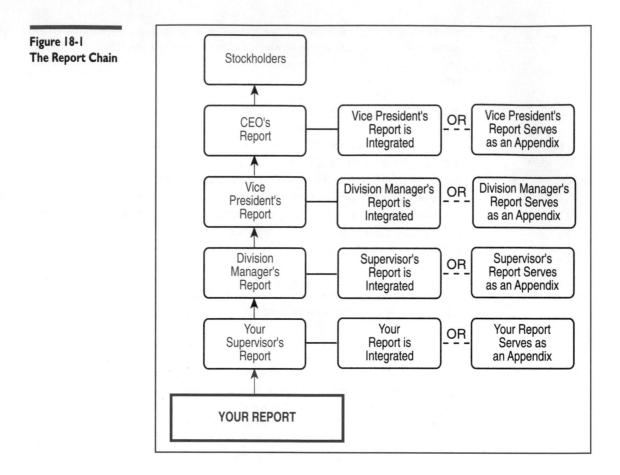

If you're writing a report for an entire project or an activity which involves several employees, you need information from those involved with aspects other than yours. One way to obtain this information is to interview the other employees. However, a more efficient way is to request that each worker submit a report to you. You can then synthesize this information for your own report.

Gathering Information

The information for these reports is based on your field experience. If you keep notes on everything you do, you can refer to your notes for the information you need when you're ready to write your report. If you haven't kept notes, you need to remember everything you've done.

In addition, you should collect journal entries, memos, and letters you've written, as well as charts, drawings, or other related documents. These may contain information your readers need. You can attach these documents to a report as appendices if you think some of your readers may want to see them in their entirety.

You should also review the original plans or proposals for a project or activity so you can compare what you've done with what you said you would do, since readers want to know the relationship between the present status and long-range goals of a project or activity. They need to ascertain whether a project or activity is on target, keeping within its budget, meeting its deadlines, and achieving its goals.

Point of View and Focus The point of view should relate to that of a decision maker, since the reader will use the information to determine the future course of action for the project or activity. The focus should revolve around the completion of the project or activity. Thus, when you discuss a budget, you not only need to discuss how much has been spent, but also whether this will mean the project will be over or under the budget projected for the entire activity. The reader can then determine if additional monies have to be found, if the project should be halted because there are not sufficient funds to complete it, or if monies can be deducted from it and given to other projects.

Style Language and sentence structure should reflect the primary readers' needs. The report should be concrete rather than abstract, and should contain specific facts about a project or activity. However, you should provide a general overview, rather than a detailed description. Readers want to know the major problems of a project, but not details. If you think readers may want additional information, place it in an appendix. The tone should be impersonal and the sentences concise, so that readers can scan the information easily. Conventions include the use of numeration or bullets to list the information within each section, so readers can recognize the various chunks of information easily. Because these reports can serve as legal documents, you may want to keep them formal.

Visual Text, Format, and Graphics Readers expect certain information to be included in each of the various types of interim reports. Use headings to indicate these major categories to help readers locate the information. If your report involves several components of a project, use subheads to help readers quickly identify these subcategories and to recognize they are moving from one aspect to another as they skim through a report.

Because most interim reports are no more than three to five pages, they are often written in a letter or memo format. If they are not, they should be sent with a cover letter or memo (see Chapter 20).

Since the reader is usually familiar with the project or activity, graphics are often unnecessary. However, if there is a deviation from original plans, then a graphic may help readers perceive a new design, model, or trend.

1. Keep notes of your activities for reference.
2. Collect journal entries, memos, letters, etc., for reference, as well as for possible inclusion in an appendix.
3. Review plans and goals for projects and activities for reference.
4. Focus text on projected plans and goals.
5. Relate text to point of view of the decision maker.
6. Be concrete, rather than abstract.
7. Provide a general, rather than detailed, discussion.
8. List items when possible.
9. Use headings and subheadings.
10. Include graphics if changes to a product or process will alter previous graphics. Indicate changes on graphic.

PROGRESS REPORTS

Progress reports are also known as status reports. They are concerned with specific projects and are usually submitted periodically to provide readers with information about the progress of a project or activity toward its goals. Funding agencies usually require that organizations to which they've provided monies report on funded projects on a semi-annual or annual basis. Regulatory agencies require that institutions over which they have jurisdiction report periodically on long-term activities. And almost all governing boards and stockholders require organizations to provide an annual report, presenting an overview of the company's present accomplishments and future plans.

Context

Readers use the information in these reports to make decisions concerning a project's continuation, funding, manpower, and interface with other departments and divisions. Readers may decide to terminate, rather than continue, a project. They may also decide to budget more funds or, if a project isn't using all of the monies originally appropriated for it, they may give some of the funding to another project. The information in the report may indicate that it is time for other divisions, e.g., the marketing or legal divisions, to become involved with the project. Based on the progress of a project, readers may decide to push back or move up a deadline. Readers also use these reports to determine if projects are on target, and to learn of any problems that might require attention. If there are problems, they may make recommendations. Discovering problems at this point rather than waiting until the entire project is complete can save both time and money.

Textual Decisions

Because these reports are reader-initiated and the reader is usually familiar with the topic, a statement of the purpose or background information is unnecessary.

The major sections of a progress report reflect the information readers need to know: what is completed (past), what is going on (present), and what still needs to be done (future). Within these categories, the various aspects of a project are discussed according to their order of importance. Thus, progress reports, sequenced according to a modified chronology that combines a chronological pattern with an inverted pyramid (most important first), usually use the following format:

A. **Introduction** A general statement of the progress of the project. The statement synthesizes what you have accomplished to date, and what you still need to do to accomplish your long-range goals.

B. **Work Completed/Accomplishments** A description of the work completed since the last report.

C. **Present Status** A description of the work in progress as of the writing of the report.

D. **Concerns (optional)** A discussion of problems that have occurred or are anticipated in the future. If the schedule or budget is not being met, the information should be included here. This information can either be integrated with the information in sections C and E, or discussed in a separate category. By creating a separate category for concerns, you send up a red flag notifying readers that a problem exists.

E. **Future Plans/Goals to be Met** A listing of work still to be done, and a schedule for completing the project. If changes in the budget or schedule are planned, the information should be included here. In addition, include all plans involving other divisions.

Figure 18-2 depicts the dual organizational pattern found in progress reports. Notice those aspects that are parallel and those that differ as you study it.

Model The progress report modeled in Figure 18-3 was submitted by a manufacturing engineer to a supervisor, who then submitted the report to another supervisor. The report is signed by the manager, not the engineer, although the engineer wrote it. The report will be integrated with reports of other projects to create a division report. This will be submitted to a vice president who will use it not only to make decisions relating to the project, but also to evaluate the division. Divisions that fall behind schedule or go over their budgets are viewed less than favorably. Eventually, if the trend continues, a shake-up in personnel can occur with some employees being let go. On the other hand, divisions that maintain a schedule and stay within a budget are looked upon favorably, and managers are often promoted. In addition, bonuses are often given out at the end of a year to employees in a division with a high rating. Thus, the writer wants to make the project look as good as possible.

**Figure 18-2
Conventional
Organizational
Pattern for
Progress Reports**

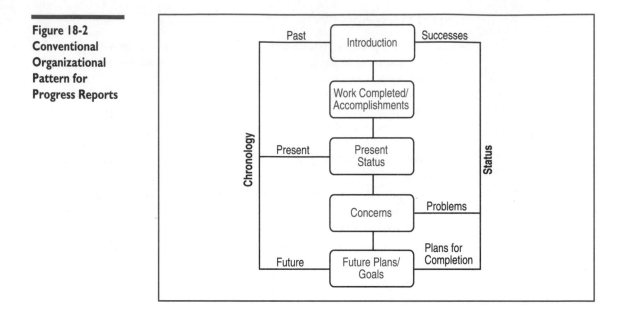

The engineer is aware that this is one of several project reports which the supervisor will receive. Therefore, the report is brief and to the point. Furthermore, the supervisor isn't interested in a detailed description of what has happened, but only in the major events as they relate to budget, schedule, and other departments and projects.

The information for the report comes from monthly reports submitted by each of the department managers involved with the project. The engineer synthesizes the information from these various departments to write this report. As you read, consider how the writer adheres to the conventions.

**STRATEGY
CHECKLIST**

1. Organize and sequence your information as follows:
   • Work completed/accomplishments
   • Present status
   • Concerns
   • Future plans/goals
2. Sequence the report in a modified chronology. Discuss each of the chronological sections from the most to the least important aspects.
3. Be specific and brief.
4. Use an impersonal tone.
5. Use lists and bullets.
6. Use third person
7. Use headings to indicate the major categories.

**Figure 18-3
Progress Report
Model**

Major Category:
Chronological &
Goal-oriented
(implies past).
Subcategory (by place).
  Relates to interface
  with Personnel Div.
  Data relates to
  status of schedule.

Subcategory.
  Relates to schedule
  & interface with
  other plants.

Major Category:
Chronological.

Relates to personnel.

Major Category:
  Goal-oriented.
  Relates to predeter-
  mined goal.
Major Category:
Chronological.

Monthly Project Summary   Month: January, 19XX
Manufacturing Engineering   Project Manager: XX
Norcross Prepared by: XX

<u>SLEEVING</u>

<u>Accomplishments</u>

<u>Tulpehocken</u>

Last MN1500 conversion kit in Tulpehocken was completed end January and is available for production start up when cells and operators are available. The final MN 1400 conversion kit in Columbus is now scheduled for completion the end of the 1st Qtr. This has been rescheduled by the plant at their request. All other equipment at both locations remains operational.

<u>Columbus</u>

Two MN1500 cell accumulators were installed in line on each side of the cell washer on line 6 in Tulpehocken. The plant reports that these are functioning satisfactorily. Information has been forwarded to them at their request for the purchase of 10 more cell accumulators. Modification to the remaining units will be completed as production schedules allow.

<u>Present Status</u>

- Equipment is currently in restart up after holiday shutdown. All units are coming back on-line satisfactorily.
- L.M. has been reassigned to L. Group. A. K. is assuming responsibility for sleeving in A. Group.

<u>Concerns</u>

- Columbus 1300/1400 preheats remain incapable of achieving the design 300 P.P.M. production rate...

<u>Plans Next Month</u>

- Retest high frequency induction preheating project.
- Issue sleeving equipment manual.
- Determine cause of problem with Columbus 1300/1400 preheats.
- Begin production after holiday shutdown.

<u>Budget</u>

(Qtr.)	1st	2nd	3rd	4th
Travel	$1,000	$1,000	$1,000	
Proj. Material	1,500			
Test Material	500			
Contract M.H.	2,000			
Overtime	0			

PERIODIC REPORTS

Periodic reports provide a picture of a division's progress toward an organization's goals. They are similar to progress reports. However, periodic reports are concerned with *all* activities that have occurred in a division rather than with a specific project.

Context

Periodic reports are written for the same types of readers as progress reports, though in most instances readers are internal to the organization. Periodic reports also have similar purposes—to provide readers with information for making decisions. In the case of periodic reports, readers are concerned with making decisions related to the functioning of a division or organization.

Textual Decisions

The organizational pattern for periodic reports, like progress reports, is analytical; it is broken down into major categories. However, because the content of a periodic report involves all activities rather than a single project as in a progress report, the organizational pattern differs slightly from that of a progress report. In a periodic report, the categories are organized according to the work in a department or division, rather than to the aspects of a project. Thus, a report for a specific division is divided according to the various projects.

Each of the major sections can be organized either chronologically or in order of importance, just as they were in the progress report. Or they can be categorized according to the various activities occurring.

Model The report shown in Figure 18-4 is a periodic report written for an internal reader by the division manager of the Facilities Engineering Department. These reports are filed quarterly. As you read, compare the content, organizational pattern and sequence, and format of this report with the progress report in Figure 18-3.

STRATEGY CHECKLIST

1. Categorize the major sections according to the major areas of work.
2. Sequence the information within a category either chronologically or by order of importance.
3. Be specific and brief.
4. Use an impersonal tone.
5. Use lists and bullets.
6. Use third person.
7. Use headings to indicate the major categories.

Figure 18-4
A Periodic Report

Major Category
(by organization).
  Subcategory (by
  geographic
  location).
  Sub-subcategory
  (by activity).

Status Report - 3rd Quarter 19xx   October 7, 19xx
Facilities Engineering Department

1.00 Corporate

  .01 Columbus (Hog Heaven Road)

    1. Can Making Expansion—Engineering work by consultant is underway,
       with drawing and bid package anticipated by November.
         Coordination meetings have been held with consultant and plant person-
       nel to plan for maintaining traffic and engaging operations during con-
       struction.
    2. Press Room Dust and Environmental Control—Approval of RAPC has
       been received and orders prepared for equipment. Deliveries of major dust
       collecting equipment are anticipated by February 1. Installations and
       change out of existing equipment is being coordinated with the Can Mak-
       ing Expansion project to keep interferences to a minimum.
    3. Compression Air Expansion—The building addition has been completed
       and the new compressor equipment is installed and running. Plant person-
       nel have handled this project on a daily basis, with the Facilities group
       providing piping installation, equipment layouts, and inspection assis-
       tance.
    4. Hazardous Waste Bulk Handling Facility—Preliminary quotes have been
       received for CPAR preparation. Several changes in location and tying-in
       with a proposed extension to the shipping dock are being evaluated prior
       to preparing the CPAR…

Subcategory.

  .02 Columbus (Exeter Rd.)

    1. Proposed Facility Expansion—Due to staffing and other priorities, no
       work has been performed on costing or layout…

Major Category.

2.0 International…

TRIP REPORTS

While progress and periodic reports are submitted at specific intervals, trip
reports are written as necessary. Trip reports provide information on the relevance of a single activity to the long-range goals of a project or organization.

Context

Because of the rapid advancement of knowledge in every field, you will probably be asked to attend a variety of staff development programs so that you can
keep up-to-date in your field. Some of these programs will be sponsored by pro-

fessional organizations, others by outside consultants. When you attend a professional conference or program by an outside consultant, you may be expected to report on the program.

In addition to your supervisor (who will probably be your primary reader), you may be asked to submit your report to your peers or subordinates if you believe they should have access to the information you acquired while attending a program.

Your supervisor's purpose in reading a trip report is not only to find out what you learned, but to determine whether or not the program was worthwhile. If you attended a professional conference, the supervisor wants to know whether to send you to a similar conference again. If you attended a staff development program, the supervisor wants to determine whether to send other employees. The purpose may also be to inform other employees about the topic.

Textual Decisions

Based on the purpose, the document should include a summary of your own impressions of the conference, including a discussion of whether or not you feel it was worthwhile for you to have gone, whether it would be worthwhile for you to attend a similar or follow-up program, and whether you would recommend the program for other personnel. The report should also include a summary of the information presented at the program. If the report is to be sent to other employees, then the information should be detailed. You may also want to include in an appendix any handouts you received.

The report should be organized in three sections, according to readers' three major concerns:

1. Overview of the program and your response to it.
2. Specific information obtained.
3. Recommendations relating to the value of the program for future attendance or for other employees.

Section 2 should be categorized by topic from most to least important rather than chronologically, since readers aren't interested in your itinerary, but in the information itself.

To help your supervisor determine the relevance of the program, you should include your own view of the program as it relates to your project's or organization's long-range goals.

The report can be written either formally or informally. You should use first person in the first and third sections when referring to yourself. If your purpose is to provide your supervisor with information to determine the value of the program, the second section should be brief. However, if the purpose is to provide employees with the information presented in the program, then it should be as detailed as is necessary for your readers to understand the information. You may list the major points, discuss the information in in-depth explanatory paragraphs, or attach an appendix with handouts, etc., from the program.

Model The excerpt in Figure 18-5, from a trip report by an assistant vice president of a national health service company, was sent not only to the writer's supervisor, but also to various division managers who needed to know the information. The writer, who expects her employees to submit detailed reports of workshops for other employees to read, comments, "If I pay $500 [for a registration fee] for an employee to attend a workshop, I want to spread that money as far as possible." As you read, notice how the writer adheres to the conventions.

Figure 18-5
A Trip Report

Numerical outline.
Personal response.

Uses first person.

Assumes persona
of company.
Lists.
Bullets.

1.0 OVERVIEW

AAA Consulting presented a seminar for management information systems executives on November 28, 1988. The seminar topic was "Trends in Information Systems Technology" and its purpose was to provide a status report on three areas of emerging technology: BEDSIDE AUTOMATION, IMAGE SYSTEMS, and NETWORK ARCHITECTURE. (There was also a "soft sell" of a new decision support product which I have not covered in this report.)

Detailed information from each of the speakers can be found later in this document. This Executive Overview focuses on my impressions of the material and the impact these technologies may have on our company.

BEDSIDE AUTOMATION

Of the three technologies presented, only bedside automation can be considered stable enough for practical implementation at this time. The benefits of bedside automation appear to be increased productivity, enhanced job satisfaction, and improved quality of care rather than actual hard dollar savings.... Our company is not in a position (either financial or technological) to implement a full-blown bedside automation system at this time. However, several less sophisticated products may offer cost justifiable solutions to our incidental overtime problems and our heavy dependence on expensive agency nurses...Areas on which we will focus are:

- Understanding our institution's nursing needs—In what area of patient care can we achieve greater productivity?
- Analyzing our competition—What type of leading edge technology will help us attract and retain highly qualified staff?...

IMAGE SYSTEMS

The technology for storing and retrieving printed documents is currently available from a variety of reliable vendors. However, there is still a large cost associated with this limited utilization of image systems (laser disk) technology. It will be several more years before maximum functionality is achieved (integration of relational database technology and enhanced document management) at a reasonable cost. Image Systems technology is not an MIS priority in the foreseeable future....

2.0 INTRODUCTION TO SEMINAR

2.1 The Issues

According to AAA Consulting, the following issues will impact on the health care industry's utilization of new information technology in the next 2-10 years.

continued

Figure 18-5
A Trip Report
continued

- Business, competitive and regulatory pressures will continue to affect the industry in the 1990's.
- The use of technology must be better aligned with business strategies and profitability goals.
- Leading-edge productivity tools must be harnessed to deliver systems more quickly.
- Complex issues related to systems consolidation and integration must be better understood as stand-alone hospitals merge into multi-entity hospital corporations.
- Corporate strategies must be developed to manage cost/quality trade-offs while maintaining or enhancing profitability and quality care.

2.2 <u>The Regulatory Outlook</u>

AAA Consulting identified the following areas of vulnerability.
- Teaching costs
- Capital costs
- PPS refinements
- Loss of Medicare waiver
- Loss of PIP…

STRATEGY CHECKLIST

1. Provide an overview of the program and a summary of your impressions, including recommendations for similar or follow-up programs for yourself or your peers.
2. Present specific information gained at the conference if readers are expecting to acquire information from your report.
3. Use the following format:
   - Overview of program and summary of your impressions.
   - Details of information presented.
   - Recommendations for future attendance or programs.
4. Present the overview and summary from your personal point of view. The presentation of information should be impersonal.
5. Organize the section for presenting information according to the topics covered and sequence the data from most to least important.
6. Use first person when referring to yourself.
7. Introduce each of the three sections with a heading.

DESIGNING TEMPLATES

Because interim reports require the same categories of information each time you write them, and because you usually write a number of these reports over a period of time, you may want to create a template so that you don't have to format your text each time you write one. A template is a formatted sheet that contains the category labels, but not the data. With the advent of word processors, templates are used more and more often, since it's easy to call up a template, copy it, and place the necessary information in the appropriate category.

Templates can be used any time you are required to report the same types of information to a reader. They can also be used in situations in which many persons report the same types of information to you. They provide a basic form that can be duplicated or printed for dissemination to all those expected to submit information.

Follow these steps to create a template:

Step 1   List the kinds of information you need.

Step 2   Look for relationships among these pieces of information.

Step 3   Label the categories of related information.

Step 4   Use the labels you determined in step 3 as labels for the major categories of your template.

Step 5   Use the information you've listed within each category as subcategory labels.

Let's follow a manager in the Health and Safety Division of a university as he develops a template for engineers in his division.

All projects involving asbestos removal must be reported to the United States Environmental Protection Agency (EPA). A report must also be filed with the Director of Physical Plants. Both of these reports are written by the Asbestos Manager, who obtains the necessary information from the engineers who do the job. To get this information, the Asbestos Manager requires division engineers to keep a record of their work. To make certain that the record contains all of the information required for both reports, the Asbestos Manager decides to develop a template.

Step I

(a) The Asbestos Manager lists all of the information required by the EPA in its guidelines.

Name of project	Project size
Type of project	Emergency
Work performed by	Work supervised by
Date work began	Date work ended
Location of work	
on campus in bldg	
Reason for work	
Description of work	

Method used for abatement
Materials used for abatement:

HEPA Respirators	HEPA vacuum	Neg. Air Machine
Disp. Coveralls	Encapsulant	Surfactant
Lag Kloth	Mini-enclosure	Signs placed
Area sealed off	Reports maintained	
Air Samples		
Disposition of ACM		

(b) The manager then lists the information which is not included on the EPA list, but is required by the Director of Physical Plants.

Date, time and day of week of complaint
Complainant's name   Complainant's phone number
Recommendation for future action

(c) Finally, the manager adds information that is not required by either the EPA or the director, but which the manager needs to know for carrying out the work.

Material cost

Steps 2 and 3

The manager places all related information together and labels it. These labels become the headings for the major categories.

Project Data

Name of project	Project size
Type of project	Emergency?

Work Data

Date, time, and day of week of complaint

Complainant's name	Complainant's phone number
Work performed by	Work supervised by
Date work began	Date work ended

Location of work
  on campus   in bldg
Reason for work

Description of Work

Method used for abatement
Materials used for abatement:

HEPA Respirators	HEPA vacuum	Neg. Air Machine
Disp. Coveralls	Encapsulant	Surfactant
Lag Kloth	Mini-enclosure	Signs placed
Area Sealed off	Reports maintained	

Air Samples
Disposition of ACM
Recommendation for future action
Material cost

Steps 4 and 5

Using the category labels and the list of information, the manager develops the template in Figure 18-6. As you study it, determine how the writer
moved from steps 2 and 3 to the final template.

**Figure 18-6
A Template**

ASBESTOS ABATEMENT PROJECT RECORD

DESIGNATOR	DATES
LOCATION	PROJECT
CONTRACTOR	PHONE

PERSON/DEPT. REQUESTING REMOVAL
REASON FOR REQUEST
DATE REQUEST SENT   DATE PROJECT AUTHORIZED
PERSONNEL ASSIGNED
JOB DESCRIPTION
DESCRIPTION OF CORRECTIVE ACTION
AIR SAMPLE DATA
      Personnel sample
      Area sample
      Clearance sample
LABOR   Time for each position   Amt. ACM abated
Pipe insulation
      Proj. designer   Acoustical
      Proj. manager   Tile
      Industrial hygienist   Other
      Supervisor
      Workers
MATERIAL

Item	Cost	Total	Item	Cost	Total
Respirator			Neg. Air Mach		
Half face			Filter 1		
Full Face			Filter 2		
Papr			HEPA filter		
Filters			HEPA vacuum		
Papr filters			Paper bag		
Cleaning agent			Plastic bag		
Disp. Clothes			Paper filter		
Goggles			Cloth filter		
Glove bags			Fine filter…		

TOTAL COST:

Notice that the Asbestos Manager expands on some items, thus ensuring that the writer doesn't omit any information. The more detailed the template, the less problem there will be when it's time to write the report. Time won't be wasted tracking down information; it will be in the reports.

STRATEGY CHECKLIST

1. Develop a template for a preformatted report when many readers need to look at the same kind of information, or when many writers must submit the same kind of information to readers.

2. Follow the steps for synthesizing information to determine the headings and subheadings in your template. Use the category labels as your headings and subcategory labels as your subheadings.

CHAPTER SUMMARY

Interim reports are written to keep readers informed about the status of a project, activity, or department/ division of an organization in relation to its long-term goals.

Progress reports provide readers with updated information on a particular project or activity; periodic reports update readers on all activities within a department/division. The organizational pattern for a progress report is analytical, examining the status of a project at various points of time (past, present, future), as well as the accomplishments and the problems.

Periodic reports are subdivided according to the various projects and each subdivision is then organized like a progress report, either according to the accomplishments and concerns of a project or chronologically.

Templates facilitate writing interim reports.

PROJECTS

As you undertake a project, be sure to engage fully in all three processes—planning, drafting, revising. You may want to use the Planning Worksheet in Appendix B, as you gather and organize your information. As you prepare to revise, you may also want to use the Evaluation Checklist (page 264-265), and/or the Document Design Criteria Checklist in Appendix B.

Collaborative Projects

1. Class template

The technical writing faculty at your university wants to reduce the number of student absences and late papers in classes. To accomplish this, the faculty needs to determine why students are absent and why they don't turn in papers on time. They have asked you and two other students to collect the data.

a. Work in groups of three.

b. Create a template for students to fill out to explain (1) each class absence, and (2) each late paper.

c. Fill out the template for all of your absences and late papers this term.

Option: To write an informational report based on the results of the data collected from all students in the class. See Chapter 19, Final Reports, Collaborative Projects: Long Term.

2. **Progress report**

Your instructor requires a monthly progress report for all group projects.

a. Write an individual progress report for a project in which you are involved in your technical writing class. The purpose of the report is to keep your instructor up-to-date on the project so he/she can determine your role in the project and whether you're on target, and can evaluate your efforts as part of the group.

b. Write a group progress report based on the individual reports of each member of your group. Your instructor needs this report to determine whether or not your group is on target and whether you'll be able to finish by the deadline.

3. **Periodic report**

In an article in the journal of the Society of Technical Communicators, the following list of "on-the-job communications" was recommended for inclusion in an undergraduate course for aerospace majors.[1]

- Oral presentations
- Abstracts
- Use of information sources
- Conference/meeting papers
- Technical reports
- Journal articles
- Letters
- Technical specifications
- Literature reviews
- Memoranda
- Technical manuals
- Newsletter/paper articles

The chairperson of your department has requested that all technical writing instructors submit a monthly report on their course to determine what on-the-job communications are being taught, and to compare their curriculum with the list of recommended topics. Your instructor has asked you to work with two other members of your class to write the report for the past month.

Individual Projects

1. Write a progress report for a project you are doing or a term paper you are writing in one of your classes. The report is for the instructor of the course, who wants to determine whether you're on target and will make the deadline.

---

[1]Barclay, Rebecca O., et al. 1991. "Technical Communication in the International Workplace: Some Implications for Curriculum Development." *Technical Communication*. 38. 324-35.

2. Write a progress report for your technical writing instructor, reporting on your progress this term in acquiring skills in technical writing. Your instructor needs the information to determine whether or not the syllabus should be altered for the remainder of the term.

3. Analyze the progress report in Figure 18-7.

    a. Read the report, shown in Figure 18-7, of a construction project. What indications do you have that the writer has a hidden agenda? What is it?

    b. Reread the progress report in Figure 18-3.

    c. Compare the report in Figure 18-3 with the report in Figure 18-7. Consider the differences and the similarities in terms of audience, purpose, organizational pattern, point of view, and style.

4. Prepare a periodic report for your parents, informing them of your previous month's activities at college. They will use the report to determine whether or not to increase your monthly allowance.

5. Develop a template that students at your university can fill out to evaluate one of the following:

    *   Their courses
    *   Their campus residence
    *   Campus activities
    *   Campus sports
    *   Student/faculty relations
    *   Campus atmosphere

Figure 18-7
A Progress Report

Jacksonville Site
11/20/19xx

Construction progress

Site: Gradework is in progress. Looks as though traffic exiting the property will have adequate view of oncoming traffic.

The backglow preventer in the fire water supply line has been reworked to correct the sinking and misalignment problems reported earlier.

Asphalt paving is in progress on the front parking area and drives. Patterson plans to relocate their construction trailers to the asphalt areas within the next week to permit finishing gradework along the street.

The storm water runoff retainage pond is being rebuilt to repair damage to north wall from heavy rains and washout.

Building: All building tilt-up panels and screen wall panels are erected.

All steelwork is erected and tied to walls. The roof deck is complete. The mezzanine is complete and the concrete floor has been poured.

Metal studs are in place for the office walls. Interior plumbing is in progress, with electrical scheduled to start very soon. The exterior window wall arrived on site today.

Pits have been poured for the dock levelers. Pouring of the perimeter floor slab is in progress.

Review items

1. Fire protection—Factory Conglomerated is not satisfied with the water flow test obtained in October. The Chicago office gave approval to proceed (Copy attached for your information), but the Atlanta office refuses to approve the drawings until a second test is obtained. I've forced the issue with the Atlanta office and they will run a test by 11/22 with the sprinkler contractor so that both will have the same information. I've impressed on Factory Conglomerated that we cannot stand a delay, and that communication must be done as quickly as possible.

Factory Conglomerated has given approval to proceed with ordering of the fire pump, which will be a 10 to 12 week delivery item. The sprinkler contractor is to notify me of the delivery date as soon as firm....

2. Roofing—Lesser Engineering has been retained as our roofing inspector and has already made contact with P. to set up a pre-roofing meeting. For some unknown reason, the roofing contractor went ahead and started roofing without telling the inspector. We discussed this and the misunderstanding has hopefully been resolved. However, in reviewing the roofing that had been placed, it was brought to P.'s attention that sloppy workmanship was evident.

Critical Schedule Items

—Make application for electrical utility service and transformer installation as soon as possible

—Complete roofing by 12/9

—Start lighting/electricals by 12/2 and complete by 12/23

—Complete storefront/windows by 12/9

—Punch list warehouse interior by 1/6

—Deliver/install/test fire pump by 1/31

Final Reports

FINAL REPORTS ARE WRITTEN to document the results of a project or activity. While an interim report is written during a project's or activity's life, a final report is written at the conclusion. An interim report discusses whether or not a project or activity may achieve a goal, while a formal report indicates whether or not the goal has been achieved.

Final reports take many forms. Information reports provide the results of data-gathering activities; feasibility reports indicate the results of an analysis of a situation; evaluation reports present the results of an assessment of personnel, organizations, projects, or situations; and research reports document the results of research studies or product development projects.

These reports provide readers with historical records of projects and activities as well as information for making decisions for future actions. Clients may use the data provided in an information report to determine how to proceed with their own projects. Managers may use the results in a feasibility report to determine whether or not to initiate a project, or how to proceed with one already begun. Regulatory agencies may use the results in an evaluation report to determine whether or not to shut down a company. Developers may use the results in a research report to determine how to improve a product for their company.

A report may be written by a single person, or it may be written as a collaborative effort by members of a team working on the same project. This collaboration may take several forms. Team members may all work together on an entire document, or each person may be responsible for a specific part of a report, with a supervisor or a designated person finally putting the parts together. Or all of the persons involved may simply submit information to a single person, who then writes the report.

GENERAL CONTEXT AND CONVENTIONS

We'll look first at general guidelines for writing final reports. Then we'll examine the conventions for specific types of reports, including information, feasibility, evaluation, and research and development reports.

Context

Readers You will write for all types of readers, ranging from internal supervisors and subordinates to external clients or peers. Readers may be experts, generalists, or novices, and their attitudes will vary. They may be receptive to your report, skeptical, or even hostile, if your message affects them negatively.

Reports are often read by multiple readers, including secondary, intermediary, and peripheral readers. A company vice president, in the position of a primary reader, may read only the executive summary of an evaluation report to learn its recommendations, and then turn it over to a secondary reader, the manager of a division who is responsible for implementing the recommendation. In all probability your supervisor, an intermediary reader, will read your report before it is sent out. Reports related to federal projects are in the public domain and are available to peripheral readers such as the news media. Evaluation reports, especially, can be used as legal documents and read by attorneys if suits concerning the topic evaluated are involved.

Regardless of their organizational position, all readers assume the role of decision makers as they read reports. If they are reading a feasibility study as shown in Figure 19-3, they need to determine whether or not to initiate a program. If they are reading an evaluation report (Figure 19-4), they need to decide whether or not to continue or to eliminate the project under discussion. If they're reading a research report (Figure 19-5), they need to determine whether or not to accept the findings, and, if they do, how to integrate the new knowledge with their own to improve their research or products.

Different types of readers read documents differently. A CEO may read only the executive summary in one document, but skim through the entire text of another. If a report discusses several divisions, managers of those divisions probably read only the sections involving them. Some readers scan a report or a section, searching for specific information; other readers study an entire section.

Purpose and Situation Your primary purpose in writing a report is to provide readers with information. However, you have a secondary purpose, that of persuading readers to use the information, or to interpret it, in a specific way. All reports are persuasive regardless of whether you attempt to persuade readers overtly, as you do in a proposal, or implicitly. For example, if you have evaluated a project, you will want to persuade readers to accept your results, conclusions, and recommendations. Results of a study are stated as claims, which are supported by evidence in the form of sources for your information, methods and procedures for carrying out your study, and statistical analyses of your data. To persuade readers to accept your results, you must convince them that your sources, methods, and analyses are valid according to criteria established by their discourse communities. Some discourse communities will accept results only if the method involves a comparison group or an ANOVA (a statistical formula) has been used for the statistical analysis.

Conclusions and recommendations are also stated in terms of claims. Your conclusions must be supported with evidence from your results. To persuade readers to accept your recommendations, you must follow a logical line of reasoning that leads from your conclusions to your recommendations (Figure 19-1).

Reports are both reader- and writer-initiated.

Reader-initiated. Most reports are reader-initiated. You write reports in response to a request for specific information. Your reader may want a recommendation concerning the feasibility of a project as shown in Figure 19-3, an evaluation of a product (Figure 19-4), or simply information to use in a project (Figure 19-2).

Writer-initiated. Information and evaluation reports are initiated by the writer when it is important for a supervisor or client to have information.

Research reports published in professional journals are writer-initiated. Researchers publish articles discussing the results of their research so that their peers and others involved with a particular topic can use the new information in their own research.

**Figure 19-1
Reader's Line of
Reasoning in
Reading a Report**

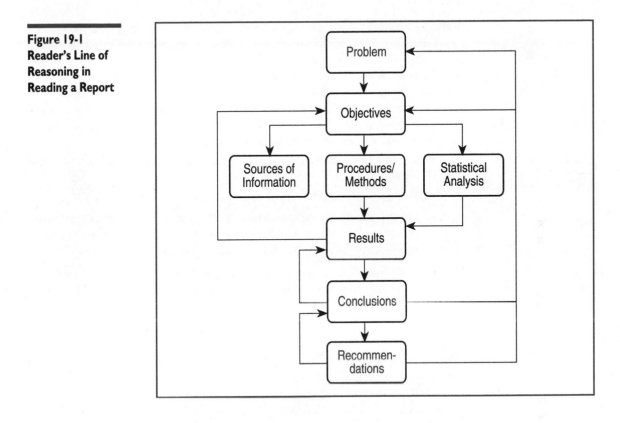

Textual Decisions

Content, Organizational Pattern and Sequence Most of your
information comes from your own field work. You may occasionally include
information from written sources as well as from observations, interviews, and
surveys in the introduction and discussion sections if your readers need the information to accept your claims, or if they need background to understand your
problem or interpret the data.

The organizational pattern of a final report is analytical, consisting of the
major areas involved in a study: problem, objectives, methods/procedures, results,
conclusions, and recommendations. You need to begin by introducing readers to
the problem so they have a frame of reference within which to understand the
information that follows. Readers also need to know the objectives of the project
you are discussing so they can determine whether the information you present and
the recommendations you make are relevant to the problem. As we have discussed,
readers need a description of how a project is implemented or a research study
conducted, to determine whether your results are valid according to their community's criteria. Once you provide readers with this information, they are ready to

learn the results of your information-gathering or investigations, and the conclusions and recommendations you draw from the results. Because readers need evidence that you have conducted a valid research study or that your evaluation of a project or situation is reliable before they accept your results, reports follow an inductive organizational pattern. Your evidence should precede your claims.

Conventions have evolved for organizing and sequencing final reports, based on these needs. The organizational pattern usually consists of the major categories below in the sequence presented here.

- Introduction—Includes rationale, statement of need, problem, and/or background.
- Purpose—States the purpose or objectives of the report.
- Description of project/procedures/methodology— Describes the project and/or explains the procedures or methodology used to obtain the data.
- Results—Presents the results neutrally, and often quantitatively.
- Conclusion/Discussion—Presents a synthesis of the data, and evaluates the results in relation to the objectives or purpose.
- Recommendations—Offers suggestions for continuing/discontinuing/ changing/extending the project or activity.

Point of View and Focus The point of view should be related to that of the decision maker. Readers must decide whether or not to accept the conclusions and recommendations of a report. Therefore, all of the information presented should help readers understand how you arrived at your conclusions and recommendations.

A report should focus on its objectives. A report explains why you selected these objectives (introduction), how you went about meeting them (procedures/methodology), whether or not you met them (results), why you believe you arrived at these results (discussion), and what you plan to do about these results (recommendations).

Style The specific style of a final report depends heavily on the discourse community in which you are writing. Use a style guide accepted by your organization or professional discourse community, or review your firm's previous reports to determine the specific conventions you should follow.

Visual Text, Format, and Graphics Readers expect and need headings in reports to locate specific aspects of a report and to know when they are moving from one major section/category to another. Subheadings should also be used in long reports to give an airier appearance, and to provide readers with a visual image of the subparts of a major section/category. The use of lists, bullets, and numeration depends on the discourse community. You need to examine previous reports and/or style guides used by the particular discourse community in which you are writing.

The length of a report varies from a single page to several hundred pages, depending on the topic and the reader's needs. Reports can be incorporated in correspondence, or stand alone.

Graphics should be used to both complement and supplement the verbal descriptions. Graphics are especially helpful if a document discusses new products, machinery, or configurations of parts.

Global Locators and Supplements Reports that are not part of a memorandum or letter often need to be accompanied by a variety of other documents. A cover letter or letter of transmittal should be sent with all reports so readers know why they have received them. If a report is reader-initiated, a transmittal form may be included in the guidelines. Place a title page at the beginning. If a report is over five pages, include an executive summary or abstract; if it is over ten pages, include a table of contents and lists of figures and tables so readers can quickly and easily locate information in which they're interested. You may also want to include appendices to provide readers with additional information to support your discussion. Other supplements to a report may include a glossary and a list of references. (See Chapter 20 for a complete discussion of global locators and supplements.)

STRATEGY CHECKLIST

1. Write for the point of view of decision makers.
2. Keep interim reports, notes, lab journals, etc., for reference.
3. Use an analytical organizational pattern, i.e., break the information down into major categories.
4. Adapt the following sequence of categories to the appropriate subgenre:
   - Introduction
   - Purpose
   - Description of project/procedures/methodology
   - Results
   - Discussion
   - Conclusion
5. Focus on the goals of the program or activity.
6. Use the conventions of your discourse community.
7. Use headings and subheadings.
8. Use graphics when appropriate.
9. Use lists. Indicate items in a list with numeration or bullets. Use hanging indentation.
10. Use document locators for long reports.
11. Use appropriate document supplements.

INFORMATION REPORTS

Many of the reports you write fall into this category. Information reports are considered final reports because they are written at the conclusion of your data-gathering efforts and present the results of those efforts. These reports are similar to those you have been writing throughout your academic career. However, instead of presenting information for the purpose of a grade, you are presenting information so readers can make a decision and/or use the data in their own work.

Context

All types of readers read these reports, and they have a variety of reading behaviors. You need to keep in mind the strategies to use when writing for multiple readers as you write these reports. For example, before writing legislation to reduce municipal waste, state legislators need information concerning the composition of landfills in terms of weight and volume. The CEO of a firm manufacturing plastic containers needs the same information to argue against legislation affecting plastic food packaging.

Although your purpose is to inform, you may also want to persuade readers to do something or make a specific decision. In these cases, select information that supports your position, or make a recommendation based on your information.

The writer-initiated report on municipal solid waste (MSW) in Figure 19-2 (beginning on page 540) is written not only for representatives of state and local governments, but also for interested business administrators and concerned citizens. The MSW report is written in a political atmosphere; many citizens are apathetic concerning the issue of solid waste, while many companies do not want to spend the time or money to initiate programs to reduce it. The purpose of this report is to persuade decision makers to reduce MSW by providing information on the components of the waste stream. As you read the document, consider how the writers lead up to the purpose. You should also notice how the writers select content that supports their claims.

Textual Decisions

Content The information readers need may require you to gather data from written sources, field sources, or a combination of both. While you may include historical data (see the methods section in Figure 19-2), you need to make certain that your information is as up-to-date as possible, so readers' decisions are based on the most current knowledge available in the field.

Organizational Pattern and Sequence Based on readers' needs, the conventional organizational pattern for information reports includes the following categories in the sequence presented:

Introduction. The introduction delineates the purpose of a report and informs readers why they are reading it. This section also includes background information readers need to understand the remainder of the report.

The purpose or objective of a report is included in this section and is usually placed at the very beginning or end of the section for emphasis. It also provides a frame for understanding the data.

As you study the introduction to the MSW report in Figure 19-2, consider the information the writer includes in the background.

Procedures/methodology.
This section describes the procedures and methods used to gather information. Readers use these descriptions to determine whether or not methods are valid, according to their organizational and professional communities. If they do not believe appropriate methods have been used, they will not accept the results of a report.

As you read the methodology section in Figure 19-2, consider how the writers attempt to persuade readers that the methodology presented is valid.

Body of the report.
This section provides a synthesis of the information you have gathered, and helps readers make a decision. The section should be organized according to categories of information that are related to the objectives, so readers can easily follow your discussion.

The EPA report in Figure 19-2 uses an analytical organizational pattern and divides the information into three major sections: characterization of MSW by weight, characterization of MSW by volume, and projections of MSW generation and management.

Each section is then further divided. The section on MSW weight is divided into two categories: materials and products. These categories are further subdivided. The materials category is organized according to types of materials, i.e., paper, glass, aluminum, ferrous metals, etc., while the products category is organized according to durable and nondurable goods. If we were to create an informal outline of the body of the report, it would look like this:

A. MSW by weight
    1. Materials
        a. paper
        b. glass
        c. aluminum
        d. ferrous metals
        e. ...
    2. Products
        a. Durable goods
            • major appliances
            • furniture and furnishings
            • rubber tires...
        b. Nondurable goods
B. MSW by volume
C. Projections

**Figure 19-2
An Information
Report**

United States
Environmental Protection
Agency

Solid Waste and
Emergency Response
(OS-305)

EPA/530-SW-90-042
June 1990

EPA

Characterization of
Municipal Solid Waste in the
United States: 1990 Update

continued

Figure 19-2
An Information
Report
continued

CHARACTERIZATION OF
MUNICIPAL SOLID WASTE
IN THE UNITED STATES, 1960 TO 2010

TABLE OF CONTENTS

...

iii

continued

Figure 19-2
An Information
Report
continued

Subsection derived
from background
section of report
(see p. 1 of
report).

EXECUTIVE SUMMARY

Many areas of the United States currently face serious problems in safely and effectively managing the garbage they generate. As a nation, we are generating more trash than ever before. At the same time, we are finding that there are limits to traditional trash management practices. As the generation of municipal solid waste (MSW) continues to increase, the capacity to handle it is decreasing. Many landfills and combustors have closed, and new disposal facilities are often difficult to site. As a result, many communities face hard choices when weighing trash management options. Some communities end up paying premium prices to transport their garbage long distances to available facilities. Others try to site facilities nearby and encounter intense public conflict. Of course, not all communities face such problems; numerous communities have found creative solutions through source reduction and recycling programs. Still, for much of the nation, the generation and management of garbage presents problems that require our focused attention.

Identifying the components of the waste stream is an important step toward solving the problems associated with the generation and management of garbage. MSW characterizations, which analyze the quantity and composition of municipal solid waste stream, involve estimating how much MSW is generated, recycled, combusted, and disposed of in landfills. By determining the makeup of the waste stream, waste characterizations also provide valuable data for setting waste management goals, tracking progress toward these goals, and supporting planning at the national, state, and local levels. For example, waste characterizations can be used to highlight opportunities for source reduction and recycling and provide information on any special management issues that should be considered.

Features of This Report

This report is the most recent in a series of reports released by the U.S. Environmental Protection Agency (EPA) to characterize MSW in the United States. It characterizes the national waste stream based on data through 1988 and includes:

- Information on MSW generation from 1960 to 1988.
- Information on recovery for recycling, composting, and combustion from 1960 to 1988.
- Information characterizing MSW by volume as well as by weight.
- Projections for MSW generation to the year 2010.
- Projections for MSW combustion through 2000.
- Projections (presented as a range) for recovery and recycling through 1995.

Unlike previous EPA characterization reports, this report does not include long-range projections for materials recovery. This is due to the significant uncertainties in making those projections. For example, rapid changes are now taking place at the federal, state, and local level that may have profound effects on such projections. In addition, shifts in consumer attitudes and behaviors, industry practices and efforts, and technological advances will greatly influence recovery and recycling. The potential impact of all of these changes is very difficult to predict.

Readers should note that this report characterizes the municipal solid waste stream of the *nation as a whole*. The information presented here may not, therefore, correlate with individual state or local estimates of waste generation and management.

Methodology

Subsection is a
condensed version
of Methodology
section of report
(see p. 1 of
report).

There are two primary methods for conducting a waste characterization study. The first is a site-specific approach in which the individual components of the waste stream are sampled, sorted, and weighed. Although this method is useful for defining a local waste stream, extrapolating from a limited number of studies can produce a skewed or misleading picture if used for a nationwide characterization of

ES-1

**Figure 19-2
An Information
Report**
continued

This subsection is
a summary of the
results of the
study.

waste. Any errors in the sample or atypical circumstances encountered during sampling would be greatly magnified when expanded to represent the nation's entire waste stream.

The second method, used in this report to estimate the waste stream on a nationwide basis, is called the "material flows methodology." EPA's Office of Solid Waste and its predecessors in the Public Health Service sponsored work in the 1960s and early 1970s to develop the material flows methodology. This methodology is based on production data (by weight) for the materials and products in the waste stream, with adjustments for imports, exports, and product lifetimes.

Report Highlights

This report underscores the problems we face in municipal solid waste management: the generation of MSW continues to increase steadily, both in overall tonnage and in pounds per capita. In addition, the report indicates that materials recovery for recycling and the combustion of MSW have been increasing in recent years, while discards to landfills have decreased. Major findings include the following:

- **In 1988, 180 million tons, or 4.0 pounds per person per day of MSW were generated.** After materials recovery for recycling, discards were 3.5 pounds per person per day. Virtually all of these discards were combusted or sent to a landfill.

- **Without source reduction, the amount of waste generated in 1995 is expected to reach 200 million tons, or 4.2 pounds per person per day. By 2000, generation is projected to reach 216 million tons, or 4.4 pounds per person per day. The per capita figure for the year 2000 is a 10 percent increase over 1988 levels.**

- **Based on current trends and information, EPA projects that 20 to 28 percent of MSW will be recovered annually by 1995.** Exceeding this projected range will require fundamental changes in government programs, technology, and corporate and consumer behavior.

Provides a brief
glossary.
Places glossary in a
box so readers
who already know
terms can skip
over them.
Box is used for
subordinating
information.

DEFINITIONS

Municipal solid waste includes wastes such as durable goods, nondurable goods, containers and packaging, food wastes, yard wastes, and miscellaneous inorganic wastes from residential, commercial, institutional, and industrial sources. Examples of waste from these categories include appliances, newspapers, clothing, food scraps, boxes, disposable tableware, office and classroom paper, wood pallets, and cafeteria wastes. MSW does not include wastes from other sources, such as municipal sludges, combustion ash, and industrial nonhazardous process wastes that might also be disposed of in municipal waste landfills or incinerators.

Generation refers to the amount (weight, volume or percentage of the overall waste stream) of materials and products as they enter the waste stream and before materials recovery, composting, or combustion (incineration) takes place.

Recovery refers to materials removed from the waste stream for the purpose of recycling and/or composting. Recovery does not automatically equal recycling and composting, however. For example, if markets for recovered materials are not available, the materials that were separated from the waste stream for recycling may simply be stored or, in some cases, sent to a landfill or incinerator.

Discards include the municipal solid waste remaining after recovery for recycling and composting. These discards are usually combusted or disposed of in landfills, although some MSW is littered, sorted, or disposed of on site, particularly in rural areas...

ES-2

continued

**Figure 19-2
An Information
Report**
continued

Introduction.
Problem.

Organizing idea.
Purpose.
Organizing idea.

Provides purposes
for readers in dif-
ferent types of
organizations.

Methodology.
Forecast.

Uses comparison/
contrast to relate
to readers' logic.
Provides counter
arguments.
Relates to authori-
ty readers consid-
er valid. Uses his-
torical data.

1. Introduction and Methodology

Background

Many areas of the United States currently face serious problems in safely and effectively managing the garbage they generate. As a nation, we are generating more trash than ever before. At the same time, we are finding that there are limits to traditional trash management practices. As the generation of municipal solid waste (MSW) continues to increase, the capacity to handle it is decreasing. Many landfills and combustors have closed, and new disposal facilities are often difficult to site. As a result many communities face hard choices when weighing trash management options. Some communities end up paying premium prices to transport their garbage long distances to available facilities. Others try to site facilities nearby and encounter intense public conflict. Of course, not all communities face such problems; numerous communities have found creative solutions through source reduction and recycling programs. Still, for much of the nation, the generation and management of garbage presents problems that require our focused attention.

Identifying the components of the waste stream is an important step toward solving the problems associated with the generation and management of garbage. MSW characterizations, which analyze the quantity and composition of the municipal solid waste stream, involve estimating how much MSW is generated, recycled, combusted, and disposed of in landfills. By determining the makeup of the waste stream, waste characterizations also provide valuable data for setting waste management goals, tracking progress toward those goals, and supporting planning at the national, state, and local levels....

Methodology

There are two primary methods for conducting a waste characterization study. The first is a site-specific approach in which the individual components of the waste stream are sampled, sorted, and weighed.

Although this method is useful for defining a local waste stream, extrapolating from a limited number of studies can produce a skewed or misleading picture if used for a nationwide characterization of waste. Any errors in the sample or atypical circumstances encountered during sampling would be greatly magnified when expanded to represent the nation's entire waste stream.

The second method—the one used in this report—is called the "material flows methodology." EPA's Office of Solid Waste and its predecessors in the Public Health Service sponsored work in the 1960s and early 1970s to develop the material flows methodology. This methodology is based on production data (by weight) for the materials and products in the waste stream, with adjustments for imports, exports, and product lifetimes...

1

EVIDENCE

CLAIM

CLAIM

EVIDENCE

EVIDENCE

CLAIM
(Implicit:
This is the
best
method.)

**Figure 19-2
An Information
Report**
concluded

**Subsection of
report.**
Introductory state-
ment to this section
of report.
Forecasts informa-
tion to be discussed.

First subcategory.

Second subcategory.

Third subcategory.

---

2. Products in Municipal Solid Waste...

Durable Goods

Durable goods generally are defined as products having a life of
three years or more, although there are some exceptions. In this report,
durable goods include major appliances, furniture and furnishings, and
rubber tires, lead-acid automotive batteries, and miscellaneous
durables (e.g., small appliances, consumer electronics) (See Tables 12
through 14). These products are often called "oversize and bulky" in
municipal solid waste management practice, and they are generally
handled in a somewhat different manner than other components of
MSW. They are often picked up separately and may not be mixed with
other MSW at the landfill, combustor, or other waste management
facility.

Durable goods are made up of a wide variety of materials. In order of
tonnage in MSW in 19xx, these include: ferrous metals, wood, plastics,
rubber and leather, glass, other nonferrous metals, (e.g., lead, copper),
textiles, and aluminum.

Generation of durable goods in MSW totaled 24.9 million tons in
19xx (almost 14 percent of total MSW generation). After recovery for
recycling, 23 million tons of durable goods remained as discards.

Major Appliances. Major appliances in MSW include refrigerators,
washing machines, water heaters, etc. They are often called "white
goods" in the trade. Generation of these products in MSW has
increased very slowly; it was estimated to be 2 million tons in 1988
(less than 2 percent of total). In general appliances have increased in
quantity but not in average weight over the years.

Some ferrous metals are recovered from shredded appliances,
although this quantity is not well documented. Recovery was estimated
to be 200,000 tons in 19xx, leaving 2.8 million tons of appliances to be
discarded.

Ferrous metals are the predominant materials in major appliances,
but other metals, plastics, glass, and other materials are also found.

Furniture and Furnishings. Generation of furniture and furnishings
in MSW has increased from 2.1 million tons in 1960 to 7.5 million
tons in 19xx (about 4 percent of total MSW). No significant recovery
of materials from furniture was identified.

Wood is the largest material category in furniture, with ferrous metals second. Plastics, glass, and other materials are also found.

Rubber Tires. About 70 percent of the rubber used in the United
States is used in the manufacture of rubber tires. Generation of rubber
tires increased from about one million tons in 1969 to 2.2 million tons
in 19xx (about one percent of total MSW). Generation was higher in
the 1970s and early 1980s, but the trend to smaller and longer wearing
tires has lowered their quantities. Small amounts of rubber are recovered for recycling (an estimated 5 percent in 19xx)...

25

Characterization of Municipal Solid Waste in the United States: 1990 Update.
U.S. Environmental Protection Agency

The subcategory on durable goods is presented in Figure 19-2. As you read, note how the writer uses an analytical organizational pattern to further subdivide this subcategory into chunks of related information.

Conclusion/summary/recommendations. This section wraps up the data by relating the results to the purpose or recommendations. The section may be omitted if the purpose of the report is strictly informative. However, if readers have requested recommendations based on the information, or if you want to persuade readers to react in a specific way to the information presented, then include this section. Make certain it is directly related to the stated purpose and also to the specific information presented, so readers can follow your line of reasoning and accept your conclusions and recommendations.

While the EPA report cited in Figure 19-2 does not include this section, a writer-initiated report on leaking underground storage tanks does. The purpose of the report is not only to inform interested citizens about the problem of leaking underground storage tanks (USTs), but also to persuade them to "motivate local UST owners and operators to adhere to regulations, and to ensure that they perform cleanups properly." Like the EPA report, this report is written within a political context. While many citizens may want to prevent leaking storage tanks, the recommendations in this report force companies to spend additional monies and many are not willing to do so. As you read, consider how the writer attempts to persuade readers to accept the recommendations.

Conclusion

EVIDENCE
Summarizes previous information.
Appeals to readers' fears.
Cites authority as evidence.

A large fraction of the country's citizens are likely sometime to experience the effects of a leaking UST system. About half the country relies on groundwater as a source of drinking water. Leaking UST systems are ubiquitous throughout the country.

Unfortunately, EPA's final UST regulations are only a small first step to protect groundwater. Its "detection rather than prevention" regulatory approach is likely to lead to an erosion of groundwater quality, making it less available as a drinking water source. EPA readily admits that with respect to petroleum contamination, "no correct action technology is capable of bringing water quality back to national standards." Nevertheless, EPA's regulations do not require petroleum tanks to have secondary containment with monitoring between the two barriers, the most effective means known for preventing UST system leaks.

CLAIM

Recommendation.

Given the inadequacy of EPA's regulations, it is essential for states and localities to adopt a "prevention before detection" regulatory approach to UST system leaks. Secondary containment is particularly important where groundwater is heavily used or especially vulnerable, where there are nearby surface water intakes, and where water comes primarily from private supplies. Citizen action to encourage states and localities to require secondary containment for new petroleum UST systems is essential (several states and counties already have a secondary containment requirement for new petroleum USTs). Equally important, people need to educate state and local legislators and regulators on how they may avoid the many other deficiencies in EPA's regulations. (See Appendix 2)

Organizing idea.

> Individuals have several means to prevent UST leaks or lessen the environmental harm from leaks:
> 1. Comment on state UST rulemakings;
> 2. Comment on EPA approvals of state UST programs;
> 3. Be aware of local UST system characteristics and report problems to implementing agencies;
> 4. Participate in corrective action plans;
> 5. Use publicity and education; and
> 6. Consider legal actions...
>
> Environmental Defense Fund, *Citizen Action: An Ounce of Prevention*, (1990).

Style Readers expect the data in information reports to be true and accurate. Therefore, the tone should be formal to imply a professional stance, and third person should be used to reflect objectivity.

Because information reports are usually written for generalists or novices, technical terms should be used only when necessary, and when used, they should be defined as in Figure 19-2. Sentence structure should be straightforward Active voice is preferred.

STRATEGY CHECKLIST

1. Determine the information to include based on readers' needs.
2. Use the following organizational pattern and sequence to present the information:
   - Introduction
   - Objective/purpose of report
   - Procedures/ methods for gathering information
   - Body of report
   - Conclusions/summary/recommendations
3. Synthesize information you have gathered in the body of the report.
4. Organize the information in the body of the report according to the readers' purpose and the objectives of the report.
5. Use a formal, neutral, and impersonal tone.

FEASIBILITY REPORTS

Feasibility reports, which are almost always reader- initiated, present the results of investigations to determine whether or not a project should be undertaken, a product developed, or an event initiated.

For example, before a government agency constructs a new road, it conducts a feasibility study to determine whether or not the road should be built. The agency needs to know whether present traffic warrants a new road and whether future traffic trends will sustain the need. The agency also needs to know how the new road will affect residents and businesses. The document in Figure 19-3 (page 550) reports the results of that study.

Context

In determining whether or not to initiate, continue, or change a project, product, or situation, readers want to know whether others have attempted to solve similar problems in a similar fashion, what the results of these attempts have been, and whether more effective or cost-efficient alternative solutions exist.

Primary readers of feasibility studies are often generalists, administrators interested in initiating a project that others, often experts in the role of secondary readers, actually carry out.

Primary readers, who determine whether or not a project should be initiated, often read only the executive summary of a report. If they do look at the main text of the report, they often read the recommendations first. They may skim through the rest of the document to search for information in which they are especially interested, or to determine whether a study was thorough and conclusions warrant the recommendations. Or they may never read beyond the recommendations.

Reports may also be read by secondary readers, those persons who will be responsible for carrying out the project. These readers study the description of the proposed project, and the conclusions, to learn the strengths and weaknesses of the project.

If the report is for a client, intermediary readers, such as a supervisor, study the report before it is sent out to make certain the study is thorough, conclusions valid, and recommendations appropriate. If the study involves a public problem, peripheral readers may read it. While the news media may skim through the introduction and recommendations, citizens who are directly affected by it may study it carefully, and if they decide to protest the project, they may use the report as a legal document.

Because feasibility studies may affect people other than those initiating a specific project, these documents are often politically sensitive. People who are negatively affected by a proposed project will probably react negatively to a report. For example, if a road is recommended, the residents whose homes may be in the way may be hostile to the document.

Textual Decisions

Content   Some of the information readers want may be found in written sources. You may also need to get data through field sources. Use recent information so your readers' decisions are based on the current situation. You may want to do a small, pilot project to determine the results. If the project impacts many people, you may want to survey them, and you may want to conduct focus groups to learn what others who have had similar problems would recommend.

The historical data included in the introduction to the report in Figure 19-3 was compiled from various documents of the Federal Aviation Administration. The remainder of the information in the body of the road feasibility report was obtained by conducting observations, land surveys, and other field studies.

Organizational Pattern and Sequence Because readers may respond negatively to the recommendations in a feasibility report, recommendations are placed at the end so readers can understand and accept the writer's conclusions.

Based on readers' needs, the conventional organizational pattern for a feasibility report includes the following categories in the sequence indicated.

Introduction. This section states the problem or need that a potential project will address, and the purpose for the report.

In the introduction to the report on the feasibility of altering an access road to an airport (Figure 19-3), notice that the purposes for the report are specified in the form of objectives, and appear at the conclusion of the introductory section.

Procedures/Methodology. This section describes the procedures and methods used to determine whether or not the proposed project meets the needs or solves the problem. If readers do not believe the procedures and methods are appropriate, according to their professional and organizational communities, they will not accept the results.

In the methods section of Figure 19-3, consider how the writer attempts to persuade readers that the methodology is valid. Notice, too, that the writer does not explain each method in detail, since readers are not research specialists.

Results. This section provides a neutral, formal description of the results of the study in relation to the purpose. The section is usually divided into two major categories, present conditions and alternatives for changing those conditions. The latter category is usually organized analytically, with each alternative listed and discussed in terms of its good and bad features.

Notice how the results section of the airport road access study adheres to the conventions.

Conclusions/Recommendations. This section presents recommendations for going ahead, revising, or discarding a project or plan. It is directly related to both results and objectives.

In the recommendations section of the report on the access road to the airport, the writer attempts to persuade readers to accept a specific recommendation rather than allow them to arrive at their own conclusions. As you read, consider the strategies the writer uses to persuade readers to accept the recommendation.

Style Because the primary reader for a feasibility report is usually a generalist, jargon should be eliminated. Technical terms should be used sparingly, and defined when used.

Feasibility reports are usually written in a formal tone. The introduction, procedures/methods, and results sections should be written impersonally and in neutral terms. However, in the discussion/conclusion and recommendations sections, you can be personal, basing the interpretations on your own experience and knowledge. These latter two sections often use first person plural to refer to the organization conducting the study.

**Figure 19-3
A Feasibility Study**

Background.
Historical data.

Purpose.

Objectives.

Methodology.
Organizing idea.
General description. No details.

Results.
Present conditions.

HIGHWAY CIRCULATION AND ACCESS NEEDS FOR THE GREATER PEORIA REGIONAL AIRPORT

INTRODUCTION

The Greater Peoria Airport has experienced considerable growth over the past decade as evidenced by the fact that the annual volume of enplaned passengers increased from 141,400 to 375,000. The consequences of this growth are evident: increased employment at the airport, higher roadway traffic volumes, more development in the vicinity of the airport, etc. To meet these growing traffic demands some roads have already been improved in the area, most particularly the widening of Middle Road and the construction of I-474 and the Maxwell Road Spur.

To further assess highway and traffic needs within the vicinity of the airport, Peoria County Highway Department and the Greater Peoria Airport Authority recently completed a detailed study of this area (Exhibit 1). This study had the following objectives:

1. Analyze the opportunity for improved vehicular access to and around the airport

2. Evaluate the need for roadway improvements to accommodate future traffic growth

3. Determine the most feasible alignment for relocating Smithville Road around the South end of the airport to accommodate the extension of Runway 4-22...

METHODOLOGY

Existing transportation conditions within the study area were inventoried to establish a data base with which to make a quantitative assessment of transportation needs. This database included: interviews with appropriate public officials; an evaluation of existing land use conditions; an inventory of the present transportation system; traffic and accident data describing the present operating patterns within the area; and new traffic data describing the present operation.

Using this data, the consultant evaluated existing transportation conditions to identify present traffic problems. Traffic forecasts through the year 2000 were also prepared for this data....

RESULTS

Existing Conditions

Data shows that traffic volumes have not increased over the past two years. Along some roads, in particular Cameron Lane, Maxwell and

CLAIM
↓
EVIDENCE

CLAIM
(Implicit:
these
are valid
methods)
EVIDENCE

CLAIM I

continued

**Figure 19-3
A Feasibility Study**
continued

Alternatives.
Relates to
Objective 2.
Background.
Relates to
Objective 1.
Forecasts alterna-
tives.

Alternative 1.
Description.
Good features.
Bad features.

Alternative 2.
Description.
Good features.
Bad features.

**Recommenda-
tion.**
Relates directly to
results.

Recommendation
relates directly to
Objective 1.

Smithville Road, volumes actually decreased significantly. This reduction in flow can probably be attributed to the opening of I-474. Some
motorists which used these county roads as a means of bypassing the
southwest portion of the urban area were attracted to this interstate
facility when it opened...

EVIDENCE

[Potential] Improved Roadway Access to Greater Peoria Airport

Airport traffic volumes are expected to double over the next 20 years.
Counts show that, at the present, 80% of the motorists enter and exit
via Middle Road. This road, and to some degree, Maxwell Road service considerable residential development. Consequently the feasibility
of developing another entryway to the airport as a means of avoiding
these residential areas was studied.

Two roadway alignments for connecting the airport entrance to Middle Road were considered. Both of these alignments would bypass
most of the vehicular and pedestrian traffic in the subdivision thereby
reducing the potential for hazardous conditions to develop in the future
as airport traffic increases. The plan and profile for each alignment is
shown in Appendix B.

CLAIM 2

EVIDENCE

Alternate AA intersects the present alignment of Middle Road at
Edwards Street. It continues west through the center of the subdivision
along the back property lines of the homes facing Middle Road and
intersects Maxwell Road slightly north of the airport entrance. Its
alignment avoids the pump station recently built near Maxwell Road
but does take two homes: one home adjacent to North Airport Road
and a second which has recently been built near Station 20. This
entrance road is estimated to cost $875,000, including right-of-way.

Alternate BB which generally passes between the subdivision and the
airport would cost approximately $600,000 to construct; $700,000 if a
secondary road is built to service the homes adjacent to the airport
from the rear. Use of this secondary road would eliminate the need for
residential driveways onto the airport entrance road, thereby avoiding
these traffic conflicts. However, a number of garages and driveways
would have to be reoriented towards the back of the residential lots.
Homeowners would be reluctant to make these changes because of the
cost involved...

RECOMMENDATION

The plan and profile sheets show that Alternate BB can be implemented without taking any homes. Consequently, this alternative will
be easier to implement than the other alignment. Furthermore, cost
studies show Alternate BB is less expensive, particularly if the secondary road is omitted. For these reasons, Alternate BB is the better
alignment. When the need for a new airport road arises, this alignment
should be selected for implementation.

EVIDENCE

CLAIM

Crawford, Murphy, & Tilly, Inc., 1991. Highway Circulation and Access Needs
for the Greater Peoria Regional Airport, Peoria, IL

**STRATEGY
CHECKLIST**

1. Include information detailing attempts to solve similar problems, if applicable, and describing alternative solutions.

2. Use the following organizational pattern and sequence to present the information:

   • Introduction—establish problem or need

   • Description of procedures/methods used to study proposed project

   • Results

   • Recommendations

3. Use a formal, neutral, and impersonal tone.

4. Use strategies for writing for a nonsympathetic audience if you are aware that your recommendations may negatively affect some people.

EVALUATION REPORTS

An evaluation report is presented to a reader at the conclusion of an assessment of a situation, person, or project, so that a reader can make a decision. You may be asked to make a recommendation in your report, or you may only be asked to provide the assessment information that allows readers to make their own decisions, as the writer for the report shown in Figure 19-4 (page 555) was asked to do. The writer, an automotive consultant, was contracted by an insurance company to determine the cause of a fire.

Because evaluations of projects often involve a team of evaluators, you may write these reports collaboratively. Once your team has completed its investigation, all members write a draft of their own observations. You then meet to share observations and to come to a consensus on an overall assessment of a project. The team leader may write the final document or assign someone else to do so. The report is submitted to all team members for final approval before it is sent to the primary reader. Team members look it over for accuracy and check that conclusions agree with those made during the team meeting.

Context

You will address a variety of readers, ranging from experts to novices, from administrators to regulators to users, from internal managers to external clients. However, they are all decision makers.

Evaluation reports are almost always reader-initiated. Readers request personnel or consultants to make an assessment and report the findings. Readers use the information to determine how to proceed in a situation, in relation to a person, or with a project. The insurance company that commissioned the report shown in Figure 19-4 will use the information to determine whether to pay for the damage, or to subrogate (transfer a claim to another party) a manufacturer for a faulty product.

Evaluation reports are usually read by multiple readers. The primary reader, a manager, will read the executive summary. He may also read the recommendation section. The person or persons being evaluated, or who are involved with the situation or project being evaluated, act as intermediary readers. They usually study the entire report to determine your accuracy in presenting the facts, as well as your bias in the evaluation. They read the report with skepticism at best, and hostility if they find your point of view does not coincide with theirs. If actions are to be taken, the report may be read by secondary readers who study those sections pertinent to their particular responsibilities. If the situation or project involves the public, then peripheral readers, such as the news media, may skim through the findings and the recommendations. These reports can be used in lawsuits to substantiate problems. If the automotive consultant finds that the fire was caused by a faulty product, then the manufacturer will also be a primary reader for the report. And if the manufacturer questions the findings, then the report may be used as evidence in a lawsuit and read by attorneys.

Evaluation reports are, by their very nature, politically sensitive, because they can affect people negatively. If you evaluate a project and find it isn't meeting its objectives, the project may be halted, or workers who are unnecessary to the process may be removed. If the consultant for the insurance company discovers the automotive fire was caused by a faulty mechanism, the manufacturer could be sued.

Evaluation reports are implicitly persuasive; the writer must persuade readers to accept the findings. The consultant's findings may eventually be used to persuade a court to find a manufacturer guilty in a product liability case.

Textual Decisions

Content Readers want to know if people are achieving their goals, if products are functioning effectively, if situations are appropriate, and if not, why not. Therefore, most of the information in an evaluation report is gathered from field resources as you investigate the specific person, product, or situation. You may need to observe a situation, interview people involved, and possibly test different solutions. You may need to use written sources to determine criteria for judging a project or to learn of alternatives. The automotive consultant combines his previous experience and knowledge with observations of the actual vehicle and a series of laboratory tests to evaluate the situation leading to the automotive fire.

Organizational Pattern and Sequence Based on readers' needs and purposes, the conventional organizational pattern and sequence includes the following categories:

Introduction. This section describes the situation, project, or person, and the reason for the evaluation. This section is often short, since the primary reader usually knows the background. (See Figure 19-4.)

Procedures/methodology.   This section describes the procedures and methodology used in the evaluation. Sometimes the section is omitted and the information is integrated into the results section.

If a single method is used, a special section is usually preferable, so readers can assess the validity of the method before reading the results. However, if various methods are used to evaluate different aspects of a situation or product, then the discussion of the method should be integrated with the discussion of the results. Thus, readers can easily see the relationship between a method and the resulting data, and can determine whether the results are based on procedures considered appropriate by their discourse communities.

This section is usually short. The primary reader is probably not an expert in evaluation techniques. Readers only need to know the general method or procedure.

In the letter investigating the car fire (Figure 19-4), the writer uses a deductive organizational pattern. He provides a brief overview of his methodology, and then combines the specific procedural steps he took with his findings for each step. Consider how he uses a cause/effect organizational pattern to explain his procedure and discuss his findings. Notice also how he attempts to persuade the reader to accept his methodology by indicating that he is doing exactly as the reader requested.

Results/evidence/findings.   This section describes the situation, project, mechanism, or person being evaluated. The section is often very detailed, so that readers can accept the conclusions. It is organized into categories of related information that are based on the purpose of the report and the subject of the evaluation.

In the text in Figure 19-4, notice how the writer uses visual text and graphics to enable the reader to read and understand the findings easily. The photographs not only complement the verbal description, but also provide additional support to the writer's claims. This can be especially important if the document is used in a legal suit later on.

The writer uses an analytical organizational pattern, dividing the section into the main mechanical components involved. He has photographed each component, and numbered each photograph. The numerical designations serve as markers to cue the reader.

Conclusions.   This section presents a discussion of the results of the evaluation, based on the information in the previous section and the writer's personal interpretation of these results. The conclusions relate directly to the purpose.

Notice that in the conclusion of the insurance report in Figure 19-4, the writer organizes the section as a modified chronology, using a cause/effect pattern within each chronological chunk.

**Figure 19-4
An Evaluation
Report**

Follows conventional
letter format.

DR. CHARLES W. PENDLETON   **Automotive Consultant
Component Failure Investigations**

November 19, 19xx

Mr. George Balta

USF&G

3100 N. Knoxville

Peoria, IL 60603

Subject: Claim # MP0694839579

Dear George:

Purpose.
Main organizing idea.

This is the report that you requested of my investigation of the fire damaged 1988 Chevrolet Caprice police car at Meyer Wrecking Service in Monmouth, Illinois. You indicated via telephone that I should inspect the vehicle to ascertain why an electrical fire caused extensive damage to the dash area of the vehicle and try to pinpoint the cause.

You suggested that I might want to inspect a new police car and consult with the service personnel at Miles Chevrolet in Decatur first. The purpose would be to identify where and how additional wiring and components were installed by their technicians on the vehicle in question before its delivery to the Monmouth Police Department.

CLAIM
(Implicit:
These
procedures
are valid.)
EVIDENCE

Methodology.
Analytical pattern.
Categories divided by
car.
Car 1.
General description of
procedures.

I visited Miles Chevrolet in Decatur on Friday, November 4, 19xx, and familiarized myself with the additional circuits, relays, circuit breakers, etc., needed to properly equip a modern patrol car...

Results.
Organized analytically
by finding.
Finding.
  Detailed description.

Results

1. This police vehicle is similar to the fire damaged vehicle. (See Photo 1.) It was provided by Miles Chevrolet for inspection. The vehicle had all of the wiring and components needed for delivery to a police department. Some of this special police equipment was provided by the factory and some installed at Miles.

Finding.
  Detailed description.

2. The "lock-out" device was installed by Miles Chevrolet personnel. (See Photo 2.) Four wires from this device run to the ignition switch. The switch when "on" allows the engine to keep running while the ignition keys are removed and while the vehicle doors are locked. Next to it is a factory installed throttle control. The wires for the "lock-out" were routed over the steering column in an orderly fashion.

Finding.
  Detailed description.

3. The fuse box had additional wiring added as indicated by the three arrows. (See Photo 3.) These were factory installed for the vehicle's spot light and other special equipment...

Car 2.
Procedures.
General description.
Findings.
Detailed description.

On Saturday, November 5, 19xx, I inspected the damaged Caprice at Meyer Wrecking Service in Monmouth in the presence of Cpt. Frank Piper of the Monmouth Police Department...

I checked under the hood to trace the electrical problem from the source (battery or alternator) to the problem area. Everything was in order except the electrical wiring around the wiper motor. The wiper motor is mounted to the engine firewall. The wires from the motor to the wiper assembly were burned, especially the ground which had its insulation burned off...

Conclusions.
Follows cause/effect
organizational pattern.

Conclusions

The police officer was driving the car with the wipers in the "low" position when the wipers stopped and the officer noticed a red glare and

EVIDENCE

continued

Figure 19-4
An Evaluation
Report
continued

Incident 1
 (wipers stopped).

Incident 2
 (stall condition).

Incident 3
 (overheating).

Incident 4.

Relates to purpose.
Closing.

Photos.

smoke in front of the windshield. An interior electrical malfunction had occurred. The wiper motor had stopped because the relay rods had contracted the plastic wiring loom that runs behind the wiper motor. This interference prevented the wiper motor from turning, causing a stall condition.

Once the motor was stopped, the current approached the stall limit. The stall current in low speed operation is 20-25 amps in a fused 25 amp circuit. Normally the copper ground strap at the wiper assembly can handle this current easily if connected tightly. However, I feel that the ground strap attachment bolt was not tight and consequently caused a high resistance. With nearly 25 amps at 15 volts trying to flow through the straps, a considerable amount of heat and burning took place at the mounting bolt and strap. This caused the red glow the officer reported.

Since the motor was stalled, the high current overheated and melted the wiper terminal board, the wire connector, and its wires. (The wire size for the "low" circuit is small because it normally doesn't carry over 4 amps.) This shorting allowed an electrical overload of the "low" speed control wire that grounds the wiper switch on the steering column turn signal stalk.

Since this small diameter wire is shorted to a larger feed wire and since there is virtually no resistance, it will catch on fire anywhere from the turn signal handle to the wiper motor connection short. When the insulation ignites around the shorted wire, the heat and burning is transmitted to other adjacent wires in the loom and more fire and short circuiting occurs which is the reason the fire damage was so severe in the dash area…

I feel the fire was caused by the above series of events. If further information is needed, please contact me.

Respectfully submitted,

Charles Pendleton, Ph.D.
Automotive Consultant
CWP:bp
Enc.

CLAIM

EVIDENCE

CLAIM

EVIDENCE

CLAIM

EVIDENCE

CLAIM

Photo 1

Photo 2

Photo 3

Recommendations. This section offers suggestions for correcting the problem. These suggestions are often divided into short- and long-term corrective actions. They are usually not detailed.

This section is optional. Sometimes readers do not want recommendations. The insurance company did not want the automotive consultant to recommend whether to pay or to subrogate, so the section is omitted.

The following recommendation section is excerpted from a report by a consultant who was requested by a manufacturer to evaluate the cause of a problem in one of its products. A "through crack" was found on a $2\frac{1}{2}$-inch-long section of a $\frac{3}{4}$-inch outer diameter brass condenser tube, which the investigator determined was caused by corrosion fatigue.

The manufacturer will use the results of the report to determine if the problem lies in its own processes or in the material. If the former, the manufacturer needs to take corrective action; if the latter, the problem can be turned over to the supplier.

The writer provides recommendations for preventing similar problems. Notice how the writer supports his recommendations with factual data.

Relates to conclusion. Recommendation.	Cyclic tensile stresses produce corrosion fatigue cracking. Therefore, to slow and stop cracking requires decreasing or eliminating cyclic tensile stresses. This in turn requires that tube vibration and/or thermal gradients be minimized.

Style Technical language is used to avoid any misunderstanding, since the report may be read and interpreted by experts if it is used in a lawsuit. While the language is technical, technical terms are defined for readers who are generalists. Sentence structure is straightforward and short. The style is formal, neutral, and impersonal. Many evaluation reports use the passive voice. However, in many cases this is unnecessary and causes awkward sentence structure.

STRATEGY CHECKLIST

1. Use the following organizational pattern and sequence to present the information:
   - Introduction—reason for evaluation
   - Procedures and methodology used in the evaluation
   - Results/evidence/findings
   - Conclusions
   - Recommendations
2. Use a neutral, impersonal, and formal style.

RESEARCH AND DEVELOPMENT REPORTS

Research and development reports provide information on the results of a research or development project. Readers use the results to improve their own research; prove that a prior study is valid by replicating it; develop or improve products or situations; invest in the resulting products; improve the quality of life; and expand their knowledge.

Unless you are in a research and development division, you will rarely be involved in writing a formal research report. However, you may be part of a project, and you may be expected to write a report at the completion of the project for your supervisor, and for any agencies funding the project or expecting to use the results. Readers may use the information in their own work, or they may use it to determine support for future projects proposed by you or your organization. Furthermore, the information may be used by the marketing or public relations department of your organization to prepare publicity materials.

Context

Research reports are read mainly by experts who usually study the abstract first. Then, if they are especially interested in the topic, they read the report in its entirety. They are often as interested in the methods and procedures as they are in the results and recommendations. They read a report to evaluate its validity, and to determine if procedures, methods, and statistical analyses are valid and reliable, according to the criteria of their discourse communities. Experts also determine if conclusions appear to be logically deduced.

Generalists interested in reading the report are usually researchers who may be involved in the general field, or who are involved with the development, production, or utilization of products related to the project or the subject being studied. They are usually concerned only with the conclusion and recommendation sections. Their purpose in reading the report is to evaluate the findings and to determine if they can adapt them to their own work. They often skim through the introduction, then study just the final two sections, i.e., conclusions and recommendations.

Novices usually read a research report to locate specific information related to their own interests. They are not usually interested in the details of how the information has been obtained and analyzed. Rather, they are looking for a general understanding of the topic and the key findings. Therefore, they are primarily concerned with the introduction, conclusion, and recommendation sections, which they may skim until they locate the particular information they need.

These reports may be politically sensitive. Whether or not a project is re-funded, or personnel associated with a project receive funding for new projects, may depend on the results documented in a report. In addition, some people may not approve of a project. Furthermore, experts are almost always skeptical, unwilling to accept new results unless procedures and methodology are valid and reliable, and all claims are supported according to the conventions of that discourse community. Thus, the writer needs to persuade readers to accept the conclusions.

The document in Figure 19-5 (page 562) reports on a study to estimate the volume of municipal solid waste and selected components in trash cans and landfills. It represents a collaborative effort by consultants, university specialists, and government employees. The study was administered and reported by consultants hired by the Council for Solid Waste Solutions, which is associated with the plastics industry. While the Council is the primary reader, secondary readers, such as government representatives, manufacturers, and interested citizens, will be given copies of the report to use in making decisions related to solid waste disposal in general, and to the manufacture and purchase of plastic products in particular.

The purpose of the study is to provide persons concerned with solid waste disposal with a valid database for characterizing the various components comprising MSW by volume. Because of trends to reduce the use of plastics, the Council has an additional purpose. It wants to persuade manufacturers to continue using plastic in products, and the general public to continue buying plastic products, by convincing them that plastics comprise only a small amount of solid waste; that other materials contribute to the problems of solid waste disposal to a much greater extent. Notice how the writer leads up to the purpose of the report.

Textual Decisions

Content   The information in a research report is based on your field research. In addition, written sources serve as authoritative evidence to support your claims in your discussion and conclusion sections. Research communities require that you relate your results to results of similar research studies, and that you interpret your results in the light of the general field as is done in the MSW report in Figure 19-5. You should cite the most recent research available. If your report is to be read by generalists and novices, you may want to provide some historical information in the introduction.

Organizational Pattern and Sequence   Based on readers' needs, research and development reports include the general categories and follow the same sequence as that described for general reports on pages 535-537.

Introduction.   This section states the problem or need addressed in the report. It establishes the reason the project has been undertaken, and provides sufficient background for readers to understand the remainder of the report. The information may include written sources, describe observations you've made, cite statements by people you've interviewed, and refer to results of surveys or focus groups you've conducted.

Academic and scientific reports for experts often include a review of the literature that summarizes the most recent information on the topic. The following introduction to a report, which was published in a professional journal, provides such a review. It is extremely short because of the limited number of studies conducted on this topic. A longer review would be necessary for a subject that has been studied more. As you read, notice how the writer cites the written sources:

EVIDENCE
  Introduction.
  Organizing Idea.
  Problem.
  Background.
  Review of the
  literature.
CLAIM

Of the South American canids, perhaps the least studied is *Dusicyon sechurae*, the Sechuran or Peruvian desert fox. Smaller than its mountain dwelling congener, *D. culpaeus*, it inhabits the coastal desert and lower western slopes of the Andes (Grimwood, 1968). Previous reports are limited to winter diet (Huey, 1969) and an account of the behavior of captive female pups (Birdseye, 1856). Herein we describe the activity pattern and diet of D. *Sechurae* before and after El Niño rains.

Cheryl S. Asa and Michael P. Wallace, "Diet and Activity Pattern of the
Sechuran Desert Fox (Dusicyon Sechurae)," Journal of Mammalogy, 1990.

Notice that because the MSW report in Figure 19-5 is intended for generalists as well as experts, most of this reference material is summarized in the introduction, rather than discussed in detail.

Objectives. This brief section specifies the purpose of the study. It clarifies the logic of the following sections, all of which are directly related to the objectives.

In a scientific or academic report for experts in a discourse community, this section is usually a single paragraph. The conventions for this section include using full sentences and beginning the sections with the phrase, "The objective of this study is/was…"

The following text was included in a report of a research project conducted under the auspices of the Alabama Agricultural Experiment Station. Notice how the statement complies with scientific conventions:

The objective of this project is to develop "lean" (90-95 percent) ground beef products with significantly reduced fat levels, which are as acceptable to the consumer in the same form as current ground beef items.

Dale L. Huffman, Ph.D. & W. Russell Egbert. Advances in Lean Ground Beef Production.
Alabama Agricultural Experiment Station.

In reports that may be read by generalists and novices, this section is often integrated with the introductory section, just as it is in the MSW report.

Procedures/methodology. This section describes the procedures and methods used in a study. The section needs to be sufficiently detailed to persuade readers that the procedures and methods are appropriate. Readers will use criteria from their discourse communities to judge the validity of the methods and procedures. In a scientific or academic research paper for experts, this section should be sufficiently detailed for another researcher to replicate the study.

As you read this section in the report shown in Figure 19-5, you should recognize that the writer is attempting to persuade readers that the procedures are valid according to criteria accepted by the scientific community. Notice that the information is detailed so readers not only accept the procedures as appropriate and sufficient, but so they can replicate the study.

Results.   This section presents a quantitative description of the results of a study in relation to its objectives.

In Figure 19-5, the writer relates the information in this section to the objectives.

Discussion/conclusion.   This section helps readers understand the data in the results section by synthesizing the data and then interpreting it in relation to the objectives. The claims made in this section must be supported by the data in the results section. You can use adjectives and adverbs to emphasize points and to achieve implicit objectives.

In the discussion of MSW in Figure 19-5, the writer not only comments on the database of solid waste by volume, but also on the specific findings of the volume of plastic in relation to the volume of other MSW components. As you read, consider how the writer supports the claims with evidence from the results section. You should also compare the content and style of this section with that of the results section.

Recommendations.   This section provides readers with suggestions for additional or continued research in the specific area, as well as for possible development of products based on the results. While this section is optional, it is included in almost all scientific and academic studies. It is also included in reports of projects that should be continued. However, it is not usually included in reports of projects that have been completed, Thus, it is not included in the MSW report.

The following text is excerpted from a scientific study to determine how juglone, an extract of walnuts, would affect the growth of wheat seeds. Notice how the writer follows the conventions for the recommendation section.

EVIDENCE

↓

CLAIM
Recommendations.

The ecological significance of experimental data shows simply that, if placed in a moist, moderate to high concentration of juglone, wheat will exhibit reduced germination and growth. This alleopathic effect may not have been exhibited if the wheat was grown in an environment that was dry or that involved several times the original seed number. Additional studies involving such variables would need to be implemented to test the existing alleopathic effects of juglone.

ESTIMATES OF THE VOLUME OF MSW...

Introduction

Introduction. Problem.	The current intense nationwide interest in municipal solid waste management, which began to accelerate in 1986, has stimulated the demand for factual information of all kinds regarding MSW (Municipal Solid Waste). One of the primary needs is reliable information on the contributions of various materials and products to the total quantities of MSW that must be managed.

Introduction.
Problem.

The current intense nationwide interest in municipal solid waste management, which began to accelerate in 1986, has stimulated the demand for factual information of all kinds regarding MSW (Municipal Solid Waste). One of the primary needs is reliable information on the contributions of various materials and products to the total quantities of MSW that must be managed.

Background.

Municipal solid waste can be measured by weight and volume. In practice, some landfill operators charge on a volume basis (cubic yards). However the incoming wastes are measured, landfill lifetime is based on the volumes of waste that are received, compacted, and covered for the long-term disposal. The volume measurement is thus very important to solid waste management planners, whether they are dealing with landfilling or the alternatives: source reduction, recycling, composting, or burning in waste-to-energy incinerators.

EVIDENCE

Measuring the weight or volume of mixed municipal solid waste provides no insight into the contribution of the individual components—products made of paper, plastics, metals, glass, etc., in the MSW. There are two ways to estimate the weight percentages of MSW components. The first is to sample, sort, and weigh the various components at the landfill or elsewhere. The second is to perform a materials flow methodology which is based on national production data for the MSW components, adjusted for imports/exports and other factors. MSW sampling studies have been done at numerous locations. In addition, there is a widely-used national database utilizing the materials flow methodology to characterize the components in MSW by weight for the years 1960 to 2000; this database has been developed, updated and refined by Franklin Associates, Ltd. for the U.S. EPA (and others) over a period of many years.

Refers to database.
Authorities accepted
by scientific discourse community.
Summarizes information; no specific
details or discussion
of published literature.

However, there has been no systematic database characterizing volumes of the various components of MSW. As a result, many estimates have been made and published, and decisions regarding solid waste management have been made without any real scientific basis. This report presents the results of a study sponsored by the Council for Solid Waste Solutions, which presents the first comprehensive systematic characterization of the relative volumes of the components of MSW. The report describes the development of an experimentally derived set of conversion factors which enable data from the MSW-by-weight database to be converted to a volume database. Results of the analysis are also presented...

CLAIM

Purpose.
Forecast.

Procedures

Procedures.
Review of the literature.

The first step of the analysis was to search the historical literature for volume data and other information. Then, telephone and personal interviews were conducted across the country to find all available information on this subject...

CLAIM

EVIDENCE

Organization and
personnel.

In order to develop a consistent, scientific, and more reliable database for the volume of materials in solid waste, an experimental program was developed as a joint project between FAL [Franklin Associates, Ltd.] and the University of Arizona. The experiments were carried out by the staff of The Garbage Project of the Department of Anthropology, Bureau of Applied Research in Anthropology...

Figure 19-5 A Research Report

Methodology.

Methodology

The goal was to sort wastes obtained from household trash bags picked up from the curb into nine categories, and to compress and crush the samples taken from each category in order to develop a reproducible compaction database which could be used to develop trash can and landfill densities. Weight to volume relationships were obtained by finding the sample density (pounds per cubic yard) under a wide range of conditions. Similar experiments were conducted on materials obtained from landfill excavations in order to establish the validity of the experimental procedures...

Equipment.

A hydraulic compression machine designed to compress trash was used by the University of Arizona to carry out these studies.

Calculations.

Using the waste material density values supplied by the Garbage Project (Appendix B) as well as other data sources, density factors were determined for 23 material and product categories in trash cans and landfills...

Perhaps the best validation of the plastics values is The Garbage Project historical database...The average weight percent of plastics was 5.7 and the volume percent was 12.2, leading to a landfill ratio of 2.1. These...results compare well with our results of 6.2 percent by weight and 16 percent for volume for MSW...Our values are higher, but because of the low number of samples for The Garbage Project data, these values appear to be within experimental ranges....Although error analysis of a complex set of numbers with widely varying sources and accuracy is not straightforward, we believe that the results and conclusions presented in this study are accurate to *better than* ± 20 percent...

Results.
Relates to primary purpose.

Results

Applying our volume factors to MSW *excluding* durables results in the calculations summarized in Table 7...

Graphics used to summarize data concisely.

TABLE 7

TRASH CAN AND LANDFILL VOLUME FOR MSW (EXCLUDING DURABLES) - 1986

	Discards (mil tons)	Weight % of Discards	Average Trash Can Density (lb/cuyd)	Trash Can Volume (mil cuyd)	Trash Can Volume (%)	Average Landfill Density (lb/cuyd)	Landfill Volume (mil cuyd)	Landfill Volume (%)
PACKAGING								
Glass Containers	10.7	8.8	654	32.7	1.7	2,816	7.6	2.8
Steel Containers	2.7	2.2	212	25.5	1.3	557	9.7	3.6
Aluminum	1.1	0.9	54	41.0	2.1	310	7.1	2.6
Paper and Paperboard	20.4	16.8	44	935.6	48.4	764	53.4	19.6
Plastics	5.6	4.6	53	210.4	10.9	356	31.5	11.5
Wood	2.1	1.7	600	7.0	0.4	792	5.3	1.9
Other Misc. Packaging	0.2	0.2	200	2.0	0.1	1,000	0.4	0.1
Packaging Subtotal	42.8	35.2	68	1,254.2	64.9	744	115.0	42.1
NONPACKAGING PRODUCTS								
Nondurable Paper	29.7	24.4	170	349.3	18.1	798	74.4	27.3
Nondurable Plastic	2.0	1.6	69	58.0	3.0	313	12.8	4.7
Apparel	1.8	1.5	48	75.0	3.9	435	8.3	3.0
Other	2.0	1.6	133	30.1	1.6	392	10.2	3.7
Nonpackaging Subtotal	35.5	29.2	139	512.4	26.5	672	105.7	38.7
NONPRODUCT WASTES								
Yard Wastes	28.3	23.3	500	113.2	5.9	1,500	37.7	13.8
Food	12.5	10.3	500	50.0	2.6	2,000	12.5	4.6
Other	2.6	2.1	2,500	2.1	0.1	2,500	2.1	0.8
GRAND TOTAL	122	100	126	1932	100	892	273	100
PAPER AND PLASTIC SUBTOTALS (PACKAGING + NONPACKAGING)								
Paper	50.1	41.2	78	1,284.9	66.5	784	127.8	46.8
Plastic	7.6	6.2	57	268	13.9	343	44.3	16.2

Note: For more detail see Appendix Tables A-7 and A-8.

continued

Figure 19-5 A Research Report *continued*

Relates to secondary
purpose.

Conclusion.
Discusses results
relating prior weight
data to the new vol-
ume data.

> An important calculation that can be made from Table 7 is the vol-
> ume percent to weight percent ratio. For plastics, these are 2.2 in the
> trash can and 2.6 in the landfill (excluding durables)…
>
> **Conclusion**
>
> The traditional use of weight factors to characterize solid waste dif-
> fers greatly from the volume perspective. The ratios of the volume per-
> cent to weight percent show this clearly. In trash cans, glass, metal, and
> other packaging (primarily wood) have ratios less than one, which
> means that they occupy little space in the trash cans. These three cate-
> gories together account for 29 percent of the weight, but only 7 percent
> of the trash can volume. Paper clearly dominates the trash can volume,
> accounting for three-fourths of the total, with the very bulky nature of
> corrugated containers being a major factor. However, at the landfill,
> this changes markedly. Corrugated and other paper products become
> wet and compact much better than many other components, resulting in
> lowering of volume percent to less than one half…

CLAIM

EVIDENCE

Figure 19-5 A Research Report *concluded*

Franklin Associates, Ltd. with the Garbage Project for the
Council for Solid Waste Solutions.

Style In research reports that are written primarily for experts, technical terms
accepted by the discourse community should be used and do not need to be
defined. If you expect generalists and/or novices to read the report, then define
those technical terms that they may not know in a glossary at the end of the main
text (see Chapter 20). The MSW report uses almost no technical terms.

Research reports are written in a formal tone. They should be neutral and
impersonal, allowing readers to decide whether or not the study and ensuing
results are valid. Numerical data follow specific conventions, depending on the
discourse community in which the study is being conducted. Scientific data
should be shown to the third significant figure (e.g. 2.003). Recorded graphs
should be folded or cut to $8^1/_2$-inch size.

Because various discourse communities perceive research reports as verbal
representations of objective scientific inquiry, they have created certain conventions which writers are expected to follow if their work is to be taken seriously.
For example, guidelines for publishing research articles in the IEEE Journal
include:

- Confine the writing style [of the abstract] to the passive voice.
- Do not write one-sentence paragraphs.
- Do not use contractions.
- Write in third person; not first or second person.

Conventions among discourse communities differ. While the IEEE guidelines require the passive voice in the abstract, other communities recommend the

active voice. Conventions also change over time. Some communities, especially those in the social sciences, are beginning to allow the use of the first person in research studies, though IEEE and many of the physical sciences continue to bar it. You should refer to style guides for the particular community in which you are writing to determine the specific conventions followed.

STRATEGY CHECKLIST

1. Use the following organizational pattern and sequence to present the information:
   - Introduction: problem or need being addressed
   - Objectives
   - Procedures/methodology
   - Results
   - Discussion/conclusion
   - Recommendations
2. Write in a formal, neutral, and impersonal tone. Use the conventions of your specific discourse community.

CHAPTER SUMMARY

Final reports provide readers with an historical record of a project as well as information for making decisions. The length of a report varies from a single page to several hundred pages, depending on the topic and the readers' needs. Reports can be incorporated in correspondence or stand alone. Most reports are reader-initiated.

You may write a report yourself or, if you are working on a project involving several people, you may collaborate with them on the report.

All readers, regardless of their organizational positions, assume the role of decision makers as they read reports. Therefore, your primary purpose in writing a report is to provide readers with information for making decisions.

Reports are also persuasive; you need to persuade readers to accept your results by indicating you have used sources of information, methods, procedures, and statistical analyses that meet criteria established by the readers' communities. The claims in your conclusion and recommendation sections must be based on evidence from your results section.

Most of the information is gathered from field sources.

Final reports are analytical and contain the following categories:

- A rationale, statement of need, and description of the problem/or background.
- A statement of the purpose or a listing of objectives.
- A description of the project and/or an explanation of the procedures/methodology used.
- Neutral and impersonal presentation of the data collected.
- Discussion of the results.

- A summary of the report, conclusions, and/or recommendations based on the purpose.

The style of a report depends on the discourse community in which it is being written.

Visual text should be used, and graphics should both complement and supplement the information to facilitate readers' comprehension of the material.

- **Information reports** provide readers with information needed to make decisions or for their own work.

- **Feasibility reports** present the results of investigations so readers can determine whether or not a project should be undertaken.

- **Evaluation reports** provide readers with assessments of situations, persons, or projects so that readers can make a decision related to that situation, person, or project.

- **Research reports** provide information concerning experimental research or project development for other researchers and developers, who may use this report in their work if they decide the research is valid and relevant.

PROJECTS

As you undertake a project, be sure to engage fully in all three processes—planning, drafting, revising. You may want to use the Planning Worksheet in Appendix B, as you gather and organize your information. As you prepare to revise, you may also want to use the Evaluation Checklist (page 264-265), and/or the Document Design Criteria Checklist in Appendix B.

Collaborative Projects: Short Term

1. **Feasibility report for improving students' preparation for writing in technical communities**

   Few students have experience writing or reading technical documents before they enter the workplace. Several strategies might be used to prepare students for writing in technical communities. These include:

   - Internships in technical industries that involve actual on-the-job writing.

   - Include technical writing in the secondary school curriculum.

   - Mentors who are already in industries who could help students draft documents for technical writing classes, and also discuss the context of their own documents and writing processes.

   - Partnerships with industry in which students are assigned to an employee and engage in writing the documents the employee writes. They may either write a separate document or collaborate with the employee.

   You may think of other possibilities.

You have been asked by your department chairperson to serve on a committee to determine the feasibility of instituting one or several of these strategies.

a. Work in groups of five.

b. Each group member should select one alternative to study. You need to discuss the idea with those people who would be involved if the strategy were to be put into practice. For example, you would want to talk to a high school English teacher and a principal if you were investigating the idea of teaching technical writing in high school. You would also want to talk to a division manager of a local industry if you were considering the mentor program, or a graduate of your college who is now working in a local industry that might be appropriate for the school/industry partnership. Once you have gathered your information, determine whether or not the alternative is feasible.

c. Discuss your findings with your group. Use collaborative debating to determine which alternatives are feasible and which are not.

d. Write your report. You will need to determine how the report will be written. You may select one person to write the entire report, with each of the others simply submitting information, or each person may write a component of the report, such as the main body, the appendices, budget, time-line, etc., or each person may be responsible for a section of the report, such as the introduction, present status, etc.

Option: *To be continued*, Chapter 20, Document Locators and Supplements, Collaborative Projects: Long Term.

2. **Information report for pharmaceuticals**

You work in the marketing division of a pharmaceuticals firm. It is presently considering developing drugs for several medical problems, including the following:

- Cystic fibrosis
- Muscular dystrophy
- Cerebral palsy
- Anorexia nervosa
- Bulimia nervosa
- Lead poisoning
- Sickle cell anemia
- Tay Sachs disease
- Apnea
- Insomnia
- Skin cancer

- Carpal Tunnel Syndrome
- Psoriasis

Before making any decisions, the firm wants to know what is already being done in relation to each of these conditions, so it doesn't re-invent the wheel or compete with a firm that is already engaged in studying the same medical condition. You are a member of the team that has been asked to research the information available on one of these medical problems. The team will then submit a report on its findings to the division manager.

a. Work with two other members of your class. Select one of the medical problems listed above.

b. You will need to use the library to obtain background information on the medical condition you have selected. If there is a medical college located in your community, you will probably want to look up information in the journals in its library. You may also want to contact your nearest hospital for brochures that they may have on the condition you are investigating. If there are any doctors in your community who specialize in this problem, you may want to contact them for additional information. Pharmacists may also have information on drugs that are being used in treatment. You may also write to drug companies to find out what they are doing in relation to the problem. You and your team members need to determine the sources each of you will investigate.

c. Once you have gathered the information, determine the content to include, and the organizational pattern and sequence of the information. You also need to determine what you and each of the other members of the team will do in terms of writing the report.

d. Write the information report.

Option: *To be continued,* Chapter 20, Document Locators and Supplements, Collaborative Projects: Long Term.

3. **Background for an R & D report**

You work in the research and development division of a firm. You and two other researchers have been working with two senior researchers on a project related to one of the following fields:

- Ceramic dental crowns
- Dental bonding
- Painless hypodermic needles
- Ceramic composites for automotive parts
- High temperature ceramic superconductors
- Photonic devices
- National aerospace plane

- LIMPET (linear induction machine programmed electric turbine) for electrical power
- Positive emission tomography scanning
- Power modulator to convert from DC to AC
- Electromagnetic fields from power lines
- Gene therapy
- Passive smoking
- Gamma rays for food preservation
- Ocean thermal energy conversion

Your research is completed and your team has been asked to write a report for the marketing division so that it can begin to develop a sales strategy for your product. While the senior researchers will write up the actual research, you and the other two researchers have been asked to write the background section of the report.

   a. Work with two other members of your class. Select one of the areas above.

   b. You will need to use the library to obtain the background information on the area you have selected. You may also want to use other written and field sources. Once you have determined the sources for your information, you and your team members need to decide which of these sources each of you will investigate.

   c. Once you have gathered the information, begin to write the background section of the report. You need to determine the content to include, and the organizational pattern and sequence of the information. You must also decide what you and each of the other members of the team will do in terms of writing the report.

Collaborative Projects: Long Term

1. **Class template continued from Chapter 18, Collaborative Projects, Exercise 1.**

   a. Have the other students in your class fill out the Absence and Tardy paper form.

   b. Tabulate the results.

   c. Analyze the results. What reasons were listed most often? least often? for absences and/or tardy papers? Rank the reasons from most to least listed.

   d. Write an evaluation report of your findings to meet your instructor's purpose.

   Option: *To be continued,* Chapter 20, Document Locators and Supplements, Collaborative Projects: Long Term.

Individual Projects

1. The background section of a research report to a governmental agency or an academic institution often includes a subsection devoted to a search of the literature. This subsection summarizes the information on a topic that can be found in written sources.

    a. Select a topic related to your major field of study. Limit the topic. For example, if you select the topic of ceramics, you might limit it to ceramics used in dentistry, in light planes, or in automotive parts.

    b. Use the library to obtain information on the topic you have selected. (Review Chapter 6, Gathering Information, for locating written sources of information in a library.)

    c. Once you have gathered the information, you need to write the section. Determine the content to include and the organizational pattern and sequence of the information.

    d. Write the review of literature subsection for a research report on the topic you have selected. Cite your sources in the style and format accepted by your professional discourse community.

2. Evaluate how your community is helping to protect the ozone layer. Base your evaluation on the following criteria:

    •   Recycling centers for CFC reclamation have been established.

    •   CFCs are recovered when refrigerators and automobiles are disposed of.

    •   Oily cans of Freon (R-12) with high-quality shut-off valves are sold over-the-counter in auto parts stores.

    •   New housing construction does not use rigid foam insulation containing CFCs. Houses should be insulated with such materials as fiberglass, fiberboard, gypsum, foil-laminated board, or cellulose.

    •   Hospitals have phased out ethylene/CFC sterilant mixtures.

    You may want to call your local or state Environmental Protection Agency office, and/or public works department, local contractors, auto parts stores, and hospitals to find out what your community is doing.

    Write a report of your evaluation for your city council to use in determining whether or not they should pass ordinances or institute services to protect the ozone layer.

3. The public relations department of your university wants to develop a booklet to recruit up-and-coming young faculty. The booklet will describe the kinds of research and development activities conducted at your institution. The writer of the booklet has requested that you provide information concerning the kinds of projects and activities in which faculty and graduate

students in your department are involved. The writer plans to devote about 750 words to each department. Write an information report describing the R & D projects and activities going on in your department.

4. The chairperson of the department in which you are majoring is required to submit an annual report to the college dean, describing all the research projects and activities in which students are engaged. The dean uses the information to determine the amount of research grant money to budget to the various departments. Your chairperson has requested that all instructors provide a report of the research and development activities conducted by their students. The instructors, in turn, have requested that each student write a report of a project or activity in which they engaged.

Write such a report for inclusion in your instructor's report.

CHAPTER 20

Document Locators and Supplements

▶ GLOBAL LOCATORS
    Pagination
    Tables of Contents
    Lists of Graphics

▶ SUPPLEMENTS
    Cover Correspondence
    Title Page
    Summaries
    References
    Glossary
    Appendix

▶ GLOBAL FORMAT

▶ PROJECTS
    Collaborative Projects: Short Term
    Collaborative Projects: Long Term
    Individual Projects

MOST LONG PROPOSALS AND REPORTS are read by multiple readers who have differing interests in and purposes for reading a document. To help these readers find the information they need, long reports and proposals provide global locators, such as pagination, tables of contents, and lists of tables and figures. In addition, a variety of supplementary documents are included, such as cover correspondence, a title page, abstracts, executive summaries, glossaries, references, and appendices.

Locators and supplements cannot be written until the main text of a document is complete, since they are based on the information in the document. However, you can begin to think about these elements while working on your document.

- As you gather information, record the information you need for a list of references.

- Set aside material for the appendix as you find it.

- Keep a list of words that should be included in a glossary as you draft.

Many word processing programs have special codes that help you create glossaries, tables of contents, lists of tables and figures, and indices as you draft.

Guidelines for these document aids are available for many organizational and professional discourse communities. You should check to see if guidelines are available for the specific community for which you are writing. In addition, if a document is reader-initiated, guidelines for writing locators and supplements may be provided.

If guidelines are not available, use the following conventions for drafting locators and supplements.

GLOBAL LOCATORS

Pages in a document are numbered, and then a table of contents and a list of tables and figures are developed to include the appropriate page numbers. These are placed at the beginning of a document so readers can locate information quickly and easily.

A CEO may want to study the recommendations of a report. A division manager may only be interested in reading the aspects of a report that are specifically related to his division. A government official may want to compare the various levels of water pollution throughout her district. The CEO and division manager can locate the information they need in the table of contents, while the official can locate the information she needs in a list of tables.

Pagination

Number each page of a document at the bottom of the page in the center. Number the pages consecutively, using Arabic numerals.

The pages preceding the main document, (title page, table of contents, lists of tables and figures, forward, preface) are not considered part of a document and are therefore numbered separately. Use small Roman numerals (e.g., i, ii, iii, iv) to number these pages.

Table of Contents

A table of contents helps readers determine whether a document contains the information in which they are interested, and then locate the page on which the information can be found.

A table of contents can serve as a general outline of a document. By looking at a table of contents, readers should be able to perceive the overall organizational pattern of a document as well as the organization of specific sections. For this reason, a table of contents should be detailed and explicit, presenting several levels of headings. Three are usually sufficient.

Headings and subheadings listed in a table of contents should accurately reflect the headings and subheadings in a document in terms of wording as well as correct page number.

Format Headings and subheadings in a table of contents need to be formatted to indicate their hierarchy in a document. Use the same visual text and format that is used in the document. If headings are in all capital letters in a document, then they should be printed in all capitals in the table of contents. If headings and subheadings are numbered in a document, then they should be listed with the corresponding number in the table of contents. Use indentation to indicate levels. The further toward the right a subhead is indented, the lower in the hierarchy it is situated.

Place the title of the page, "Table of Contents," at the top. Use all capital letters.

List the headings and subheadings and report supplements in the order in which they are found in the document.

To facilitate readers' locating the proper page of a section, you may want to use dotted lines from the end of the heading or subheading to the page number.

Lists of Graphics

Graphics are listed either as tables or figures. Any graphic that isn't a table is considered a figure. Like the table of contents, lists of tables and lists of figures help readers locate specific information.

Graphics should be listed with their corresponding page number, in the order in which they occur in the document. Center the title "List of Tables" or "List of Figures" at the top of the page.

The tables in Figure 20-1 are included in a report to the U.S. Congress. Because there are so many items, the list is subdivided by chapter to facilitate readers' locating the table or figure they need. Notice that without the dots from the heading to the page number it is more difficult to locate the appropriate page number and there is greater chance of locating an incorrect one.

STRATEGY CHECKLIST

1. Paginate. Number pages prior to main text with Roman numerals. Use Arabic numerals for main text.
2. Include a table of contents and lists of tables and figures.

U.S. Environmental Protection Agency, Characterization of Municipal Solid Waste in the United States: 1990 Update.

Figure 20-1 Conventional Lists of Tables and Figures

SUPPLEMENTS

Many readers have neither the time nor the need to read an entire report or proposal. Often they are only interested in certain aspects of a document. A Chief Executive Officer (CEO) may only need to know the main recommendations of an evaluation report before passing the document to a secondary reader who will actually oversee the corrective actions. A cover letter and executive summary or abstract can provide primary readers, such as CEOs, with sufficient information.

Secondary and peripheral readers may need more details or explanations than primary readers. Appendices, glossaries, and reference lists can provide this information. While a CEO, by skimming through an executive summary, can learn the main problems and the evaluator's recommendations for solving those problems, an engineer may need a detailed analysis to implement the recommendations. These details can be provided in an appendix. An interested citizen may need definitions of technical terms that are part of the primary reader's everyday vocabulary. These terms can be defined in a glossary.

This section examines the conventions for cover correspondence, title pages, abstracts and executive summaries, glossaries, references, and appendices.

Cover Correspondence

You need to include a cover letter or memo so readers know why they are receiving a document and what it is about. Cover letters or memos, sometimes referred to as letters of transmittal, should precede a title page and should be short, no longer than a page.

Sometimes readers read only the cover letter before sending a document to a secondary reader. A CEO at a company may be the person addressed on an evaluation report, but she may direct the report to the vice president in charge of the area that was evaluated, expecting the VP to provide a brief summary of the report's findings. Or an addressee may simply act as a gateway to direct a document to appropriate readers. Without looking at the document, the director of a division of a funding agency may direct proposals to outside reviewers, using the information included in the cover letter to determine the appropriate persons.

If a document is reader-initiated, the reader may provide a form to serve as a letter of transmittal. The information requested on this form consists of the main information readers need to understand the major concepts of the attached information: name, address, and type of organization; title and type of project, amount requested, duration of project, and principal participants.

If you do not have a transmittal form, use the guidelines below to write your cover correspondence.

Content, Organizational Pattern and Sequence   A cover letter or memo should contain the following information in the order listed:

- **The purpose of the letter:** Explain that the enclosed report or proposal is either in response to the reader's request or is writer-initiated. If it is writer-initiated, explain why.

- **The topic discussed in a nutshell:** Provide a synthesis of the proposal or report. If it is a proposal, use the main organizing ideas from your problem and solution sections. If it is a report, use the main organizing idea from the introductory section. If it is an information report, include a list of the major topics covered.

- **Additional comments:** You may want to discuss one or two specific aspects of the document in which the reader may be interested or which you believe might persuade the reader to accept your recommendations or support your proposal. If you are trying to persuade readers to do something or think in a certain way, you may want to include some background or a brief summary of the proposed project or activity.

- **Conventional courtesy closing:** Regardless of whether it is a proposal or report, close by offering to provide the reader with additional information or answers to questions. Comment that you hope the reader finds the information of interest. If appropriate, indicate that you look forward to a response, and that you appreciate the reader's assistance.

If you expect your readers to provide a response to the document, request it just before the closing.

Style, Format The style should be the same as that used in the accompanying report.

Use the conventional format of a letter or memo. The cover letter should be no longer than a single page.

The letter in Figure 20-2, submitted to the board of a local airport authority, accompanied a feasibility report on noise. Consider how the writer follows the general conventions for correspondence as well as the specific conventions for this type of letter.

Title Page

Readers use a title page to orient themselves to a document. Sometimes, by the time secondary or peripheral readers see a document, the cover letter has been removed and the title page is their only introduction to the document. Through

**Figure 20-2
Cover Letter**

Heading.

Salutation.
Statement of
purpose, topic.

Background.
Additional
information.

Instructions for a
response.

Closing.

Crawford, Murphy & Tilly, Inc.
Consulting Engineers
600 North Commons Drive Suite 107
Aurora, Illinois 60504

CMT

Dear Recipient:

This preliminary draft document contains the analysis and recommendations of a F.A.R. part 150 Noise Compatibility Study for the Greater Peoria Regional Airport. This study is a voluntary action undertaken by the Airport authority to evaluate existing and potential future noise conditions at the airport.

In conjunction with the preparation of the study, a Notice of Opportunity for a Public Hearing was published on December 14, 19xx. This document is to be made available to the public for 30 days following publication of this notice. In order that this document be available for all interested individuals, please keep this copy of the report on file at this location for public review. This copy will be picked up by a representative of our office after January 16, 19xx.

If you or interested individuals have any questions, you may contact the following individuals.

Mr. Tom Miller, Ass't Dir.
Greater Peoria Regional Airport
1900 S. Maxwell Road
Peoria, IL 61607
(309) 697-8272

Mr. Greg Heaton
Crawford, Murphy, & Tilly, Inc.
2750 W. Washington
Springfield, IL 62702
(217) 787-8050

Thank you for your assistance.

Yours truly,

Greg Heaton

this page they discover the specific topic to be discussed, how current the document is, who has authored the document and sponsored the project under discussion, and for whom the document is primarily intended.

Content, Organizational Pattern and Sequence The title page contains: the title, date, author, sponsoring agency, and addressee.

Title. The title is extremely important. It provides readers with the basic frame in which to read the remainder of the report. Therefore, use a title that accurately represents the topic. To assure accuracy, the title should relate directly to the main organizing idea.

The title should be a short phrase rather than a complete sentence. In fact, it may be a truncated version of the main organizing idea.

Titles for research reports are often long, since they must specify the particular aspect of an area under consideration. They are also often divided into two parts. The first provides a specific description of a topic; the second part gives a general description.

Specific   **ACID MINE DRAINAGE FROM SULPHIDE ORE-BODIES:**
General   **Kinetics and Mechanisms of Oxidation and Dissolution**

Titles should also reflect the language of the primary reader:

Expert   **REGULATION OF INSECT CUTICLE-DEGRADING ENZYMES FOR THE MYCOINSECTICIDE BEAUVERIA BASSIANA**

Generalist **BIOCONTROL OF INSECTS USING FUNGI**

Novice   **AN ATTRACTIVE ALTERNATIVE TO THE USE OF CHEMICAL PESTICIDES**

Generalist/ **BIOCONTROL OF INSECTS USING FUNGI:**
Novice   **An Attractive Alternative to the Use of Chemical Pesticides**

Date. Use the date the report is expected to be published.

Author and/or sponsoring agency. List all authors and/or your organization, as well as any other organization involved with the project.

Addressee. Include the name of the primary reader, the organization the reader represents, and the city and state where the reader is located.

Visual Text and Format The title of the document should be in all capital letters. If the title is in two parts, the first part should be in all capital letters, the second part in capital and lower case letters. If the title is longer than a line of type, do not split phrases.

Incorrect **BIOCONTROL OF INSECTS USING FUNGI: An Attractive Alternative to the Use of Chemical Pesticides**

Correct **BIOCONTROL OF INSECTS USING FUNGI: An Attractive Alternative to the Use of Chemical Pesticides**

The title should be centered and placed one-third of the way down the page. If the title is longer than a line of type, center the remaining lines as shown in the example above. Do not number the title page. However, consider it as the first page of your prefatory information.

Center the remaining information on the title page. Provide sufficient spacing between each of the main categories for the reader to perceive the different chunks at a glance. Study how the writer of Figure 19-2 adheres to the conventions on the title page of the report.

Summaries

A summary is placed after the table of contents and the lists of tables and figures in a document. It provides a brief synopsis of the contents so that if readers do not have the time or interest to read an entire document, they can still learn the main points. Readers may also use summaries to determine whether or not they are interested in the specific topic being discussed.

There are two types of document summaries: abstracts and executive summaries. Some discourse communities prefer abstracts, others prefer executive summaries. Guidelines for reader-initiated documents usually specify whether to include an abstract, an executive summary, or both. If the guidelines do not specify, check previous documents written for that discourse community to determine which of the two to use.

Abstracts are usually written for research and development reports and proposals, and the primary reader is usually an expert. Executive summaries are usually written for other types of reports and proposals, and are usually written for decision makers. Sometimes the primary reader is an expert.

Abstracts Abstracts precede the main text and provide brief syntheses of the information contained in documents. Readers use an abstract to learn the major concepts involved in a proposal or report. If they want additional details, they then turn to those sections of the document in which they are specifically interested. Abstracts also provide readers with a frame in which to read a report.

Cataloguers use abstracts to index a document. For this reason abstracts should contain the key words readers use to look up a report in a reference index.

Content, organizational pattern and sequence. An abstract usually contains the main organizing ideas of each section of the accompanying document. In a proposal, an abstract contains the main organizing ideas for the problem and solution sections respectively. It also includes the major methods used to achieve the solution, and the major reasons the solution is advantageous. An abstract for a report includes the objectives and the main organizing ideas for the methodology, conclusion, and recommendation sections.

Information in an abstract is presented in the same sequence as it is presented in the accompanying report or proposal. An abstract should not contain any information that is not included in the accompanying document.

Style, format. The language of an abstract should reflect that of the accompanying report. If the report uses technical terminology, then the abstract should also do so. In many discourse communities the abstract is written in the passive voice. It is always written in an impersonal and formal style.

Abstracts should be short, usually no more than 250 words. Introduce the text for the abstract by placing the word "ABSTRACT" in capital letters at the top center of the first page of the main document. Single-space the text.

The abstract shown in Figure 20-3 summarizes a proposal to a university's toxic substances research program. Consider how the writer adheres to the conventions.

Executive Summaries Executive summaries serve the same purpose as abstracts: they provide readers with a summary of the information contained in an accompanying document. However, because the reader is a decision maker rather than a peer in the role of a learner, the focus is different. Because readers are specifically interested in the solutions, conclusions, and recommendations of a document, summaries are usually longer and provide more detail. Primary readers may never read beyond the executive summary. Therefore, an executive summary should contain all the information a primary reader must know before turning the document over to a secondary reader.

**Figure 20-3
Abstract for a
Research Proposal**

Statement of
problem.
Proposed solution.
Description of
project.
Technical term-
inology.

ABSTRACT

The combustion of fuels produces oxides of nitrogen (NOXx) that lead to air pollution and acid rain. We propose a post-combustion exhaust clean-up process that will efficiently and economically remove the NOx from the exhaust for a wide variety of combustion processes.

In this process, selective gas phase NOx reductions would be performed in conjunction with a commercially available regenerative heat exchanger. This process is particularly advantageous for use with toxic incinerators because the process is decoupled from the combustion process and from the acid scrubbing of the flue gas from toxic incinerators.

A Novel Technique for Removal of NOx from Products of Combustion of Toxic Compounds by Robert W. Dribble, Department of Mechanical Engineering, University of California, Berkeley

Content, organizational pattern and sequence, style. Executive summaries contain not only the main organizing ideas of each of the major sections of a report, but also the major points in each section. Because primary readers are most interested in the solutions discussed in a proposal and the recommendations in a report, executive summaries should contain a list of the major problems or findings, as well as the solutions proposed to solve the problems or the recommendations suggested for enactment.

Like abstracts, executive summaries precede the main text. The information is presented in the same sequence it is presented in the accompanying document. The style should also reflect that of the text. If a document attempts to persuade readers to do or think in a certain way, then summaries should reflect that same point of view. Executive summaries should never include information that is not included in the main report.

Visual text, format, and graphics. Executive summaries can range from two to ten double-spaced, typewritten pages. Use headings and subheadings that parallel those of the accompanying document to introduce the major sections, so readers scanning the document can locate them easily as well as recognize when the topic changes. Use graphics similar to those in the accompanying documents to supplement and complement the readers' understanding of the text. (See the Executive Summary for the report in Figure 19-2.)

References

If you use information from written or field sources, you need to refer to these sources in your text. You not only need to cite the sources of direct quotations, but also the sources from which you have elicited data of any kind. Readers often need to know these sources if they are to accept your argument. Sources should be cited for the following reasons:

- To give credit to those persons responsible for the information. This is not just a courtesy: failure to do so may be considered plagiarism. You are guilty of plagiarism if you use another person's words, data, or ideas without giving that person credit, but rather allow the reader to assume the words or ideas are yours.

- To provide readers with information for determining the acceptability of your information, based on criteria established by their discourse communities. By observing the date the information was written, readers can determine how current it is. Unless you are writing about historical data, readers want the most current information available.

   If readers recognize the source as someone in their organizational or professional community, they can determine the biases of the data. In addition, if the source is a well-known authority in their discourse community, readers are more apt to accept the data as accurate.

- To provide readers with information for reading the original written document, or for discussing the information with the primary source. Readers may want to read or hear more about the particular topic, or they may want to check the data.

You provide readers with information about your sources through in-text citation and reference lists.

In-text Citation When you quote or refer to a source in your text, either directly before or after you refer to an idea or to data from that source, cite the source. "In-text citation" can be provided in several ways:

- Place the source of the information in parentheses immediately after presenting the data.

Parenthetical.

> Of the three main thermochemical processes, gasification is the most widely commercialized (Erickson 1986b).

- Integrate the source of information into the text itself either by citing the source at the beginning (a), or the end (b), of the sentence.

Place source at beginning.

> (a) Klass (1988) estimates that biomass consumption in 1988 was about 3.11 EJ, although Rinebolt (1989a) estimates current consumption to be 4.0 EJ.

Place source at end.

> (b) Direct combustion of wood and wood products in 1987 generated 2.87 EJ, according to Klass (1988).

Notice that each of these citations includes the name of the author of the source from which the information was obtained, and the date of publication, the basic information necessary for readers to determine the validity of the data. This type of in-text citation follows the "author/year" style, used by some technical discourse communities. We will discuss other types of in-text citation under the section on conventions.

Reference Lists While in-text citations provide readers with basic information about a source, these citations don't provide sufficient information for readers to look up or locate the primary source. This additional information is provided in a list of citation sources at the end of the main body of a document. Readers can use the information in an in-text citation to look up the source in which they are interested in a reference list. The list of references must contain sufficient data to enable readers to locate the primary text.

The following information is provided in the list of references for the sources cited in the previous example. As you study the list, notice the diverse types of sources the writer uses.

> Klass, Donald L. 1988. "The U.S. Biofuels Industry." Report presented at the International Renewable Energy conference. Honolulu. September 18-24. Chicago: Institute of Gas Technology.

Komanoff, Charles. 1988. Personal communication.

—————1981. *Power Plant Cost Escalation*. New York: Komanoff Energy Associates.

Rinebolt, David C. 1989a. National Wood Energy Association. Testimony before the House of Representatives Science, Space and Technology Committee. March 2...

—————1989b. Personal Communication...

While there are two references for Komanoff, the dates of publication differ. Because dates are included in the in-text citations, readers can determine which of the two sources in the reference list they want. However, both of the Rinebolt references occur in 1989. To help readers distinguish between these two sources, they are designated as "a" and "b."

Conventions The conventions that govern the specific information to be included, the sequence of the data, and the punctuation for both in-text citations and the list of references, depend on the discourse community for which you are writing. You may be familiar with the conventions associated with English literature, commonly referred to as the name and page style. These conventions are outlined in a style guide by the Modern Language Association.

Technical documents follow different sets of conventions, depending on the specific discourse community for which they are written. These conventions are known by the style of their respective citations: author/year, or number.

The following list indicates the style used by various technical discourse communities.

- agriculture   author/year
- archaeology   author/year
- astronomy   author/year
- biology   author/year
- botany   author/year
- chemistry   number
- computer science   number
- geology   author/year
- health, medicine, nursing number
- home economics   author/year
- mathematics   number
- physical education   author/year
- physics   number
- psychology   author/year
- zoology   author/year

Author/year style.   In the in-text citation, always include the name of the author and the year the source was published. If you use a direct quotation, also include the page number for the quotation, for example, *p. 215.*

Title the list of citations at the end of your text "References." All sources cited in your text must be included in your list of references. Do NOT include sources that you may have read, but which you have not cited in your text.

Your references should be listed alphabetically by the author's last name. List several sources by a single author chronologically, with the most recent source listed first. When both sources were published in the same year, list the sources alphabetically, according to the author of the article or book.

The basic style and format for the author/year style is listed below.

Author(s).

- The author's last name, first name/ initials, and middle initial, followed by a period.

Date of publication. Article or chapter title.

- The year of the publication or speech in parentheses, followed by a period.

- The name of the article or chapter, followed by a period. Capitalize only the first word of the title and the first word of the subtitle.

Journal or book title, publication information.

- The name of a book or journal should be underlined. If you are using desktop publishing, use italics and omit the underlining. Place a period at the end.

- The place of publication or where the speech was delivered, followed by a colon.

- The name of the publisher or organization where the speech was made, followed by a period if it is a book, by a comma if it is a journal.

- The volume number in Arabic numerals if it is a journal, followed by a comma. Underline the number.

- Page numbers of an article in a journal or chapter in a book, followed by a period. Do NOT precede the numbers with "p."

The list is a general set of guidelines for the disciplines that use the author/year system. However, each discipline has its own conventions, which may differ somewhat from those above. For example, while the list of sources in psychology is titled "References," the list of sources in anthropology/archaeology is titled "References Cited," and the list in biology/botany/ zoology is titled "Literature Cited." Because of such variations in the form and style, you should purchase a stylebook for the discourse community in which you are writing to be sure you are following the appropriate conventions.

Figure 20-4, which appears in a brochure on energy sources, follows the conventions of the author/year system. Notice how the writer adheres to the conventions in both the text and the list of references. Read the text in "A" and then see how easily you can look up the specific references in "B."

A

As designs and performance have improved, the cost of wind power has declined dramatically. The levelized cost of electricity generated by modern wind turbines is now estimated to be 7-9¢/kWh, down from over 25¢/kWh in 1981, while the average price per installed kilowatt of intermediate-size wind turbines declined from $1,300-$2,000 in 1981 to $950-$1,100 in 1988 (AWEA, 1988a). One advantage of wind power is that operations-and-maintenance (O&M) costs can be quite low. O&M costs at operating wind farms range from 0.8 to 2¢/kWh and average 1.2¢/kWh (AWEA, 1988b).

26

B

References

American Wind Energy Association (AWEA). 1989a. *Wind Energy Has Come of Age in California.* Washington, DC: AWEA.

---------- 1989b. "Response of American Wind Energy Association to 'Wind Energy Development in California.'" Washington, DC: AWEA.

---------- 1988a. "Wind Energy Costs Drop." Collection of viewgraphs. Washington, DC: AWEA.

---------- 1988b. *Windletter.* Issue #7.

Bain, Don, 1988. "Pacific Northwest Wind Energy Planning." Oregon Department of Energy.

Bath, Thomas D. 1989. Manager, Analysis and Evaluation Office, Solar Energy Research Institute. Testimony before the Senate Committee on Energy and Natural Resources, March 14.

81

From *Cool Energy,* Union of Concerned Scientists

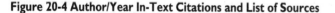

Figure 20-4 Author/Year In-Text Citations and List of Sources

Number system. The numbering system differs markedly from the author/year and year/page style, and creates more work for the reader. Instead of information identifying the author, date of publication, or page, a number is placed after the information cited. The reader must turn to the list of references and look up the corresponding number to acquire the information about the source.

A general set of guidelines for using the numbering system follows:

- Number the citations sequentially in the text. The number can be placed either in parentheses or brackets, or set as superscript.
- List the sources numerically in the list of references, which is titled "Works Cited."

As with the author/year style, the number system varies according to the discipline. You should obtain a stylebook for the discipline in which you are writing to be certain you are using the appropriate conventions.

Figure 20-5, which appears in a report on the transfer of toxic chemicals, follows the conventions of the number system. Notice how you can locate the source in the reference list, based on the number in the text.

A	**B**
Example 1	**Reference List**
Sender: Formosa Plastics, Point Comfort, Texas	
Receiver: PPG Industries, Lake Charles,	[1]53 FR 4500, "Toxic Chemical Release Reporting:
** Louisiana**	Community Right-to-know," Feb. 16, 1988, p.
	4516.
Formosa Plastics, a major Texas plastics manufactur-	[2]EPA, Environmental Fact Sheet: "New regulations
er, sends up to 14 million pounds of ethylene dichlo-	proposed for hazardous waste burners," EPA/530-
ride (EDC) distillation bottoms (a production	SW-89-063; 54 FR 43718, Oct. 26, 1989.
residue) off-site each year[3]—but the company	[3]Formosa Plastics, letter to the U.S. Environmental
reported transferring only 750 pounds off-site under	Protection Agency, June 19, 1987.
the toxics-release inventory (TRI). The unreported	[4]Louisiana Department of Environmental Quality,
shipments are sent by highway to PPG Industries in	internal memorandum, Dec. 13, 1990.
Lake Charles, Louisiana where a minimum of 35% is	[5]Texans United, *The Formosa Plastics Story: Report*
allegedly recycled in a feedstocks remanufacturing	*of Environmental Investigation*, 42pp., 1990.
process,[4] and the remainder is incinerated.	[6]Tulane School of Public Health and Tropical Medi-
Tests reveal high EDC concentrations in groundwa-	cine, *Assessment of Environmental Data from the*
ter and storage ponds at the Formosa Plastics site in	*Calcasieu System*, 1990.
Texas,[5] and in soils and waterways near PPG Indus-	
tries in Louisiana.[6] EDC (or 1,2-dichloroethane) is a	
highly flammable and toxic chemical that may cause	
cancer, birth defects and other health problems.	
8	19

The "Recycling" Loophole in the Toxics Release Inventory: Out of Site, Out of Mind. Paul Orum, Working Group on Community Right to Know; Ann Moest, Environmental Defense Fund/ Environmental Information Exchange; David Allen, National Toxics Campaign Fund/Industrial Pollution Prevention Project; Ed Hopkins, Ohio Citizen Action

Figure 20-5 Number System In-Text Citations and List of Sources

Glossary

A glossary is a list of words with their definitions. When you are writing for generalists and novices and it is necessary to use more than a few technical terms,
use a glossary to provide definitions for unfamiliar terms. It is also advantageous
to use a glossary when you are writing for a general audience that includes
experts, generalists and/or novices. By defining the terms in a glossary, rather
than in the running text, experts do not have to read information they already
know.

A glossary is usually placed at the back of the main text before the list of
references.

Content, organizational pattern and sequence. Define the terms
according to the strategies discussed in Chapter 9. Alphabetize the list.

Style. Use a noun phrase immediately after the term to define it. To create a
noun phrase, place the word "is" after the term to be defined, and complete the
sentence. Then cross out the word "is" and replace it with a dash (see Figure 20-6).

Use parallel grammatical structure for all definitions. Additional explanations for a term should be in complete sentences.

Visual text, format. Visual text should be used to emphasize the terms so readers can easily locate the word they want. Place the term in boldface or italic type, or underline it. Use a dash or a colon to separate the word from its definition. Single-space each definition, and use a double space between terms.

The glossary in Figure 20-6 appears in a feasibility study on reducing noise pollution at a midwestern airport for members of the local airport authority. As you study it, notice the types of definitions used and how the writer adheres to the conventions.

Appendix

The appendix is the final section of a report. It is a catch-all for various types of information. An appendix may contain one or more pieces of information, including a single-page map or table, a two-page model of a questionnaire, or a complete report of 50 pages.

Information that secondary or peripheral readers may need to know, but primary readers do not, should be placed in an appendix; e.g., a list of suppliers for certain products, or a detailed description of procedures used in a feasibility study. In addition, documents that are summarized in the main text because they are too long or detailed to be included verbatim may be placed in an appendix, e.g., tabulations of survey responses.

Appendix A

GLOSSARY

A-Weighted Sound Level—A sound pressure level which has been frequency filtered or weighted to quantitatively reduce the effect of the low frequency noise. It was designed to approximate the response of the human ear to sound.

Approach Light system/ALS—An airport lighting facility which provides visual guidance to landing aircraft by radiating light beams in a directional pattern by which the pilot aligns the aircraft with the extended centerline of the runway on his final approach for landing.

Attenuation—Acoustical phenomenon whereby a reduction in sound energy is experienced between the noise source and receiver. This energy loss can be attributed to atmospheric conditions, terrain, vegetation, and man-made and natural features.

Base Leg—A flight path at right angles to the landing runway off its approach end. The base leg normally extends from the downwind leg to the intersection of the extended runway.

Crawford, Murphy & Tilly, Inc., 1991, F.A.R. Part 150 Noise Compatibility Study for the Greater Peoria Regional Airport, Peoria, IL

Figure 20-6 A Glossary

If a report is a compilation of several other reports, the original report may be placed in an appendix for readers who want to know all of the original data. Detailed lists of organizations, persons, or materials may be placed in an appendix for readers who may need this information. Sample data sheets and questionnaires that readers may want to see or copy can also be placed in an appendix.

Place each category of information in a separate appendix. Appendices are listed using alphabetic designations, such as Appendix A, Appendix B, etc. Each of the appendices should be listed in the Table of Contents of the document.

In the main text readers need to be informed that additional information on a topic is available. Following the discussion of a topic in the text, cite the appropriate appendix. Place the citation in parentheses.

The examples in Figure 20-7 demonstrate the variety of information found in an appendix. The appendices are included with a report on the transfer of toxic waste. As you scan these examples, determine why the writer decided to place the data in an appendix, rather than the main text.

GLOBAL FORMAT

Locators and supplements may be included with any document to aid readers in reading and understanding a text. The sequence for these aids follows the sequence presented here:

1. Cover correspondence
2. Title page
3. Summary/abstract
4. Table of Contents
5. Lists of Tables and Figures
6. Main body of the document
7. References
8. Glossary
9. Appendix

STRATEGY CHECKLIST

1. Provide a cover letter, memo, or letter of transmittal to explain to readers why they are receiving a document.

2. Include a title page to provide the reader with a statement indicating what the document is about.

3. Use an abstract or executive summary to introduce a long document. Check your discourse community to determine which one you need.

4. Include a glossary if you have used many technical terms that some of your readers may not know.

5. Include a list of references if you have cited sources.

6. Use an appendix to include information that is extraneous to some readers' needs, but that other readers may want or that the primary reader may like to read.

A

Appendix A: Superfund Sites Related to Solvent Recycling

At least 112 toxic contamination sites on the federal Superfund list involved recycling as a site activity. Of these, 47 involved waste-oil processing, 36 resulted from solvent recovery, 25 involved battery recycling and 14 were caused by drum recycling. (Some sites involved more than one recycling activity.) Superfund sites with solvent recovery as a site activity are listed below.

State	Site Name	Location
California	Industrial Waste Processing	Fresno
Connecticut	Sovents Recovery Service New England	Southington
Delaware	Chem-Solv, Inc.	Cheswold
Florida	Gold Coast Oil Corp.	Miami
Florida	City Industries, Inc.	Orlando...

B

Appendix D: Health and Environmental Effects Matrix

This matrix presents data from a preliminary search of known health effects assembled by EPA consultants, with additional data on stratospheric ozone depleters. The assembled information does not assess the severity of the effect, the appropriateness of the study method, or the presence of conflicting test results. Further testing would be likely to produce additional entries on the matrix.

CAR *Carcinogen*—includes chemicals known or suspected of causing cancer in humans or laboratory animals.

MUT *Mutagen*—chemicals that have the potential to produce changes in genetic material that can be passed on to the next generation.

DEV *Developmental Toxin*—chemicals that can cause birth defects, miscarriage, growth retardation, mental retardation or learning disorders.

REP *Reproductive Toxin*—chemicals that can damage the ability to reproduce.

NEU *Neurotoxin*—chemicals that can cause adverse effects on the nervous system, including the brain, spinal cord, and nerves.

ACU *Acute Toxin*—chemicals that can cause serious health effects or death from short-term exposure.

CHR *Chronic Toxin*—includes chemicals that can cause adverse health effects (other than cancer) from long-term exposure, such as liver, lung or kidney damage.

Rank	Chemical	CAR	MUT	DEV	REP	NEU	ACU	CHR
1	Toluene			•	•			
2	Xylene (mixed isomers)			•	•			•
3	Methyl ethyl ketone			•	•	•		•
4	Acetone							•
5	Methanol				•			
6	Aluminum (fume/dust)	•						
7	Ethylene						•	
8	Naphthalene			•			•	
9	Dichloromethane	•					•	
10	Methyl isobutyl ketone					•		

The "Recycling" Loophole in the Toxics Release Inventory: Out of Site, Out of Mind. Paul Orum, Working Group on Community Right to Know; Ann Moest, Environmental Defense Fund/ Environmental Information Exchange; David Allen, National Toxics Campaign Fund/Industrial Pollution Prevention Project; Ed Hopkins, Ohio Citizen Action

Figure 20-7 Appendices

**CHAPTER
SUMMARY**

Because long proposals and reports are read by different types of readers with different types of purposes, global locators and document supplements are included with the main text.

Global locators help readers find specific information quickly and easily. These include: pagination, tables of contents, and lists of tables and figures.

Number pages of the main text consecutively, using Arabic numerals. Number prefatory pages separately, using small Roman numerals.

A table of contents serves as a global locator as well as a global outline of a document. It should include chapters, sections, and subsections of a report. The chapter titles, and the section headings and subheadings, should reflect those used in the main text and provide the page number on which they begin.

Lists of tables and figures should be used to provide readers with locators for finding graphics.

Document supplements provide for the differing reading behaviors of various readers. Primary readers may only want a general overview of a document's contents. They can obtain this by scanning cover correspondence, title pages, abstracts, and executive summaries. Secondary or peripheral readers may need to know some information that primary readers don't need. This information can be located in a glossary, reference list, or appendix. Supplements can provide additional information to support a writer's claims, which may be too long to include except in summary form in the main text. This information can also be included in an appendix.

A cover letter should be short, and include:

- the purpose of the letter
- a brief summary of the topic to be discussed
- additional comments to persuade the reader to accept your recommendation or support your proposal
- conventional courtesy closing

A title page should contain:

- title
- date of publication
- author
- sponsoring agency
- addressee

Both abstracts and summaries precede the main text of a document and provide brief syntheses of the information contained in a document. If readers don't have time to read an entire document, they can obtain the essence of it from these supplements.

An abstract is usually no more than 250 words, and is usually found preceding research proposals and reports. An executive summary may be several pages in length, and is more detailed than an abstract. It will include the main findings or problems in a report, and the recommendations or solutions in a proposal. The organization and sequence of the information and the style of the text in both of these should reflect that of the main document.

A glossary is used to provide secondary and peripheral readers with definitions of terms that primary readers know. The words should be listed in alphabetical order.

A list of references provides readers with information for locating and reading primary sources cited in the main text.

An appendix is the final section of a document, and contains information that secondary or peripheral readers may need which primary readers don't. It provides detailed information that is only summarized in the main text, and which includes information too cumbersome to include in the main text, but necessary as support for the writer's claims.

PROJECTS

Collaborative Projects: Short Term

1. **Writing cover letters for a document.**

    Work in trios. You and your partners are technical editors for the organization doing the report on the volume of components in municipal solid waste in Figure 19-5. The report is to be sent to the Council on Solid Waste Solutions, an organization associated with the plastics industry which sponsored the project. The report will also be sent to federal, local, and state government agencies and representatives, manufacturers, and interested citzens.

    a. Review the report in Figure 19-5 on pages 562-564.

    b. Determine whether to write a single cover letter for all readers or separate cover letters for readers of each of the different kinds of agencies.

    c. Write the cover letter. You may want to collaborate on the letter(s), or if you decide to write separate letters, each member of the trio should be responsible for writing the cover letter to a particular type of reader.

    d. If you decide to write separate letters, compare the cover letters and discuss the similarities and differences in terms of content, organization and sequence, point of view and focus, and style. If there are differences, determine why you made the different decisions.

2. **Summarizing an Executive Summary**

    Work in pairs. Consider that you and your partner are both assistants in the Department of Engineering for a city with a population of about 150,000 people. You are presently involved in the problems of the city's landfills, which are filling up. A report on municipal solid waste has just been released by the Environmental Protection Agency. (The report appears in

Figure 19-2, pages 540-545.) You haven't time to read the entire report, only the summary. Based on what you read in the summary, you believe your supervisor should be apprised of the information.

 a. Read the Executive Summary in Figure 19-2.

 b. Collaborate on writing a memo to your supervisor concerning the report. You will probably want to indicate whether or not the person should take some kind of action.

3. **Providing for readers' global needs**

 Work with a partner. Each of you should select a report which you have written since entering college. Swap reports. Assume the role of editors.

 a. Determine which locators and supplements should be included in your partner's document so readers can quickly and easily locate information and comprehend the writer's message.

 b. Return your reports.

 c. Add the necessary locators and supplements to your document.

Collaborative Projects: Long Term

- **Solid waste** continued from Chapter 14, Collaborative Projects: Short Term Exercise 1.

- **Manual for a software program** continued from Chapter 16, Collaborative Projects Exercise 1.

- **Tutorial for a software program** continued from Chapter 16, Collaborative Projects Exercise 2.

- **Class template** continued from Chapter 18, Collaborative Projects, Exercise 1.

- **Improving students' preparation for writing in technical communities** continued from Chapter 19, Collaborative Projects Exercise 2.

- **Information report for pharmaceuticals** continued from Chapter 19, Collaborative Projects: Short Term Exercise 2.

 a. Determine which locators and supplements you should include in your document.

 b. Add the necessary locators and supplements to your document.

Option: *To be continued* in Chapter 21, Oral Presentations, Collaborative Projects.

Individual Projects

1. Create a global List of Figures for Part I of this textbook.

2. If the report in Figure 19-4 had been submitted as a separate document instead of as part of a letter, the writer would have needed to write an Executive Summary. Write the Executive Summary.

3. For the report in Figure 19-3, write a cover letter to be sent to the Board of Directors of the Greater Peoria Airport Authority. The Board is comprised of local citizens who are interested in aviation. Most members are generalists, a few are experts.

4. You are a member of the firm of Crawford, Murphy, & Tilly, Inc. Write a cover letter for the report in Chapter 19, Figure 19-3 to be sent to the Greater Peoria (IL) Airport Authority. Consider you are sending the report today.

5. Using a report that you have written since entering college, draft the global locators and supplements for the report. Consider the instructor who received the report as your primary reader.

CHAPTER 21

Oral

Presentations

▶ TYPES OF PRESENTATIONS

     Impromptu Presentations
     Prepared Presentations

▶ WRITTEN VERSUS ORAL DISCOURSE

▶ PREPARING ORAL PRESENTATIONS

     Analyzing Context
     Making Decisions
     Designing Visuals
     Outlining Extemporaneous Presentations
     Example

▶ MAKING THE PRESENTATION

     Rehearsal
     Vocalization
     Eye Contact
     Body Language
     Visuals

▶ THE WRITTEN REPORT

     Full Reports
     Presentation Reports
     Intermediate Reports

▶ PROJECTS

     Collaborative Projects: Short Term
     Collaborative Projects: Long Term
     Individual Projects

I N THIS CHAPTER WE WILL DISCUSS ORAL COMMUNICATION. Instead of learning strategies for drafting written documents, as we have been doing, you will acquire strategies for preparing oral presentations.

    Throughout your career you will need to make oral presentations. You may think of these presentations as being formal speeches in which you stand before a large group, talking about a topic you have researched. However, oral presentations in the workplace actually take a variety of forms. While you may occasionally be asked to give a "formal speech," most of the time you are simply asked to present an idea you are proposing, to report on work you have been doing, or to discuss the results of a project on which you have been working. The presentation may involve ten to twelve people internal to your organization, or it may involve three or four members of a client's organization. Your presentation may take place in a conference room around a long rectangular table, and may be scheduled for anywhere from five minutes to an hour. A short presentation to a small group of people can be as nerve-racking as a long one if approval for your entire project rests on your presentation.

In this chapter we will look at various types of oral presentations, as well as at the differences between written and oral texts. We will also look at strategies for preparing and presenting oral presentations. Finally, we will look at the written documents that may result from oral presentations.

TYPES OF PRESENTATIONS

There are two types of oral presentations, impromptu and prepared.

Impromptu Presentations

An impromptu presentation is one that occurs on the spur of the moment. You do not know ahead of time that you will be expected to speak, and have made no preparations. The topic obviously must be one with which you are familiar.

You may be called on to make such a presentation during a meeting in which a topic with which you are familiar is brought to the attention of those attending. The group may want additional information on the topic, and you may be requested to provide it. You will need to gather your thoughts together quickly, determining the information your listeners need and the order in which to present it. This is often difficult to do. An impromptu speech may quickly become disorganized and unfocused. Even if you can line up your presentation quickly, there is always the chance that you will become sidetracked as you begin to speak, and wander off on a tangent, leaving your listeners trying to figure out where you're going and what the relationship is between the information they're hearing and the purpose of your presentation. In addition, you do not have much time to think about listeners' biases, questions, or arguments, nor to prepare appropriate comments to avoid negative responses to your information.

You need to keep the following strategies in mind as you talk if you are to make a clear and concise presentation which your audience accepts.

- Make your presentation short and to the point to avoid becoming sidetracked. If your audience wants additional details, they can ask for them during a question-and-answer period following your remarks.

- Present factual information. Don't embellish with adjectives or insert personal opinion. Audiences tend to respond more positively to factual discussions.

Because of the pitfalls inherent in impromptu presentations, you should always prepare a presentation if you know you will be called on to speak.

Prepared Presentations

If you know you will be expected to make a presentation, you can determine the information, organizational pattern and sequence, and style ahead of time.

Prepared presentations assume two forms: the prewritten presentation, and the extemporaneous (outline) presentation.

Prewritten Presentations For the prewritten presentation, the speaker writes out the complete text and then either memorizes or reads it. Memorization is a difficult technique at best. Because even a five-minute presentation is a lot to memorize, only people who memorize quickly and easily should use this method. If you forget a part midway through, you might get caught staring desperately at your audience trying to recapture your script. In addition, a memorized script often sounds mechanical and boring. This is also true of a paper which is read. While the conventions of a professional meeting may require a research paper be read to a group of peers in the discourse community, an audience of generalists and novices prefer a more extemporaneous approach.

Extemporaneous Presentations The extemporaneous presentation, which uses an outline to guide the speaker, provides a compromise between the impromptu and the prewritten speech. It is prepared ahead of time, but, instead of speaking from a complete script, you talk from an outline. The presentation appears less stilted and more sincere, and at the same time well organized and focused. Most presentations in the workplace should be in this form. The remainder of this chapter is concerned with the extemporaneous presentation.

WRITTEN VERSUS ORAL DISCOURSE

Preparing to give an oral presentation is similar to preparing the draft of a written document. Just as you consider the context in which you write, you must consider your listeners, and the purpose and situation in which you speak. And just as the information about which you write usually will emanate from your own work, so will the topics on which you speak. You gather, organize, and revise your information just as you do for written documents, and you employ the same kinds of rhetorical techniques to describe, instruct, and persuade readers. However, your audience cannot reread what you have written and, in many cases, cannot question you or ask you to repeat or clarify your discussion. Therefore, it is even more crucial that you organize information so listeners can follow it easily, that you use techniques to help them remember major concepts, and that you provide for people who may be more oriented towards reading than listening.

Because oral presentations provide listeners with a vocalization of the text as well as a visual image of the speaker, the presentation of the text is as important as the content. Your voice must convey the message clearly and with proper emphasis. At the same time, your facial expressions and body movements must focus the listener on the topic.

Because listeners have more difficulty understanding and accepting an oral message, you need to learn some strategies that relate directly to oral discourse. These relate to the preparation and presentation of oral text.

PREPARING ORAL PRESENTATIONS

This section points out the similarities between preparing written and oral discourse, and provides strategies for overcoming problems specifically related to oral presentations.

Analyzing the Context

Oral presentations are made before a variety of audiences, for various purposes, and under different situations. A pre-proposal presentation, in which you present an idea to obtain feedback for improving your concept, may be a relatively informal presentation that occurs around a circular table and involves only three or four peers, all of whom are in the same discourse community as yourself. Other presentations, such as an exit meeting following a plant or division evaluation, may still involve a small group, but the listeners may be managers who are generalists in your field, and your presentation may be made in a more formal setting around a long, rectangular table. Still other presentations, such as those you might be asked to make to a local Rotary Club or the YMCA, may be quite formal and involve standing before a large audience of novices.

Oral presentations may be both reader- and writer-initiated. You may request a meeting to present a proposal, or a local community group may request that you speak.

Your listeners may or may not want to take notes. If you wish them to do so, you should notify them ahead of time so they bring paper and pens. If they are taking notes, you need to speak slowly; be prepared to repeat key points so that they can get them down accurately. A visual aid that lists the key points or details can further help the audience in notetaking.

However, listeners prefer to concentrate on a presentation and they therefore avoid taking notes. They would rather have the speaker provide handouts that summarize the important information and which include the charts, tables, etc., they need to remember or use.

Making Decisions

Once you have analyzed your audience and determined your purpose and situation, you are ready to make decisions similar to those you made for written documents. These decisions are directly related to the context in which you will speak, just as they were to the context in which you wrote.

Content   You gather information in the same way as you do for a written document. In many cases, the information can be found in documents you have already written. The criteria for the information to include is based on your readers and your purpose, just as it is in a written document.

Organizational Pattern and Sequence   You use the same criteria for determining the organizational pattern and sequence that you use in drafting written text.

Because listeners have more difficulty following an oral presentation, you should begin with an introduction that establishes the purpose of the presentation, the topic, and a list of subtopics in the sequence in which they will be discussed. The introduction provides listeners with a frame of reference and prepares them for the information to follow.

You should conclude with a summary of the major points discussed to refresh listeners' memory and to synthesize the information presented.

Style   Because it is more difficult to grasp words that are heard but not seen, it is vitally important to use words with which the audience is familiar. You should also avoid using words with which you are not familiar, and which are difficult or tricky to pronounce. If you must use such words, write them out phonetically (the way they sound) and indicate the accented syllable so that when you read them, you read them correctly. Your dictionary can provide this information.

cicatrix   si ka trix

Armageddon   Ar ma gued don

Series of words containing sounds that irritate or draw your listeners' attention away from the meaning of the words should also be avoided. Alliteration, assonance, and sibilance[1] in your prose can cause problems.

Difficult   The disc-to-piston pin was propelled past the plaster post and stripped at the point of engagement.

Smooth   The pin in the piston was thrust beyond the column and the threads were stripped at the point of engagement.

---

[1]Alliteration—the repetition of consonant sounds. Assonance—the repetition of vowel sounds. Sibilance—the repetition of the "s" sound

The string of P's causes the first sentence to be difficult. If you read it aloud, you can see how the P's cause a problem.

Sentences should be short. Try to keep them to a maximum of 25 words. It is difficult to deliver a long sentence without running out of breath. In addition, listeners have difficulty remembering the beginning of a long sentence. Whereas a reader's eyes can take in two lines of type at a glance, listeners hear only one word at a time. Simple and compound sentences are easier to follow than complex sentences. It is even more important in speaking than in writing that you do not interrupt the subject-predicate-object sequence. Most listeners cannot remember the subject of a sentence if several phrases separate it from its predicate.

Because you are speaking directly to your audience, even if that audience numbers in the hundreds, you should use first and second person. An audience perceives the anonymous pronoun "one" as superficial.

Designing Visuals

Because most people are visual as well as aural learners, visuals help your audience understand your message. As you learned in Chapters 12 and 13, visuals help the audience perceive chunks of information and understand it. In addition, visual text can provide an outline of the presentation to help listeners locate where they are in a speech.

There are many types of visuals, including handouts; easels, greaseboards, and chalkboards; slides and transparencies; video cassettes; and LCD displays. You may use one or all of these visuals during a presentation, depending on your audience's needs and your situation (Table 21-1).

Handouts Handouts, written documents related to the topic of a presentation, are prepared ahead of time and distributed to the audience. Handouts can fulfill several functions. They can:

- **Provide information** to which the audience needs to respond during a presentation. For instance, a speaker who is discussing two designs for a proposal product may want the audience to look at and compare the designs.

  Informational handouts should be distributed to the audience before a presentation begins.

- **Outline or summarize a presentation.** Listeners can use an outline or summary before, during, and after a presentation. By scanning the handout prior to the presentation, the audience can perceive the frame for the upcoming speech, and can more easily perceive the relationship between the focus and the details which are presented.

  Listeners can refer to these handouts during a presentation the way theatergoers refer to a program for a play. The outline or summary helps listeners locate where the speaker is in relation to the entire presentation.

VISUALS

	Preparation	Distribution/ Presentation	Audience Use	Advantages	Disadvantages
Handouts	Prior to presentation	Prior to presentation	• Frame of reference • Road map • Supplement • Review • Future reference	• Includes as much information as necessary • Allows eye contact • Inexpensive	None
Easels, Boards	None	During presentation	• Visual reinforcement • Complement	• Responds to audience's immediate needs	• Can't include much information • Equipment not readily available • Requires space, appropriate room arrangement • Handwriting must be legible • Difficult to use while speaking
Slides, Transparencies	Prior to presentation	During presentation	• Visual reinforcement • Complement	• Allows eye contact • Legible visuals • Attractive graphics • Accessible to audience	• Expensive (slides) • Time consuming • Equipment not readily available • Can't include much information
Video Cassettes	Prior to presentation	During presentation	• Visual reinforcement • Complement • Supplement	• Demonstrates action, interpersonal relationships	• Expensive • Equipment not readily available
LCD Displays	None	During presentation	• Visual reinforcement	• Demonstrates computer software	• None

Table 21-1 A Guide for Using Visuals

These handouts can also allow listeners to review key points of a presentation later. For example, after listening to a presentation reporting on a study of noise pollution in an urban area, a listener may want to examine the causes to determine which ones relate to his specific industry.

These handouts should be distributed before a presentation.

• **Supplement information** provided in the presentation. Listeners may want additional information that the presenter does not have time to discuss. The audience, in listening to the presentation on the noise pollution study, may want specific information relating to the methodology used to measure pollution, but the presenter may not have time to explain the methods in detail.

These handouts should be distributed after the presentation.

- **Provide data for future reference.** Listeners may need to use some of the data after the presentation. The presentation on the noise pollution study may include a table of the decibel levels for each of the major industries in the area. Listeners may want to refer to this data after the presentation to learn the level for their respective industry, and to compare their industry's level to others.

These handouts should be distributed *before* the presentation and can be referred to *during* and *after* the speech.

Label and number handouts in the order in which they are discussed in the presentation. Handouts should be easy to read. They should be double-spaced and in black ink on white paper. In addition, make certain that the print on all copies is sufficiently dark to be read easily.

Handouts can be drafted either during or after you write your presentation.

Easels, Greaseboards, and Chalkboards

These boards allow you to record information for the audience to see as you speak. However, they require certain preconditions:

- The equipment must be available in the room in which you will be speaking.
- The equipment must be positioned so everyone can see it. In a large room, people in the rear often have difficulty seeing easels or boards.
- Sufficient space must exist in the front of a room to place an easel. In addition, you need sufficient room to move around the equipment.
- Your handwriting must be legible and large enough to be read easily by everyone.

Using these boards can be difficult. You need to:

- Speak and write simultaneously. Your listener doesn't want to wait while you write.
- Speak loud enough to be heard, especially if you are turned toward the easel or board while you write.
- Maintain eye contact with the audience while you write. Try not to turn your back on the audience. You can accomplish this by writing sideways, so that you can continue to look at the audience while you write.

An easel pad is approximately 20 x 30 inches. Chalkboards and greaseboards in meeting rooms are not very large either. If you write large enough for everyone to read the words easily, you will not be able to place much information on a page of an easel pad or on a board. Therefore, these boards should be used to record only brief bits of information, such as lists, key words and phrases, definitions of unfamiliar words, and numerical computations.

Because of the problems associated with these boards, they should only be used to display information that results from direct communication between you and the audience during your presentation or a question-and-answer period. Information in your presentation that you want the audience to see should be placed on a slide or overhead transparency prior to the presentation.

Slides and Transparencies These visuals must be prepared ahead of time. Like the easel and boards, you need to be certain these visuals can be used in the room in which you are meeting, and that the equipment for showing them is available. Since they don't rely on your handwriting, slides and transparencies allow you to maintain eye contact with your audience. You can glance quickly at the screen for a cue for your next point, then look at the audience while you discuss it. Many places have rheostats, which allow you to dim the lights sufficiently for the slide or transparency to be clear and bright and at the same time allow sufficient light for your audience to take notes easily.

While professional graphic artists can create beautiful visuals, you can also create very readable and attractive transparencies with a computer and laser printer. In addition, because these visuals can be projected on a screen over the heads of the audience and because the screen can be enlarged, the audience does not have difficulty seeing them. Use capital and lower case lettering, except for titles, and be careful that the lettering is easy to see if you use colors. Keep in mind that some members of your audience may be "color blind," and unable to differentiate between red and green. Use large lettering, and an easy-to-read typeface.

Slides are more expensive, difficult, and time-consuming to make than transparencies. Therefore, they should only be used for showing visuals that cannot be shown on a transparency, such as photographs or illustrations. Transparencies may be used for projecting visual text, charts, and tables. However, for groups of over 100 persons, slides should be used for these also, since slides are easier to see.

As with the easel and boards, transparencies and slides should be easy to read. They should not be crowded with information, and chunks of information should be surrounded by white space. A transparency or slide should contain a headline for a text, not the text. An entire paragraph should not be transferred to these.

Transparencies and slides, like the easel and boards, should be used only for listing information the audience needs to remember, key words and phrases you want to emphasize, and definitions of unfamiliar words the audience needs to know to understand your discussion.

The transparency in Figure 21-1 is used in a program to persuade county residents to support a stormwater program. The transparency emphasizes a major reason citizens should support the program.

Transparencies or slides can also be used to provide the audience with "organizers" for following your presentation. Titles, headings, and subheadings can be listed on a transparency to help listeners locate where they are in your speech. Transparencies and slides may include the main organizing ideas for the various sections of your presentation, as well as the organizing ideas for the

**Figure 21-1
Emphasizing a
Point with a Visual**

COORDINATED, UNIFORM REGULATIONS PREVENT FUTURE DAMAGES

**Figure 21-2
Helping the
Audience Predict
What They Will
Hear with a
Transparency**

STORMWATER MANAGEMENT PLAN

I. Problem
II. Solution
   A. Legislation
      1. Establishment of stormwater management committee
      2. Objectives
      3. Policies
   B. Plan
      1. Major elements
      2. Short/Long term results

major categories in each section. These enable the listener to remember the main point as you discuss the details related to it.

The transparency shown in Figure 21-2 prepares the audience for listening to the presentation on the stormwater program.

Visual text that appears on a transparency should be derived directly from the text or outline of the presentation and, therefore, should not be written until after the text or outline of the presentation is completed.

Video Cassettes These visuals should be used only when a topic requires an audience to see action, or to see and hear some form of interpersonal communication. Making a cassette is more expensive than making the other visuals and often the equipment to run it is not readily available.

LCD Displays These displays are extremely useful for programs involving computers because they permit an audience to see what is on a computer monitor. The equipment, which is connected to a computer at one end and an overhead projector at the other, projects the image on the monitor to a large viewing screen that the audience can easily see. This visual aid is especially useful for providing instructions in software use. As with the other types of visuals, you need to be certain that the appropriate equipment is available, and that it can be used in the room in which you will give your presentation.

Outlining Extemporaneous Presentations

If you were to draft an entire speech, you would do so in the same way you would draft a written document. But, because we have already asserted that reading a completely written-out text is not a good way to make a presentation, we will only discuss the use of outlining in preparing a presentation.

Functions   The outline in an extemporaneous speech serves several functions:

- **It serves as a truncated script.** Therefore, the outline should include all the information you want to present in an abbreviated form. It should include all the major categories and subcategories of a topic, as well as lists of information, examples, and quotes that you may not remember. It should also indicate when you plan to use visuals, to refer to handouts, and when you want a response from the audience.

- **It provides cues to topics you plan to discuss.** A single word or phrase in an outline should serve as a trigger for an entire discussion of a subject. Therefore, an outline should include key words for each major category and subcategory you plan to include in your presentation.

- **It serves as a road map to keep you on track.** Topics and subtopics should be in a logical order. You may want to insert transition words at the beginning of each new category or subcategory. These are useful for indicating to your audience the relationship between the subject you are about to discuss and the previous one.

Scripting   If you organize your information according to one of the methods discussed in Chapter 7, creating an outline is easy. Use the major categories and subcategories as the basis of your outline. Fill in additional information that you want to be certain to remember. Sometimes, even after you do this, you're not sure exactly how the presentation fits together. Some people find they need to write out the entire presentation as a complete written document, then derive the outline from their text. You may find you need to write a complete draft several times before you are comfortable drafting only an outline.

Once you have drafted your outline, you need to print it out so that it is easy for you to read as you speak. Some people use note cards (5 x 8 inch cards are easiest to see), others use the pages of a legal pad, while still others use transparencies.

Your outline should be easy to read, and allow you to quickly find your place.

- Print the letters, using capital and lower case. Don't use cursive script.
- Use large lettering.
- Separate each category with plenty of white space so you can see when you come to the end of one category and are ready to begin another.
- Place a single subcategory on a page so you can easily keep your place.

If you were to make a presentation on preparing a speech, you might use the example below. It follows a formal outline; you may or may not want to make your outline quite as formal. Notice that it is derived directly from the headings and subheadings in this section of the chapter. The outline also includes lists from the various sections as well as examples and, in parentheses, specific data and quotes the presenter wants to remember.

I. PREPARING ORAL PRESENTATIONS

   (30 minutes)   Time

   A. Introduction—purpose, topic, sequence   3 min.

   B. Analyzing the Context   3 min.

       1. Audience—all types, sizes

       2. Purpose—reader-/writer-initiated

       3. Situation

       Ex. pre-proposal to exit mtg.

   C. Making Decisions   3 min.

       1. Content—same as for writ. docs.

       2. Org. Pattern & Sequence—same as for writ docs but always...

          a. Begin with intro—purpose, topic in nutshell, sequence of info

          b. Conclude with wrap-up. Review main pts.

       3. Style   3 min.

          a. Familiar words

          b. Short sentences (25 wds) (subj-pred-obj)

          c. Avoid alliteration

             (The disc-to-piston pin was propelled past the plaster post and
             stripped at the point of engagement.)

          d. Personal (you, I)...

Timing One of the major aspects of an oral presentation that you need to address is time. Often, you are told how much time you will have. You don't want to come up short, but neither do you want to go over your limit. It can be embarrassing to give a ten-minute speech when you've been allotted a half-hour. But it can also be aggravating to the audience, as well as to other speakers, if you speak for 45 minutes when you've only been given 20 minutes. This is one major reason you need to practice your speech. By practicing, you get an idea of the amount of time it will take, and you know in advance whether you need to include more information or delete some.

It often helps to indicate the amount of time you plan to spend on each major aspect of your presentation, as the speaker does in the example above. Determine the amount of time according to the importance of the information. The introduction and conclusion should always be short (they should take no more than 20 percent of the total time allowed). Spend the majority of your time on the most important aspects. Also, plan to spend extra time on concepts that may be difficult for your audience to understand. When you present your speech, place your watch on the table or podium in front of you, so you can refer to it periodically to determine if you are running over or under your time. If you find you are doing either, make a mental note to expand on a point or eliminate a category in order to remain within your limits.

Example

Figure 21-3 contains the script for a formal slide presentation made by a guest speaker at a conference. The presentation is prewritten so that the speaker moves from slide to slide in the proper sequence. Because the writer is speaking to a group only on the periphery of his discourse community, he does not use technical terminology. The tone of the speech is informal, and parts are even humorous.

As you read, notice that visual prompts are included. In addition, concepts that should be emphasized are underlined to cue the speaker.

Figure 21-3
A Scripted Speech

Introduction.

Humor.
Indicates emphasis with underlinings.

Forecast.
Problem/solution organization pattern.
Problem 1.
Visual directions.

OILED AFTERMATH: LESSONS FROM THE ALASKA AND PERSIAN GULF OIL SPILLS; WHERE DO WE GO FROM HERE?

A Speech Presented at The International Environmental
Film Festival, Boulder, Colorado
by Albert M. Manville II, Ph.D.
Senior Staff Wildlife Biologist, Defenders of Wildlife
Washington, DC

Good afternoon filmmakers, videographers, ladies and gentlemen. Welcome to the panel on oil and the environment—this is a precursor to the St. Patties Eve Green party tonight.

I am Senior Staff Wildlife Biologist for Defenders of Wildlife, a national wildlife conservation organization with over 80,000 members, supporters and activists.

In case you were wondering, a wildlife biologist is a game warden who wears a necktie—not to be confused with an ecologist who is also a game warden but wears a sport shirt and doesn't work on weekends. And wildlife management is the process whereby men and women are paid for things that animals do on their own, while wilderness is an area with only one television station and a considerable distance from the nearest bar.

Now that we've cleared that up, back to the subject at hand. To get a better idea about the problems related to an oil spill, let's first take a trip to south-central Alaska, look at the problem, then talk about solutions to the EXXON VALDEZ and the Persian Gulf spills, including things we can all do.

[dim lights please]
[focus slide]

continued

Figure 21-3
A Scripted Speech
continued
Presents slides.

1) The shortfalls of the use of petroleum have never been made more clear than with the March 24, 1989 disaster of the EXXON VALDEZ in Prince William Sound—now nearly 2 years ago.

2) On Good Friday 1964, Alaska experienced a massive earthquake—a natural disaster of unprecedented proportions of 8.3 on the Richter scale. Exactly 25 years to the day on Good Friday 1989, Alaska experienced a manmade disaster of unprecedented proportions when at least 11 million gallons of North Slope crude began pouring into one of our nation's most sensitive ecosystems. Nature recovered rapidly from the disaster of its own making, but nature still has a long way to go to recover from this disaster of man's making.

3) What resulted in a 3-square-mile spill on Good Friday soon spread to a 100-square-mile spill by the following week. By May 10, 1989, over 10,000 square miles of Prince William Sound, the Gulf of Alaska, and Cook Inlet had been affected.

Uses lists.

4) • One of the richest marine ecosystems in the world, Prince William Sound alone supported a $100 million fishing industry;

   • one of the most diverse wildlife habitats, Prince William Sound supported the largest bird migration in the world, contained a pre-spill population of 10,000-12,000 sea otters, at least 10 pods of killer whales, endangered humpback and California gray whales, hundreds of Steller's sea lions and harbor seals, and many other species of wildlife;

   • and was one of the most beautiful places on Planet Earth...

Relates to audience's values. Uses 1st person plural to include audience in his ideas. Directions.

26) Although the immediate blame for this spill surely falls on Exxon, our demand for petroleum—and for the many products made from petroleum, including plastic—plays into this whole disaster. We must start becoming more energy <u>conscious</u> and resource <u>cautious</u>. Think about what's happened to Alaska when you use petroleum, electricity, automobiles, or plastic.

[lights on please]

Gives video cue. Shows video.

<u>Show 7:20 video tape</u>: "Oiled Aftermath; the EXXON VALDEZ retrospective"

Reminder.

Special thanks to:
   Lisa Swann, Defenders' public relations director
   Chris Cram, her husband, of Potomac Television
Much of the video footage and slides I shot; Potomac Television kindly provided the rest.

Problem 2. Discussion.

Let's talk briefly about the PERSIAN GULF situation.

During the week of January 23rd, Saddam Hussein's forces apparently unleashed an act of ecoterrorism—the first conflict in which ecoterrorism played a major role in a combatant's battle plan...

...But cleaning up the mess is <u>not</u> the real solution. The simple fact of the matter: America is addicted to oil. And we are paying dearly to feed our habit—oil spills, polluted air, and global warming. But we <u>can</u> prevent future wars over energy resources.

continued

Figure 21-3
A Scripted Speech
concluded

While some people advocate drastic changes in our lifestyles to conserve energy, the fact is we don't have to make major sacrifices to save our planet. The <u>SOLUTION</u>:

- combine sensible conservation measures with more efficient energy use. Here are a few examples:

- cars already exist that get 50 or more MPG; there, however, has been little effort to raise the <u>national fuel economy standard</u>;

- energy efficient lightbulbs already exist; utility companies, however, are <u>not</u> promoting them;

- our government should be informing citizens about ways to conserve energy;

- we need new sources of <u>renewable</u> energy—solar, wind, biomass, water—to provide ample safe, clean, and economical energy.

In short, we need an effective national energy policy. This does <u>not</u> mean opening up more land for oil and gas drilling, especially in the Arctic National Wildlife Refuge.

Solution.
Recommendations.

In conclusion, do things to help the environment. Here are a few examples:

- buy and use fewer plastic products—ask for paper bags rather than plastic, buy margarine in sticks rather than plastic tubs, use canvas shopping bags when you shop, and reuse grocery bags;

- support recycling efforts for paper, plastic, glass and metals; if you don't have a recycling center, request that your city council establish one;

- recycle your own trash—reuse plastic and paper bags, compost vegetable wastes, grass clippings, and leaf litter; recycle used motor oil;...

Strong conclusion
directed at audi-
ence. Use of 2nd
person.

The decision is yours. Your voice and your actions can and do make a difference. We all <u>must</u> make this the decade for the environment. Thank you very much.

Defenders of Wildlife

MAKING THE PRESENTATION

Many people become nervous if they have to make a speech. But there is no need to do so, because in almost all cases you are speaking knowledgeably about a project in which they have been involved. You know your subject well. If you have written a good outline, you don't have to worry that you will ramble or forget a major point, because your notes will keep you on course. But there are some strategies you should follow to ensure that your presentation is effective.

Rehearsal

Rehearse your speech ahead of time. There are several different ways to practice a speech:

- Close the door to your office and go through the entire speech out loud. This kind of practice gives you the opportunity to discover words you trip over or don't know how to pronounce, sentences that are too long, and an estimate of the time for discussing each aspect.

- Audio-tape your rehearsal, then replay it. As you listen to the replay, critique the sound of your presentation. Listen for words you mumble or for places when you drop your voice too low to be heard. Check that your tone emphasizes the words and phrases you want to stress. Determine whether the tone and pitch of your voice varies, or if you drone monotonously. Hear whether you speak very fast or very slowly. Note your problems and try again.

- Video-tape your rehearsal, then replay it. As you watch the replay, critique the appearance of the presentation, as well as the sound. Listen for the problems noted above. In addition, watch to see if you maintain eye contact with the camera which, in this case, is your audience. Look for irritating facial expressions, like twitching your nose, which are distracting. You also want to be careful you don't have any distracting body movements, like tossing your hair back or tapping your fingers.

Vocalization

Your voice carries your message. The audience must be able to hear you and understand your words. In addition, your voice must keep the audience interested; it should not lull them to sleep. And, finally, your voice needs to serve in the same capacity as visual text does in written prose. By stressing words and ideas and by altering your pitch, you can indicate important ideas, humorous or ironic aspects and questions.

- Speak loudly enough for people in the back row to hear you. If you have a naturally soft voice or are in a large room, use a microphone. After a few minutes of speaking, make a visual check of the back row to see if people seem to be straining to hear.

- Speak slowly enough for listeners to understand you, but not so slowly that the audience can fill in the words before you pronounce them.

- Speak distinctly. Pronounce each word clearly. Don't let your voice fade out at the end of a sentence or paragraph.

- Speak with authority. Let your audience know you are knowledgeable.

- Use your tone to stress important ideas or phrases or to indicate humor.

- Try not to use fillers, such as "uh" or "hmm," or "you know." If you have to pause to look at your notes or see where you're going, simply keep quiet. A moment of silence while you gather your thoughts will be accepted by audiences.

Eye Contact

Eye contact—looking directly at your audience—is very important, especially if you are attempting to persuade your listeners to accept your ideas. Think of your presentation as a conversation. In ordinary conversation, you usually look at the other person. In this situation, you should look at your listeners.

If you are in a small group, you should shift your gaze from one person to another so that all your listeners feel included. If you are making a presentation to a large group, you should shift your gaze from one side of the room to the other to take in all those present.

You may need to glance periodically at your notes or at a visual aid. Once you determine the next category on which to speak, look back at your audience. If you are using visuals, such as easels or greaseboards, stand to the side so that you can point to them and still maintain eye contact with your audience. If you are using a transparency, look at the hard copy or at the actual transparency on the projector, rather than at the screen, so you don't have to turn your back on your audience.

Body Language

Your body, like your voice, helps deliver your message. You need to project confidence as well as authority. Stand straight and avoid nervous movements. If you are speaking before a large group, it helps to have a podium in front of you where you can not only place your notes, but also rest your hands. Do not fiddle with your notes.

Visuals

Check your visuals carefully before your presentation. Make certain all equipment is available, plugged in, and working. There is nothing worse than building a presentation around visuals, only to discover, when you are ready to speak, that there is no projector available. Also, don't waste the audience's time while you look for a plug or try to figure out how to turn on a projector.

If you are using transparencies or slides, make certain they are lined up so they don't appear upside down or backward on the screen. If you are using an LCD display, check that the proper software is available. If you are using easels, chalk, or greaseboards be sure you have sufficient pens or chalk for writing on them, and an eraser.

Display your visuals only when you begin to speak about the topic they depict. Leave them on the screen while you discuss the topic, then remove them. Always turn off an overhead or slide projector if you are not showing a slide. Keep as many lights on in the room as possible so you can see your audience and they can see to write notes. Some rooms allow you to dim the lights directly over the screen, leaving the rest of the lights in the room on.

THE WRITTEN REPORT

Oral presentations serve more as executive summaries of information than as full reports. Often, supervisors or clients want a written report in addition to the oral presentation. They need the written report for their records and for a reference when they are back in their offices. Furthermore, they may want more details than you have time to provide in an oral presentation.

Readers usually use these written reports to obtain specific information. Therefore, they may skim through a report, searching for the information they want, then scan the particular section or paragraph in which they are interested. For this reason, long reports should always contain global locators.

Reports emanating from oral presentations can assume the form of a full report, a presentation report, or an intermediate report.

Full Reports

We discussed full reports in Chapter 19. Knowing ahead of time that your supervisor or client wants a full report, you should write it, and then draw your outline for your presentation from it.

Presentation Reports

You may not have sufficient time to draft a final report. Clients want information as quickly as possible, and consultants often make oral presentations prior to writing a report. The presentation report has evolved to provide clients or management with a written record of an oral presentation as quickly and efficiently as possible. It is comprised of copies of the transparencies used in an oral report.

Intermediate Reports

Clients and supervisors often need more information than is contained in the transparencies of an oral presentation but, having heard the presentation, they do not need a fully-drafted text. The intermediate report is a compromise between the full report and the presentation report. It is an elaboration on the transparencies used during a presentation. Using an outline format, numeration and bullets, and truncated sentence structure, the intermediate report provides readers with the information included in the presentation plus relevant details. It is easier and quicker to write, as well as easier to scan and read, than a full report.

Figure 21-4 contains an exerpt from an intermediate report on a marketing survey of drug treatment by medical professionals. The report expands on the presentation.

**Figure 21-4
A Page from an
Intermediate
Report**

II.   CURRENT USE OF INJECTABLE ANALGESICS

A.   <u>Speed of adoption</u>

Anesthesiologists are the boldest prescribers of new injectable analgesics.

1. Forty-two percent would try a promising new agent in this class either in clinical trials before launch or when it is first released ("early adopters").

   • Less than one-fourth would wait until the new product had been tried by many MDs or widely used ("late adopters").

Strategic Marketing Corporation, Philadelphia, PA

STRATEGY
CHECKLIST

1. Make extemporaneous presentations if possible.

2. Use an outline to guide your discussion.

3. Vary your tone so it isn't monotonous.

4. Speak clearly and authoritatively, emphasizing important points. Try to speak naturally, neither too slowly, nor too precisely. Speak loudly enough for people to hear you.

5. Use facial and body expressions to emphasize your main points. Be careful that your facial and body expressions do not detract from your speaking.

6. Establish eye contact with your audience.

7. Open with an introduction that provides a frame for the remainder of your discussion.

8. Conclude with a summary of your main points.

9. Use first and second person.

10. Use the following visuals to enhance your discussion:
     • handouts
     • easels, greaseboards, and chalkboards
     • slides and transparencies
     • video cassettes
     • LCD displays

11. Gauge your time. Do not speak either over or under your allotted period.

12. Rehearse your presentation.

13. Accompany your speech with one of the following written reports:
     • full report
     • presentation report
     • intermediate report

**CHAPTER
SUMMARY**

There are two types of oral presentations: impromptu and prepared. Prepared presentations can be either prewritten or extemporaneous. Extemporaneous presentations appear less formal and are preferred by listeners.

Preparing oral presentations is similar to drafting written documents. However, because listeners cannot reread the text or question the speaker, and because some listeners may be visually rather than aurally oriented, speakers must be careful to provide "organizers" at the beginning of a presentation and cues throughout the speech to help listeners follow the discussion. Important information needs to be displayed visually, and summarized at the conclusion of the presentation. Language and sentence structure should be simplified and personalized.

Use visuals in conjunction with your discussion. A variety of visuals can be used to supplement a presentation, including handouts, easels, grease and chalk boards, slides, transparencies, video cassettes, and LCD displays.

To prepare for an extemporaneous presentation, draft an outline. Outlines serve as truncated scripts that cue you to the topics to discuss and keep you on track. Place your outline on note cards, legal pads, or transparencies, and determine your timing for major categories.

When you make your presentation, your voice, eye contact, and body language are as important in presenting your message as your words.

Determine if your audience wants a written report of the work discussed in your presentation. If so, develop a presentation report by using your transparencies as an outline, and filling in basic information your readers need to know.

PROJECTS

As you undertake a project, be sure to engage fully in all three processes—planning, drafting, revising. You may want to use the Planning Worksheet in Appendix B, as you gather and organize your information. As you prepare to revise, you may also want to use the Document Design Criteria Checklist in Appendix B.

Collaborative Projects: Short Term

1. **Obtaining support for a class project**

    A local financier has offered to finance a program for one of the following community problems, if college students will donate their time.

    - Pick up recyclable goods that the local waste company does not collect from residences.
    - Manage a soup kitchen on campus for the homeless.
    - Rehabilitate structures for use by the homeless.
    - Find housing for the homeless on the campus and surrounding area.

- Conduct a survey of vehicles transporting toxic chemicals through your community.
- Conduct a feasibility study for increasing recycling efforts in your community.
- Work with illiterate parents.
- Tutor children with AIDS.

You and four other classmates have decided to participate.

a. Work in groups of five. As a group, decide on one of the problems on which you'd like to work. You would like to have the entire class support your group's proposal.

b. Collaborate to make a five-minute extemporaneous pre-proposal oral presentation to your class, informing them of your decision and attempting to persuade them to adopt it as a class project.

c. After all groups have made presentations to the class, vote on the project the class should support.

2. **An impromptu speech**

You have just received a half-hour notice to make a three-minute presentation to a small group of people interested in a topic that you have been studying.

a. Select a paper you have been working on in one of your classes.

b. Use the half-hour to prepare your presentation.

c. Meet in groups of five. Take turns making your presentations. Go in alphabetical order, based on your last names. The last person also assumes the role of chairperson. The first person assumes the role of assistant chair. The chair's major responsibility is to indicate to the speakers when they have spoken for three minutes and need to wrap up their presentation. The assistant chair keeps the time during the chairperson's presentation.

3. **Construction and landscaping improvements**

Select an area on the campus or in the community for construction or landscaping improvements.

a. Work with five people. Develop a five-minute prewritten presentation to urge the city council/county board/campus administration to make the improvements.

b. Make the presentation to the other members of the class.

Collaborative Projects: Long Term

- **Communicating about a campus/community problem** continued from Chapter 4, Collaborative Projects: Long Term Exercise 1.

- **Proposing microcomputer laboratories** continued from Chapter 4, Collaborative Projects: Long Term Exercise 2, Projects 1, 2, or 3.

- **Gathering information for a feasibility report: diaper laundry service** continued from Chapter 6, Collaborative Projects: Short Term Exercise 2.

- **Gathering data for an information report: cleaning supplies** continued from Chapter 6, Collaborative Projects: Short Term Exercise 3.

- **Organizing data for a proposal: recycled paper** continued from Chapter 7, Collaborative Projects: Short Term Exercise 2.

- **Organizing data from a food and music survey** continued from Chapter 7, Collaborative Projects: Short Term Exercise 3.

- **Solid waste** continued from Chapter 14, Collaborative Projects Exercise 1.

- **Improving students' preparation for writing in a technical community** continued from Chapter 19, Collaborative Exercises: Short Term Exercise 1.

Develop and make a five-minute oral presentation of your proposal or report to your class. Invite campus administrators, community leaders, etc., who might be interested in the project to attend.

- **Class template** continued from Chapter 18, Collaborative Projects Exercise 1.

Develop and make a five-minute oral presentation to your class of your findings.

- **Information report for pharmaceuticals** continued from Chapter 19, Collaborative Projects: Short Term Exercise 2.

Develop and make a five-minute oral presentation to a group of five students. The group should assume the role of the managers of various divisions who would be involved in this project if the organization decides to go with it.

Individual Projects

1. Prepare a prewritten, five-minute oral presentation to inform the other members of your class about a recent theory or discovery in your field in which you are interested. Make your presentation.

2. Prepare an extemporaneous five-minute oral presentation to recruit high school seniors to your college.

3. Select a topic on which you feel strongly. Prepare an extemporaneous five-minute oral presentation to convince members of your class to feel similarly.

4. You have been hired by a fast food chain as a campus representative. Select the chain you wish to represent. Visit at least one of its restaurants to learn what is being done in terms of reducing solid waste. Using the information from the report in Chapter 19, Figure 19-5, and the letter in Chapter 5, Figure 5-10, develop a three-minute, prewritten oral presentation, presenting your fast food chain's efforts to be ecologically and environmentally responsible. Your purpose is to persuade students that your fast food chain is an environmentally responsible organization, and that they should select it over other fast food restaurants. You will be making this presentation at a large number of sorority and fraternity meetings, dorm meetings, and club meetings, and at a student council meeting.

A Brief Reference Guidebook

This Guidebook should serve as a brief overview of the most basic grammatical and conventional rules pertinent to technical writers. While it cannot replace a a comprehensive handbook, it can help with the major aspects of sentence structure, grammar, punctuation, and mechanics that are of concern in writing for technical communities. Refer to Chapter 11 for advice on readability.

SENTENCE COMPLETENESS

Sentences make complete statements by using three basic structures. A **simple sentence** contains a single independent clause with a core subject and verb, often with various modifiers:

Simple: Hazardous waste [subject] abounds [verb] in a house.

A **compound sentence** contains two or more independent clauses connected by a coordinate conjunction such as *and, but, nor*:

Compound: Hazardous waste abounds in a house, *but* most residents don't think of such material as hazardous.

A *complex sentence* contains an independent clause and at least one dependent clause, with a connecting word such as *which, whom, that, after, although,* etc.:

Complex: The hazardous wastes *that* abound in a house may include furniture stripper and paint remover.

In addition to these basics, a **compound/complex sentence** is a combination structure involving both a compound sentence (with two or more independent clauses) and a complex sentence (with at least one dependent clause):

Compound/Complex: Hazardous wastes, *which* abound in a house, include furniture stripper and paint remover, *but* most residents don't think of these as hazardous.

Sentences may also contain **verbal phrases** serving as components. These include **infinitive phrases** serving as nouns and using a base verb form frequently preceded by an infinitive marker, *to*; **gerundive phrases** serving as nouns and using the verb form plus an *-ing* ending; and **participial phrases** serving as adjective modifiers and using the verb form plus an *-ing* ending:

Infinitive phrase: Household hazardous wastes are hard *to identify*

Gerundive phrase: *Identifying household hazardous materials* can be difficult.

Participial phrase: Lead paint *chipping off the walls* is deadly.

These verb-form constructions, commonly used as nouns or modifiers within sentences, cause problems if they are left by themselves without a main verb; they cannot stand alone as complete statements or sentences.

Avoiding Fragments

To avoid writing an incomplete thought punctuated as if it were a sentence, check sentences to be sure they are complete:

1. Find a main verb, expressing action or status. Disqualify verbals and verbal phrases like those above.

2. Locate a subject for the verb: either a noun, a pronoun, or any of the phrases above serving as a noun.

3. Look for connecting words (eg. which, whom, that, after, although, etc.:) which may create a dependent clauses that is not properly linked to a simple or compound sentence (see above).

If you find a fragment —

A. Revise subordinate clauses by converting them to a simple independent clause (eliminating the connecting word), or join the dependent clause to form a complex sentence (see above).

B. Revise phrases to include them as a component of a sentence.

C. Revise any repeating structures to incorporate them into one of the basic sentence structures above.

Eliminating Comma Splices and Fused Sentences

Sentences of explanation with expanded examples, or sentences with transitional words such as *therefore, consequently*, or *for example*, may signal a risk for a **comma splice** — two sentences insufficiently connected with a simple comma— or a **fused sentence** — two sentences with no boundary between them. When complete sentences are strung together, not bounded by proper connecting words or punctuation, they don't form any of the basic structures above because their boundaries or connections are missing.

Fused sentence: Hazardous wastes may be hard to locate therefore many believe their houses are hazard-free.

Comma splice: Hazardous wastes may be hard to find, therefore many believe their houses are hazard-free.

There are several ways to correct this error:

Make two sentences: Hazardous wastes may be hard to find.
Therefore many believe their houses are hazard-free.

Use a semicolon to separate the clauses:
Hazardous wastes may be hard to find; therefore many believe their houses are hazard-free.

Separate clauses with a comma and coordinating conjunction:
Hazardous wastes may be hard to find, and that's why many believe their houses are hazard-free.

Make one clause dependent to create a complex sentence:
Hazardous wastes may be so hard to find that many believe their houses are hazard-free.

Sentence Completeness and Style

In formal discourse addressed to managers, experts, or peers in a scientific or technical community, complete sentences are a necessity for clear communication, using one of the basic three or four patterns above. Beyond this, however, some myths about sentence style persist:

- *Myth:* Some may feel that for "factual" statements in technical discourse, sentences should begin with subject first, rather than with a phrase or subordinate clause. In fact, opening phrases or clauses help break up the monotony of a subject-verb-object rhythm, and are widely used in many technical communities.

- *Myth:* Some feel sentences should never begin with coordinating conjunctions *and, or, but*, or *nor*. Such sentence openings are seldom found in formal discourse to peers and experts, but in less formal discourse to novices, as well as in advertising and literary contexts, sentences may begin with these words if the meaning is clear.

To eliminate household hazards, you can use a variety of alternative substances, including water- instead of oil-based paint. *And,* of course, you can do what Tom Sawyer's Aunt Polly did — use whitewash instead of paint.

SENTENCE PARTS

Subject-Verb Agreement

A predicate verb must agree in number and person with its subject. This normal pattern is a problem only in certain confusing situations when extra caution and remedial action is required.

For singular subject followed by a phrase containing a plural noun. Use a singular predicate; disregard the phrase.

>   One of the rooms *has* lead paint chips.

For single subject linked to another subject by a prepositional phrase indicating addition. Use a singular predicate.

>   The living room along with the bedroom *has* several layers of paint.

For an indefinite pronoun as a subject. Use a singular predicate with the following:

- each
- either
- neither
- one

- everyone
- everybody
- no one
- no body

- anyone
- anybody
- someone
- somebody

Use a plural predicate with the following:

- several
- few
- both
- many

For the following, use either a singular or plural predicate depending on the noun it refers to:

- some
- all
- any
- none
- most

Singular: Some of the paint is five layers deep.
Plural: Some of the chips have lead.

For a compound subject connected by "and." Use a plural predicate.

>   The living room and the kitchen *have* lead-based paint.

For a compound subject connected by "or" or "nor." Use a singular predicate.

> Neither the bedroom nor the study *has* lead-based paint.

For a compound subject including both a singular and a plural noun connected by "or" or "nor." The predicate should agree with the closest noun.

> Neither the painters nor the owner *is* legally responsible for using lead-based paint fifty years ago.

For a collective noun as a subject, such as faculty, species, committee, or class. The predicate can be either singular or plural, depending on whether the noun is considered as a single unit or as separate individuals.

Singular:	The committee *is* writing the proposal. (The committee as a whole is putting together the proposal.)
Plural:	The committee *are* voting on the issue. (Each member of the committee is voting.)

For a singular subject and a plural predicate nominative or vice versa. The predicate agrees with the subject, not the predicate nominative.

> The problem with old paint is the chips.

For a sentence beginning with "There is" or "Here is." The predicate agrees with the noun immediately following the verb *to be*.

Singular:	There *is* lead in the paint chips.
Plural:	There *are* paint chips with lead from old paint.

For a numerical value as a subject. The predicate is singular if the value is considered a single unit; the predicate is plural if the value is considered in terms of its individual parts.

Singular:	Three points above the mean *is* considered a significant increase.
Plural:	Ten percent of the participants sampled *believe* in being proactive.

For a title as a subject. Use a singular predicate.

> "Dances with Wolves" *continues* to be a classic film.

For a relative pronoun as a subject, i.e. *who, which, that*. The predicate is determined by the pronoun's antecedent (the noun to which it refers).

One of the rooms which were tested for lead *has* tested positively.

Pronouns and Antecedents: Reference and Agreement

A pronoun renames and refers to a noun or indefinite pronoun located somewhere nearby in the same or a neighboring sentence; every pronoun needs to have an **antecedent** to which it refers in order to get its specific meaning. To make the reference clear, observe these rules:

Keep pronouns close to their antecedents. Too many nouns between a pronoun and its antecedent can be confusing (eg. EPA or OSHA officers when dealing with an angry client or a troubled group must keep *their [?]* tempers cool.) Rewriting to close the gap will help: (eg. EPA or OSHA officers must keep their tempers cool when...).

Keep pronouns referenced to specific things or persons. Avoid using vague pronoun constructions such as *you know, they say, it figures,* and be sure to make *this, that, which,* or *it* refer to specific nouns. Check for sentences with vague or implied antecedents (eg. In a thunderstorm *it [?]* creates a lightning discharge); rewrite them to make a specific antecedent (eg. If an electric *charg*e builds during a thunderstorm, *it* creates lightning.)

To make an accurate reference, a pronoun must agree with the number, person, and gender of its antecedent. Some sentence situations may make this agreement unclear and may require some corrective action. Here are some of these situations:

Agreement when there is a compound antecedent. When two or more antecedent nouns are joined by certain conjunctions, the pronoun should be plural if the connecting conjunction is *and*; it should be singular if the connecting conjunction is either *or* or *nor*.

Singular: Either the EPA or OSHA, whichever *is* willing to assume responsibility, should initiate the program.

Plural: Both the EPA and OSHA, which *are* responsible for our safety, should be involved in the decision.

Agreement when the antecedent is a common indefinite pronoun. The pronoun should be singular when referring to indefinite antecedents: each, anyone, everyone, etc.; (see the section above on subject-verb agreement).

Choosing an appropriate gender for the personal pronoun referring to one of these is no problem when the antecedent is intentionally limited to an all-male or female group (eg. The sisters shared lunch; each passed her soda around). However, in most other cases these singular indefinite antecedents create a dilemma for the non-sexist writer, who must either use the somewhat awkward *he or she* construction or rewrite the sentence. Rewriting options commonly include these:

Plural constructions: They all passed their *sodas* around.

Passive voice, de-emphasizing the actors: *Sodas* were passed around.

See Chapter 11 for a discussion on avoiding sexist language.

Pronoun Use and Style

A writer who uses first person (*I, we*) or second person (*you*) can create a personal tone or a feeling of "I-you" direct contact. If this is appropriate to your audience community, use these pronouns from the beginning of the document. They should not be inserted abruptly into the middle or end of the text, creating confusion.

Don't mix your use of person in a sentence or a paragraph.

Mixed:	If *you* check *your* house, *you* may find *you* are discarding the same kinds of hazardous wastes that industry does. *One* must be careful not to add to the problem.
Correct:	If *you* check *your* house, *you* may find *you* are discarding the same kinds of hazardous wastes that industry does. *You* must be careful not to add to the problem.

Second person is used to speak directly to the audience as in example "a." It should not be used simply as a figure of speech as in example "b."

Correct:	(a) *You* should check *your* house for toxic substances and dispose of them properly.
Incorrect:	(b) Toxic substances can be found in *your* everyday household cleaners. These wastes include ammonia; drain, oven, and toilet cleaners; and mothballs.
Revised:	(c) Toxic substances can be found in everyday household cleaners. These wastes include ammonia; drain, oven, and toilet cleaners; and mothballs.

Verb Use: Active and Passive Voice

Verbs in the active voice name the subject of the clause as actor or protagonist of the act or assertion which the verb expresses; when the verb is transitive, anything acted upon becomes the object of the verb:

Active: The EPA inspector (subject) found (active verb) several violations (object).

Active voice is the clearest and simplest mode for making direct statements, and is preferable for most general exposition.

Verbs in the **passive voice** are useful to emphasize the object acted upon, and also to avoid the active voice's emphasis on the actor. Passive verbs are formed with the past participle verb form and a form of *to be*, with the original subject relegated to a *by* phrase:

Passive: Several violations *were found* **by** the EPA inspector.

If the *by* phrase is dropped, the object of the verb gets even more emphasis and the actor is eliminated from view :

Passive: Several violations *were found.*

Because the passive voice leaves the actor less specific and is sometimes wordy, some readers strongly object to its frequent use. Yet scientific documents may often use passive voice extensively, especially when the body of the text focuses emphatically on the material acted upon in the procedure, rather than on the technicians or experimenters. However, even scientific documents often use active voice in the introduction and conclusion sections or whenever it becomes important for the performer of the procedure to accept responsibility for a decision. If passive voice becomes a deliberate way to avoid responsibility for a statement or to give a false impression of objectivity, its use becomes irresponsible. It should not predominate when it is important that the performer be identified, as in a proposal.

Verb Use: Some Troublesome Forms

Verbs frequently confused: transitive and intransitive forms.

Some verbs with similar sounds and related meaning are commonly confused and given incorrect roles in a sentence. A *transitive verb* takes a direct object:

The inspector [subject] *sets* [transitive verb] *standards* [object].

An **intransitive verb** takes no action upon any object:

Now the inspector[subject] *sits* [intransitive verb with no object].

The transitive verb *set* is incorrect in such a sentence.

Similarly the transitive verb *lay* takes an object (*Engineers lay pipelines.*), but the intransitive verb *lie* takes no object (*Here the pipes lie.*); in the latter sentence transitive *lay* is incorrect.

The transitive verb *raise* takes an object (*Legislators raise taxes.*), but the intransitive verb *rise* takes no object (*Taxes rise.*), and in the latter sentence the transitive *raise* is incorrect.

The split infinitive. The infinitive — the base verb form — is frequently preceded by the infinitive marker *to*. Some traditional readers are disturbed by any verbal modifier that "splits" the marker (e.g. *to*) from the base form (e.g. *go*), as in *Star Trek's* famous phrase "to boldly go where no one has ever gone before..." While some readers would be more comfortable with "to go boldly" or another phrasing, certainly most readers would be confused by a longer splitting modifier (eg. "to decisively and forever boldly go..."). On the other hand a very brief splitting modifier may be widely heard and read without discomfort. As in the example (to boldly go...), the closely paired verb and modifier may in fact achieve special impact. Often the split formation may prevent more awkward phrasing in a sentence full of complex modifiers. (See Modifiers below.)

SENTENCE STRUCTURE

Modifiers: Placement and Misplacement

In general, modifiers refer most clearly when they are placed *immediately before or after to the word they modify*, but their placement must also avoid breaking up the basic integrity of the sentence statement. In certain situations, especially with lengthy modifiers, placement calls for care to prevent confusion.

Reposition a lengthy modifier that splits a subject and its verb, or splits a verb and its object or complement. Usually a lengthy modifying phrase or clause will be less disruptive if moved to the beginning or end of the main statement.

Confusing: Scientists are, because of their habitual and healthy skepticism, often hesitant to declare a discovery.

Repositioned: Because of their habitual and healthy skepticism, scientists are often hesitant to declare a discovery.

Use care with limiting modifiers. **Limiting modifiers** can change meaning significantly — *only, almost, nearly, even, simply*. They must be placed with special care. Notice the different meanings:

> *Nearly* 75 percent of the 200 sites were contaminated.

> Seventy-five percent of the *nearly* 200 sites were contaminated.

Reposition modifiers that describe two elements simultaneously. In certain phrasings a modifier's placement can create ambiguity:

> People who objected to the plan *quietly* talked of their dismay.

Repositioning the modifier will clarify meanings, either as "People who objected *quietly*..." or "...talked *quietly* of their dismay."

Correct "dangling" modifiers. A modifier is said to "dangle" when the word it modifies is not clearly visible in the sentence. This is likely to happen either when a long introductory phrase or clause is not followed up with a word it can modify, or when a sentence with a passive-voice verb (see above) provides no mention of the performer of the action. Correct the problem by rewriting sentences to provide a word that can be clearly modified.

> Dangling: After doing the inspection, a report was filed.

> Corrected: After doing the inspection, she filed a report.

Parallelism

Words, phrases, and clauses that are paired together or compared must share an equivalent or parallel grammatical form to achieve sentence clarity and consistency. This principle operates within sentences whenever coordinate or correlative conjunctions are used or whenever comparisons or contrasts are made.

Use parallel words, phrases, and clauses with coordinate and correlative conjunctions. When elements are joined or paired by coordinate conjunctions — *and, but, for, or, nor, so, yet*— or when they are contrasted with correlative conjunctions — *either/or, neither/nor, both/and, not only/but also* — then the linked elements must be the same kind of grammatical unit.

> Parallel adjectives: NOT: Hazardous waste is a danger and contaminating.
>   BUT: Hazardous waste is *dangerous* and *contaminating*.

> Parallel phrases: NOT: There is a need for inspecting very quickly yet to work methodically.
>   BUT: There is a need to inspect very *quickly* yet to work *methodically*.

Use parallel structures with compared and contrasted elements. Elements are contrasted or compared using such expressions as *rather than, on the other hand, like, unlike,* or *just as/ so too*. The compared elements must share an equivalent grammatical form.

Not:	Engineers have described the pollution as destructively influencing water chemistry rather than as a civic danger.
But:	Engineers have described the pollution as a *destructive influence* on water chemistry rather than as a *civic danger*.

Use parallel grammatical forms for entries in a list or outline. To emphasize the logical connections among items in outlines and lists, make sure the elements are of equivalent grammatical form.

Not: Items needing attention:
- air pollution
- making inspections
- to find help
- how to find equipment

But: Items needing attention:
- *identifying* air pollution
- *making* inspections
- *finding* help
- *finding* equipment

Sentence Styles

A writer chooses from common styles of sentence structure according to the purpose and emphasis of the document. The common sentence styles are — loose, periodic, and cumulative. Each has a distinctive emphasis and flavor.

In loose construction, the subject and predicate are presented at the beginning of a sentence, followed by a string of phrases or clauses while in periodic construction the phrases or clauses are at the beginning of a sentence and the subject and predicate are at the end. A cumulative sentence is similar to a loose sentence in that the subject and predicate are presented in the beginning of a sentence. In a cumulative sentence, the phrases are usually participial phrases.

Loose:	The EPA inspector found violations in the landfill in terms of hazardous waste from previous years being dumped in blatant disregard of regulations now in effect.
Periodic:	From the dumping of hazardous waste prior to regulations being enacted, the landfill was contaminated by toxic chemicals.
Cumulative:	The EPA inspector found the landfill seeping with hazardous wastes, oozing through underground springs, contaminating and killing a world unconscious of its existence.

PUNCTUATION TO MARK SENTENCE STRUCTURE

End Punctuation

Periods. Periods are used to mark the end of a sentence. A period is always placed inside a quotation that ends a sentence, and inside a parenthesis only if the parenthetical remark is a complete sentence. Periods are used with most standard abbreviations — Mr., Mrs., Ave., Dr., and also with Ms. (though it is not an abbreviation). A period is NOT used with acronymns or acronymn abbreviations — FBI, NATO, IBM, FDIC.

Question marks. A direct question is marked with an end question mark, and after a quoted question within quotation marks. However, an indirect question (such as "He asked when this might happen.") uses no question mark.

Exclamation marks. Exclamations, except after a direct exclamatory quotation within marks, would not normally be used as part of informative technical writing in most communities. Its overuse generally conveys a breathless, less than serious tone to a statement.

Semicolons

Use a semicolon, not a comma, to link independent clauses.
Semicolons can link clauses that are intended to be closely related; commas cannot do this without creating the comma splice error; (see above, "Sentence Completeness.") The choice between separating clauses into sentences with a period versus linking with a semicolon depends on your intention about their relationship. If you spoke each clause aloud, you would find your tone or timing between the two would differ. Thus in examples (a) and (b) there is a difference in meaning and in sound:

(a) You should use borax with vinegar as a cleanser. It isn't toxic.

(b) You should use borax with vinegar as a cleanser ; it isn't toxic.

Similarly, independent clauses can be linked by semicolons — not by commas— immediately before the following conjunctions and conjunctive expressions: *however, therefore, moreover, nevertheless, furthermore, instead, consequently, accordingly, of course, for example.*

Use borax with vinegar as a cleanser ; of course, it isn't toxic.

Use a semicolon to join a series of clauses or phrases that require commas within them.

The flags of the United States, Canada, and Mexico respectively contain the colors red, white, and blue ; red and white ; and red, white, and green.

Semicolons are placed outside of end quotation marks.

Colons and Dashes

Colons. Colons are used after a complete sentence to announce a list (or another statement) that appears separate from the announcing sentence.

> Alternatives for household toxic substances include the following: baking powder to clean hard surfaces, diluted vinegar to clean glass, and cedar chips to preserve fabrics.

> You can use a number of alternatives to household toxic substances: baking powder, vinegar, and cedar chips replace toxics in several cases.

A colon should follow a complete sentence, NOT a fragment that lacks critical parts.

> Incorrect: The colors are: red, white, and blue.

Correct this error by completing the sentence before the colon.

> Correct: These are the colors: red, white, and blue.

Dashes. A dash may serve, more informally than a colon, to announce a list (or another statement) that appears separate from or as an interruption to an opening statement.

Dashes are also used to set off and emphasize a statement added to a complete sentence. The added statement may be a complete sentence, but a phrase or a clause.

> We walked in — we were totally exhausted — and collapsed.

Emphatic interruptions with dashes should be used sparingly in most technical writing. A dash is indicated on a typewriter or computer by a double hyphen.

Commas

The purpose of commas is to avoid confusion within a sentence by setting off or separating words, phrases, or clauses. When they are incorrectly found joining totally independent sentences, they create a comma splice error; (see "Sentence completeness" above). The correct uses of commas can be described under the following categories.

Commas separate words, phrases, or clauses.

Use a comma before a coordinating conjunction (and, but, or, nor, for, so, yet) in a compound sentence. Use a comma.

However, it is not necessary to use a comma if the two clauses are short.

Long   Homes built over twenty years ago may have been painted on the inside as well as outside with lead paint, but homes built today do not use paint containing lead.

Short   Old homes have lead paint but new homes do not.

Use a comma before items in a series of three or more words, phrases, or dependent clauses. You may or may not use a comma before the "and" in the series, depending on the style guide used by the discourse community for which you are writing. Some guides require it, others do not. The present trend in technical writing is to use the comma. In any case, be consistent throughout your text.

Household toxic wastes include nail cleaners, polish, glue, and disinfectant.

Do NOT use a comma to separate only two items in a pair joined by and.

Commas separate introductory and transitional units.

Use a comma following an introductory phrase or clause preceding the subject of a sentence.

After tossing the paint thinner down the drain, she realized she had disposed of a toxic chemical.

Use a comma to follow these transitions:

- however
- therefore
- on the other hand
- of course

If the transition is included within an independent clause, place a comma in front of the transition also. However, if the transition separates two independent clauses, a semi colon, not a comma, should precede the transition.

Within a clause: There is no program for residents to discard household toxic waste. Such wastes, however, should be discarded in the same way as commercial wastes.

Between clauses: There is no program for residents to discard household toxic waste; however, there needs to be.

Use a comma to set off quotations. If a clause introduces a quote, the comma comes immediately after the introductory clause and prior to the quotation marks as in example "a." If an explanatory clause follows a quote, a comma is placed inside the quotation marks as in example "b.". The latter rule was determined by a printer who thought the comma inside the quotations marks was more aesthetically pleasing. If you are in doubt, use the printer's criteria to decide what to do.

(a) According to the Household Products Disposal Council, "Don't dispose of more than one chemical at a time."

(b) "When hazardous household products are misused or improperly stored or disposed of, residents are the ones most likely exposed to the dangers," according to officials of the Environmental Protection Agency.

Commas set off nonessential or interrupting units.

Commas are often used to set apart clauses or phrases that are non-essential, that is, not limiting to a basic sentence's meaning. Such units, (also called non-restrictive), do not restrict meaning but merely add more information. Commas should NOT be used with any essential (restrictive) clause or phrase that limits meaning with an important modifier.

Essential modifier: The paint *that was chipped from the walls contained lead* [The modifying clause means we're limiting discussion to lead paint on the walls; paint from the floor and elsewhere is not under discussion.]

Non-essential: The paint, *which was chipped from the walls*, contained lead. [Here we're discussing all the paint, which incidentally was removed by chipping.]

Because the non-essential additional information is seen as an interruption to the basic statement, it is set off or surrounded by commas. However, if on re-reading you decide that such information is essential and should set a limiting restriction on the meaning intended, you should signal this by eliminating any commas around the clause or phrase.

OTHER PUNCTUATION FOR QUOTATIONS AND WORD FORMS

Apostrophes

Apostrophes are used to mark certain plural forms, show possession, and to show the omission of letters in contractions.

Apostrophes and personal pronouns. Apostrophes should NEVER be used to show possession with personal pronouns, which have their own forms in the possessive case: your, yours, (eg. your book; the book is yours), its, whose, our, ours, etc. These forms must be systematically distinguished from non-possessive pronouns which use apostrophes in contractions with to be verbs: it's (it is), who's (who is), you're (you are), etc.

Apostrophes for contractions. While very formal documents have traditionally refrained from using contraction forms (can't, won't, it's), more and more documents in most technical communities can be found using a contraction style. If in doubt, check previous similar documents in the discourse community for which you're writing.

Possession with multiple nouns. When a pair or cluster of words functions as a single unit, or when joint possession is indicated, add an 's to the last word: *Bill and Mary's car*; *brother-in-law's car*. If possession is individual to each person named, each gets an 's: *Both Bill's and Mary's notes are meticulous*.

With certain plurals. An 's is used in the plural to name letters, numerals, or words identified as words: *dot your i's, three 2's, two and's*. NEVER use apostrophes to form regular plurals of nouns.

Quotation Marks

Use quotation marks (" ") to enclose a direct quote from a written source or from a person you've interviewed or surveyed. Even a single word should be placed in quotation marks if you are borrowing the word from another source. However, do not use quotations marks if you are paraphrasing a written source or person.

> According to the Environmental Protection Agency, "*Government agencies aren't adequately funded to handle the problem*" of residential toxic waste.

If a quotation is long, it should be separated from the text and placed in a block quotation. The maximum number of lines before a quotation should be separated from its text and the number of spaces a block quotation should be indented from the left and right margins varies among discourse communities. Check a style guide for the community in which you are writing to determine the exact format. Do not use quotations marks to set off a quote in block form.

Quotation marks have also been used to specify a word, indicate a word is slang or jargon, or indicate a word is being used in a special way as the example below indicates. With the introduction of desktop publishing, various typographical markers are being used to achieve the same effect.

> Even though cleansers may not be "*industrial strength*," they can be dangerous to your health.

Ellipses

Ellipses (...) are used to indicate omissions in quotes. Often you don't want to include an entire paragraph in a quote, but only two or three parts of it that relate to your specific discussion. You use ellipses to indicate to readers that parts have been omitted. If the omission comes in the middle of a sentence, use three dots. If the omission begins at the end of a sentence, use three dots after the period.

"Just because a cleanser isn't "industrial strength' doesn't mean it's not hazardous....We suggest they [residents]... hold on [to their toxic waste] until there's a collection in their community."

Parentheses

Parentheses ( ), as columnist Russell Baker notes, whispers information. It is used to include unimportant or subordinate information which readers don't need to know but which you feel they might want to know. The information serves as an aside. The present trend is to eliminate parentheses. If the information is sufficiently important to be included, place it between commas. If it isn't important enough, eliminate it.

In technical documents, parentheses may be used to include brief definitions if there is a possibility readers won't know a term.

One of the toxic chemicals found In a home is Sodium Hydrochlorite (bleach).

Brackets

Brackets [ ] are used to indicate additional information within quotation marks.

"Just because a [dishwashing] cleanser isn't 'industrial strength,' don't [sic] mean it ain't hazardous."

The writer inserts the word "dishwashing" into the quote to clarify the type of cleanser for the reader. The brackets indicate the word is not part of the original quote.

The word "sic" is Latin for "as such." Placing the word in brackets following a grammatical error indicates to the reader that the error has been made by the speaker, not the writer of the document.

MECHANICS

Italics

Italics are rendered in sloped type in typography, or with an underline on most typewriters or word processors. In sentences, italics may be used for emphasis or mechanical purposes. They can call specific attention to a word or letter that may need specific emphasis to impact meaning:

While commas are useful, they may *not* be used alone to join independent clauses.

Overuse of italics devalues such emphasis and makes writing appear overexcited and unconvincing. Most writing for technical communities makes very sparing use of italics for emphasis. However, such special markers are used, especially for warnings, in instructions, procedures, and documentation.

Mechanical uses. Italics have three main mechanical uses.

1. Italics identify key technical terms or numerals that will be defined or referred to as words or numerals:

    The term *pollutant* will be designated as types 1 and 2.

2. Italics identify foreign words or expressions not yet assimilated into English, as in the Latin expression *post hoc* (after the fact). Note that many terms originally foreign have been assimilated and don't use italics, as in the Latin expression alter ego (someone's stand-in). For any given word your dictionary or the style sheet for your discourse community is the final arbiter.

3. Italics are key elements in bibliographical conventions. Both in text and in a reference listing, italics designate titles for the following:
    • books, long poems, plays
    • movies, broadcast shows, works of art, or long musical works
    • individually named transport craft: ships, trains, aircraft
    • newspapers, magazines, and periodicals

With newspapers it's conventional not to italicize the word *the*, even if this is part of the paper's title. With technical journals of all kinds, it's best to consult a bibliography from your discourse community to determine how titles are given.

Abbreviations and Acronyms

In using abbreviations, be careful that the readers in your discourse community know what the abbreviations stand for, especially for acronymns or in designating the names of technical journals. When in doubt, write out the full word, at least at the first occurrence. In formal documents, use as few abbreviations as possible.

Abbreviations, names, and titles. In formal documents, names should not be abbreviated. When using civil or military titles, write out the full title if you use only a person's surname. However, you may abbreviate a title if you use a person's full name.

Senator Kennedy
Sen. Robert Michel

Social titles, i.e. Mr., Ms., Dr., are always abbreviated.
Company designations are abbreviated, i.e. Co., Inc., Ltd.

Abbreviations for places, times. Spell out the names of cities and states when discussing them in a text. However, when the names of states are part of an address, they should be abbreviated according to the post office's two letter designation.

Names of the months and days of the week should be spelled out.

Acronyms. Government and industry have a tendency to use the initials of titles to create acronyms that serve as abbreviations; i.e., Institute for Nuclear Power Operations — INPO. If you are a member of that discourse community, you may use the acronym so often you forget what the initials actually stand for. However, when you write a formal document or to anyone outside your discourse community, you need to write out the entire name. If you plan to use the name again in your text, place the acronym in parentheses following the full designation.

Capitalization

In addition to designating the beginning of each sentence, capitals are used in titles for words of significance. The first and last words of a title are capitalized, including articles. Prepositions of five letters or more are also capitalized. However, short prepositions and the articles *a, an, the* are NOT capitalized in titles.

Proper noun capitalization. Capitals are conventionally used for any noun that refers to a *particular* person, place, or object that has been given an individual (proper) name or title. This includes the following:

1. Names of people or groups of people.
2. Religions, religious titles, and nationalities.
3. Languages, places (including those in addresses), and regions. Capitalize *the South* (a region) but NOT a compass heading as in *We headed south.*
4. Adjectives formed from proper nouns (French perfume) and titles of distinction that are part of proper nouns.
5. Names of days, months, holidays, historical events, periods.
6. Particular objects (*Hope Diamond*) and brand-name products.
7. Abbreviations for words that are themselves capitalized.
8. Acronymns for companies, agencies, established abbreviations.

Numbers

In written sentences, numbers that begin sentences or numbers that can be expressed in one or two words (seventy-six, three-fourths) are normally written out. When time is expressed with the term *o'clock* the number is written out (ten o'clock BUT 10 a.m.). As you proofread, it's important to be sure that the short numbers have been treated consistently throughout the document.

Treatment of numbers and abbreviations must conform to the standards set by the leading journals and conventions of your discourse community.

Model Worksheet and Checklist

PLANNING WORKSHEET

1. Determine Purpose
   Yours:

   Your Reader's:

2. Analyze Your Audience:

3. Gather and List Information:

Data	Label	**Category Relevance**		
		R (Relevant)	? (Possible)	I (Irrelevant)

4. Find Relationships and Label Categories:

5. Write Sentences: Several sentences for each category relating two or more pieces of information in that category:

6. Write an Organizing Idea for Each Category:

7. Write a Main Organizing Idea:

8. Determine Relevance (Return to step 3):

9. Sequence Categories (Return to step 3):

DOCUMENT DESIGN CRITERIA CHECKLIST

1. Does the document appear dense?   Yes ( ) No ( )

2. Do headings and subheadings introduce sections and subsections?   Yes ( ) No ( )

    a. Are the headings and subheadings organized hierarchically, using   Yes ( ) No ( )
       appropriate margins indentations and/or typography?

    b. Do the headings and subheadings serve as truncated main organizing   Yes ( ) No ( )
       ideas (M.O.I.s) or organizing ideas (O.I.s) respectively?

3. Does information need to be boxed because of its importance or because it   Yes ( ) No ( )
    is extraneous to readers' needs?

4. Does information need to be listed?   Yes ( ) No ( )

    a. Are listed items parallel spatially and grammatically?   Yes ( ) No ( )

    b. Do listed items grammatically complete the initiating sentence?   Yes ( ) No ( )

    c. Are long lists separated into sections?   Yes ( ) No ( )

    d. Does spacing allow readers to perceive individual steps?   Yes ( ) No ( )

    e. Are items identified by numbers or bullets?   Yes ( ) No ( )

5. Do chunks of information or steps in a list need to be set off by white space?   Yes ( ) No ( )

6. Are margins sufficient?   Yes ( ) No ( )

7. Is text spaced to provide an airy appearance and good readability?   Yes ( ) No ( )

8. Is typeface, size, and style, and line length legible?   Yes ( ) No ( )

9. Do global markers indicate various parts of the document?   Yes ( ) No ( )

10. Do tables or graphs provide visual representations of numerical concepts?   Yes ( ) No ( )

11. Do charts provide visual representations of hierarchical, directions, or   Yes ( ) No ( )
    time relationships?

12. Do illustrations inform or persuade readers or clarify concepts about   Yes ( ) No ( )
    mechanisms, objects, processes, procedures, or theories?

13. Are the appropriate conventions used with graphics?   Yes ( ) No ( )

If you have marked "No" for any of these questions, you may need to include graphics or visual text or alter your format. Return to your document to determine whether or not you do.

BIBLIOGRAPHY

Chapter 1

Barclay, Rebecca O., et al. 1991. "Technical Communication in the International Workplace: Some Implications for Curriculum Development." *Technical Communication* 38:324-35.

Kent, Thomas. "On the Very Idea of a Discourse Community." *College Composition and Communication* 1991: 425-45.

Rosenblatt, Alfred, et al., eds. July, 1991. "Concurrent Engineering: Special Report." *Spectrum*. New York: Institute of Electrical and Electronics Engineers. 22-37.

Society For Technical Communication. 1991. "Technical Communicators Part of Team at Artisoft." *Intercom*. 1.

Chapter 2

Broadhead, Glenn J., and Richard C. Freed. 1986. *Variables of Composition: Process and Product in a Business Setting.* Carbondale, IL: Southern Illinois University Press.

Faigley, Lester, et al. 1985. *Assessing Writers' Knowledge and Processes of Composing.* Norwood, NJ: Ablex.

Flower, Linda, and John Hayes. 1984. "Images, Plans and Prose: The Representation of Meaning in Writing." *Written Communication* 1: 120-60.

Flower, Linda, and John R. Hayes. 1981. "A Cognitive Process Theory of Writing." *College Composition and Communication* 32: 365-87.

Kaufer, David S., John R. Hayes, and Linda Flower. 1986. "Composing Written Sentences." *Research in the Teaching of English* 20: 121-40.

Shahani, Chandru J., and William K. Wilson. 1987. "Preservation of Libraries and Archives." *American Scientist* 1987: 240-51.

Spivey, Nancy Nelson. 1990. "Transforming Texts: Constructive Processes in Reading and Writing." *Written Communication* 7: 256-85.

Witte, Stephen P. 1987. "Pre-Text and Composing." *College Composition and Communication* 38: 387-425.

Chapter 3

Allen, Jo. 1989. "New Directions in Audience Analysis." In *Technical Writing: Theory and Practice*, ed. Bertie E. Fearing and W. Meats Sparrows, 53-62. New York: Modern Language Association.

Barabas, Christine. 1990. *Technical Writing in a Corporate Culture.* Norwood, NJ: Ablex Publishing Corporation.

Blyler, Nancy. 1990. "Rhetorical Theory and Newsletter Writing." *Technical Writing and Communication* 20: 139-52.

Blyler, Nancy Roundy. 1991. "Reading Theory and Persuasive Business Communications: Guidelines for Writers." *Technical Writing and Communication.* 383-96.

Brostoff, Anita. 1981. "The Functional Writing Model in Technical Writing Courses." In *Courses, Components, and Exercises in Technical Communication*, ed. Dwight W. Stevenson, 65-73. Urbana, IL: National Council of Teachers of English.

Dobrin, David N., ed. 1989. *Writing and Technique.* Urbana, IL: National Council of Teachers of English.

Duin, Anne. 1988. "How People Read: Implications for Writers." *The Technical Writing Teacher* 15: 185-94.

Faigley, Lester, et al. 1985. *Assessing Writers' Knowledge and Processing of Composing.* Norwood, NJ: Ablex Publishing Corporation.

Huckin, Thomas N. 1983. "A Cognitive Approach to Readability." In *New Essays in Technical and Scientific Communication: Research, Theory and Practice*, ed. Paul V. Anderson, R. John Brockman, and Carolyn R. Miller, 90-108. Farmingdale, NY: Baywood Publishing Company, Inc.

Kirsch, Gesa. 1990. "Experienced Writers' Sense of Audience and Authority: Three Case Studies." In *A Sense of Audience in Written Communication*, ed. Gesa Kirsch and Duane H. Roen. Newbury Park, CA: Sage Publications, Inc.

Kirsch, Gesa, and Duane H. Roen, eds. 1990. *A Sense of Audience in Written Communication.* Newbury Park, CA: Sage Publications, Inc.

McCormick, Kathleen, Gary Waller, and Linda Flower. 1987. *Reading Texts: Reading, Responding, Writing.* Lexington, MA: D. C. Heath and Company.

Mulcahy, Patricia. 1991. "Cognitive Links in Reading ans Writing for Technical Communicators." In *Studies in Technical Communication* ed. Brenda Sims, 103-120. University of North Texas Press.

Paivio, Allan, and Ian Begg. 1981. *Psychology of Language.* Englewood Cliffs, NJ: Prentice-Hall, Inc.

Schriver, Karen A. 1992. "Teaching Writers to Anticipate Readers' Needs: A Classroom Evaluated Pedagogy." *Written Communication.* 9: 179-208.

Spilka, Rachel. 1988. "Studying Writer-Reader Interactions in the Workplace." *The Technical Writing Teacher* 15: 208-22.

Spyridakis, Jan H. and Michel J. Wenger. 1992. "Writing for Human Performance: Relating Reading Reasearch to Document Design." *STC Technical Communication.* 39: 202-218.

Weaver, Constance. 1988. *Reading Process and Practice.* Portsmouth, NH: Heinemann Educational Books.

Ziv, Nina Dansky. 1989. "Reading Technical Writing." In *Technical and Business Communications: Bibliographic Essays for Teachers and Corporate Trainers* ed. Charles H. Sides, 39-52. Urbana, IL: National Council of Teachers of English.

Chapter 4

Allen, Jo. 1989. "New Directions in Audience Analysis." *Technical Writing: Theory and Practice* ed. Bertie E. Fearing and W. Meats Sparrow, 53-62. New York: Modern Language Association.

Barabas, Christine. 1990. *Technical Writing in a Corporate Culture.* Norwood, NJ: Ablex Publishing Corporation.

Blyler, Nancy. 1990. "Rhetorical Theory and Newsletter Writing." *Technical Writing and Communication* 20: 139-52.

Coney, Mary B. 1992. "Technical Readers and their Rhetorical Roles." *IEEE Transactions on Professional Communications.* 35: 58-63.

Cooney, Barry D. 1989. "Japan/America: Culture Counts." *Training & Development Journal:* 58-61.

Ede, Lisa, and Andrea Lunsford. 1984. "Audience Addressed/Audience Invoked: The Role of Audience in Composition Theory and Pedagogy." *College Composition and Communication* 35: 155-71.

Faigley, Lester, et al. 1985. *Assessing Writers' Knowledge and Processing of Composing.* Norwood, NJ: Ablex Publishing Corporation.

Hein, Robert G. 1991. "Culture and Communication." *Technical Communication.* 125-26.

Hildebrandt, H. H. 1973. "Communication Barriers between German Subsidiaries and Parent American Companies." *Michigan Business Review:* 11.

Huckin, Thomas N. 1983. "A Cognitive Approach to Readability." In *Essays in Technical and Scientific Communicaiton: Research, Theory, Practice* ed. Paul V. Anderson, R. John Brockman, Carolyn R. Miller, Farmingdale, NY: Baywood Publishing Company. 90-108.

Matalene, Carolyn. 1985. "Contrastive Rhetoric: An American Writing Teacher in China." *College English* 47: 789-808.

Mathes, J. C., and Dwight W. Stevenson. 1985. *Designing Technical Reports: Writing for Audiences in Organizations.* Indiannapolis: ITT Bobbs Merrill Educational Publishing Company.

Raven, Mary Elizabeth. 1991. The Art of Audience Adaption: Dilemmas of Cultural Diversity. Proceedings of the 38th International Technical COmmunication Conference.

Spilka, Rachel. 1990. "Orality and Literacy in the Workplace: Process- and Text-Based Strategies for Multiple-Audience Adaptation." *Journal of Busines and Technical Communication* 4:44-67.

——. 1988. "Studying Writer-Reader Interactions in the Workplace." *The Technical Writing Teacher* 15: 208-22.

Tan, Wen-lan. 1987. Think in Advance; Plan for the Future. Unpublished paper. Illinois State University, Normal, IL

Subbiah, M. New Dimension to the Teaching of Audience Awareness: Cultural Awareness. *IEEE Transactions on Professional Communication.* 35: 14-18.

Varner, I. I. (1987). "Internationalizing Business Communication Courses." *Bulletin of the Association of Business Communication* 50: 7-11.

Weaver, Constance. 1988. *Reading Process and Practice.* Portsmouth, NH: Heinemann Educational Books.

Chapter 5

Barabas, Christine. 1990. *Technical Writing in a Corporate Culture.* Norwood, NJ: Ablex Publishing Company.

Blyler, Nancy Roundy. 1989. "Purpose and Professional Writers." *The Technical Writing Teacher* 16:52-67.

Bryan, John. 1992. "Down the Slippery Slope: Ethics and the Technical Writer as Marketer." *Technical Writing and Communication.* 22: 73-88.

Brockman, R. John, and Fern Rook, eds. 1989. *Technical Communication and Ethics.* Washington, D.C.: Society for Technical Communication.

Crowhurst, Marion, and G. Piche. 1979. "Audience and Mode of Discourse Effects on Syntactic Complexity in Writing at Two Grade Levels." *Research in the Teaching of English* 13:101-09.

Dombrowski, Paul M. 1991. "The Lessons of the Challenger Investigation." *IEEE Transactions in Professional Communication* 211-16.

Driskell, Linda. 1989. "Understanding the Writing Context in Organizations." In *Writing in the Business Professions* ed. Myra Kogen. Urbana, IL: National Council of Teachers of English.

Dulek, Ronald E. 1991. "The Challenge to Effective Writing: The Public Policy Maker's Multiple Audiences." *IEEE Transactions on Professional Communications* 34: 224-27.

Fink, Steven. 1986. *Crisis Management.* New York: American Management Association.

Flower, Linda. 1989. "Rhetorical Problem Solving." In *Writing in the Business Professions* ed. Myra Kogen. Urbana, IL: National Council of Teachers of English.

Gilman, Joel. 1990. "Product Liability and the Duty to Warn." *Intercom* 36: 7-14.

Gorlin, Rena A., ed. 1990. *Codes of Professional Responsibility.* 2nd ed. Washington, D.C.: Bureau of National Affairs.

Killingsworth, M. Jimmie, and M.K. Gilbertson. 1986. "Rhetoric and Relevance in Technical Writing." *Journal of Technical Writing and Communication* 16: 287-97.

Killingsworth, M. Jimmie, and Dean Steffens. 1989. "Effectiveness in the Environmental Impact Statement." *Written Communication* 155-80.

Kleimann, Susan D. 1991. "The Complexity of Workplace Review." *Technical Communication.* 520-26.

Madigan, Chris. 1985. "Improving Writing Assignments with Communication Theory." *College Composition and Communication* 36:183-90.

Markel, Mike. 1991. "A Basic Unit on Ethics for Technical Communicators." *Technical Writing and Communication* 21: 327-50.

Michaelson, Herbert B. 1991. "Commentary: Unethical Blunders by Authors." *Technical Writing and Communication* 21: 25-28.

Mortensen, Peter L. 1990. "Understanding Readers' Conceptions of Audience: Rhetorically Challenging Texts." In *A Sense of Audience in Written Communication* ed. Gesa Kirsch and Duane H. Roen. Newbury Park, CA: Sage Publications.

Ornatowski, Cezar M. 1992. "Between Efficiency and Politics." *Technical Communication Quarterly* 1: 91-103.

Redish, Janet. 1989. "Writing in Organizations." In *Writing in the Business Professions* ed. Myra Kogen. Urbana, IL: National Council of Teachers of English.

Ross, Susan, and Bill Karis. 1991. "Communicating in Public Policy Matters: Addressing the Problem of Non-Congruent Sites of Discourse." *IEEE Transactions on Professional Communications* 34: 247-53.

Sanders, Scott P. "Culture." 1991. *IEEE Transactions on Professional Communications* 34: 7.

Seltzer, Jack. 1983. "The Composing Process of an Engineer." *College Composition and Communication* 34:178-87.

Shimberg, H. Lee. 1989. "Ethics and Rhetoric in Technical Writing." In *Technical Communication and Ethics* ed. R. John Brockman, Fern Rook. Washington, D.C.: Society for Technical Communication. 59-62.

Smith, Herb. 1990. "Technical Communications and the Law." *Technical Writing and Communication.* 20: 307-20.

Weinstein, Alvin S., et al. 1978. *Products Liability and the Reasonably Safe Product: A Guide for Management, Design, and Marketing.* New York: John Wiley and Sons.

Winsor, Dorothy. 1993. "Owning Corporate Texts." *Journal of Business and Technical Communication.* Forthcoming.

Chapter 6

Ary, Donald, Lucy Cheser Jacobs, and Asghar Razavieh. 1972. *Introduction to Research in Education.* 2nd ed. New York: Holt, Rinehart, and Winston.

Babbe, Earl. 1990. *Survey Research Methods.* 2nd ed. Belmont, CA: Wadsworth.

Bledsoe, Joseph C. 1973. *Essentials of Educational Research.* Athens, GA: Optima House.

Hastings, Elizabeth, and Philip Hastings. 1989. *Index to International Public Opinion.*

——. *Index to International Public Opinion.* Westport, Conn.: Greenwood Press.

Institute For Nuclear Power Operations. 1986. Lesson Plans for Evaluators. Atlanta, GA.

Krueger, Richard A. 1988. *Focus Groups: A Practical Guide for Applied Research.* Newbury Park, Ca: Sage Publications.

Monroe, Alan. *FlexText-POL Manual.* 1991. Springfield, IL: Text Analysis Service Corp.

——. *PALS Manual.*

Zimmerman, Juliet G., and Robert N. Zelnio. 1985. "Listening Is the Key to More Productive Focus Group Sessions." *Medical Marketing and Media.* 84-88.

Chapter 7

Barker, Thomas T. 1989. "Word Processors and Invention in Technical Writing." *The Technical Writing Teacher.* 16: 126-35.

Broadhead, Glenn J., and Richard C. Freed. 1986. *The Variables of Composition: Process and Product in a Business Setting.* Carbondale, IL: Southern Illinois University Press.

Dobrin, David N. 1989. *Writing and Technique.* Urbana, IL: National Council of Teachers of English.

Faigley, Lester, et al. 1985. *Assessing Writers' Knowledge and Processes of Composing.* Norwood, NJ: Ablex Publishing Corporation.

Faigley, Lester, and Stephen P. Witte. 1983. "Topical Focus in Technical Writing." In *New Essays in Technical and Scientific Communication: Research, Theory, and Practice* eds. Paul V. Anderson, et al. 59-68. Farmingdale, NY: Baywood Publishing Company, Inc.

Flower, Linda. 1989. "Rhetorical Problem Solving." In *Writing in the Business Professions.* ed. Myra Kogen. 3-36. Urbana, IL: National Council of Teachers of English.

Kellogg, Ronald T. 1987. "Writing Performance: Effects of Cognitive Strategies." *Written Communication* 4: 269-98.

Selzer, Jack. 1989. "Arranging Business Prose." In *Writing in the Business Professions.* ed. Myra Kogen. 37-64. Urbana, IL: National Council of Teachers of English.

Spivey, Nancy Nelson. 1990. "Transforming Texts: Constructive Processes in Reading and Writing." *Written Communication* 7: 256-87.

Van Nostrand, A. D., C. H. Knoblauch, and J. Pettigrew. 1982. *The Process of Writing: Discovery and Control.* 2nd ed. Boston: Houghton Mifflin Company.

Chapter 8

Bizzell, Patricia, and Bruce Herzberg. 1989. "Knowledge and Argument: An Example from English Studies." In *Argument in Transition: Proceedings of the 3rd Summer Conference.* ed. David Zarefsky, M.O. Sillars, & J. Rhodes. Annandale, VA: Speech Communication Association.

Brockriede, Wayne, and Douglas Ehringer. 1971. "Toulmin on Argument: An Interpretation and Application." in *Contemporary Theories of Rhetoric.* ed. Richard L. Johannesen. New York: Harper & Row, 1971. 241-55.

Connor, Ulla. 1990. "Linguistic/Rhetorical Measures for International Persuasive Student Writing." *Research in the Teaching of English.* 67-87.

Corbett, Edward P. C., ed. 1987. *The Technical Writing Teacher.* Vol. 14. St. Paul: Association of Teachers of Technical Writing.

French. J.R.R., and B. Raven. 1968. "The Bases of Social Power." in *Group Dynamics: Research and Theory.* ed. Cartwright & Zander. New York: Harper & Row.

Gilsdorf, J. W. 1992. "Writing to Persuade." *Writing and Speaking in the Technology Professions: A Practical Guide.* 105-10.

Grice, H. P. 1975. "Logic and Conversation." in *Syntax and Semantics 3: Speech Acts.* Ed. P. Cole and J.L. Morgan. New York: Academic Press. 41-58.

Hart, Roderick P. 1990. *Modern Rhetorical Criticism.* Glenview, IL: Scott Foresman/ Little Brown Higher Education.

Hewes, Dean E., et al. 1990. "Interpersonal Communication Research: What Should We Know." in *Speech Communication: Essays to Commenorate the 75th Anniversary of the Speech Communication Association.* Ed. Gerald M. Phillips & Julia T. Wood.Carbondale, IL: Southern Illinois University Press. 130-80.

Kelman, H. C. 1961. "Processes of Opinion Change." *Public Opinion Quarterly.* 57-78.

Maslow, A. H. 1954. *Motivation and Personality.* New York: Harper.

McDonald, Daniel. 1989. *The Language of Argument.* sixth ed. New York: Harper & Row.

Miller, Carolyn. "Fields of Argument: A Special Topoi." in *Argument in Transition: Proceedings of the 3rd Summer Conference.* ed. David Zarefsky, M.O. Sillars, & J. Rhodes. Annandale, VA: Speech Communication Association. 147-58.

——. 1985. "Invention in Technical and Scientific Discourse: A Prospective Survey." in *Research in Technical Communication: A Bibliographic Sourcebook.* ed. Michael G. Moran and Debra Journet. Westport, Conn.: Greenwood Press, 1985. 117-62.

Moore, Patrick. 1990. "Using Case Studies to Teach Courtesy Strategies." *The Technical Writing Teacher.* 8-25.

——. 1990. "Using Case Studies to Teach Courtesy Strategies." *The Technical Writing Teacher.* 8-25.

Neman, Beth. 1993. *Teaching Writing.* 2nd ed.

Riley, Katherine. 1988. "Conversational Implicature and Unstated Meaning in Professional Communication." *The Technical Writing Teacher.* 94-104.

Rosen, Leonard J., and Laurence Behrens. 1992. *Handbook.* Boston: Allyn & Bacon.

Rottenberg, Annette T. 1991. *Elements of Argument.* 3rd ed. Boston: Bedford Books of St. Martin's Press.

Toulmin, S. E. 1958. *The Uses of Argument.* Cambridge: Cambridge University Press.

Chapter 9

Curry, Jerome. 1989. "Introducing Realism into the Technical Definition Assignment." *The Technical Writing Teacher.* 123-25.

——. 1992. "Technical Instruction and Definition Assignments: A Realistic Approach." *Journal of Business and Technical Communication.* 116-22.

Chapter 10

Boiarsky, Carolyn. 1984. "A Model for Analyzing Revision." *Journal of Advanced Composition.* 65-78.

D'Aoust, Debbie. 1990. "Instructional Software for the Introductory Composition Classroom: One Teacher's Review and Recommendations." *Computers and Composition.* 109-27.

Dobrin, David N. 1989. *Writing and Technique.* Urbana, IL: National Council of Teachers of English.

Faigley, Lester, et al. 1985. *Assessing Writers' Knowledge and Processes of Composing.* Norwood, NJ: Ablex Publish Corporation.

Flower, Linda. 1989. "Rhetorical Problem Solving: Cognition and Professional Writing." In *Writing in the Business Professions.* Myra Kogen, ed. Urbana, IL: National Council of Teachers of English, 1989. 3-36.

Markel, Mike. 1990. "The Effect of the Word Processor and the Style Checker on Revision in Technical Writing: What Do We Know and What Do We Need to Find Out?" *Technical Writing and Communication.* 329-42.

Murray, Donald. 1985. *A Writer Teaches Writing.* 2nd ed. Boston: Houghton Mifflin Company, 1985.

Selzer, Jack. 1989. "Composing Processes for Technical Discourse." In *Technical Writing: Theory and Practice.* eds. Bertie E. Fearing and W. Keats Sparrow. New York: Modern Language Association, 1989. 43-50.

Sims, Brenda R., and Donna DiMaggio. 1990. "Using Computerized Textual Analysis with Technical Writing Students: Is It Effective." *The Technical Writing Teacher*. 61-68.

Unikel, Graham. 1988. "The Two-Level Concept of Editing." *The Technical Writing Teacher*. 49-55.

Chapter 11

American Psychological Association. 1983. *Publication Manual of the American Psychological Association*. 3rd ed. Washington, D.C.: American Psychological Association.

Chafe, Wallace. 1988. "Punctuation and the Prosody of Written Language." *Written Communication*: 395-426.

Colomb. Greg. 1987. Disciplinary Secrets and the Apprentice Writer. Paper Presented at University of Chicago Symposium, Chicago, IL.

Cooper, Allene. 1988. "Given-New: Enhancing Coherence Through Cohesiveness." *Written Communication*: 354-67.

Ferguson, K. Scott, and Frank Parker. 1990. "Grammar and Technical Writing." *Technical Writing and Communication*. 357-69.

Fulkerson, Richard. 1988. "Technical Logic, Comp-Logic, and the Teaching of Writing." *College Composition and Communication*: 436-52.

Hall, Dean G. 1990. "Sex-Biased Language and the Technical Writing Teacher's Responsibility." *Journal of Business and Technical Communication*. 69-80.

Huckin, Thomas N. 1983. "A Cognitive Approach to Readability." *New Essays in Communication: Research, Theory, and Practice*. ed Paul V. Anderson, R. John Brockman, & Carolyn Miller. Farmingdale, NY: Baywood Publishing Company, Inc. 90-108.

Markel, Mike. 1990. "The Effect of the Word Processor and the Style Checker on Revision in Technical Writing: What Do We Know, and What Do We Need to Find Out?" *Technical Writing and Communication*. 329-42.

Riley, Katherine. 1988. "Speech Act Theory and Degrees of Directness of Professional Writing." *The Technical Writing Teacher*: 1-30.

Rock, Fern. "Positively Stated." 1991. *Technical Communication*. 134-35.

Selzer. Jack. 1983. "What Constitutes a "Readable" Technical Style?" in *New Essays in Technical and Scientific Communication: Research, Theory, and Practice*. ed. Paul A. Anderson, R. John Brockman, Carolyn R. Miller. Farmingdale, NY: Baywood Publishing Company, Inc. 71-89.

Sloan, Gary. 1990. "Frequency of Errors in Essays by College Freshmen and by Professional Writers." *College Composition and Communication*: 299-308.

Steinberg, Erwin R. 1986. "A Pox on Pithy Prescriptions." *College Composition and Communication*: 96-100.

University Of Chicago Press, comp. 1982. *The Chicago Manual of Style*. 13th ed. Chicago: University of Chicago Press.

Vaughn, Jeannette. 1989. "Sexist Language-Still Flourishing." *The Technical Writing Teacher*: 33-40.

Williams, Joseph M. 1989. *Style: Ten Lessons in Clarity and Grace*. 3rd ed. Glenview, IL: Scott, Foresman and Company.

Winsor, Dorothy A. 1990. "Engineering Writing/Writing Engineering." *College Composition and Communication*: 58-70.

Woodward, John B. 1990. "Numerical Analysis as Technical Communication." *Technical Communication*. 221-24.

Chapter 12

Barton, B. En F., and Marthalee S. Barton. 1989. "Trends in Visual Representation." In *Technical and Business Communication*. ed. Charles H. Sides. Urbana, Il: National Council of Teachers of English. 95-136.

Barton, Ben F., and Marthalee S. Barton. 1990. "Postmodernism and the Relation of Word and Image in Professional Discourse." in *Sudies in Technical Communication: Proceedings of the 1990 CCCC and NCTE Meetings*. ed. Brenda R. Sims. Conference on College Composition and Communication and the National Council of Teachers of English, 1990. 27-46.

Gross, Alan G. 1990. "Extending the Expressive Power of Language: Tables, Graphs, and Diagrams." *Technical Writing and Communication*. 221-36.

Hickman, Dixie Elise. 1985. *Teaching Technical Writing: Graphics*. Texas: Association of Teachers of Technical Writing.

Kostelnick, Charles. 1990. "Typographical Design, Modernist Aesthetics, and Professional Communication." *Journal of Business and Technical Communication*. 5-24.

McKinley, Lawrence C. 1990. "Writers and Artists Should Be Friends." *Technical Communications*. 451.

Petersson, Rune. 1989. *Visuals for Information: Research and Practice*. Englewood Cliffs, NJ: Educational Technology Publications.

Tufte, Edward R. 1990. *Envisioning Information*. Cheshire, CN: Graphic Press.

—. 1983. *The Visual Display of Quantitative Information*. Cheshire, CN: Graphic Pess.

Winn, Bill. 1987. The Role of Graphics in Training Documents: Towards an Explanatory Theory of How They Communicate. In "Creating Usable Manuals and Forms: A Document Design Symposium." Carnegie-Mellon University Communications Design Center.

Chapter 13

Barton, B. En F., and Marthalee S. Barton. 1989. "Trends in Visual Representation." In *Technical and Business Communication*. ed. Charles H. Sides. Urbana, Il: National Council of Teachers of English. 95-136.

Barton, Ben F., and Marthalee S. Barton. 1990. "Postmodernism and the Relation of Word and Image in Professional Discourse." in *Sudies in Technical Communication: Proceedings of the 1990 CCCC and NCTE Meetings*. ed. Brenda R. Sims. Conference on College Composition and Communication and the National Council of Teachers of English, 1990. 27-46.

Barton, Ben F., and Marthalee S Barton, eds. 1990. *The Technical Writing Teacher*. Vol. 17. No. 3, Minnappolis: Association of Technical Writing Teachers.

Boyarski, D. 1987. *Helping Developers See the User's Point of View: A Graphic Designer's Perspective*. Pittsburgh: Communications Design Center.

Grove, Laurel. 1991. Document Design for the Visually Impaired. The Engineered Communication: Designs for Continued Improvement. IPC Proceedings. Vol. 2. New York: IEEE Professional Communications Society, 1991. 304-09.

Halio, Marcia Peoples. 1990. "Student Writing: Can the Machine Maim the Message." *Academic Computing*. 16-19.

Hickman, Dixie Elise. 1985. *Teaching Technical Writing: Graphics*. Texas: Association of Teachers of Technical Writing.

Kostelnick, Charles. 1990. "Typographical Design, Modernist Aesthetics, and Professional Communication." *Journal of Business and Technical Communication*. 5-24.

Kostelnick, Charles. 1989. "Visual Rhetoric: A Reader-Oriented Approach to Graphics and Designs." *The Technical Writing Teacher*. 77-88.

Pavio, Allen, and Ian Begg. 1981. *Psychology of Language*. Englewood Cliffs, NJ: Prentice Hall.

Petersson, Rune. 1989. *Visuals for Information: Research and Practice*. Englewood Cliffs, NJ: Educational Technology Publications.

Sullivan, Patricia. 1990. "Visual Markers for Navigating Instructional Texts." *Technical Writing and Communication* 1990. 255-68.

Tufte, Edward R. 1990. *Envisioning Information*. Cheshire, CN: Graphic Press.

—. 1983. *The Visual Display of Quantitative Information*. Cheshire, CN: Graphic Pess.

U.S. Department of Education. Office of Educational Research and Improvement. 1986. *Redesigning and Testing a Work Order Form*. Washington, D.C.: ERIC Document Reproduction Sevice.

Weaver, Constance. 1988. *Reading Process and Practice*. Portsmouth, NH: Heinemann Educational Books.

Winn, Bill. 1987. The Role of Graphics in Training Documents: Towards an Explanatory Theory of How They Communicate. In "Creating Usable Manuals and Forms: A Document Design Symposium." Carnegie-Mellon University Communications Design Center.

Chapter 14

Barclay, Rebecca O., et al. 1991. "Technical Communication in the International Workplace: Some Implications for Curriculum Development." *Technical Communication*. 324-35.

Bocchi, Joseph S. 1991. "Forming Constructs of Audience: Convention, Conflict, and Conversation." *Journal of Business and Technical Communication* 1991.

Dragga, Sam. 1991. "Classification of Correspondence: Complexity Versus Simplicity." *The Technical Writing Teacher*. 1-14.

Gerson, Steven M., and Sharon J. Gerson. 1991. "Using Templates to Teach Technical Correspondence.". 570-72.

LaQuey, Tracy. 1989. "Networks for Academics." *Academic Computing*. 32-39.

McCahill, Mark P. 1989. "Communications: Network Applications." *Academic Computing*. 40-52.

Milliman, Ronald E. 1991. "An Experiment Designed to Maximize the Profitability of Customer Directed Post-Transaction Communication." *Technical Writing and Communication*. 397-410.

Richardson, Malcolm. 1987. "A Memo-Writing Assignment to Freeze the Blood." *The Technical Writing Teacher*. 97-98.

Riley, Katheryn. 1988. "Speech Act Theory and Degrees of Directness in Professional Writing." *The Technical Writing Teacher*. 1-30.

Rogers, Priscilla S. 1990 "A Taxonomy for the Composition of Memorandum Subject Lines: Facilitating Writer Choice in Managerial Contexts." *Journal of Business and Technical Communication*. 21-43.

Turner. 1988. "'E-Mail' Technology Has Boomed But Manners of Its Users Fall Short of Perfection." *Chronicle of Higher Education*. 1.

U.S. Postal Service. 1982. National Council of Teachers of English. *All About Letters*. Revised ed. Washington, D.C.: U.S. Postal Service.

United Express. 1990. "America's Executives May not Have the "Write Stuff"" *Destinations*. 18.

Chapter 15

Caterpillar, Inc. *Writing Your Resume*. Peoria, IL: Caterpillar, Inc.

Charney, Davida H., Jack Rayman, and Linda Ferreira-Buckley. 1992. "How Writing Quality Influences Readers' Judgments of Resumes in Business and Engineering." *Journal of Business and Technical Communication*. 38-74.

Croft, Barbara L. *Getting a Job: Resume Writing, Job Application Letters, and Interview Strategies*. Merrill Publishing Co., 1989.

Morgan, Meg. 1990. "Teaching the Resume as a Technical Document." *Technical Communication*. 436-38.

Norman, Rose. 1988. "'Resumex': A Computer Exercise for Teaching Resume Writing." *The Technical Writing Teacher*. 162-66.

Topf, Mel A. 1987. "Jop Application Correspondence: Integral to the Technical Communication Course." *The Technical Writing Teacher*. 114-17.

Vaughn, Jeannette W., and Nancy Darsey. "Negative Behavior Factors in the Employment Interview: Interviewer Opinions and Observations." *The Technical Writing Teacher* 1987: 208-18.

Williams, Charles. *Selection Criteria for New Hires*. Peoria, IL: Caterpillar, Inc.

Chapter 16

American Institutes for Research. Carnegie Mellon University. 1981. *Guidelines for Document Designers*. Washington, D. C.: American Institutes for Research.

Dobrin, David N. 1991. "Alphabetic Software Manuals: Notes and Comments." *Technical Communication*. 89-100.

Document Design Center. American Institutes for Research. Carnegie Mellon University. 1989. *Document Design: A Review of the Relevant Research*. Washington, D.C.: American Institutes for Research.

Duffy, Thomas M., et al. 1988. *Writing Online Information: Expert Strategies*. Pittsburgh, PA: Communications Design Center.

Duffy, Thomas M., and Robert Waller, eds. *Designing Usable Texts*. Educational Technology Series. Orlando, FL: Academic Press, 1985.

Duffy, Thomas, et al. 1988. *Creating Usable Manuals and Forms: A Document Design Symposium*. Pittsburgh, PA: Communications Design Center.

Eissler, Margaret. 1991. "Hypertext: A Communicator's View." *Intercom*. 3-7.

Hammond, J. S., et al. 1990. "Developing Standards: A DOE Case Study." *The Engineered Communication: Designs for Continued Improvement*. New York: IEEE Professional Communication Society. 180-90.

Horton, Wiliam. 1991. "Is Hypertext the Best Way to Document Your Product? An Assay for Designers." *Technical Communication*. 20-32.

Horton, William. 1990. "The Wired Word: Designing Online Documentation." *Technical Communication*. 444-47.

IBM, ed. *IBM Jargon and General Computing Dictionary*. 10th edition. IBM, London: 1990.

Johnson, Pamela R. 1990. "Employee Handbooks: An Integration of Technical Writing Concepts." *The Technical Writing Teacher*. 1-7.

—. 1991. "Employee Handbooks: An Integration of Technical Writing Concepts-Part II." *The Technical Writing Teacher* 1991: 60-68.

Major, John H. 1992. "What Should You Write: A User's Guide, Tutorial, Reference Manual, Or Standard Operating Procedure." *Writing and Speaking in the Technology Professions: A Practical Guide*. New York: IEEE. 121-24.

Mirel, Barbara, Susan Feinberg, and Leif Allmendinger. 1991. "Designing Manuals for Active Learning Styles." *Technical Communication.* 75-87.

Montgomery, Tracy T. 1991. "Negotiating Corporate Culture: An Exercise in Documentation." *The Technical Writing Teacher.* 75-80.

Redish, Janice C., and David A. Schell. "Writing and Testing Instructions for Usability." *Technical Writing: Theory and Practice.* Ed. Bertie E. Fearing and Keats W. Sparrow. New York: Modern Language Association, 1989. 63-71.

Sankar, Chetan S., and Walter H. Hawkins. 1991. "The Role of User Interface Professionals in Large Software Projects." *IEEE Transactions on Professional Communications.*n 94-100.

Spuridakis, Jan H., and Carol H. Isakson. 1991. "Hypertext: A New Tool and Its Effect on Audience Comprehension." *The Engineered Communication: Designs for Continued Improvement.* New York: IEEE Professional Communication Society. 37-44.

Weinschenk, Susan. 1991. "Will Online Documentation Save Us." *Intercom.* 3.

Young, Al R. 1990. "The Elephant and the Blind Men: A Look at Functionality." *Technical Communication.* 136-39.

Chapter 17

Allen, Lori. 1991. "Proposal Tips for the Engineer: Why the Facts Aren't Enough." *The Engineered Communication: Designs for Continued Improvement.* New York: IEEE Professional Communication Society. 173-77.

Barakat, Robert A. 1991. "Developing Winning Proposal Strategies." *IEEE Transactions on Professional Communication.* 130-39.

——. 1989. "Storyboarding Can Help Your Proposal." *IEEE Transactions in Professional Communications.* New York: IEEE Professional Communication Society. 20-25.

Georges, T. M. 1992. "Fifteen Questions to Help You Write Winning Proposals." *Writing and Speaking in the Technology Professions: A Practical Guide.* New York: IEEE Press. 149.

McIsaac, C. M., and M. A. Aschauer. 1990. "Proposal Writing at Atherton Jordan, Inc.: An Ethnographic Study." *Management Communication Quarterly.* 527-60.

Pfeiffer, William S. 1989. *Proposal Writing: The Art of Friendly Persuasion.* Columbus, OH: Merrill Publishing Co.

Chapter 18

Barabas, Christine. 1990. *Technical Writing in a Corporate Culture.* Norwood, NJ: Ablex Publishing Corporation, 1990.

Chapter 19

Garay, Mary Sue. 1989. "Clinic Case: A Case to Elicit Point-Making During Revision of Survey-Based Reports." *The Technical Writing Teacher.* 155-68.

Hager, Peter J., and H. J. Scheiber. 1990. "Reading Smoke and Mirrors: The Rhetoric of Corporate Annual Reports." *Technical Writing and Communication.* 113-30.

Stocker, Deborah J. 1991. "An Engineered Approach to Report Production." *The Engineered Communication: Designs for Continued Improvement.* New York: IEEE Professional Communication Society. 169-73.

Werner, Warren W. 1989. "An RFP for Research Reports." *The Technical Writing Teacher.* 120-22.

Chapter 20

Curtis, Donnelyn, and Bernhards. Stephen A. 1991. "Keywords, Titles, Abstracts, and Online Searches: Implications for Technical Writing." *The Technical Writing Teacher.* 142-61.

Curtis, Donnelyn, and Stephen A. Bernhardt. 1991. "Keywords, Titles, Abstracts, and Online Searches: Implications for Technical Writing." *The Technical Writing Teacher.* 142-61.

Manning, Alan D. 1990. "Abstracts in Relation to Larger and Smaller Discourse Structures." *Technical Writing and Communication.* 69-90.

Vaughan, David K. "Abstracts and Summaries: Some Clarifying Distinctions." *The Technical Writing Teacher* 1991: 132-41.

Chapter 21

Doty, D. 1989. "Preparing a Desktop Presentation." *The Page.* 10-11.

Dzujna, C. C. 1989. "Visual Presentations to Hold Audience Interest." *The Office.* 85-86.

Zaremba, A. 1989. "Q and A: The Other Part of Your Presentation." *Management World.* 8-10.

INDEX